NETWORK FUNCTIONS AND PLASTICITY

NETWORK FUNCTIONS AND PLASTICITY

Perspectives From Studying Neuronal Electrical Coupling in Microcircuits

Edited by

JIAN JING

ELSEVIER

ACADEMIC PRESS

An imprint of Elsevier

Academic Press is an imprint of Elsevier
125 London Wall, London EC2Y 5AS, United Kingdom
525 B Street, Suite 1800, San Diego, CA 92101-4495, United States
50 Hampshire Street, 5th Floor, Cambridge, MA 02139, United States
The Boulevard, Langford Lane, Kidlington, Oxford OX5 1GB, United Kingdom

Library of Congress Cataloging-in-Publication Data
A catalog record for this book is available from the Library of Congress

British Library Cataloguing-in-Publication Data
A catalogue record for this book is available from the British Library

ISBN: 978-0-12-803471-2

For information on all Academic Press publications visit our website at https://www.elsevier.com/books-and-journals

Publisher: Mara Conner
Acquisition Editor: Natalie Farra
Editorial Project Manager: Kathy Padilla
Production Project Manager: Kirsty Halterman and Karen East
Designer: Maria Inês Cruz

Typeset by TNQ Books and Journals

To my family and friends for their love and friendship.

CONTENTS

LIST OF CONTRIBUTORS

G.B. Awatramani
University of Victoria, Victoria, BC, Canada

J.P. Bacon
University of Sussex, Brighton, United Kingdom

A. Bédécarrats
Université de Bordeaux, INCIA, CNRS UMR 5287, Bordeaux, France

M.V.L. Bennett
Albert Einstein College of Medicine, Bronx, NY, United States

J.M. Blagburn
University of Puerto Rico, San Juan, PR, United States

E.C. Cropper
Icahn School of Medicine at Mount Sinai, New York, NY, United States

A. Draguhn
Heidelberg University, Heidelberg, Germany

J. Golowasch
New Jersey Institute of Technology and Rutgers University-Newark, Newark, NJ, United States

M. Gray
New Jersey Institute of Technology and Rutgers University-Newark, Newark, NJ, United States

S. He
ShanghaiTech University, Shanghai, China

J. Jing
Nanjing University, Nanjing, Jiangsu, China; Icahn School of Medicine at Mount Sinai, New York, NY, United States

W.-C. Li
The University of St Andrews, St Andrews, United Kingdom

X. Li
New Jersey Institute of Technology and Rutgers University-Newark, Newark, NJ, United States

N.S. Magoski
Queen's University, Kingston, ON, Canada

F. Nadim
New Jersey Institute of Technology and Rutgers University-Newark, Newark, NJ, United States

R. Nargeot
Université de Bordeaux, INCIA, CNRS UMR 5287, Bordeaux, France

J. O'Brien
McGovern Medical School and The University of Texas Health Science Center at Houston, Houston, TX, United States

A.E. Pereda
Albert Einstein College of Medicine, Bronx, NY, United States

P. Phelan
University of Kent, Canterbury, United Kingdom

I. Rabinowitch
Fred Hutchinson Cancer Research Center, Seattle, WA, United States

J.C. Rekling
University of Copenhagen, Copenhagen, Denmark

C.P. Ribelayga
McGovern Medical School and The University of Texas Health Science Center at Houston, Houston, TX, United States

W.R. Schafer
MRC Laboratory of Molecular Biology, Cambridge, United Kingdom

S.-H. Shi
Memorial Sloan Kettering Cancer Center, New York, NY, United States

M.-F.M. Shih
Stony Brook University, Stony Brook, NY, United States (Present address)

R.D. Traub
IBM T.J. Watson Research Center, Yorktown Heights, NY, United States; Columbia University, New York, NY, United States

S. Trenholm
Friedrich Miescher Institute for Biomedical Research, Basel, Switzerland

J. Turecek
Harvard Medical School, Boston, MA, United States

K.R. Weiss
Icahn School of Medicine at Mount Sinai, New York, NY, United States

J.P. Welsh
Seattle Children's Research Institute, Seattle, WA, United States; University of Washington, Seattle, WA, United States

M.A. Whittington
University of York, York, United Kingdom

C.-L. Wu
Chang Gung University, Taoyuan, Taiwan; Linkou Chang Gung Memorial Hospital, Taoyuan, Taiwan

PREFACE

The ultimate goals of neuroscience are to understand the human brain and mind, and to ameliorate and possibly cure neuropsychiatric disorders. Two fundamental developments in modern biology have made these two goals attainable. One well-known development applies to all of biology, ie, the theory of evolution, first developed by Darwin in the 1850s. Humans, and our brains, come about not through divine intervention but through millions of years of natural evolution. Thus, we can understand our brains, at least partly, by both studying the brains of close relatives, such as gorillas and monkeys, and distant relatives, such as mice, nematodes, insects, crustaceans, leech, and molluscs. Indeed, many advances in modern biology, including the discovery of cells and neurons, genes, synapses, and the ionic basis of the action potential, have demonstrated that all animals, including humans, utilize the same components and basic mechanisms.

Another development, less well-known, is more specific to neuroscience. In fact, it is safe to say that one of the more important developments in contemporary neuroscience is the realization that different brain functions are localized to a specific brain region and/or regions (Jing, 2009; Kandel et al., 2013; Llinas, 2001; Mashour and Alkire, 2013). All behaviors are mediated by a specific neural circuit and/or interactions between circuits. These include relatively simple behaviors, such as reflexes, and rhythmic behaviors, such as locomotion and feeding. This principle also applies to behaviors or brain functions that are more complex, such as language processing, emotions, and consciousness. Understanding how neural circuits drive behavior continues to be one of the major components of modern neuroscience. Indeed, starting from the mid-20th century, neuroscientists have targeted a variety of neural circuits in a number of animal model systems and have made tremendous progress. More recently, there has been a reduction in the number of model systems studied as a consequence of the more advanced approaches that have become available to study mammalian and human brains. Nevertheless, the model system approach continues to be useful for two reasons. First, many insights gained in simpler circuits have proven to be applicable in more complex circuits. Second, if we cannot completely understand simpler circuits, how would we expect to understand more complex ones? For example, if mammalian or human circuits have some fundamental difference that makes them more difficult to understand than simpler circuits, we would only grasp this difference by elucidating simpler circuits first. This book is an attempt to synthesize current research in circuitry studies using model systems, by both looking back at progress that has been made and looking forward to what is expected in the future.

All neural circuits are formed by a defined group of neuronal elements usually composed of multiple, distinct cell types (consult the Glossary for a list of useful terms in circuit research). In general terms, the functions of a given circuit can be considered in the following way: it receives input, processes the information, and generates a specific output. The activity of neural circuits is determined by (1) the intrinsic properties of circuit neurons and (2) by the electrical and chemical connections they make. Thus one major goal of circuitry studies is to characterize these network properties. Additionally, both the intrinsic properties and the connections of neural circuits are often plastic. Indeed, this feature may ultimately contribute to learning and memory in animals. Thus a second major goal of circuitry studies is to characterize this plasticity. Both goals are represented in the title of the book.

In this book, we seek to elucidate circuit function from three perspectives, which are aimed, in part, to complement related books (Byrne et al., 2014; Dere, 2012; North and Greenspan, 2007; Orlovsky et al., 1999; Shepherd, 2004; Shepherd and Grillner, 2010; Stein et al., 1997). First, there is a growing realization that more in-depth/details in a particular circuit are necessary to achieve the ultimate understanding. Indeed, this line of thinking has motivated attempts to establish connectomes in some model systems. Thus, for this book, we have assembled a group of experts who work on tractable, well-defined circuits in both invertebrate and vertebrate model systems, many of which have been studied intensively over decades and are more likely to offer a full understanding. The investigators utilize a variety of methods, including molecular, genetic, physiological, biophysical, pharmacological, systems, and computational. Second, we emphasize functional understanding of each given circuit. Specifically, rather than cataloging circuit properties, we try to delineate them that are most relevant to specific functions of the circuit under study.

Third, we focus on the roles of electrical coupling/gap junctions (see Chapter 8 by Drs. Pereda and Bennett for a historical account of the discovery of electrical transmission in invertebrates and vertebrates, and the Glossary for relevant terms) in a circuit. Electrical coupling is initially identified based on electrophysiological recordings. Gap junctions refer to the morphological basis of electrical coupling and are composed of innexin proteins in prechordates and connexin proteins in chordates. Innexins and connexins are not homologous. In chordates, proteins that are homologous to innexins are present and termed pannexins but typically do not mediate electrical coupling. Recently, there are suggestions that innexins should be called pannexins to simplify the terminology (Shestopalov and Panchin, 2008). Although often considered secondary to chemical synapses and therefore likely overlooked in some instances, electrical coupling is present in a number of circuits including many circuits in the mammalian nervous systems. As can be seen from this book, when elucidated in simpler neural circuits, electrical coupling can be vital to functional network output. Indeed, the chapters in the book have clearly illustrated that electrical coupling is critical for sensory and motor functions, neural computations, decision-making, regulation of network activity, learning and memory, and circuit development. In addition, we also consider what "special" properties electrical coupling/gap junctions may bring to a circuit and use these findings to evaluate how chemical synapses may function differently in a circuit, thus offering multiple perspectives toward understanding circuit functions and ultimately the behavior they mediate.

Specifically, most of the chapters in this book focus on one or more circuits within different animal model systems. The first chapters mainly describe the invertebrate neural circuits of *Caenorhabditis elegans*, *Drosophila*, crustacean, leech, and molluscs (eg, *Aplysia*). Subsequent chapters describe circuit function in vertebrates. Each chapter focuses on either network functions or plasticity (Table 1). Chapter 1 (Rabinowitch and Schafer) describes a network motif: a hub-and-spoke gap junction circuit in *C. elegans* that mediates the integration of mechanosensory information to control nose touch avoidance behavior. Most interestingly, the authors describe the novel synthetic biology approach that was used to demonstrate the roles of electrical coupling in network function. Chapter 2 (Phelan, Bacon, and Blagburn) presents an integrated view of the *Drosophila* giant fiber escape system with a focus on the roles of electrical coupling/gap junctions. In particular, the authors demonstrate the roles of rectifying synapses in the circuit. Chapter 3 (Shih and Wu) describes the roles of gap junctions between two *Drosophila* modulatory neurons in a specific form of classical conditioning, using a variety of approaches. Chapter 4 (Nadim, Li, Gray, and Golowasch) explores theoretical and experimental evidence for the role of electrical coupling in small oscillatory networks of crustacean, leech, and other model systems. Chapter 5 (Jing, Cropper, and Weiss) describes how electrical coupling in multiple but specific sites may play a role in rhythm generation in a local circuit and the propagation of activity between projection neurons and remote circuit elements in the feeding and locomotor networks of gastropod molluscs, such as *Aplysia*, *Lymnaea*, *Clione*, *Tritonia*, and *Pleurobranchaea*. The chapter illustrates that a thorough understanding of the circuit localization of electrical coupling provides novel insights into its functions. Chapter 6 (Nargeot and Bédécarrats) summarizes learning-induced plasticity in electrical coupling and intrinsic excitability in identified neurons of the feeding circuit in *Aplysia californica*. In addition, it provides an extensive review of relevant work in other model systems including vertebrates. Interestingly, the plasticity in electrical coupling and excitability in *Aplysia* underlie different aspects of learning-induced changes in rhythmic motor output, ie, more regularity and higher frequency, respectively. Chapter 7 (Magoski) provides in-depth, valuable information on the role of electrical coupling in neuroendocrine neurons in a variety of species, including invertebrates and vertebrates.

Chapter 8 (Pereda and Bennett) presents a historical, current, and future account of electrical coupling studies in fishes. Indeed, fish research provided one of the first evidence of electrical coupling/gap junctions in vertebrates. Chapter 9 (Trenholm and Awatramani) describes critical roles of electrical coupling in a variety of functions including visual processing, neural computations, and disease states in the retina. Chapter 10 (Ribelayga and O'Brien) describes extensive evidence that sheds light on how intricate interactions of neuromodulators and signaling pathways can modulate dynamic

Table 1 Chapter Summary

Chapter Number and the Authors	Model Systems	Types of Networks	General Roles	More Specific Roles (see Glossary and Index)	Methods
1. Rabinowitch and Schafer	*Caenorhabditis elegans*	Sensory	Network operations	Nose-touch circuit, network motif, coincidence detection, innexin	Genetic, physiological, computational
2. Phelan, Bacon, and Blagburn	*Drosophila*	Sensory motor	Network operations	Escape circuit, rectifying synapses, heterotypic channels, innexin	Genetic, physiological, biophysical
3. Shih and Wu	*Drosophila*	Memory	Plasticity	Classic conditioning, modulatory neurons, heterotypic gap junctions, innexin	Genetic, physiological
4. Nadim, Li, Gray, and Golowasch	Crustacean, leech, and other systems	Motor	Network operations	Feeding, rectifying synapses, rhythm generation, oscillation, innexin	Physiological, computational
5. Jing, Cropper, and Weiss	*Aplysia*, *Lymnaea* and other gastropod molluscs	Motor	Network operations	Feeding and locomotion, heterogeneous elements, rhythm generation, functional modules	Physiological, pharmacological
6. Nargeot and Bédécarrats	*Aplysia*, other model systems	Motor	Plasticity	Feeding, reward, operant conditioning, decision making	Physiological, pharmacological, biophysical
7. Magoski	Invertebrates and vertebrates	Endocrine	Network operations	Network activity, hormone, peptidergic neurons, innexin	Physiological, pharmacological
8. Pereda and Bennett	Vertebrates (fishes)	Sensory and motor	Network operations	Synchrony, mixed chemical and electrical connections, connexin, neuromodulation	Genetic, physiological, computational

9. Trenholm and Awatramani	Vertebrates	Sensory (retina)	Network operations	Visual processing, neural computation, retinal degeneration, connexin	Physiological, pharmacological, computational
10. Ribelayga and O'Brien	Vertebrates	Sensory (retina)	Plasticity	Neuromodulation, signaling pathways, circadian and light-adaptive control, dopamine, connexin	Physiological, pharmacological
11. Li and Rekling	Vertebrates	Motor (spinal cord, brainstem)	Network operations	Chewing, locomotion and respiration, rhythm generation, connexin	Physiological, pharmacological
12. Welsh and Turecek	Mammals	Motor (inferior olive)	Plasticity	Oscillation, synchrony, motor learning, connexin	Physiological, pharmacological
13. Traub, Whittington, and Draguhn	Mammals	Hippocampus, cortex	Network operations	Very fast oscillations, gap junctions between axons of pyramidal neurons, connexin	Physiological, pharmacological, computational
14. He and Shi	Mammals	Neocortex	Development	Lineage-dependent electrical synapse formation, formation of cortical microcircuits, connexin	Genetic, physiological, pharmacological

properties of electrical coupling between sensory circuit elements and thereby contribute to circadian and light-adaptive control of information processing in retina. Chapter 11 (Li and Rekling) reviews the roles of electrical coupling in vertebrate rhythm generation, including locomotion and respiration. Chapter 12 (Welsh and Turecek) describes recent experiments indicating the modifiability of electrical coupling/gap junctions and its functional consequence on the network activity of the mammalian inferior olive. Chapter 13 (Traub, Whittington, and Draguhn) describes the roles that gap junctions between axons of pyramidal neurons in cortex and hippocampus play in one form of network oscillations. Chapter 14 (He and Shi) describes extensive and elegant experimental evidence that demonstrates the role of lineage-dependent electrical synaptic formation in generating functional cortical microcircuit during development.

This book is necessarily a snapshot of the current progress in a given field due to space limitations, but some general conclusions/principles regarding circuit functions can already be drawn from the book. I encourage you to read through chapters that interest you, and I am confident that you will not be disappointed. Here, as an editor, I can offer the following take-home messages. First, each circuit in a model system offers its own distinct advantage, and therefore each investigator tends to focus on different aspects of electrical coupling/gap junctions. Thus, students who work on circuits in different model systems would benefit from learning from each other. For example, when three or more neurons are electrically coupled, a genetic organism like *C. elegans* is advantageous in identifying gap junctions between each pair of neurons and therefore specific network motifs such as hub-and-spoke gap junction circuit (Chapter 1). Such motifs would be more difficult to ascertain in other animal models. So the question arises: is the network motif described in *C. elegans* present in other (eg, molluscan) model systems? Conversely, are functional modules defined by electrical coupling in the molluscan feeding circuit (Chapter 5) ubiquitous in other (eg, *C. elegans*) model systems? In addition, is the elaborate neuromodulation of electrical coupling described in the retina (Chapter 10) present in other circuits? More generally, the different research focuses highlight and reinforce the utility of diverse circuits with varying degrees of complexity.

Second, a major question is whether electrical coupling/gap junctions are present in a circuit. Although chemical synapses may be the dominant form of neuron-to-neuron communication in the adult nervous systems of mammals (Chapters 11 and 14 but see Chapter 12), a recurrent theme is that the presence of electrical coupling/gap junctions is often underestimated. As described in Chapter 1, this appears to be the case in the only complete connectome (*C. elegans*) currently available. Chapter 5 showed that electrical coupling can be overlooked between projection neurons and remote circuit elements, even in the *Aplysia* model system where electrical coupling is usually easy to detect between identified neurons. Thus when electrical coupling is not found in a particular circuit, one might ask how conclusive the data are (see Chapter 11 about adult motor circuits in vertebrates). The answer can be tricky when electrical coupling may only be present in a small subset of neurons in mammalian circuits where a large number of circuit elements are involved (Chapter 13). Another issue that may make the identification of electrical coupling complicated is the fact that chemical and electrical synapses can coexist between two neurons (Pereda, 2014), as is also shown in this book (eg, Chapters 2–5 and 8). One important issue is what are the extent and functions of this type of synapses that may occur in mammalian circuits. Moreover, although electrical coupling differs from chemical synapses in that it often allows electrical signals to transmit bidirectionally, rectifying electrical coupling and the underlying heterotypic gap junctions (eg, Chapters 2–5 and 8) do enable effective unidirectional signal transmission as chemical synapses do. Thus, it is also important to determine the circuit localization and functions of rectifying electrical coupling.

Third, electrical coupling, wherever it is present, is an essential component in a circuit, whether driving network activity or contributing to plasticity and development. For example, electrical coupling plays various roles in sensory processing and neural computation in both invertebrates and vertebrates (Chapters 1, 9, and 10). Moreover, neural networks are not just passive machines that are only driven by inputs from the outside world. Rather, they generate intrinsic network activities, many of which are rhythmic and underlie behavioral and mental activities (Buzsaki, 2006; Grillner, 1985; Llinas, 2001; Marder and Calabrese, 1996; Stein et al., 1997; Traub and Whittington, 2010). Indeed, electrical coupling plays a major role in the generation of these rhythmic, oscillatory activities in some circuits (Chapters 4–6, 11–13). In these rhythm-generating circuits, electrical coupling is often present in neurons that fire at the same phase (but see Chapter 4). Interestingly, within a single phase, electrical coupling may be present in subgroups of neurons, which may form distinct functional modules (Chapter 5). Plasticity in electrical coupling may also contribute to specific forms of classical conditioning (Chapter 3) and operant conditioning (Chapter 6). The important lesson is that just like chemical synapses, electrical synapses in a circuit can contribute specifically to aspects of learning and memory. Electrical coupling may also be important during development so as to form functional circuits in some mammalian circuits but not in others (Chapter 14).

Gap junction proteins, especially connexins and pannexins in chordates, are also present and play diverse roles in nonneuronal cells, eg, glial cells. This subject is beyond the scope of this book and readers are referred to Dere, 2012 and Chapter 9 of Byrne et al., 2014 for more information.

Finally, I would like to thank all the chapter authors for their hard work and marvelous contributions. Some of you, as well as my friend Dr. Nikolai Dembrow, also contributed by providing valuable comments to other chapters, the Preface and/or the Glossary. I would like to thank coauthors of my chapter, Drs. Klaudiusz R. Weiss and Elizabeth C. Cropper, also my mentors and long-time collaborators, for their help with the research that forms the basis for the chapter. This book is partly based on a 2013 Society for Neuroscience symposium (titled "Electrical Coupling and Microcircuits: Network Operation and Plasticity") organized by the editor, held at San Diego, California, USA. Thanks to Natalie Farra, Senior Acquisition Editor at Elsevier, for the initial suggestion for the book, and encouragement and trust throughout the process, and Kathy Padilla, Editorial Project Manager, Kirsty Halterman and Karen East, Production Project Managers, for editorial assistance. With efforts from all of you, the book becomes a reality. It is the editor's hope that the book will contribute significantly to the advance in the circuitry studies of electrical coupling/gap junctions, and ultimately, to the understanding of our brains.

Jian Jing
Nanjing, China
October, 2016

REFERENCES

Buzsaki, G., 2006. Rhythms of the Brain. Oxford University Press, New York, NY.

Byrne, J.H., Heidelberger, R., Waxham, M.N. (Eds.), 2014. From Molecules to Networks. Academic Press, London, UK.

Dere, E. (Ed.), 2012. Gap Junctions in the Brain: Physiological and Pathological Roles. Academic Press, London, UK.

Grillner, S., 1985. Neurobiological bases of rhythmic motor acts in vertebrates. Science 228 (4696), 143−149.

Jing, J., 2009. Command systems. In: Squire, L.R. (Ed.), Encyclopedia of Neuroscience, vol. 2. Academic Press, Oxford, UK, pp. 1149−1158.

Kandel, E., Schwartz, J.H., Jessell, T.M., Siegelbaum, S.A., Hudspeth, A.J. (Eds.), 2013. Principles of Neural Science. McGraw-Hill, New York, NY.

Llinas, R.R., 2001. I of the Vortex: From Neurons to Self. MIT Press, Cambridge, MA.

Marder, E., Calabrese, R.L., 1996. Principles of rhythmic motor pattern generation. Physiol. Rev. 76 (3), 687−717.

Mashour, G.A., Alkire, M.T., 2013. Evolution of consciousness: phylogeny, ontogeny, and emergence from general anesthesia. Proc. Natl. Acad. Sci. U.S.A. 110 (Suppl. 2), 10357−10364.

North, G., Greenspan, R. (Eds.), 2007. Invertebrate Neurobiology. Cold Spring Harbor Laboratory Press, Cold Spring Harbor, NY.

Orlovsky, G.N., Deliagina, T.G., Grillner, S., 1999. Neural Control of Locomotion. Oxford University Press, New York, NY.

Pereda, A.E., 2014. Electrical synapses and their functional interactions with chemical synapses. Nat. Rev. Neurosci. 15 (4), 250−263.

Shepherd, G.M. (Ed.), 2004. The Synaptic Organization of the Brain. Oxford University Press, New York, NY.

Shepherd, G.M., Grillner, S. (Eds.), 2010. Handbook of Brain Microcircuits. Oxford University Press, New York, NY.

Shestopalov, V.I., Panchin, Y., 2008. Pannexins and gap junction protein diversity. Cell. Mol. Life Sci. 65 (3), 376−394.

Stein, P.S., Grillner, S., Selverston, A.I., Stuart, D.G. (Eds.), 1997. Neurons, Networks, and Motor Behavior. MIT Press, Cambridge, MA.

Traub, R., Whittington, M., 2010. Cortical Oscillations in Health and Disease. Oxford University Press, New York, NY.

CHAPTER 1

Electrical Coupling in *Caenorhabditis elegans* Mechanosensory Circuits

I. Rabinowitch[1], W.R. Schafer[2]
[1]Fred Hutchinson Cancer Research Center, Seattle, WA, United States; [2]MRC Laboratory of Molecular Biology, Cambridge, United Kingdom

1. INTRODUCTION

The nematode *Caenorhabditis elegans* is in many ways an ideal organism to investigate microcircuits and their roles in behavior. It is currently the only organism with a complete physical connectome; each of its 302 neurons has been individually identified and its synaptic and gap junctional connections mapped at the level of electron microscopy. It is also highly accessible to genetic manipulation, with a sequenced genome, a short generation time, and amenability to transgenesis and gene replacement. Moreover, its transparency and compactness have made it well suited for optogenetic manipulation and recording of neural activity in behaving animals. Together, these tools make it possible to dissect how the interactions between defined neurons generate the functional properties of microcircuits, and how those properties relate to whole animal behavior.

Gap junctions form an important component of the *C. elegans* connectome. The published *C. elegans* "wiring diagram" includes approximately 900 gap junctions along with 8000 chemical synapses (White et al., 1986). Analyses using modern machine vision methods (Xu et al., 2013) suggest this is an underestimate, with over 4000 gap junctions reported in data published online (wormwiring.org). Like other invertebrates, *C. elegans* gap junctions are formed from innexins rather than connexins (Altun et al., 2009; Simonsen et al., 2014). The *C. elegans* genome contains 25 innexin genes, 20 of which are neuronally expressed (Altun et al., 2009). These show varying patterns of expression, some expressed widely and others in only a few neurons. A few have been shown to have behavioral phenotypes; for example, loss-of-function mutations in *unc-7* and *unc-9*, which are expressed in motorneurons and premotor interneurons, result in strongly uncoordinated movement (Kawano et al., 2011).

The pattern of gap junction connections in the worm nervous system has been analyzed with the goal of identifying motifs of potential functional importance. In particular, the frequencies of all possible three-neuron and four-neuron connectivity patterns have been determined and compared with their expected frequencies in a random network (Varshney et al., 2011). Overrepresented patterns, such as a triangular connection of three neurons, might represent microcircuit elements with a conserved function in computation. One

Network Functions and Plasticity
ISBN 978-0-12-803471-2, http://dx.doi.org/10.1016/B978-0-12-803471-2.00001-1

overrepresented motif observed in the four-neuron analysis is the hub and spoke, in which a single hub neuron is connected to each of the other three neurons ("spokes"). Another overrepresented four-neuron motif is the "diamond" motif whereby all neuron pairs except for one are connected by gap junctions (Varshney et al., 2011). The hub-and-spoke architecture, whereby multiple neurons are connected to one central neuron, is a recurring circuit motif in the *C. elegans* connectome (Varshney et al., 2011); indeed, larger hub-and-spoke circuits, with a single hub receiving gap junctions from a large number of "spoke" inputs, are not uncommon. For example, the hub-and-spoke circuit in which many sensory neurons of varying modalities are connected by gap junctions to interneurons called RMG has been shown to control aggregation behavior, and to modulate responses to nematode pheromones (Macosko et al., 2009; Jang et al., 2012).

We have undertaken an analysis of a simpler hub-and-spoke network involved in nose touch (Chatzigeorgiou and Schafer, 2011; Rabinowitch et al., 2013). This circuit involves a small number of input neurons of a single (mechanosensory) modality, and the behavioral output, an escape response called a reversal, is robust and easily correlated with the activity of the hub neuron. From these studies, we have aimed to uncover general principles of how hub-and-spoke circuits process information and control behavior.

2. THE NOSE TOUCH CIRCUIT

The natural habitat of *C. elegans* consists of soil and rotting fruit. With no sense of vision, it relies heavily on a range of mechanosensory cues to navigate, locate food, interact with conspecifics, and avoid threats. Such complex interaction with the environment presents several challenges to the worm's mechanosensory system. It must be able to discriminate between different textures and patterns, distinguishing, for example, between food (bacteria), soil particles, a mating partner, and a predator. In addition, its dynamic range must be extensive enough to detect both the gentlest and harshest mechanical inputs. Gap junctions might be useful building blocks for neural circuits that implement these features. We demonstrate this in the nose touch circuit, one of several neural circuits involved in mechanosensation in *C. elegans*. Other circuits include the polymodal nociceptive circuit involving the ASH neurons (Kaplan and Horvitz, 1993; Hart et al., 1995), the gentle body touch circuit (Chalfie et al., 1985), and the harsh body touch circuit (Way and Chalfie, 1989).

The nose touch circuit is important for the transduction and processing of mechanosensory information sensed by the nose, often the first body part to come into contact with the changing texture that the worm encounters as it navigates through its surroundings. The circuit comprises several classes of mechanosensory neurons. The neurons in each class share a distinct morphology, are equipped with specific mechanoreceptors, and are linked to separate downstream circuits (Fig. 1.1): Four CEP neurons extend their dendrites to the tip of the nose and require the transient receptor potential N

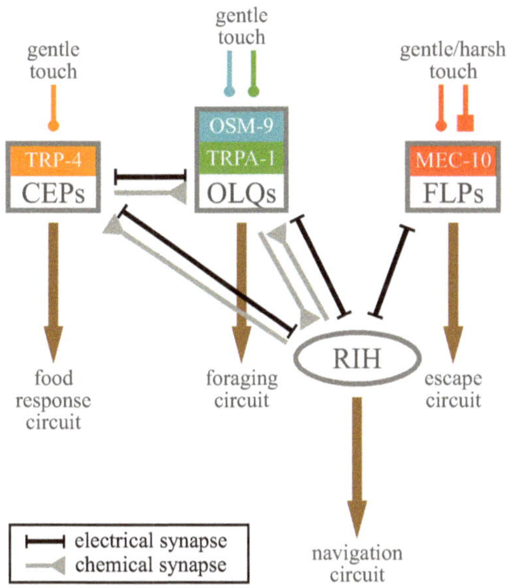

Figure 1.1 ***The* C. elegans *nose touch circuit.*** The *C. elegans* nose touch circuit consists of the CEP, OLQ, and FLP mechanosensory neurons (rectangles), each expressing specific mechanoreceptors (indicated by different colors), each contributing to a specific circuit (indicated by downward pointing arrows), and all connected by gap junctions to a single interneuron, RIH (oval), whose output contributes to the navigation circuit.

channel TRP-4 for mechanosensory transduction (Li et al., 2006; Kindt et al., 2007a; Kang et al., 2010). These neurons are dopaminergic and are involved in modifying locomotion on physical contact with food (Sawin et al., 2000). Four OLQ neurons have similar morphology to the CEPs. However, they use different mechanoreceptors, the transient receptor potential V channel OSM-9 (Colbert et al., 1997; Chatzigeorgiou and Schafer, 2011), and the transient receptor potential A (TRPA) channel TRPA-1 (Kindt et al., 2007b). These neurons are involved in controlling foraging and head withdrawal. Two FLP neurons have multidendritic processes. This is a rare morphology for *C. elegans* neurons, most of which have a simple bipolar structure. They use the degenerin/epithelial-like sodium channel (DEG/ENaC) channel MEC-10 to detect both gentle and harsh mechanical contact with the nose (Huang and Chalfie, 1994; Chatzigeorgiou and Schafer, 2011). The FLP neurons form part of an escape mechanism responsive to noxious physical stimulation of the nose.

What drew our attention to the nose touch circuit was the observation that FLP responses to gentle nose touch, as measured by calcium imaging, depended only partially on the cell-autonomous activity of the DEG/ENaC channel MEC-10, which is completely necessary for responses in FLP evoked by harsh mechanical contact with the nose, but on the other hand, non—cell autonomously, on functional OSM-9 in the OLQs (Chatzigeorgiou and Schafer, 2011). This suggested that the activity of other sensory neurons in the circuit facilitated FLP activation. The CEPs, OLQs, and FLPs are all connected via gap junctions to a single neuron RIH, an interneuron with a role in navigation (Fig. 1.1).

We found that this facilitation is conveyed by the gap junctions connecting the OLQ neurons to RIH, and the FLP neurons to RIH. Loss of OSM-9 abolished the input from OLQ that contributed to part of the FLP response to gentle touch (Chatzigeorgiou and Schafer, 2011).

3. SIMPLIFIED MATHEMATICAL MODEL OF THE NOSE TOUCH CIRCUIT

To better understand how the nose touch hub-and-spoke gap junction circuit works, we formulated a simplified mathematical model describing current flow between two input neurons and a hub interneuron (Fig. 1.2) (Rabinowitch et al., 2013). First we took into account the flow of current through the membrane of each individual neuron in the circuit. The neuron's membrane is typically modeled as a resistor and capacitor operating in parallel (eg, hub neuron in Fig. 1.2). The resistor, which represents membrane conductance G_m, corresponds to the passive ion channels embedded in the cell's membrane, through which current can leak into or out of the cell. Electrical and chemical gradients

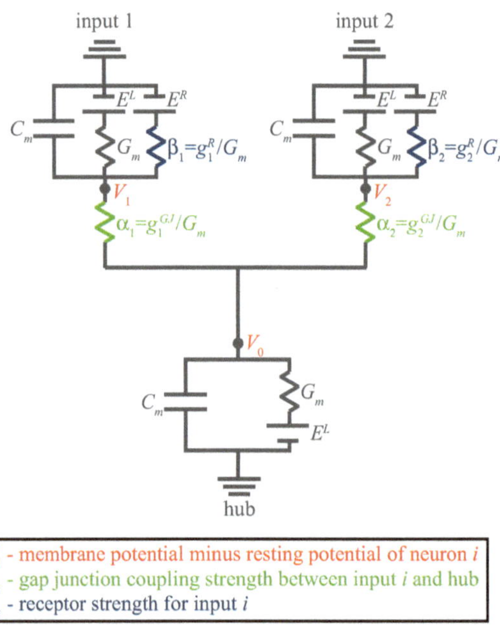

Figure 1.2 ***Model of a simplified hub-and-spoke circuit.*** Each neuron is modeled by a capacitor in parallel to a driving force and resistor in series. In the sensory input neurons an additional driving force and resistor in series, representing the sensory receptor, are connected in parallel. Gap junctions between the input neurons and the hub are represented by resistors.

formed across the membrane create a driving force, represented as a voltage source, E^L, that drives current through these membrane channels. Accumulation of electrical charge on the membrane produces membrane capacitance, C_m, represented in the equivalent circuit as a capacitor. The activity of *C. elegans* neurons is characterized mostly by graded potentials rather than by spiking temporal patterns, which allowed us to ignore discrete brief electrical events at faster timescales. In addition to these passive components, the input neurons include in parallel a sensory receptor driving force, E^R, and conductance, g^R. We express this conductance in terms of membrane conductance as $\beta = g^R / G_m$ (see input neurons in Fig. 1.2).

The gap junctions connecting between the input and hub neurons were modeled as additional resistors with conductance g^{GJ}. We define the gap junction coupling strength as $\alpha = g^{GJ} / G_m$. A system of three differential equations fully captures the flow of current in or out of each neuron. We focused our analysis on the steady-state membrane potential of the model neurons. Further details about the analysis and the assumptions underlying the model can be found in Rabinowitch et al., 2013. The model serves as an effective tool for addressing questions and deriving experimental predictions about coincidence detection and gain control in gap junction circuits. In particular, we have investigated both theoretically and experimentally two possible mechanisms for nose touch detection: lateral facilitation and shunting inhibition.

4. LATERAL FACILITATION

The activity of one input neuron might be enhanced by the activity of a second input neuron through lateral facilitation (Fig. 1.3A). Such lateral facilitation can be mediated by gap junctions by allowing current to flow from one input neuron to the other, via the hub. However, because current flows away from one neuron to facilitate the other, the activity in that first neuron should diminish, implying that lateral facilitation is asymmetric. To investigate what determines the direction and extent of lateral facilitation in the model, we compared the steady-state membrane potentials of input 1, either when both inputs 1 and 2 are activated (Fig. 1.3A), or when input 1 is activated but completely isolated from the circuit (Fig. 1.3B). A positive difference between the two conditions corresponds to a facilitation of input 1 by input 2.

As an example, we have plotted the percent facilitation of input neuron 1 as a function of receptor strength for the case whereby the coupling strengths of both gap junctions are equal, $\alpha_1 = \alpha_2 = 0.5$ (Fig. 1.3C). Notably, neuron 1 can be facilitated by neuron 2 only when the strength of receptor 1 is considerably weaker than that of receptor 2. This stands to reason because in order for neuron 2 to facilitate neuron 1 the membrane potential of neuron 2 must be higher than that of neuron 1. However, it is not just the relative level of activity between the two neurons that matters, the gap junction coupling strengths also play an important role. In Fig. 1.3D, we have plotted the percent

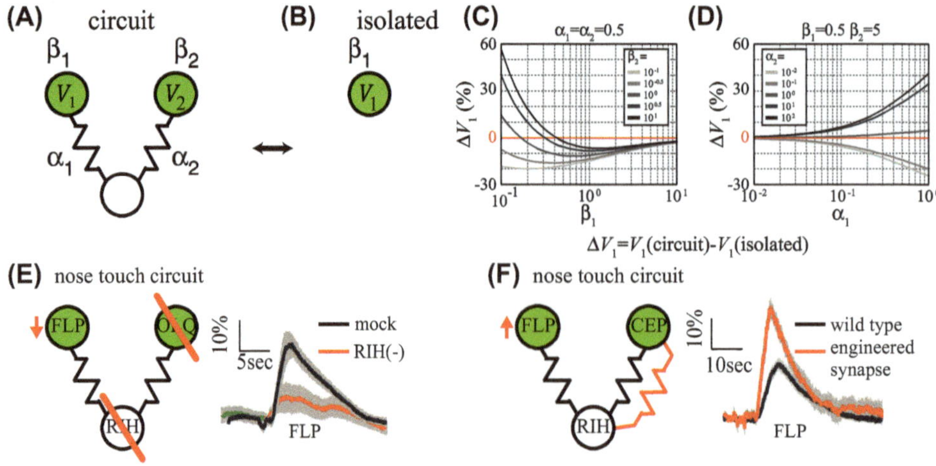

Figure 1.3 ***Lateral facilitation.*** (A) Hub-and-spoke circuit configuration, in which both input neurons are active. (B) Only input 1 is active and is isolated from the circuit. (C, D) Percentage difference between the steady-state voltage of input neuron 1 when connected to the circuit compared with when in isolation, as a function of transduction strength (C) or gap junction coupling strength (D). (E) Input neuron 1 (FLP) calcium response to gentle nose touch decreases (downward arrow in schematic diagram) when input neuron 1 is isolated from the hub-and-spoke circuit. (F) Input neuron 1 (FLP) calcium response to gentle nose touch increases (upward arrow in schematic diagram) when the electrical coupling between input neuron 2 (CEP) and the hub output neuron (RIH) is enhanced by synaptic engineering.

facilitation of input neuron 1 as a function of gap junction coupling in a circuit in which the receptor strength of neuron 1 is much weaker than that of neuron 2 ($\beta_1 = 0.5 < \beta_2 = 5$). The extent to which neuron 2 influenced the activity of neuron 1 directly depended on the coupling strength of input neuron 1 and the hub (Fig. 1.3D). Very weak coupling entailed almost no change in V_1, because very little current could make its way to neuron 1. In contrast, stronger coupling resulted in either considerable facilitation or weakening of V_1. The coupling strength between neuron 2 and the hub determined whether neuron 1 activity would be facilitated or weakened. Strong coupling of neuron 2 and the hub allows for current to flow from neuron 2 to the hub and then to neuron 1. Weak coupling between neuron 2 and the hub isolates neuron 2 and draws current away from neuron 1 into the hub, reducing the activity of neuron 1.

As described earlier, we have found that the response of the FLP neurons to gentle touch is facilitated by the OLQ neurons (Fig. 1.3E) (Chatzigeorgiou and Schafer, 2011). Consistent with our model, this might be due to relatively weakened responsiveness of FLP to gentle touch compared with the OLQ response.

Our model also predicts that strong coupling of both FLP and OLQ to the hub interneuron RIH is important for enabling the facilitation of FLP. Thus, according to our model, FLP responses might be even further facilitated if, for example, the gap junction coupling between another input neuron and the hub RIH were stronger (Fig. 1.3D). To test this prediction, we artificially increased the gap junction coupling strength between the CEP input neurons and RIH and measured the effect on FLP calcium responses. We did this by using a synaptic engineering approach (Rabinowitch et al., 2014; Rabinowitch and Schafer, 2015). We expressed the mouse gap junction protein Connexin36 in the CEP and RIH neurons, to form additional gap junctions between these neurons. Because vertebrate gap junction proteins, such as Connexin36, belong to a different family than invertebrate gap junction proteins, which are called innexins, we expected no nonspecific heterotypic gap junctions to form between the engineered Connexin36 and endogenous *C. elegans* innexins. We found that, indeed, as the model predicted, increased gap junction coupling between CEP neurons and RIH enhanced the FLP response to gentle touch to the nose (Fig. 1.3F) (Rabinowitch et al., 2013).

5. INHIBITION BY SHUNTING

Lateral facilitation consists of one active neuron contributing to the activity of another active neuron. Another form of interaction in gap junction circuits may occur when one input neuron is active and another input neuron is not. This might happen following certain stimulation patterns that evoke a specific response only in neurons that are tuned to the particular features of the stimulus, whereas other neurons remain silent. We wished to examine using our model, the effect that the inactivity of input neuron 2 might have on input neuron 1 (Fig. 1.4A). To this end we considered the percent difference in the membrane potential of neuron 1, V_1, when neuron 2 is inactive compared with when neuron 2 is removed from the circuit, ie, is ablated (Fig. 1.4A and B).

We found that for all values of gap junction coupling strength, α_1, a_2, and receptor strength, β_1, β_2, the membrane potential of input neuron 1, V_1, was always inhibited by the inactivity of input neuron 2; ie, the difference in V_1 steady-state values between silencing input 2 and ablating it was always negative. This kind of electrical suppression of membrane potential is called shunting inhibition. We demonstrated shunting inhibition experimentally, taking advantage of the fact that each neuron class in the nose touch circuit relies on different receptors for transduction (Fig. 1.1), and is therefore selectively silent in mutants lacking the relevant functional receptor. Thus, in worms with dysfunctional OSM-9 receptors, and thus silent OLQ neurons, FLP and RIH calcium responses to nose touch were diminished, unless the OLQ neurons were ablated. Similarly, in *trp-4* mutants with silent CEPs, FLP and RIH responses were reduced unless the CEPs were ablated, and in worms with defective MEC-10, and thus silent FLPs, CEP and RIH responses were attenuated, but normal if the FLPs were ablated (Rabinowitch et al., 2013).

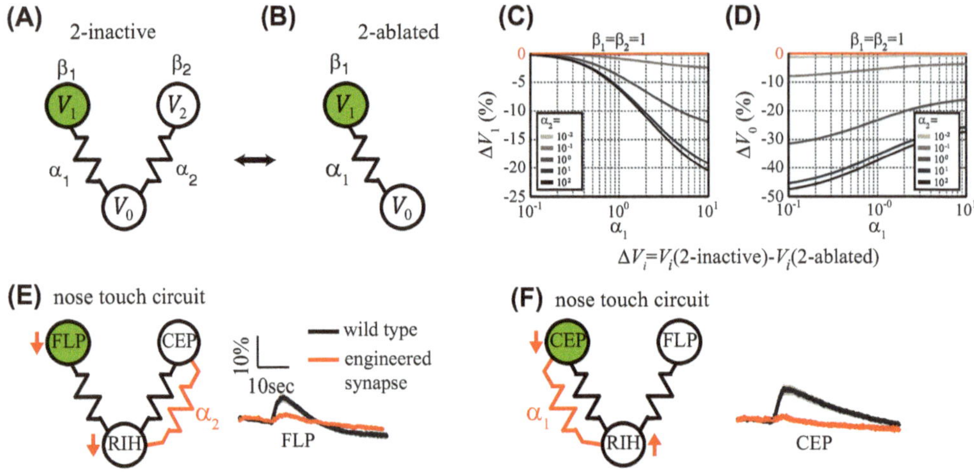

Figure 1.4 ***Inhibition by shunting.*** (A) Hub-and-spoke circuit configuration, in which input neuron 1 is active and input neuron 2 is silent. (B) Circuit configuration, in which input neuron 1 is active and input neuron 2 is ablated. (C, D) Percentage difference between the steady-state voltage of input neuron 1 (C), or the hub output neuron (D) when one input neuron is active and the other is silent compared with when one input neuron is active and the other is ablated, as a function of gap junction coupling strength. (E) Input neuron 1 (FLP) calcium response to gentle nose touch decreases (*downward arrow* in schematic diagram) when the coupling between input neuron 2 (CEP) and the hub output neuron (RIH) is enhanced by synaptic engineering. (F) Input neuron 1 (CEP) calcium response to gentle nose touch decreases (*downward arrow* in schematic diagram) when the electrical coupling between input neuron 1 and the hub output neuron (RIH) is enhanced by synaptic engineering.

We further found from the model that the degree of shunting inhibition depended strongly on gap junction coupling strength. As can be seen in the example in Fig. 1.4C and D, stronger coupling to the hub of input neuron 2 produced more shunting inhibition of V_1 as well as the hub membrane potential, V_0, because more current could get drawn away from neuron 1 through the hub into neuron 2. In contrast, stronger gap junction coupling between input neuron 1 and the hub also increased the shunting inhibition of neuron 1, whereas it reduced the inhibition of the hub. The reason is that more current drawn out of input neuron 1 could remain in the hub rather than flow into input neuron 2 (Fig. 1.4D).

We used synaptic engineering once more to test these predictions. As earlier, we examined worms with a gap junction inserted between CEP and the hub, RIH. First, we considered CEP as input neuron 2, and used *trp-4* mutants to simulate silencing of the CEPs. In this case it was α_2, the coupling strength between neuron 2 and the hub that was increased by the engineered gap junction (Fig. 1.4E). As predicted by the model, strengthening α_2 resulted in decreased calcium responses to nose touch in both FLP (input neuron 1 in this case) and RIH (the hub). Second, we switched roles between CEP and FLP. Now we used *mec-10* mutants to silence

FLP (input neuron 2). Consistent with the model, strengthening α_1 this time, led to reduced calcium responses in CEP (input neuron 1), but to increased responses in RIH (the hub) (Fig. 1.4F).

6. CONCLUSIONS AND FUTURE PERSPECTIVES

These studies illustrate how the function of a simple microcircuit can rely on the unique properties of electrical synapses. Electrical coupling between sensory neurons, either directly or indirectly through a hub interneuron, can enhance coincident responses through lateral facilitation. Conversely, in the absence of coincident activity, current can be shunted through hub neurons to inactive inputs, which act as current sinks and inhibit overall network activity. Together, lateral facilitation and inhibition by shunting allow the hub-and-spoke gap junction circuit to function as an analog coincidence detector, with inputs summing nonlinearly in the network output. Synthetic increases in gap junction coupling to the hub can increase both facilitation and inhibition, making network output depend more strongly on coincidence. Although this work concerns a specific hub and spoke circuit in the worm, the computational properties of the circuit follow straightforwardly from its architecture; thus, it seems reasonable to imagine hub-and-spoke circuits might perform similar functions in more complex brains.

The use of ectopic connexin expression to create new electrical synapses has been useful in the analysis of electrically coupled circuits in *C. elegans*, and it may have broader applications as well (Rabinowitch et al., 2014, 2016). Other invertebrates, such as *Drosophila,* also lack endogenous connexin genes, so targeted Cx36 expression should allow the creation of synthetic electrical synapses in other genetically tractable invertebrate model organisms. Moreover, the vertebrate counterparts of innexins, the pannexins, are thought to form only hemichannels and are therefore unlikely to form gap junctions with a nematode innexin partner. Thus, the converse approach of ectopically expressing worm innexins in vertebrate neurons may allow the insertion of ectopic electrical synapses in animals such as the mouse. Connexin and innexin transgenesis may not only prove to be useful tools for the analysis of neural circuit mechanisms, but also allow for neural repair or even the engineering of new functions in transformed animals. This could represent a first step toward synthetic neurobiology.

OUTSTANDING QUESTIONS/FUTURE DIRECTIONS

- Do other hub-and-spoke circuits in *C. elegans* perform coincidence detection?
- Do hub-and-spoke circuits in other animals perform coincidence detection, and how does the existence of action potentials affect their function?
- What kind of functions do other gap junction circuit motifs in *C. elegans* perform?
- Is synaptic engineering effective in model organisms other than *C. elegans*?

REFERENCES

Altun, Z.F., Chen, B., Wang, Z.-W., Hall, D.H., 2009. High resolution map of *Caenorhabditis elegans* gap junction proteins. Dev. Dyn. 238, 1936–1950.

Chalfie, M., Sulston, J.E., White, J.G., Southgate, E., Thomson, J.N., Brenner, S., 1985. The neural circuit for touch sensitivity in *Caenorhabditis elegans*. J. Neurosci. 5, 956–964.

Chatzigeorgiou, M., Schafer, W.R., 2011. Lateral facilitation between primary mechanosensory neurons controls nose touch perception in *C. elegans*. Neuron 70, 299–309.

Colbert, H.A., Smith, T.L., Bargmann, C.I., 1997. OSM-9, a novel protein with structural similarity to channels, is required for olfaction, mechanosensation, and olfactory adaptation in *Caenorhabditis elegans*. J. Neurosci. 17, 8259–8269.

Hart, A.C., Sims, S., Kaplan, J.M., 1995. Synaptic code for sensory modalities revealed by *C. elegans* GLR-1 glutamate receptor. Nature 378, 82–85.

Huang, M., Chalfie, M., 1994. Gene interactions affecting mechanosensory transduction in *Caenorhabditis elegans*. Nature 367, 467–470.

Jang, H., Kim, K., Neal, S.J., Macosko, E., Kim, D., Butcher, R.A., Zeiger, D.M., Bargmann, C.I., Sengupta, P., 2012. Neuromodulatory state and sex specify alternative behaviors through antagonistic synaptic pathways in *C. elegans*. Neuron 75, 585–592.

Kang, L., Gao, J., Schafer, W.R., Xie, Z., Xu, X.Z.S., 2010. *C. elegans* trp family protein TRP-4 is a pore-forming subunit of a native mechanotransduction channel. Neuron 67, 381–391.

Kaplan, J.M., Horvitz, H.R., 1993. A dual mechanosensory and chemosensory neuron in *Caenorhabditis elegans*. Proc. Natl. Acad. Sci. U.S.A. 90, 2227–2231.

Kawano, T., Po, M.D., Gao, S., Leung, G., Ryu, W.S., Zhen, M., 2011. An imbalancing act: gap junctions reduce the backward motor circuit activity to bias *C. elegans* for forward locomotion. Neuron 72, 572–586.

Kindt, K.S., Quast, K.B., Giles, A.C., De, S., Hendrey, D., Nicastro, I., Rankin, C.H., Schafer, W.R., 2007a. Dopamine mediates context-dependent modulation of sensory plasticity in *C. elegans*. Neuron 55, 662–676.

Kindt, K.S., Viswanath, V., Macpherson, L., Quast, K., Hu, H., Patapoutian, A., Schafer, W.R., 2007b. *Caenorhabditis elegans* TRPA-1 functions in mechanosensation. Nat. Neurosci. 10, 568–577.

Li, W., Feng, Z., Sternberg, P.W., Shawn Xu, X.Z., 2006. A *C. elegans* stretch receptor neuron revealed by a mechanosensitive TRP channel homologue. Nature 440, 684–687.

Macosko, E.Z., Pokala, N., Feinberg, E.H., Chalasani, S.H., Butcher, R.A., Clardy, J., Bargmann, C.I., 2009. A hub-and-spoke circuit drives pheromone attraction and social behavior in *C. elegans*. Nature 458, 1171–1175.

Rabinowitch, I., Schafer, W.R., 2015. Engineering new synaptic connections in the *C. elegans* connectome. Worm 4, e992668.

Rabinowitch, I., Chatzigeorgiou, M., Schafer, W.R., 2013. A gap junction circuit enhances processing of coincident mechanosensory inputs. Curr. Biol. 23, 963–967.

Rabinowitch, I., Chatzigeorgiou, M., Zhao, B., Treinin, M., Schafer, W.R., 2014. Rewiring neural circuits by the insertion of ectopic electrical synapses in transgenic *C. elegans*. Nat. Commun. 5, 4442.

Rabinowitch, I., Laurent, P., Zhao, B., Walker, D., Beets, I., Schoofs, L., Bai, J., Schafer, W.R., Treinin, M., 2016. Neuropeptide-driven cross-modal plasticity following sensory loss in *Caenorhabditis elegans*. PLoS Biol. 14, e1002348.

Sawin, E.R., Ranganathan, R., Horvitz, H.R., 2000. *C. elegans* locomotory rate is modulated by the environment through a dopaminergic pathway and by experience through a serotonergic pathway. Neuron 26, 619–631.

Simonsen, K., Moerman, D., Naus, C.C., 2014. Gap junctions in *C. elegans*. Membr. Physiol. Membr. Biophys. 5, 40.

Varshney, L.R., Chen, B.L., Paniagua, E., Hall, D.H., Chklovskii, D.B., 2011. Structural properties of the *Caenorhabditis elegans* neuronal network. PLoS Comput. Biol. 7, e1001066.

Way, J.C., Chalfie, M., 1989. The mec-3 gene of *Caenorhabditis elegans* requires its own product for maintained expression and is expressed in three neuronal cell types. Genes Dev. 3, 1823–1833.
White, J.G., Southgate, E., Thomson, J.N., Brenner, S., 1986. The structure of the nervous system of the nematode *Caenorhabditis elegans*. Philos. Trans. R. Soc. Lond. B Biol. Sci. 314, 1–340.
Xu, M., Jarrell, T.A., Wang, Y., Cook, S.J., Hall, D.H., Emmons, S.W., 2013. Computer assisted assembly of connectomes from electron micrographs: application to *Caenorhabditis elegans*. PLoS One 8, e54050.

CHAPTER 2

Neural Circuits Underlying Escape Behavior in *Drosophila*: Focus on Electrical Signaling

P. Phelan[1], J.P. Bacon[2], J.M. Blagburn[3]
[1]University of Kent, Canterbury, United Kingdom; [2]University of Sussex, Brighton, United Kingdom; [3]University of Puerto Rico, San Juan, PR, United States

1. INTRODUCTION

Animals use a wide variety of ingenious strategies, including chemical defense, camouflage, mimicry, and locomotory behaviors, to avoid predation. Marine molluscs release ink, opaline, and other secretions that, depending on precise chemical composition, are sensed by attackers as noxious or appetizing, thereby directly repelling them or distracting them while the intended prey makes its escape (Kicklighter et al., 2005; Derby and Aggio, 2011). Cephalopods, in particular, are masters of camouflage, rapidly and dynamically changing body color and pattern to match their diverse habitats (Hanlon et al., 2009; Allen et al., 2010; Zylinski and Johnsen, 2011). Many spider species dupe their predators by modifying body form and behavior to resemble ants that are unpalatable to spider-eating predators (Durkee et al., 2011; Uma et al., 2013). When deception or concealment is not an option, survival relies on the timely execution of high-speed escape behaviors. Well-studied invertebrate escape behaviors include the tail-flip response of the crayfish (Edwards et al., 1999; Herberholz et al., 2004), the jump-flight response of the locust (Simmons et al., 2010; Fotowat and Gabbiani, 2011; Card, 2012), the cockroach turn-and-run response (Domenici et al., 2008, 2009; Booth et al., 2009), and the fly jump-flight response (Allen et al., 2006; Card, 2012), which is the subject of this review.

Escape behaviors are underpinned by sensorimotor circuits adapted for speed and reliability. Common features, found in pathways from jellyfish to goldfish (Mackie, 2004; Eaton et al., 2001), are large-diameter axons, which maximize conduction velocity, and electrical synapses, which minimize delay. Wiersma's (1947) work in the crayfish was the first to demonstrate that firing of giant fibers (GFs) (interneurons that he referred to as "command neurons") triggered a fast behavioral response. The GFs relay sensory input to giant motorneurons innervating abdominal muscles, which flex to propel the animal away from threatening visual or mechanical stimuli. Furshpan and Potter (1959) subsequently discovered that the GF-giant motorneuron synapses were electrical

Network Functions and Plasticity
ISBN 978-0-12-803471-2, http://dx.doi.org/10.1016/B978-0-12-803471-2.00002-3

junctions, the first such synapses to be identified in any organism. It took a further three decades before the molecular components of arthropod electrical synapses (gap junctions) were identified (Phelan et al., 1998a,b; reviewed Phelan, 2000, 2005; Phelan and Starich, 2001). Pivotal to this were studies of the *Drosophila* Giant Fiber System (GFS), one of at least two parallel escape neural circuits in the fly (reviewed Allen et al., 2006; Card, 2012).

The GFS mediates rapid escape in response to visual and mechanosensory cues; when environmental threats are less pressing, a parallel, slightly slower but more directed, escape circuit is activated (von Reyn et al., 2014). The neural components of this alternative pathway have yet to be identified; by comparison, the GFS is well characterized. In the 1980s, a traditional forward genetic screen identified mutants in which GF-evoked escape behavior is perturbed (Thomas and Wyman, 1984); analysis of one of these mutants, *shakB*, led to the discovery of the innexin family of gap-junction genes in the 1990s (Phelan et al., 1996, 1998a,b). *shakB* gene products are differentially expressed in the neurons of the GFS and mediate transmission at rectifying and nonrectifying electrical synapses (Phelan et al., 2008).

2. THE *DROSOPHILA* GIANT FIBER SYSTEM

The *Drosophila* GFs originally came to light in Power's (1948) study of insect nervous system anatomy. He observed, using light microscopy of silver-stained preparations, a pair of large-diameter axons extending apparently uninterrupted from the brain to the periphery, branching to innervate muscles of the legs and wings. In 1980, higher-resolution electron microscopy (EM) studies by King and Wyman revealed that the GFs were not, in fact, motorneurons but interneurons that terminated within the thoracic nervous system where they established connections with polysynaptic pathways to the muscles. Since these original studies, the anatomy and physiology of the GF and its synaptic partners have been extensively studied in *Drosophila* and in larger flies, notably *Calliphora* and *Musca*. Although there are some subtle differences in cell morphology, the pattern of interconnections is very similar in all species studied. The focus is on *Drosophila*. For reference, relevant studies in other Dipterans also include Strausfeld and Bassemir (1983), Bacon and Strausfeld (1986), and Milde and Strausfeld (1990).

2.1 GFS Cellular Elements

The adult *Drosophila* nervous system (Fig. 2.1) consists of the brain and the thoracicoabdominal ganglion. The brain is a fusion of the cerebral (supraesophageal) ganglion, which encompasses the central brain and flanking optic lobes, and the gnathal (subesophageal) ganglion (Ito et al., 2014). The thoracic ganglion consists of three prominent neuromeres (*pro*, T1; *meso*, T2; *meta*, T3), the third of which is attached to the small cone-shaped abdominal ganglion (labeled as A in Fig. 2.1A). The GFs are the largest, in terms of

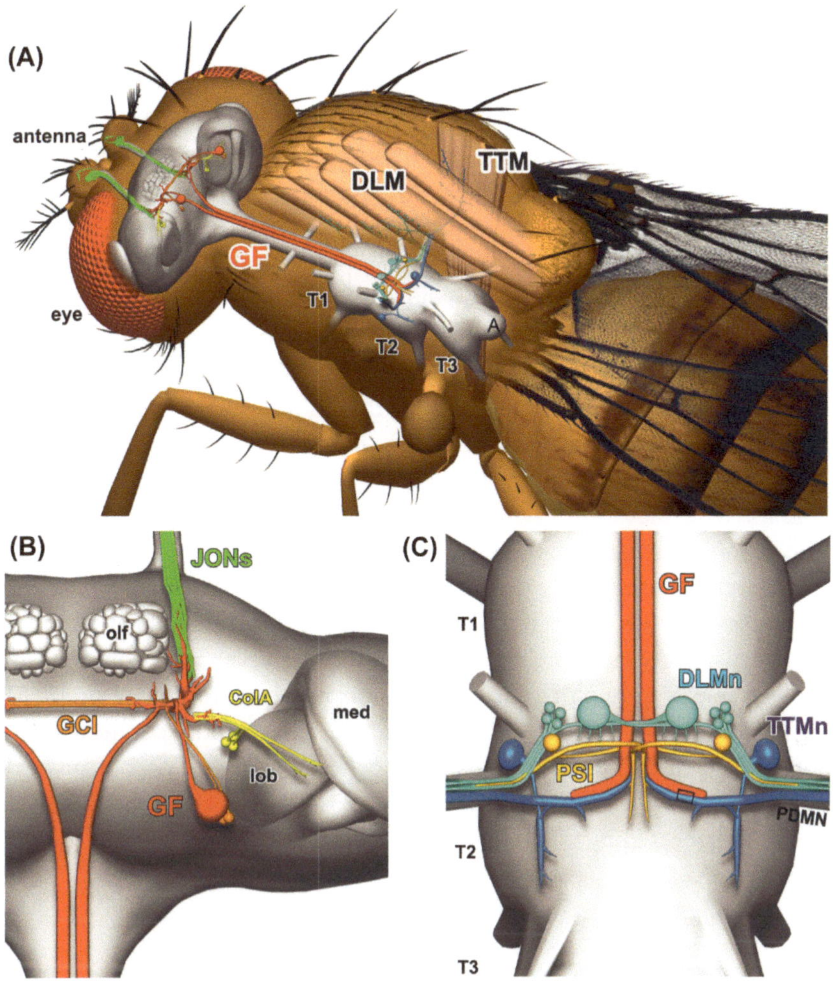

Figure 2.1 ***The*** **Drosophila** ***Giant Fiber System.*** (A) The body of the fly showing the position of the central nervous system (white) and the tergotrochanteral (TTM) and dorsal longitudinal (DLM) muscles. The major neurons of the Giant Fiber (GF) System are visible and shown at higher magnification in (B) and (C). (B) The cell bodies (one shown, *red*) of the bilaterally symmetrical GFs are located in the brain. The GF dendrites make contact with a bundle of Giant Commissural Interneurons (GCIs, *orange*), the axons of which extend across the midline interconnecting the two GFs, and with sensory neurons from the eyes (lobula columnar, ColA, *yellow*) and antennae (Johnston organ, JON, *green*). (C) The GF axons extend from the brain through the cervical connective to the thoracic ganglion. In the mesothoracic neuromere (T2), each axon bends laterally and synapses with the dendrite of the ipsilateral tergotrochanteral muscle motorneuron (TTMn, *dark blue*). Anterior to the bend, the GFs synapse at the midline with the peripherally synapsing interneurons (PSI, *dark yellow*). Each PSI extends an axon across the ganglion to terminate in the contralateral posterior dorsal mesothoracic nerve (PDMN) where it synapses with the axons of some, or all, of five dorsal longitudinal muscle motorneurons (DLMn, *pale blue*). The cell bodies of DLMn 1–4 form a cluster ipsilateral to the muscle they innervate; the large DLMn 5 cell body is contralateral to the innervated muscle. *T1, T2, T3*, pro-, meso-, and metathoracic neuromeres; *A*, abdominal ganglion; *lob*, lobula; *med*, medulla of the optic lobe; *olf*, olfactory neuropil.

axon diameter (6—8 μm), of a cluster of descending interneurons (Power, 1948; King and Wyman, 1980). Their cell bodies lie in the protocerebrum, the largest region of the cerebral ganglion. The axons extend from the brain through the cervical connective to the mesothoracic neuromere where they terminate with a sharp bend. The GF bends contact the dendrites of the tergotrochanteral muscle motorneurons (TTMns), the axons of which exit the ganglion in the ipsilateral posterior dorsal mesothoracic nerve (PDMN) to innervate the tergotrochanteral "jump" muscles (TTMs) that extend the middle legs. Just anterior to the bend, the GFs contact the peripherally synapsing interneurons (PSIs), the axons of which project across the ganglion into the contralateral PDMN where they synapse with the axons of dorsal longitudinal muscle motorneurons (DLMns) that innervate the wing depressor dorsal longitudinal muscles (DLMs) (Fig. 2.1; King and Wyman, 1980; Phelan et al., 1996; Allen et al., 2006). Other motor outputs of the GF (not shown in Fig. 2.1) have been reported but much less intensively studied than the TTMn and DLMn pathways. These include motorneurons of the dorsoventral muscles (DVMs) responsible for wing elevation (Tanouye and Wyman, 1980), the tibial levator muscles (TLMs) involved in leg extension (Trimarchi and Schneiderman, 1993), and the anterior pleural muscles that control wing opening (Tanouye and King, 1983). In the brain, each GF arborizes to produce three to four main dendritic branches that occupy at least partially distinct spatial domains. The lateral dendrites receive input from the compound eyes via at least one subgroup of optic lobe columnar interneurons (ColA cells; Mu et al., 2014) and possibly others (Mu et al., 2012). The posterior dendrite connects with mechanosensory afferents from the antennae (Kamikouchi et al., 2009; Tootoonian et al., 2012; Lehnert et al., 2013; Mu et al., 2014; Pézier et al., 2014, Fig. 2.1). Right and left dendritic arbors are interconnected by a bundle of giant commissural interneurons (GCIs; Phelan et al., 1996, Fig. 2.1).

2.2 GFS Synaptic Connections

Various experimental approaches have been applied to investigate the nature of synaptic connections in the GFS. These have revealed that all three types of synapses, electrical (signal transmitted directly via gap junctions), chemical (neurotransmitter-receptor signaling), and mixed synapses (capable of both electrical and chemical modes of transmission), are present (Fig. 2.2).

The earliest evidence for electrical synapses came from electrophysiological analyses. Tanouye and Wyman (1980) recorded activity in the TTM and DLM muscles in response to stimulation either of their respective motorneurons or of the GF. The latencies from motorneuron stimulation to muscle response in the TTM and DLM pathways were 0.66 and 0.83 ms, respectively, indicating typical chemical synapses at the neuromuscular junctions. The corresponding figures from GF stimulation to muscle response were 0.81 (GF-TTM) and 1.25 ms (GF-DLM). This implied a delay of just

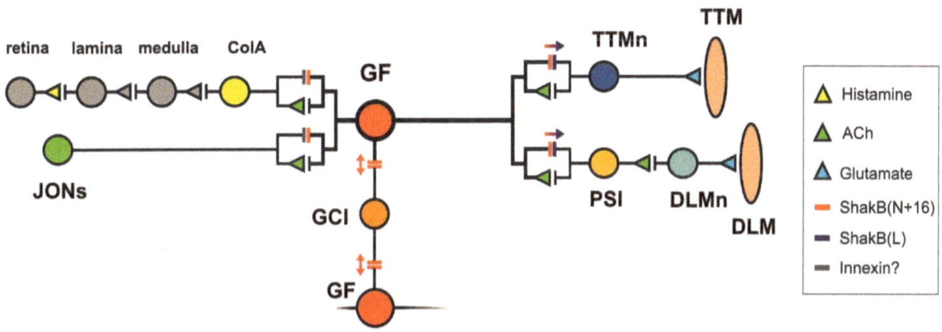

Figure 2.2 ***Synaptic connections in the Giant Fiber System (GFS).*** Schematic of the GFS showing known electrical (=) and chemical (Δ) synapses and the relevant innexin proteins and neurotransmitters, respectively. Only electrical synapses, composed of the innexin ShakB(Neural+16), have been identified between the GF and GCIs. The PSI-DLMn synapses and the neuromuscular junctions (TTMn-TTM and DLMn-DLM) are purely chemical synapses that use the neurotransmitters acetylcholine (ACh) and glutamate, respectively. All other synapses between the GF and its pre- and postsynaptic targets are mixed electrical–chemical connections. The GF-TTMn and GF-PSI electrical synapses are predicted to be heterotypic ShakB(Neural+16)/ShakB(Lethal) gap junctions; the chemical synapses between these neurons are cholinergic. At the synapses between the GF and sensory afferents from the retina (ColA) and antennal Johnston organ (JON), the innexin composition of the electrical junctions is not known; ACh is the chemical transmitter. Histamine is a transmitter of some of the optic lobe interneurons in the pathway from the retina to the ColA cells. *GF*, Giant Fiber; *GCI*, Giant Commissural Interneuron; *TTM*, tergotrochanteral muscle; *TTMn*, tergotrochanteral muscle motorneuron; *PSI*, peripherally synapsing interneuron; *DLM*, dorsal longitudinal muscle; *DLMn*, dorsal longitudinal muscle motorneuron. Neurons are color coded as in Fig. 2.1.

0.15 ms (0.81–0.66), far too short for chemical transmission, at the GF-TTMn synapse. A deduced delay of 0.42 ms (1.25–0.83) in the GF-DLMn pathway, allowed for no more than one chemical synapse in this disynaptic pathway. Aided by ultrastructural observations of close membrane appositions at sites of GF-TTMn and GF-PSI contact (subsequently confirmed to be gap junctions, Blagburn et al., 1999) and typical chemical contacts between the PSI and DLMn (King and Wyman, 1980), the authors reasoned that the GF formed electrical synapses with the TTMn and PSI and the latter, chemical synapses with the DLMns.

The presence of electrical synapses in the GFS was further confirmed by dye filling; this approach relies on cell-to-cell transfer of small molecule tracers, such as cobalt, Lucifer yellow, and neurobiotin, through gap junctions. Dyes microinjected into the GF reliably passed into the TTMn, PSI, GCIs (Phelan et al., 1996), and into a subset of antennal Johnston organ (JON) neurons (Pézier et al., 2014); the latter have also been shown to be electrically coupled to the GF (Lehnert et al., 2013). Visualizing coupling between the GFs/GCIs and the retinal ColA inputs has proved to be elusive in *Drosophila*, although there is unequivocal evidence that the GF is dye coupled to ColA afferents in *Musca* and *Calliphora* (Strausfeld and Bassemir, 1983;

Bacon and Strausfeld, 1986) and these cells are clearly contiguous in the *Drosophila* brain (Kamikouchi et al., 2009; Mu et al., 2014).

Evidence that the GFs could transmit signals chemically as well as electrically emerged, first, from physiological analysis of flies deficient in gap-junction genes. The loss of electrical synapses in these mutants unmasked a GF-evoked response in the TTM (Thomas and Wyman, 1984; Baird et al., 1990) with a longer latency than that recorded in the earlier studies in wild-type flies (Tanouye and Wyman, 1980). Subsequently, ultrastructural studies revealed chemical connections, with characteristic pre- and postsynaptic densities, neighboring close membrane appositions typical of gap junctions at sites of GF-TTMn, and GF-PSI contact (Blagburn et al., 1999, Fig. 2.3).

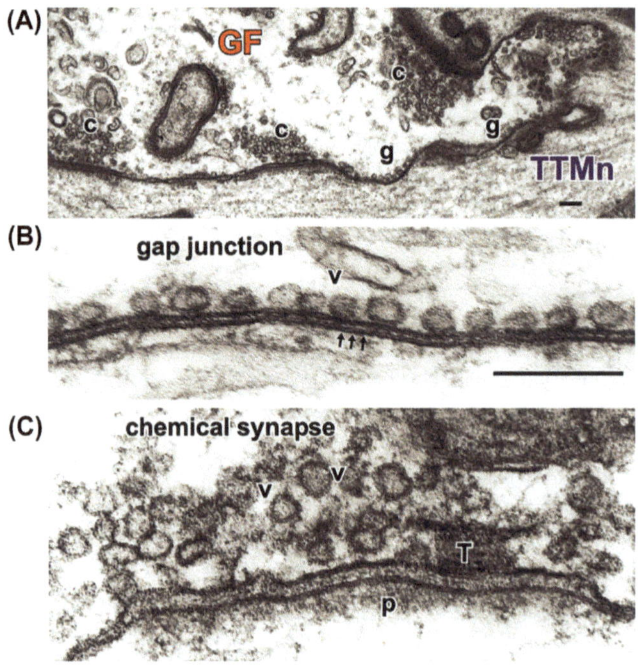

Figure 2.3 ***Mixed synapses in the Giant Fiber System.*** Electron micrographs of sites of synaptic contact between the Giant Fiber (GF, upper) and tergotrochanteral muscle motorneuron (TTMn, lower). The region corresponds roughly to the boxed region in Fig. 2.1C. (A) Low-magnification view showing chemical synapses (c) neighboring gap junctions (g) of electrical synapses. (B) Higher-magnification view of a gap junction. The junction has a 2–4 nm intercellular space and cross-striations at intervals of approximately 18 nm (arrows). A single row of large vesicles (v) is situated on the cytoplasmic side of the GF membrane. (C) Higher magnification view of a chemical synapse. The intercellular space is 13–20 nm. Clusters of small synaptic vesicles (v) and a T-bar specialization (T, an electron-dense structure characteristic of the active zone) are seen in the presynaptic GF opposite postsynaptic densities (p) in the TTMn. Scale bars: 200 nm. *(Reprinted with permission from Blagburn, J.M., Alexopoulos, H., Davies, J.A., Bacon, J.P., 1999. Null mutation in shaking-B eliminates electrical, but not chemical, synapses in the Drosophila Giant Fiber System: a structural study. J. Comp. Neurol. 404, 449–458. Copyright John Wiley and Sons.)*

Some of the synapses between the GF and its afferent inputs are also likely to be mixed electrical–chemical junctions. Consistent with the presence of chemical synapses, a presynaptic density marker localizes to sites of contact between JON auditory afferents and the GF (Pézier et al., 2014) and the GF dendrites express receptors for the neurotransmitter acetylcholine (ACh; Fayyazuddin et al., 2006). Chemical and electrical synapses have been observed in electron micrographs between the terminals of JON axons and unidentified postsynaptic neurons in the antennal mechanosensory center of the brain where the GF dendrites arborize (Sivan-Loukianova and Eberl, 2005). Anatomical confirmation of mixed synapses between ColA afferents and GF dendrites is not yet available for *Drosophila*, although such synapses have been observed electron microscopically in *Calliphora* and *Musca* (Strausfeld and Bassemir, 1983).

3. ELECTRICAL TRANSMISSION IN THE GFS: MOLECULES AND MECHANISMS

3.1 shakB: The First Invertebrate Prechordate Gap-Junction Gene

Although studies in an invertebrate, the crayfish, pioneered the discovery of the gap junction and the phenomenon of electrical synaptic transmission (Furshpan and Potter, 1959), it was vertebrate studies that led the molecular analysis of junctional communication. The first mammalian connexin (Cx) was cloned in the mid-1980s (Paul, 1986) and this gene family expanded rapidly as new members were added through the late 1980s and 1990s (Nielsen et al., 2012). During that time the molecular identity of prechordate gap junctions remained a mystery. Given the ultrastructural and functional similarities between these and vertebrate junctions, researchers reasonably anticipated conservation at the molecular level, yet efforts to isolate Cx homologues from lower organisms failed (reviewed Phelan, 2000). The sequencing of the *Caenorhabditis elegans* (*C. elegans* Sequencing Consortium, 1998) and *Drosophila* (Adams et al., 2000) genomes ultimately explained why. Cxs were simply not present. In the meantime, proteins unrelated to Cxs, including *Drosophila* Shaking B (ShakB), emerged as strong candidate gap-junction proteins (Barnes, 1994).

shakB (aka *Passover*) originally came to light in classical forward genetic screens for loci affecting sensorimotor behavior (Homyk et al., 1980; Thomas and Wyman, 1984). Wyman's group specifically screened for "nonjumping" flies with defects in a visually evoked "escape" jump, a behavior they showed was mediated by the GFS. Using the electrophysiological approach they had earlier developed in wild-type flies (Section 2.2), they showed that mutations in *shakB* impaired fast electrical transmission between the GF and the leg and wing muscles but spared the slower chemical component of synaptic transmission (Thomas and Wyman, 1984; Baird et al., 1990). Despite this early association between *shakB* and electrical synapses it was some time before the gene was explicitly linked to gap junctions. A key observation that mutations in the *shakB* gene eliminated dye coupling throughout the GFS provided a strong clue (Phelan et al., 1996). In the absence of any anatomical or developmental defects that could account for the loss of intercellular coupling in these mutants, the encoded protein(s) was either a structural component of gap-junction channels or an essential accessory protein. These possibilities were distinguished by heterologous expression in Cx-depleted paired *Xenopus* oocytes, which demonstrated unambiguously that ShakB (specifically ShakB(Lethal); see below and Fig. 2.5A–C) could form functional intercellular channels independently, analogous to the connexins (Phelan et al., 1998a). Thus, *shakB* was the first invertebrate prechordate gap-junction gene to be characterized. It is one of eight such genes in the fly genome (Stebbings et al., 2002; see also Shih and Wu, 2017 (Chapter 3)) that are part of the now large evolutionarily conserved innexin/pannexin family (Phelan et al., 1998b; Panchin et al., 2000). Innexins form gap junctions in all prechordate lineages and persist as divergent sequences, termed pannexins, in chordates (reviewed Phelan, 2005; Bond and Naus, 2014; Yen and Saier, 2007; Abascal and Zardoya, 2013).

3.2 shakB Gene Transcripts Are Differentially Expressed in GFS Neurons

shakB is a large and complex gene that produces several transcripts as a consequence of alternative promoter usage and alternative splicing. These give rise to three proteins (Fig. 2.4): ShakB(Neural) (Krishnan et al., 1993) and ShakB(Lethal) (Crompton et al., 1995; Krishnan et al., 1995), each encoded on a single transcription unit, and ShakB(Neural+16), which is encoded by multiple transcripts that differ only in noncoding sequences (Zhang et al., 1999). The transcript suffix relates to the phenotype of the first characterized mutants; *neural* transcripts were implicated in GFS function (Thomas and Wyman, 1984), whereas *shakB(lethal)* [aka *shakB(vital)*] mutations caused early larval lethality possibly by affecting gut or heart function (Crompton et al., 1995). In reality, the transcripts are widely expressed in neural and nonneural tissues.

shakB(neural) was the first to be described. Originally it was called *shaking B* (*shakB*; Homyk et al., 1980) or *passover* (Thomas and Wyman, 1984) after two allelic mutations (*shakB*^{2} and *shakB*^{Pas}, Fig. 2.4A) that disrupted gene function in the GFS. These mutations subsequently were found to eliminate the function not only of *shakB(neural)* but also of *shakB(neural+16)*. The two transcripts are nearly identical, differing only by an extreme 5′ exon that codes for a 16-amino acid N-terminal tail in the latter; the *shakB*^{2} and *shakB*^{Pas} mutations target a shared coding exon (Zhang et al., 1999, Fig. 2.4). This raised the question, addressed by phenotypic rescue studies, of whether one or both proteins were required to form electrical synapses in the GFS. Phelan et al. (2008) reintroduced the individual transcripts one at a time into the uncoupled neurons of *shakB*^{2} [*shakB(neural/neural+16)* null] mutants. *shakB(neural)* failed to effect rescue, although the protein was localized to sites of synaptic contact. The longer transcript encoding ShakB(Neural+16), on the other hand, restored functional electrical and dye coupling

(A)

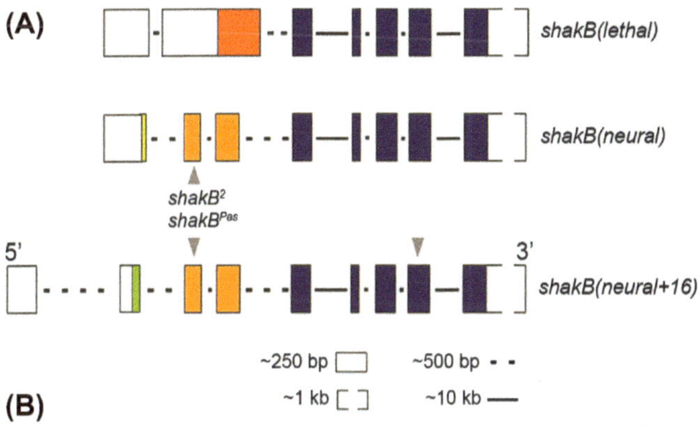

(B)

```
Dm_ShakB(Lethal)      MLDIFRGLKNL-----VKVSHVKTDSIVFRLHYSITVMILMSFSLIITTRQYVGNPIDCV  55
Dm_ShakB(Neural)      -----------------MVSHVKIDSPVFRLHTNATVILLITFSIAVTTRQYVGNPIDCV  43
Dm_ShakB(Neural+16)   -MEFLRGVYAFMQVSRSSVSHVKIDSPVFRLHTNATVILLITFSIAVTTRQYVGNPIDCV  59
                                        ***** ** ***** . **::*::**: :*************

Dm_ShakB(Lethal)      HTKDIPEDVLNTYCWIQSTYTLKSLFLKKQGVSVPYPGIGNSDGD-PADKKHYKYYQWVC  114
Dm_ShakB(Neural)      HTRDIPEDVLNTYCWIHSTYTVVDAFMKKQGSEVPFPGVHNSQGRGPLTIKHTKYYQWVA  103
Dm_ShakB(Neural+16)   HTRDIPEDVLNTYCWIHSTYTVVDAFMKKQGSEVPFPGVHNSQGRGPLTIKHTKYYQWVA  119
                      **:*************:****: . *:**** .**:**: **:*  *   ** ******.

Dm_ShakB(Lethal)      FCLFFQAILFYTPRWLWKSWEGGKIHALIMDLDIGICSEAEKKQKKKLLLDYLWENLRYH  174
Dm_ShakB(Neural)      FTLFFQAILFYTPRWLWKSWEGGKIHALIMDLDIGICSEAEKKQKKKLLLDYLWENLRYH  163
Dm_ShakB(Neural+16)   FTLFFQAILFYTPRWLWKSWEGGKIHALIMDLDIGICSEAEKKQKKKLLLDYLWENLRYH  179
                      * **********************************************************

Dm_ShakB(Lethal)      NWWAYRYYVCELLALINVIGQMFLMNRFFDGEFITFGLKVIDYMETDQEDRMDPMIYIFP  234
Dm_ShakB(Neural)      NWWAYRYYVCELLALINVIGQMFLMNRFFDGEFITFGLKVIDYMETDQEDRMDPMIYIFP  223
Dm_ShakB(Neural+16)   NWWAYRYYVCELLALINVIGQMFLMNRFFDGEFITFGLKVIDYMETDQEDRMDPMIYIFP  239
                      ************************************************************

Dm_ShakB(Lethal)      RMTKCTFFKYGSSGEVEKHDAICILPLNVVNEKIYIFLWFWFILLTFLTLLTLIYRVVII  294
Dm_ShakB(Neural)      RMTKCTFFKYGSSGEVEKHDAICILPLNVVNEKIYIFLWFWFILLTFLTLLTLIYRVVII  283
Dm_ShakB(Neural+16)   RMTKCTFFKYGSSGEVEKHDAICILPLNVVNEKIYIFLWFWFILLTFLTLLTLIYRVVII  299
                      ************************************************************

Dm_ShakB(Lethal)      FSPRMRVYLFRMRFLRVRRDAIEIIVRRSKMGDWFLLYLLGENIDTVIFRDVVQDLANRL  354
Dm_ShakB(Neural)      FSPRMRVYLFRMRFLRVRRDAIEIIVRRSKMGDWFLLYLLGENIDTVIFRDVVQDLANRL  343
Dm_ShakB(Neural+16)   FSPRMRVYLFRMRFLRVRRDAIEIIVRRSKMGDWFLLYLLGENIDTVIFRDVVQDLANRL  359
                      ************************************************************

Dm_ShakB(Lethal)      GHNQHHRVPGLKGEIQDA  372
Dm_ShakB(Neural)      GHNQHHRVPGLKGEIQDA  361
Dm_ShakB(Neural+16)   GHNQHHRVPGLKGEIQDA  377
                      ******************
```

Figure 2.4 shakB *gene transcripts encode three partially identical proteins.* (A) Schematic of transcripts encoding ShakB(Lethal), ShakB(Neural), and ShakB(Neural+16) proteins. Coding exons are represented by colored boxes, noncoding exons by white boxes, and introns by lines. ShakB(Neural+16) is encoded on multiple transcription units (one shown) which differ only by a 5′ noncoding exon. All transcripts share five 3′ coding exons (*blue*), which are spliced onto a specific complement of 5′ coding exons (*red, orange, yellow, green*); *shakB(neural)* and *shakB(neural+16)* are distinguished only by the extreme 5′ coding exon depicted in yellow and green, respectively. The site of the *shakB²* and *shakB^Pas* mutations is indicated. *(Adapted with permission from Phelan, P., Starich, T.A., 2001. Innexins get into the gap. Bioessays, 23, 388–396. Copyright John Wiley and Sons.)* (B) Sequence alignment (Clustal Omega 1.2.1) of ShakB proteins. Predicted transmembrane domains are shaded. The N-terminal 16 amino acids that distinguish ShakB(Neural+16) from ShakB(Neural) are underlined. Asterisk (*), identical residue; colon (:), residue with strongly similar properties; dot (.), residue with weakly similar properties.

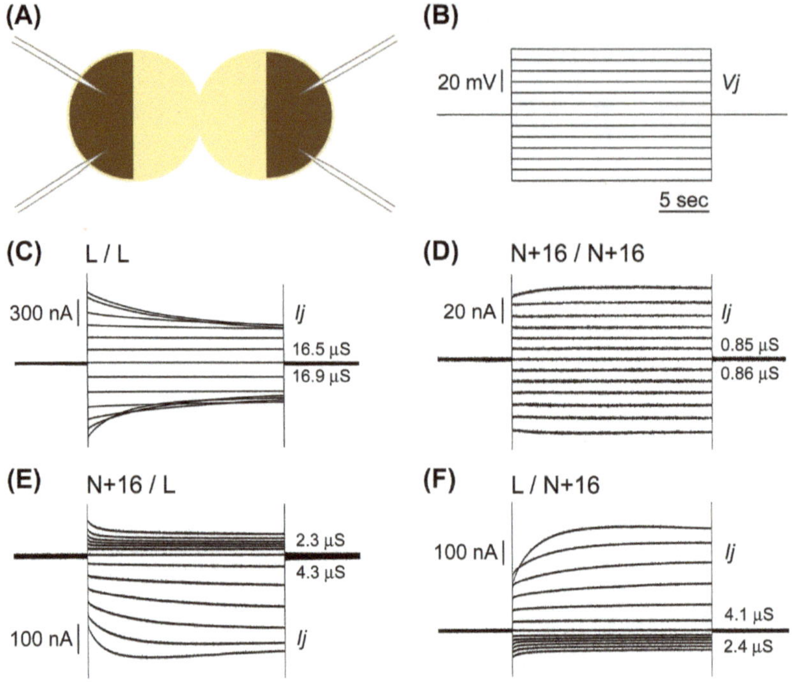

Figure 2.5 ***ShakB proteins assemble rectifying and nonrectifying junctions.*** Dual-voltage clamp recordings of *Xenopus* oocyte pairs expressing ShakB proteins. (A) Schematic of a cell pair. (B) Stimulus protocol consisting of sequential depolarizing and hyperpolarizing voltage steps imposed from a holding potential. The protocol is applied to one cell of a pair of cells clamped at the same holding potential, generating a voltage difference between the two cells (transjunctional voltage, Vj). (C—F) Junctional currents (Ij) recorded from cell pairs expressing different combinations of ShakB proteins. Induced currents are recorded in one cell in response to application of the stimulus protocol (B) to the other cell of the pair. In voltage clamp mode, induced currents are opposite in sign to the applied current, ie, traces below the baseline represent the response to depolarizing voltage steps and traces above the baseline, the response to hyperpolarizing steps. Junctional conductance (Gj, μS) is Ij/Vj; values given are maximum Gj. (C) L/L. Pairs in which both cells express ShakB(Lethal) form homotypic channels that respond symmetrically to applied voltage; Ij decreases over time with increasing Vj of either polarity. (D) N+16/N+16. Pairs in which both cells express ShakB(Neural+16) form voltage-insensitive homotypic channels; Ij is linear for all Vjs. (E, F) N+16/L and L/N+16. Pairs in which one cell expresses ShakB(Neural+16) and the other ShakB(Lethal) form heterotypic channels that respond asymmetrically to applied voltage. In (E), the ShakB(Neural+16)-expressing cell was stimulated and the response recorded from the ShakB(Lethal)-expressing cell. In (F), the ShakB(Lethal)-expressing cell was stimulated and the response recorded from the ShakB(Neural+16)-expressing cell. Depolarizing voltage steps applied to the ShakB(Neural+16)-expressing cell elicit large junctional currents (E), whereas depolarizing steps applied to the ShakB(Lethal)-expressing cell tend to close the channel (F) and vice versa for hyperpolarizing voltage steps. *(Adapted from Phelan, P., Goulding, L.A., Tam, J.L.Y., Allen, M.J., Dawber, R.J., Davies, J.A., Bacon, J.P., 2008. Molecular mechanism of rectification at identified electrical synapses in the Drosophila Giant Fiber System. Curr. Biol. 18, 1955—1960.)*

between the GF and its synaptic partners. In vitro expression corroborated these studies; *shakB(neural)*, unlike the other *shakB* transcripts, did not form open intercellular channels in paired oocytes (Phelan et al., 1998a, 2008). The precise function of this protein is as yet unknown.

Consistent with its function in the GFS, *shakB(neural+16)* RNA is expressed in the GFs and GCIs (Zhang et al., 1999; Phelan et al., 2008, Fig. 2.2). It has not, however, been detected in the TTMns and PSIs; rather there is evidence from reporter gene expression that the TTMn at least (and possibly the PSI) expresses *shakB(lethal)* (Jacobs et al., 2000; Phelan et al., 2008, Fig. 2.2). The identity of the innexins at sensory afferent-GF synapses is not known at present.

3.3 Some GFS Synapses Are Rectifying Junctions

Gap junctions are assemblies of intercellular channels each of which is composed of two multimeric hemichannels. The terms homotypic and heterotypic refer to channels in which the subunit composition of the docked hemichannels is identical or different, respectively. Both channel types are possible and probable in the GFS as a consequence of the cell-specific expression of *shakB* splice variants (Fig. 2.2). *shakB(neural+16)* is expressed in the GF (Zhang et al., 1999; Phelan et al., 2008) and is necessary and sufficient to couple this cell to the GCIs in the brain and the TTMn and PSI in the thoracic ganglion (Phelan et al., 2008). As the GCIs also express *shakB(neural+16)* (Zhang et al., 1999; Phelan et al., 2008), the gap-junction channels at GF-GCI synapses are suggested to be homotypic. Channels at the GF-TTMn (and possibly the GF-PSI) synapses, on the other hand, are predicted to be heterotypic, with pre- and postsynaptic hemichannels composed of ShakB(Neural+16) and ShakB(Lethal), respectively (Phelan et al., 2008; Jacobs et al., 2000).

How does the molecular composition of the GFS synapses influence their functional properties? This question has not yet been addressed ex vivo in the *Drosophila* nervous system. It would involve direct recording of transmission at individual synapses, which remains a technical challenge in the diminutive fruit fly. Modeling the synapses in a paired cell expression system, however, has been instructive (Phelan et al., 2008, Fig. 2.5). Cell pairs in which both cells expressed ShakB(Neural+16), as at the GF-GCI synapses, formed simple homotypic channels (Fig. 2.5A, B and D) that transmitted currents bidirectionally, a property that fits with a possible role for the GCIs in synchronizing the activity of right and left GFs. Cell pairs in which one cell expressed ShakB(Neural+16) and the other, ShakB(Lethal), mimicking the GF-TTMn synapse, formed heterotypic channels that displayed rectification, ie, they transmitted depolarizing signals in one direction from the ShakB(Neural+16)-expressing cell ("presynaptic GF") to the ShakB(Lethal)-expressing cell ("postsynaptic TTMn"). This unidirectionality was a result of differential voltage sensitivity of the molecularly distinct hemichannels;

channels opened when the presynaptic side was positive relative to the postsynaptic side and vice versa (Phelan et al., 2008, Fig. 2.5A, B, E and F). Structure–function studies (Marks and Skerrett, 2014) suggested that the ShakB N-terminus mediates voltage gating. Rectification at electrical synapses classically is associated with speed and reliability. The phenomenon was first described in the crayfish (where the GF-giant motorneuron synapse can be recorded in situ; Furshpan and Potter, 1959) and shown to depend on a differential response to voltage of the apposing hemichannels (Jaslove and Brink, 1986; Giaume et al., 1987). The work in *Drosophila* (albeit based in part on artificial synapses) confirmed and extended this by demonstrating that asymmetric voltage gating is underpinned by molecular asymmetry (Phelan et al., 2008).

Interestingly, both fly and crayfish rectifying synapses also exhibit morphological asymmetry in the form of a single row of large (~40–80 nm) juxtamembrane vesicles in the GF, but not in the postsynaptic motorneuron (Hanna et al., 1978; Leitch, 1992; Blagburn et al., 1999, Fig. 2.3B). By contrast, nonrectifying junctions in the crayfish have vesicles on both sides of the synapse (Peracchia, 1973; Leitch, 1992). These vesicles are clearly linked to gap junctions. In *Drosophila*, they are absent in *shakB* mutants that lack electrical synapses (Blagburn et al., 1999). In crayfish (at least at nonrectifying septate synapses), the vesicles are anchored to the junctional membrane by filament-like structures (Ohta et al., 2011). Although the presence of these vesicles has been widely documented, their function is unknown. Ohta et al. (2011) showed that they express the vesicular nucleotide transporter and speculate that they may store ATP, which is known to be released by innexin and pannexin unpaired channels (Dahl and Muller, 2014). If so, this would provide another means of signaling between the GF and its targets.

4. CHEMICAL TRANSMISSION IN THE GFS: TRANSMITTERS AND RECEPTORS

Knowledge of the neurotransmitters and receptors operating at chemical synapses in the GFS is still incomplete. ACh appears to be the main classical transmitter of the circuit interneurons (Fig. 2.2). Genetic manipulation of the ACh synthetic enzyme, choline acetyltransferase, was found to disrupt transmission at the PSI-DLMn synapse (Gorczyca and Hall, 1984) and the chemical component of the mixed GF-TTMn synapse (Allen and Murphey, 2007); by inference, the GF-PSI chemical synapse is also likely to be cholinergic, although this has not been directly demonstrated. The *Drosophila* nicotinic ACh receptor (nAChR) α7 subunit (Dα7) is expressed on the GF lateral and posterior dendrites that receive ColA visual and antennal inputs, respectively (Section 2.1 and 2.2; Figs. 4.1 and 4.2) and, based on mutant analysis, is required for transmission at these synapses and at the PSI-DLMn synapse (Fayyazuddin et al., 2006). Whether the functional pentameric receptor includes non-α7 subunits has yet to be determined as has the complement of nAChR subunits expressed by the TTMn and PSI. Glutamate, acting on ionotropic receptors, is the main excitatory transmitter at neuromuscular junctions in the fly (Qin et al., 2005; DiAntonio, 2006) and thus the synapses between the TTMn and DLMn motorneurons and their respective muscles (Koenig and Ikeda, 2005) are believed to be glutamatergic. These neuromuscular synapses are unaffected by genetic (Fayyazuddin et al., 2006; Allen and Murphey, 2007) or pharmacological (Mejia et al., 2013) blockade of cholinergic transmission.

5. THE GF CIRCUIT IS RESPONSIBLE FOR SHORT-MODE ESCAPE

More than 40 years ago, Levine and Tracey (1973) suggested that the "giant neuron," then thought to be a large interganglionic motor neuron, was responsible for triggering the fly escape jump and the stereotypical sequence of contractions of the TTM jump muscle (or as they termed it, "TDT" for tergal depressor of the trochanter), followed shortly by the DLM wing elevator. It was the subsequent combination of EM and detailed electrophysiological studies, carried out in the laboratory of Robert Wyman, that determined the true interneuronal nature of the GF, and first suggested the now accepted pattern of circuitry in the thorax (King and Wyman, 1980; Tanouye and Wyman, 1980; Koto et al., 1981, Section 2.1 and 2.2; Figs. 2.1 and 2.2). Having elucidated the circuitry, the Wyman laboratory developed a forward-genetic behavioral screen for disruptions in the GF escape jump circuit, using visual (light-off) stimuli, which led to the subsequent identification of the *bendless* and *passover* (aka *shakB)* mutations (Thomas and Wyman, 1982, 1984, Section 3.1). However, they were forced to use white-eyed flies for the screen, because wild-type flies do not readily jump in response to light-off visual stimuli alone.

The basic features of the GF system are conserved throughout the Diptera (Strausfeld and Bassemir, 1983) and the larger flies are more suitable for elucidating candidate sensory inputs to the GF for triggering escape (Bacon and Strausfeld, 1986). The latter study showed that both mechanosensory inputs from the antenna and small-field motion stimuli or "light-on" or "off" stimuli to the eyes were able to evoke subthreshold responses in the GF. In the 1990s, more detailed studies of neuromuscular activation were able to distinguish voluntary flight initiation (and the jump/flight evoked by olfactory stimuli) from that evoked by visual (or electrical) stimulation of the GF (Trimarchi and Schneiderman, 1995a,b).

In this century, the development of high-speed video cameras allowed a more precise dissection of these different escape sequences, which have called into question the idea of the GF being a "command neuron" that is necessary and sufficient for escape behavior (Trimarchi and Schneiderman, 1995b; Card, 2012). Hammond and O'Shea (2007) showed that a more natural looming stimulus with its accompanying air displacement (in contrast to the less ecologically relevant light-off stimulus) evokes wing elevation before leg extension, whereas previous studies had shown that GF activation first elicits leg extension (Trimarchi and Schneiderman, 1995a,b). It was subsequently shown that the fly's response to visual looming is in fact a complex series of behaviors that involve adjustments of leg and body posture, then wing elevation, prior to a controlled take-off in the opposite direction (Card and Dickinson, 2008; Card, 2012). The GF was found not to be required to initiate this process; instead, other descending interneurons were invoked (Fotowat et al., 2009). What role, then, does the GF play? This was resolved by a detailed study of the visual parameters by which these neurons are activated, in a technical tour-de-force where patch-clamp recordings were made from the cell body of the GF, in an animal still able to jump and fly (von Reyn et al., 2014). It was found that both the GF and a parallel circuit (involving as yet unidentified descending interneurons) detect visual looming, with the former having a higher threshold for excitability. Thus, when the "threat" approaches fairly slowly, the parallel circuit is activated first, postural adjustments are made, and a controlled directed jump and flight is initiated (long-mode escape). If looming is rapid, the GF threshold is reached first and an uncontrolled but more rapid emergency escape jump is made (short-mode escape; von Reyn et al., 2014). Presumably, in this latter scenario, air movement from the imminent threat also excites the antennal mechanosensory neurons, ensuring that the GF spikes in time.

6. SUMMARY

The *Drosophila* GFS is a relatively simple and experimentally accessible neural circuit. It underlies a rapid jump-flight response, part of the fly's repertoire of escape behaviors to potentially threatening visual and mechanosensory cues. This review has focused on synaptic connectivity in the GFS, in particular, on studies of electrical transmission that were instrumental in the discovery of the innexin/pannexin family of gap-junction genes and have provided insight into the molecular mechanisms of synaptic rectification. It has not been possible here to review several excellent studies of GFS development (eg, Godenschwege et al., 2002a,b; Orr et al., 2014; Pézier et al., 2014) that have contributed to knowledge of the molecules and mechanisms that regulate neuronal development and synaptogenesis.

QUESTIONS ARISING

- Although the GFS is one of the most extensively described central neural circuits in the fly, the picture is not yet complete. Questions arising from the work described here include:
  - The full complement of GFS sensory inputs, the nature of all sensory afferent to GF synapses, and the innexin proteins and neurotransmitters thereof
  - Potential functional interactions between neighboring electrical and chemical synapses

- The role of the large juxtamembrane vesicles on the presynaptic side of putative rectifying electrical synapses

- The non-GF escape circuits have not been characterized at the cellular level. The identification of these neurons will be a further important contribution to the field.

Note: Pézier et al. (2016) have demonstrated that the electrical synapses between the giant fiber and auditory Johnston organ afferents are homotypic ShakB(Neural+16) junctions.

REFERENCES

Abascal, F., Zardoya, R., 2013. Evolutionary analyses of gap junction protein families. Biochim. Biophys. Acta 1828, 4–14.

Adams, M.D., Celniker, S.E., Holt, R.A., Evans, C.A., Gocayne, J.D., Amanatides, P.G., et al., 2000. The genome sequence of *Drosophila* melanogaster. Science 287, 2185–2195.

Allen, M.J., Murphey, R.K., 2007. The chemical component of the mixed GF-TTMn synapse in *Drosophila melanogaster* uses acetylcholine as its neurotransmitter. Eur. J. Neurosci. 26, 439–445.

Allen, M.J., Godenschwege, T.A., Tanouye, M.A., Phelan, P., 2006. Making an escape: development and function of the *Drosophila* giant fibre system. Sem. Cell Dev. Biol. 17, 31–41.

Allen, J.J., Mäthger, L.M., Buresch, K.C., Fetchko, T., Gardner, M., Hanlon, R.T., 2010. Night vision by cuttlefish enables changeable camouflage. J. Exp. Biol. 213, 3953–3960.

Bacon, J.P., Strausfeld, N.J., 1986. The dipteran 'Giant fibre' pathway: neurons and signals. J. Comp. Physiol. A 158, 529–548.

Baird, D.H., Schalet, A.P., Wyman, R.J., 1990. The *Passover* locus in *Drosophila melanogaster*: complex complementation and different effects on the giant fiber neural pathway. Genetics 126, 1045–1059.

Barnes, T.M., 1994. OPUS: a growing family of gap junction proteins? Trends Genet. 10, 303–305.

Blagburn, J.M., Alexopoulos, H., Davies, J.A., Bacon, J.P., 1999. Null mutation in *shaking-B* eliminates electrical, but not chemical, synapses in the *Drosophila* Giant Fiber System: a structural study. J. Comp. Neurol. 404, 449–458.

Bond, S.R., Naus, C.C., 2014. The pannexins: past and present. Front. Physiol. 5. Article 58.

Booth, D., Marie, B., Domenici, P., Blagburn, J.M., Bacon, J.P., 2009. Transcriptional control of behavior: engrailed knockout changes cockroach escape trajectories. J. Neurosci. 29, 7181–7190.

Sequencing Consortium, 1998. Genome sequence of the nematode *C. elegans*: a platform for investigating biology. Science 282, 2012–2018.

Card, G., Dickinson, M.H., 2008. Visually mediated motor planning in the escape response of *Drosophila*. Curr. Biol. 18, 1300–1307.

Card, G.M., 2012. Escape behaviors in insects. Curr. Opin. Neurobiol. 22, 180–186.

Crompton, D., Todman, M., Wilkin, M., Ji, S., Davies, J., 1995. Essential and neural transcripts from the *Drosophila shaking-B* locus are differentially expressed in the embryonic mesoderm and pupal nervous system. Dev. Biol. 170, 142–158.

Dahl, G., Muller, K.J., 2014. Innexin and pannexin channels and their signalling. FEBS Lett. 588, 1396–1402.

Derby, D.C., Aggio, J.F., 2011. The neuroecology of chemical defenses. Integr. Comp. Biol. 51, 771–780.

DiAntonio, A., 2006. Glutamate receptors at the *Drosophila* neuromuscular junction. Int. Rev. Neurobiol. 75, 165–179.

Domenici, P., Booth, D., Blagburn, J.M., Bacon, J.P., 2008. Cockroaches keep predators guessing by using preferred escape trajectories. Curr. Biol. 18, 1792–1796.

Domenici, P., Booth, D., Blagburn, J.M., Bacon, J.P., 2009. Escaping away from and towards a threat: the cockroach's strategy for staying alive. Commun. Integr. Biol. 2, 497–500.

Durkee, C.A., Weiss, M.R., Uma, D.B., 2011. Ant mimicry lessens predation on a North American jumping spider by larger salticid spiders. Environ. Entomol. 40, 1223–1231.

Eaton, R.C., Lee, R.K.K., Foreman, M.B., 2001. The Mauthner cell and other identified neurons of the brainstem escape network of fish. Prog. Neurobiol. 63, 467–485.

Edwards, D.H., Heitler, W.J., Krasne, F.B., 1999. Fifty years of a command neuron: the neurobiology of escape behavior in the crayfish. Trends Neurosci. 22, 153–161.

Fayyazuddin, A., Zaheer, M.A., Hiesinger, P.R., Bellen, H.J., 2006. The nicotinic acetylcholine receptor Dα7 is required for an escape behavior in *Drosophila*. PLoS Biol. 4, e63.

Fotowat, H., Gabbiani, F., 2011. Collision detection as a model for sensory-motor integration. Annu. Rev. Neurosci. 34, 1–19.

Fotowat, H., Fayyazuddin, A., Bellen, H.J., Gabbiani, F., 2009. A novel neuronal pathway for visually guided escape in *Drosophila melanogaster*. J. Neurophysiol. 102, 875–885.

Furshpan, E.J., Potter, D.D., 1959. Transmission at the giant motor synapses of the crayfish. J. Physiol. 145, 289–325.

Giaume, C., Kado, R.T., Korn, H., 1987. Voltage-clamp analysis of a crayfish rectifying synapse. J. Physiol. 386, 91–112.

Godenschwege, T.A., Hu, H., Shan-Crofts, X., Goodman, C.S., Murphey, R.K., 2002a. Bi-directional signaling by Semaphorin 1a during central synapse formation in *Drosophila*. Nat. Neurosci. 5, 1294–1301.

Godenschwege, T.A., Simpson, J.H., Shan, X., Bashaw, G.J., Goodman, C.S., Murphey, R.K., 2002b. Ectopic expression in the giant fiber system of *Drosophila* reveals distinct roles for Roundabout (Robo), Robo2, and Robo3 in dendritic guidance and synaptic connectivity. J. Neurosci. 22, 3117–3129.

Gorczyca, M., Hall, J.C., 1984. Identification of a cholinergic synapse in the giant fiber pathway of *Drosophila* using conditional mutations of acetylcholine synthesis. J. Neurogenet. 1, 289–313.

Hammond, S., O'Shea, M., 2007. Escape flight initiation in the fly. J. Comp. Physiol. A 193, 471–476.

Hanlon, R.T., Chiao, C.C., Mäthger, L.M., Barbosa, A., Buresch, K.C., Chubb, C., 2009. Cephalopod dynamic camouflage: bridging the continuum between background matching and disruptive coloration. Philos. Trans. R. Soc. B 364, 429–437.

Hanna, R.B., Keeter, J.S., Pappas, G.D., 1978. The fine structure of a rectifying electrotonic synapse. J. Cell Biol. 79, 764–773.

Herberholz, J., Sen, M.M., Edwards, D.H., 2004. Escape behavior and escape circuit activation in juvenile crayfish during prey–predator interactions. J. Exp. Biol. 207, 1855–1863.

Homyk, T., Szidonya, J., Suzuki, D.T., 1980. Behavioural mutants of *Drosophila melanogaster* III. Isolation and mapping of mutations by direct visual observation of behavioural phenotypes. Mol. Gen. Genet. 177, 553–567.

Ito, K., Shinomiya, K., Ito, M., Armstrong, J.D., Boyan, G., Hartenstein, V., Harzsch, S., Heisenberg, M., Homberg, U., Jenett, A., Keshishian, H., Restifo, L.L., Rössler, W., Simpson, J.H., Strausfeld, N.J., Strauss, R., Vosshall, L.B., Insect Brain Name Working Group, 2014. A systematic nomenclature for the insect brain. Neuron 81, 755–765.

Jacobs, K., Todman, M.G., Allen, M.J., Davies, J.A., Bacon, J.P., 2000. Synaptogenesis in the giant-fibre system of *Drosophila*: interaction of the giant fibre and its major motorneuronal target. Development 127, 5203–5212.

Jaslove, S.W., Brink, P.R., 1986. The mechanism of rectification at the electrotonic motor giant synapse of the crayfish. Nature 323, 63–65.

Kamikouchi, A., Inagaki, H.K., Effertz, T., Hendrich, O., Fiala, A., Göpfert, M.C., Ito, K., 2009. The neural basis of *Drosophila* gravity-sensing and hearing. Nature 458, 165–171.

Kicklighter, C.E., Shabani, S., Johnson, P.M., Derby, C.D., 2005. Sea hares use novel antipredatory chemical defenses. Curr. Biol. 15, 549–554.

King, D.G., Wyman, R.J., 1980. Anatomy of the giant fiber pathway in *Drosophila* I. Three thoracic components of the pathway. J. Neurocytol. 9, 753–770.

Koenig, J.H., Ikeda, K., 2005. Relationship of the reserve vesicle population to synaptic depression in the tergotrochanteral and dorsal longitudinal muscles of *Drosophila*. J. Neurophysiol. 94, 2111–2119.

Koto, M., Tanouye, M.A., Ferrus, A., Thomas, J.B., Wyman, R.J., 1981. The morphology of the cervical giant fiber neuron of *Drosophila*. Brain Res. 221, 213–217.

Krishnan, S.N., Frei, E., Swain, G.P., Wyman, R.J., 1993. Passover: a gene required for synaptic connectivity in the Giant Fiber System of *Drosophila*. Cell 73, 967–977.

Krishnan, S.N., Frei, E., Schalet, A.P., Wyman, R.J., 1995. Molecular basis of intracistronic complementation in the *Passover* locus of *Drosophila*. Proc. Natl. Acad. Sci. U.S.A. 92, 2021–2025.
Lehnert, B.P., Baker, A.E., Gaudry, Q., Chiang, A.S., Wilson, R.I., 2013. Distinct roles of TRP channels in auditory transduction and amplification in *Drosophila*. Neuron 77, 115–128.
Leitch, B., 1992. Ultrastructure of electrical synapses: review. Electron Microsc. Rev. 5, 311–339.
Levine, J.D., Tracey, D., 1973. Structure and function of the giant motorneuron of *Drosophila melanogaster*. J. Comp. Physiol. 87, 213–235.
Mackie, G.O., 2004. Central neural circuitry in the jellyfish *Aglantha*: a model 'simple nervous system'. Neurosignals 13, 5–19.
Marks, W.D., Skerrett, I.M., 2014. Role of amino terminus in voltage gating and junctional rectification of Shaking B innexins. J. Neurophysiol. 111, 1383–1395.
Mejia, M., Heghinian, M.D., Marı, F., Godenschwege, T.A., 2013. New tools for targeted disruption of cholinergic synaptic transmission in *Drosophila melanogaster*. PLoS One 8, e64685.
Milde, J.J., Strausfeld, N.J., 1990. Cluster organization and response characteristics of the Giant Fibre pathway of the blowfly *Calliphora erythrocephala*. J. Comp. Neurol. 294, 59–75.
Mu, L., Ito, K., Bacon, J.P., Strausfeld, N.J., 2012. Optic glomeruli and their inputs in *Drosophila* share an organizational ground pattern with the antennal lobes. J. Neurosci. 32, 6061–6071.
Mu, L., Bacon, J.P., Ito, K., Strausfeld, N.J., 2014. Responses of *Drosophila* giant descending neurons to visual and mechanical stimuli. J. Exp. Biol. 217, 2121–2129.
Nielsen, M.S., Axelsen, L.N., Sorgen, P.L., Verma, V., Delmar, M., Holstein-Rathlou, N.H., 2012. Gap junctions. Compr. Physiol. 2, 1981–2035.
Ohta, Y., Nishikawa, K., Hiroaki, Y., Fujiyoshi, Y., 2011. Electron tomographic analysis of gap junctions in lateral giant fibers of crayfish. J. Struc. Biol. 175, 49–61.
Orr, B.O., Borgen, M.A., Caruccio, P.M., Murphey, R.K., 2014. Netrin and Frazzled regulate presynaptic gap junctions at a *Drosophila* giant synapse. J. Neurosci. 34, 5416–5430.
Panchin, Y., Kelmanson, I., Matz, M., Lukyanov, K., Usman, N., Lukyanov, S., 2000. A ubiquitous family of putative gap junction molecules. Curr. Biol. 10, R479–R480.
Paul, D., 1986. Molecular cloning of cDNA for rat liver gap junction protein. J. Cell Biol. 103, 123–134.
Peracchia, C., 1973. Low resistance junctions in crayfish. II. Structural details and further evidence for intercellular channels by freeze-fracture and negative staining. J. Cell Biol. 57, 54–65.
Pézier, A., Jezzini, S.H., Marie, B., Blagburn, J.M., 2014. Engrailed alters the specificity of synaptic connections of *Drosophila* auditory neurons with the giant fiber. J. Neurosci. 34, 11691–11704.
Pézier, A.P., Jezzini, S.H., Bacon, J.P., Blagburn, J.M., 2016. Shaking B mediates synaptic coupling between auditory sensory neurons and the giant fiber of *Drosophila melanogaster*. PLoS One. 11 (4), e0152211.
Phelan, P., Starich, T.A., 2001. Innexins get into the gap. Bioessays 23, 388–396.
Phelan, P., Nakagawa, M., Wilkin, M.B., Moffat, K.G., O'Kane, C.J., Davies, J.A., Bacon, J.P., 1996. Mutations in *shaking-B* prevent electrical synapse formation in the *Drosophila* giant fiber system. J. Neurosci. 16, 1101–1113.
Phelan, P., Stebbings, L.A., Baines, R.A., Bacon, J.P., Davies, J.A., Ford, C., 1998a. *Drosophila* Shaking-B protein forms gap junctions in paired *Xenopus* oocytes. Nature 391, 181–184.
Phelan, P., Bacon, J.P., Davies, J.A., Stebbings, L.A., Todman, M.G., Avery, L., Baines, R.A., Barnes, T.M., Ford, C., Hekimi, S., Lee, R., Shaw, J.E., Starich, T.A., Curtin, K.D., Sun, Y., Wyman, R.J., 1998b. Innexins: a family of invertebrate gap junction proteins. Trends Genet. 14, 348–349.
Phelan, P., Goulding, L.A., Tam, J.L.Y., Allen, M.J., Dawber, R.J., Davies, J.A., Bacon, J.P., 2008. Molecular mechanism of rectification at identified electrical synapses in the *Drosophila* Giant Fiber System. Curr. Biol. 18, 1955–1960.
Phelan, P., 2000. Gap junction communication in invertebrates: the innexin gene family. Curr. Top. Membr. 49, 389–422.
Phelan, P., 2005. Innexins: members of an evolutionarily conserved family of gap-junction proteins. Biochim. Biophys. Acta 1711, 225–245.
Power, M.E., 1948. The thoracico-abdominal nervous system of an adult insect. *Drosophila melanogaster*. J. Comp. Neurol. 88, 347–409.

Qin, G., Schwarz, T., Kittel, R.J., Schmid, A., Rasse, T.M., Kappei, D., Ponimaskin, E., Heckmann, M., Sigrist, S.J., 2005. Four different subunits are essential for expressing the synaptic glutamate receptor at neuromuscular junctions of *Drosophila*. J. Neurosci. 25, 3209–3218.

Shih, M.-F.M., Wu, C.-L., 2017. Gap junctions underlying labile memory. In: Jing, J. (Ed.), Network Functions and Plasticity. Elsevier (Chapter 3).

Simmons, P.J., Rind, F.C., Santer, R.D., 2010. Escapes with and without preparation: the neuroethology of visual startle in locusts. J. Insect Physiol. 56, 876–883.

Sivan-Loukianova, E., Eberl, D.F., 2005. Synaptic ultrastructure of *Drosophila* Johnston's organ axon terminals as revealed by an enhancer trap. J. Comp. Neurol. 491, 46–55.

Stebbings, L.A., Todman, M.G., Phillips, R., Greer, C.E., Tam, J., Phelan, P., Jacobs, K., Bacon, J.P., Davies, J.A., 2002. Gap junctions in *Drosophila*: developmental expression of the entire innexin gene family. Mech. Dev. 113, 197–205.

Strausfeld, N.J., Bassemir, U.K., 1983. Cobalt-coupled neurons of a giant fibre system in Diptera. J. Neurocytol. 12, 971–991.

Tanouye, M.A., King, D.G., 1983. Giant fibre activation of direct flight muscles in *Drosophila*. J. Exp. Biol. 105, 241–251.

Tanouye, M.A., Wyman, R.J., 1980. Motor outputs of giant nerve fiber in *Drosophila*. J. Neurophysiol. 44, 405–421.

Thomas, J.B., Wyman, R.J., 1982. A mutation in *Drosophila* alters normal connectivity between two identified neurones. Nature 298, 650–651.

Thomas, J.B., Wyman, R.J., 1984. Mutations altering synaptic connectivity between identified neurons in *Drosophila*. J. Neurosci. 4, 530–538.

Tootoonian, S., Coen, P., Kawai, R., Murthy, M., 2012. Neural representations of courtship song in the *Drosophila* brain. J. Neurosci. 32, 787–798.

Trimarchi, J.R., Schneiderman, A.M., 1993. Giant fibre activation of an intrinsic muscle in the mesothoracic leg of *Drosophila melanogaster*. J. Exp. Biol. 177, 149–167.

Trimarchi, J.R., Schneiderman, A.M., 1995a. Different neural pathways coordinate *Drosophila* flight initiations evoked by visual and olfactory stimuli. J. Exp. Biol. 198, 1099–1104.

Trimarchi, J.R., Schneiderman, A.M., 1995b. Flight initiations in *Drosophila melanogaster* are mediated by several distinct motor patterns. J. Comp. Physiol. A 176, 355–364.

Uma, D., Durkee, C., Herzner, G., Weiss, M., 2013. Double deception: ant-mimicking spiders elude both visually- and chemically-oriented predators. PLoS One 8, e79660.

Von Reyn, C.R., Breads, P., Peek, M.Y., Zheng, G.Z., Williamson, W.R., Yee, A.L., Leonardo, A., Card, G.M., 2014. A spike-timing mechanism for action selection. Nat. Neurosci. 17, 962–970.

Wiersma, C.A.G., 1947. Giant nerve fibre system of the crayfish. A contribution to comparative physiology of synapse. J. Neurophysiol. 10, 23–38.

Yen, M.R., Saier Jr., M.H., 2007. Gap junctional proteins of animals: the innexin/pannexin superfamily. Prog. Biophys. Mol. Biol. 94, 5–14.

Zhang, Z., Curtin, K.D., Sun, Y.A., Wyman, R.J., 1999. Nested transcripts of gap junction gene have distinct expression patterns. J. Neurobiol. 40, 288–301.

Zylinski, S., Johnsen, S., 2011. Mesopelagic cephalopods switch between transparency and pigmentation to optimize camouflage in the deep. Curr. Biol. 21, 1937–1941.

CHAPTER 3

Gap Junctions Underlying Labile Memory

M.-F.M. Shih[1,4], C.-L. Wu[2,3]
[1]Cold Spring Harbor Laboratory, Cold Spring Harbor, NY, United States; [2]Chang Gung University, Taoyuan, Taiwan; [3]Linkou Chang Gung Memorial Hospital, Taoyuan, Taiwan; [4]Stony Brook University, Stony Brook, NY, United States (Present address)

1. INTRODUCTION

Gap junctions are a type of cell−cell communication that make holochannels formed by pairs of hexameric hemichannels from apposed cell membranes, which allows direct transmission of ions and small molecules less than 1 kD in size (McCracken and Roberts, 2006). A hexameric hemichannel could be homomeric or heteromeric, depending on whether it has identical composition or different interlocking subunits. A holochannel could be either homotypic if two hemichannels are identical, or heterotypic if they are not identical. Neuronal gap junctions were identified as a type of direct intracellular communication for electrical synapses during the 1970s and had once been regarded as the primary mechanism of neural transmission until the emergence of studies of chemical neurotransmission. After that, this direct neuronal communication was considered a fast but primitive version in the nervous system of lower organisms. However, the impression that electrical synapses only provide bidirectional current flow must be revised and expanded by adding rectifying electrical synapses, where heterotypic channels are asymmetrically gated by voltage and hence favor one-directional flow (Edwards et al., 1999b; Marder, 1998; Phelan et al., 2008; see also Phelan et al., 2017 (Chapter 2)). Such synapses are gifted with some advanced physiological characteristics, like functional polarity and coincidence detection. Following the advances of molecular biology, more and more components of gap junctions have been identified, such as the *connexin/pannexin* gene family for chordates and the *innexin* gene family for invertebrates (Paul, 1986; Phelan and Starich, 2001). Intriguingly, *cx36*, a neuronal *connexin* gene (Condorelli et al., 2000), has been suggested to play a role in learning and memory (Frisch et al., 2005), and pharmacologically blocking gap junctions can induce memory impairment (Juszczak and Swiergiel, 2009), but there is still a lack of definitive evidence that a specific spatio-temporal influence of neuronal gap junctions is required in memory formation.

In the book *Memory: From Mind to Molecules* (Squire and Kandel, 2000), Squire and Kandel said that "Memory is the process by which what is learned persists across time," and from the information processing perspective it can be divided into three main

Network Functions and Plasticity
ISBN 978-0-12-803471-2, http://dx.doi.org/10.1016/B978-0-12-803471-2.00003-5

stages—acquisition, consolidation, and retrieval. The memory study originated from and was restricted to the domain of philosophy before the late nineteenth century. Thanks to the success of experimental science, the philosophical exploration of mental processes was gradually replaced by empirical studies of the mind, as psychology emerged. The last century has seen great advances in memory study with the integration of biology and psychology, especially for studies using *Drosophila melanogaster* as a model animal. Currently, neuroscientists can decipher each well-defined memory behavior with respect to gene and circuit.

Flies can be trained to form associative memory between odor and electric shock with Pavlovian olfactory conditioning. Flies have a natural aversion to electric shock, and thus it plays the role of unconditioned stimulus (US) well. When the shock is paired with delivery of a particular odor, the conditioned stimulus (CS), flies learn the association and subsequently avoid the odor (Tully and Quinn, 1985). This aversive olfactory memory can persist for several hours after one training session and is operationally divided into two forms based on its susceptibility to cold-induced anesthesia: labile anesthesia-sensitive memory (ASM) and consolidated anesthesia-resistant memory (ARM) (Fig. 3.1) (Davis, 2005; Dubnau and Tully, 1998; Quinn and Dudai, 1976). ASM is also called the middle-term memory (MTM) (Tully et al., 1994). ASM and ARM each account for about half of the total retention level when flies are tested 3 h after conditioning, a time point in the range of the intermediate-term memory (ITM) (Stough et al., 2006; Tamura et al., 2003; Tully et al., 1994).

In fly the mushroom body (MB), a neuropil composed of approximately 2000 intrinsic neurons, the Kenyon cells (KCs), in each hemisphere, is the olfactory learning and memory center (Aso et al., 2014; Davis, 2005; de Belle and Heisenberg, 1994; Margulies et al., 2005). The dendrites of KCs form the cap or calyx of the MB, and their axons project anteriorly to form the stalk-like peduncle before those axons terminate in one or more lobes, termed α/β, α'/β', and γ lobes for $\alpha\beta$ KCs, $\alpha'\beta'$ KCs, and γ KCs, respectively (Jefferis et al., 2002). KCs receive olfactory input (CS) from the antennal lobe projection neurons (PNs), which contact KC dendrites in the calyx, and manifest sparse neuronal activity in MB subsets (Akalal et al., 2010; Honegger et al., 2011; Lin et al., 2014; Wang et al., 2008; Yu et al., 2006). Two MB modulatory neurons, the anterior paired lateral (APL) neuron and the dorsal paired medial (DPM) neuron, contact all of the 2000 intrinsic MB KCs with the dense fibers covering whole MB lobes and anterior part of the peduncle, and the APL neuron has additional coverage in the MB calyx and the posterior part of the peduncle (see also Fig. 3.2) (Tanaka et al., 2008). These two neurons co-release multiple modulatory neurotransmitters, such as serotonin and octopamine (see Table 3.1). Serotonin takes center stage in the seminal findings uncovering the cellular models for simple learning and memory in *Aplysia*, which has also been shown important in other organisms, such as fruit fly, rodents, and human (Glanzman, 2008; Seyedabadi et al., 2014).

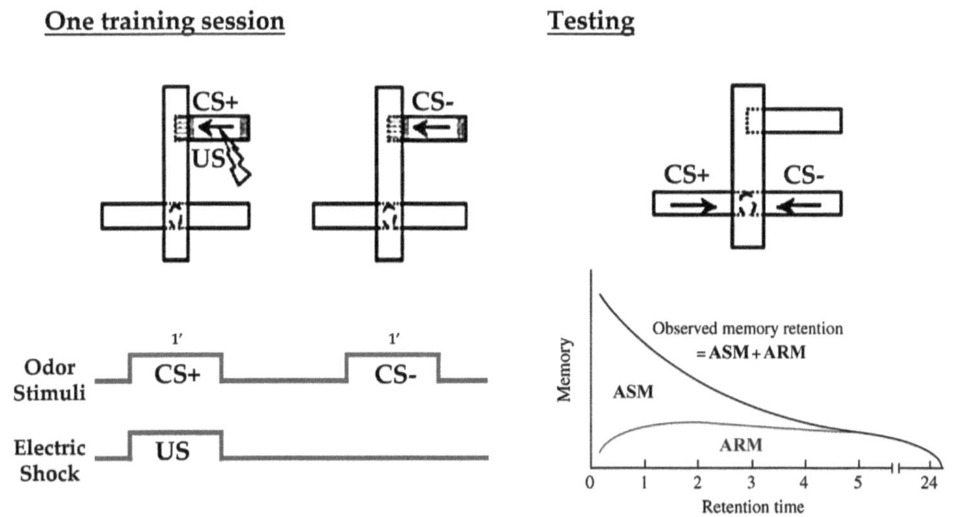

Figure 3.1 ***Protocol for aversive olfactory associative memory.*** *CS,* Conditioned stimulus; *ARM,* anesthesia-resistant memory; *ASM,* anesthesia-sensitive memory; *US,* unconditioned stimulus. *(This figure is from Fig. 1 in Shih, M.-F.M., October 22, 2013. Anterior Paired Lateral Neurons in* Drosophila *Intermediate-term Memory. (A.-S. Chiang). Institute of Biotechnologies, National Tsing Hua University. Retrieved from: http://etd.lib.nctu.edu.tw/cgi-bin/gs32/hugsweb.cgi?o=dnthucdr&s=id=%22GH000934272%22.&searchmode=basic.)*

After operational segregation of two memory components, ASM and ARM, in *Drosophila* (Quinn and Dudai, 1976), the first gene identified as critical for ASM is *amnesiac* (*amn*), which is a putative neuropeptide gene homologous to vertebrate pituitary adenylyl cyclase-activating peptide, which modulates the phosphorylation of α-amino-3-hydroxy-5-methyl-4-isoxazolepropionic acid receptor and synaptic plasticity (Toda and Huganir, 2015). *amn* mutant flies have been characterized as having normal learning but fast forgetting (Feany and Quinn, 1995; Quinn et al., 1979). The fast decline of memory retention was ascribed to the anesthesia-sensitive MTM lost (Folkers et al., 1993; Tully et al., 1994). It eventually became clear that *amn* predominantly expressing in the DPM neurons is necessary and sufficient for 1-h memory (Waddell et al., 2000), and that age-related memory impairment is identical to the *amn*-dependent MTM lost (Tamura et al., 2003), implying that the DPM neurons release AMN for ASM formation.

The *amn*-expressing DPM neuron, an MB modulatory neuron, innervates all of the lobes of the MB and part of the peduncle but does not send processes anywhere else like the calyx (Waddell et al., 2000) (see also Fig. 3.2B). This anatomical arrangement suggests that DPM and KCs are recurrently connected, ie, the DPM neuron likely receives direct KC input and in turn releases its neuropeptide transmitter onto the KC axons (Keene and Waddell, 2007). Consistent with the notion of pooling inputs from many KCs, the DPM neuron responds to both shock and a wide range of different odors (Wu

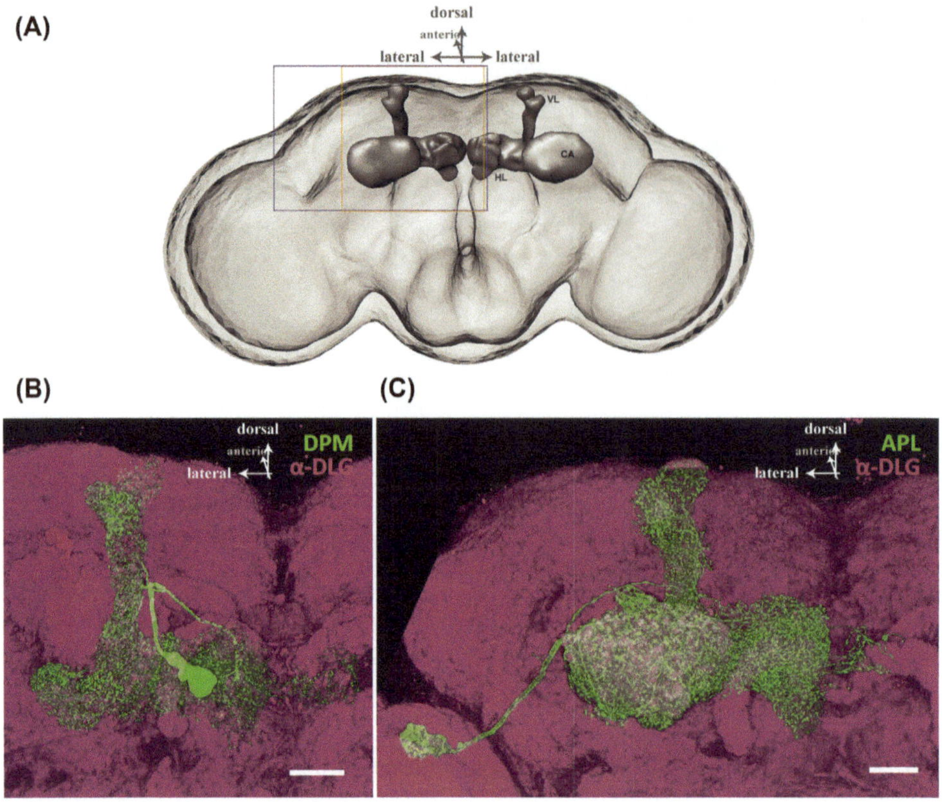

Figure 3.2 ***Single neuron morphology of the dorsal paired medial (DPM) and anterior paired lateral (APL) neurons.*** (A) A cross-section image of Drosophila CNS showing the location of the mushroom body (dark shading). Yellow box: the field of view of the image in panel (B); purple box: the field of view of the image in panel (C). (B) MARCM image of DPM neuron (green) in the left-brain hemisphere. Genotype: *hs-FLP,UAS-mCD8::GFP/y;FRT*^{g13}*,tubP-GAL80/FRT*^{g13}*,UAS-mCD8::GFP;Trh-Gal4/+*. (C) MARCM image of APL neuron (green) in the left-brain hemisphere. Genotype: *hs-FLP,UAS-mCD8::GFP/y; FRT*^{g13}*,tubP-GAL80/FRT*^{g13}*,GH146-GAL4,UAS-mCD8::GFP*. All the brains are counterstained with DLG antibody (magenta). Scale bars represent 20 μm. *CA*, calyx; *HL*, horizontal lobe; *VL*, vertical lobe; *α-DLG*, anti-discs large; *MARCM*, mosaic analysis with a repressible cell marker. *(Panel C is adapted from Fig. 3 in Shih, M.-F.M., October 22, 2013. Anterior Paired Lateral Neurons in* Drosophila *Intermediate-term Memory. (A.-S. Chiang). Institute of Biotechnologies, National Tsing Hua University. Retrieved from: http://etd.lib.nctu.edu.tw/cgi-bin/gs32/hugsweb.cgi?o=dnthucdr&s=id=%22GH000934272%22.&searchmode=basic.)*

et al., 2011; Yu et al., 2005). The APL neuron, another MB modulatory neuron, has a morphology that partly overlaps with DPM. Its cell body is located in the lateral protocerebrum where it gives rise to a primary neurite that projects medially and bifurcates to innervate the entire MB, one branch entering the calyx and another branch entering the vertical lobe (Liu and Davis, 2009) (see also Fig. 3.3C). The APL neuron also responds to the US shock (Liu and Davis, 2009). Using functional imaging to monitor

Table 3.1 Immunoreactivity of Neurotransmitters in APL and DPM Neurons

Antigen	APL	DPM
ChAT	Negative	Negative
Gad1	NA	Positive
VGAT	NA	Positive
GABA	Positive	Positive
DDC	Negative	Positive
Serotonin	Negative	Positive
Amnesiac	NA	Positive
Tyramine	Positive	NA
Tβh	Positive	NA
Octopamine	Positive	NA
TH	Negative	Negative
inx7	Positive[a]	Negative[a]
inx6	Negative[a]	Positive[a]

Cholinergic antigen: ChAT; GABAergic antigen: Gad1, VGAT, and GABA; serotonergic antigen: DDC and serotonin; peptidergic antigen: Amnesiac; octopaminergic antigen: tyramine, Tβh, and octopamine; and dopaminergic antigen: TH.
APL, anterior paired lateral; *ChAT*, choline acetyltransferase; *DDC*, dopa decarboxylase; *DPM*, dorsal paired medial; *GABA*, gamma aminobutyric acid; *Gad1*, glutamic acid decarboxylase 1; *NA*, not available; *TH*, tyrosine hydroxylase or pale; *Tβh*, tyramine β hydroxylase; *VGAT*, vesicular GABA transporter.
[a]Immunostaining against INX7 and INX6 did not provide neuronal specificity. The quality here reflects the RNAi-mediated knockdown effects on 3-h memory.

intracellular calcium changes, we found that the APL neuron is also an odor generalist responsive to at least eight tested odorants (Wu et al., 2011). The GABAergic APL neuron was reported to take part in feedback inhibition of KCs to maintain sparse, decorrelated odor coding and the odor specificity of memories (Lin et al., 2014); notably, training attenuates its response to the conditioned odor within 5 min (Liu and Davis, 2009). In contrast, *amn* neuropeptide released by DPM neurons is dispensable during the acquisition phase of learning but is critical for a delayed memory trace and a robust ITM (Keene et al., 2006, 2004; Waddell et al., 2000; Yu et al., 2005).

2. GAP JUNCTIONS BETWEEN APL AND DPM NEURONS FOR LABILE MEMORY

The overlapping projections of DPM and APL neurons, along with their similar odor response properties, prompted us to examine whether information may be exchanged between these two neurons. Whole-cell patch-clamp recordings from the DPM cell

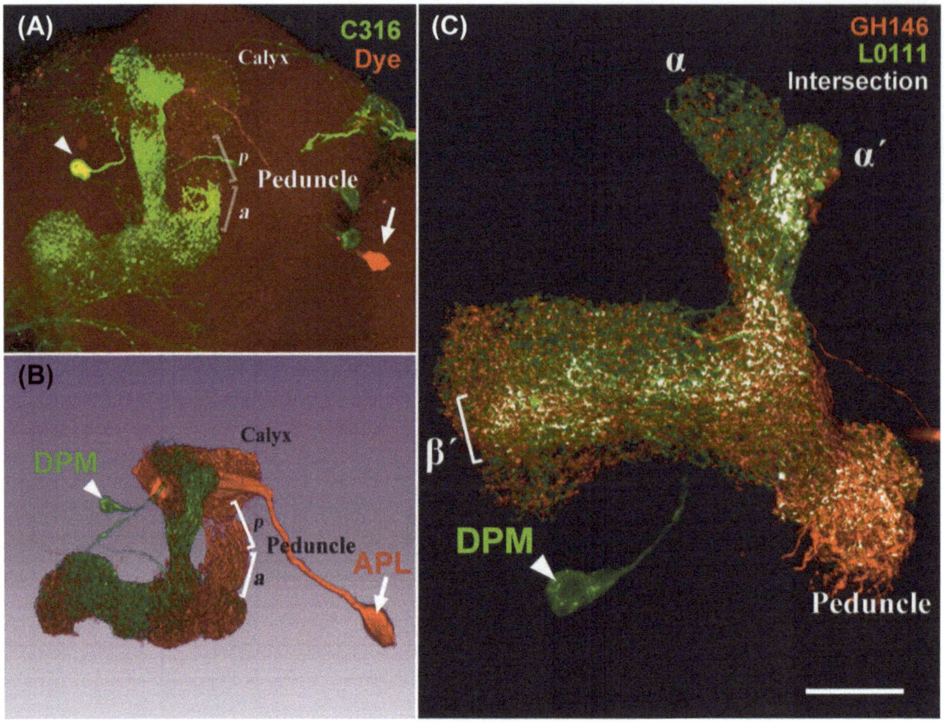

Figure 3.3 ***Dye coupling between dorsal paired medial (DPM) and anterior paired lateral (APL) neurons.*** (A) Biocytin hydrazide loaded into the soma of a GFP-labeled DPM neuron (*arrowhead*) via whole-cell patch-clamp recording appeared in the soma of APL neuron (*arrow*) and its fibers (red) projecting into the entire mushroom body (MB) including calyx and posterior (p) peduncle (*dotted line*). The DPM neuron innervates all MB lobes and anterior (a) peduncle. Alexa Fluor 546-conjugated streptavidin was used to visualize biocytin-filled neuron. Genotype shown is *UAS-mCD8::GFP;C316-GAL4*. (B) Anterior view of DPM and APL neurons labeled by GFP (green) and mKO (red), respectively. For clarity, only the regions containing APL and DPM neurons are shown in the volume image. Genotype shown is *GH146-GAL4,UAS-mKO,UAS-mKO/L0111-LexA::GAD,lexAop-rCD2::GFP;lexAop-rCD2::GFP/+*. (C) The intersection (white) between APL neuron (red) and DPM neuron (green). The image is a maximum projection of (B) after removing the calyx. Voxels occupied by APL and DPM neurons are assigned as the white channel. Genotype shown is the same as in (B). Scale bar represents 20 μm. *Arrowhead* indicates DPM cell body, *arrow* indicates APL cell body. *(This figure is from Fig. 1 in Wu, C.-L., Shih, M.-F.M., Lai, J.S.-Y., Yang, H.-T., Turner, G.C., Chen, L., & Chiang, A.-S., 2011. Heterotypic gap junctions between two neurons in the* Drosophila *brain are critical for memory. Curr. Biol. 21 (10), 848–854.http://doi.org/10.1016/j.cub.2011.02.041.)*

body were used to dye-fill the DPM neuron. In addition to labeling the DPM cell body and its processes in the lobes, staining extended through the entire peduncle, calyx to the APL cell body (Fig. 3.3A). Dye coupling indicates gap-junctional connections between neurons (Stewart, 1981). To further resolve the site of DPM-APL contact, we labeled the APL and DPM neurons with two different colors of fluorescent proteins using the dual binary systems of GAL4 and LexA (Wu et al., 2011). We found that, although both neurons innervate all lobes of the MB, they contact preferentially in the α'/β' lobes (Fig. 3.3C).

Next, we asked which gap junction genes expressed in APL and DPM neurons are critical for memory. There are eight *innexin*-encoding loci in the *Drosophila* genome (Adams et al., 2000). Taking the advantage of a *Drosophila* RNAi library (Dietzl et al., 2007) under control of the GAL4 expression system, we knocked down each of the eight *innexin* genes in both the APL and DPM neurons using a combination of two GAL4 drivers, *GH146-GAL4* and *C316-GAL4* (Wu et al., 2011). The effectiveness of each inx^{RNAi} was validated with quantitative polymerase chain reaction. We then tested flies for memory retention 3 h after associative conditioning, a phase of memory referred to as intermediate-term (Berry et al., 2008). The *inx6* and *inx7* products in APL and DPM neurons were found necessary for normal 3-h memory (Wu et al., 2011). We confirmed the specificity of knockdown using quantitative immunostaining to show that $inx6^{RNAi}$ or $inx7^{RNAi}$ expression in both *GH146-GAL4* and *C316-GAL4* neurons effectively decreases the levels of the respective proteins within the MB lobes without affecting their levels in other brain regions (Wu et al., 2011). Immunocytochemistry also indicated that INX6 and INX7 are widely expressed in the fly brain but are not always colocalized. In the central complex, strong immunopositive signals were observed for INX6 and INX7 in the fan-shaped body and ellipsoid body, respectively.

Within the MB, INX6 and INX7 colocalized preferentially in the α'/β' lobes and the anterior half of the peduncle, whereas the posterior half of the peduncle was immunonegative (Wu et al., 2011). The preferential localization of INX6 and INX7 immunopositive signals to the α'/β' lobes, together with the fact that the anatomical overlap between APL and DPM neurons is most extensive in the α'/β' lobes, strongly suggests that this is the site of gap-junctional coupling between these two neurons.

Because both APL and DPM neurons play critical but distinct roles in associative olfactory learning and memory (Keene et al., 2004, 2006; Keene and Waddell, 2007; Lin et al., 2014; Liu and Davis, 2009), we next examined the role of the APL-DPM gap junctions at various stages of memory retention. Surprisingly, we found that RNAi-mediated knockdown of either INX6 or INX7 in APL and DPM neurons impairs only ITM, whereas initial learning is normal (Fig. 3.4A–B). ITM in flies comprises two distinct components, ASM and ARM, each accounting for about half of the total retention level when tested 3 h after one session of associative conditioning (Margulies et al., 2005; Stough et al., 2006; Tamura et al., 2003) (see also Fig. 3.1). Using cold shock anesthetization at 2 h after training to abolish ASM (Tully et al., 1994), we found that ARM was normal in flies with knockdown of INX6 (Fig. 3.4A) or INX7 (Fig. 3.4B) proteins in both APL and DPM neurons. The result was verified with an independent $inx6^{RNAi}$ line (v8638) targeting a distinct sequence with a different insertion site to eliminate the possibility of an off-target effect (Wu et al., 2011).

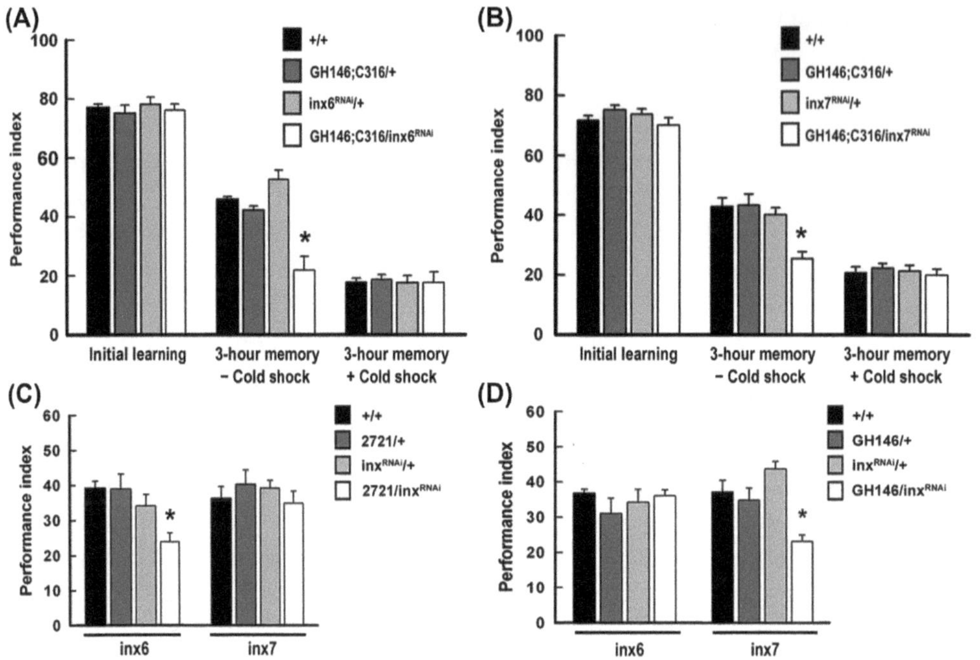

Figure 3.4 ***Knockdown of INX6 or INX7 impairs anesthesia-sensitive memory while leaving learning and anesthesia-resistant memory intact.*** (A) Effects of *inx6*^{RNAi} (v46398) expression in both anterior paired lateral (APL) and dorsal paired medial (DPM) neurons driven by *GH146-GAL4* and *C316-GAL4*, respectively. Genotypes shown are as follows: "+/+," wild type; "GH146; C316/+," *FRT*^{g13}*,GH146-GAL4,UAS-mCD8::GFP/+;C316-GAL4/+*; "inx6^{RNAi}/+," *+/UAS-inx6*^{RNAi} (v46398)*;+/UAS-Dcr-2*; "GH146; C316/inx6^{RNAi}," *FRT*^{g13}*,GH146-GAL4,UAS-mCD8::GFP/UAS-inx6*^{RNAi} (v46398)*;C316-GAL4/UAS-Dcr-2*. (B) Effects of *inx7*^{RNAi} (v103256) expression in both APL and DPM neurons driven by *GH146-GAL4* and *C316-GAL4*, respectively. Genotypes shown are as follows: "+/+," wild type; "GH146; C316/+," *FRT*^{g13}*,GH146-GAL4,UAS-mCD8::GFP/+;C316-GAL4/+*; "inx7^{RNAi}/+," *+/UAS-inx7*^{RNAi} (v103256)*;+/+*; "GH146; C316/inx7^{RNAi}," *FRT*^{g13}*,GH146-GAL4,UAS-mCD8::GFP/UAS-inx7*^{RNAi} (v103256)*;C316-GAL4/+*. (C) Effects of *inx6* and *inx7* knockdowns in the DPM neurons driven by *2721-Gal4* on 3-h memory. Genotypes shown are as follows: "+/+," wild type; "2721/+," *2721-GAL4/+;+/+*; "inx^{RNAi}/+," *+/inx6*^{RNAi} (v46398)*;+/UAS-Dcr-2* for "inx6" or *+/inx7*^{RNAi} (v103256)*;+/+* for "inx7"; "2721/inx^{RNAi}," *2721-GAL4/inx6*^{RNAi} (v46398)*;+/UAS-Dcr-2* for "inx6" or *2721-GAL4/inx7*^{RNAi} (v103256)*;+/+* for "inx7." (D) Effects of *inx6* and *inx7* knockdowns in the APL neurons driven by *GH146-Gal4* on 3-h memory. Genotypes shown are as follows: "+/+," wild type; "GH146/+," *GH146-GAL4/+;+/+*; "inx^{RNAi}/+," *+/inx6*^{RNAi} (v46398)*;+/UAS-Dcr-2* for "inx6" or *+/inx7*^{RNAi} (v103256)*;+/+* for "inx7"; "GH146/inx^{RNAi}," *GH146- GAL4/inx6*^{RNAi} (v46398)*;+/UAS-Dcr-2* for "inx6" or *GH146-GAL4/inx7*^{RNAi} (v103256)*;+/+* for "inx7." *(This figure is adapted from Figs. 3 and 4 in Wu, C.-L., Shih, M.-F.M., Lai, J.S.-Y., Yang, H.-T., Turner, G.C., Chen, L., & Chiang, A.-S., 2011. Heterotypic gap junctions between two neurons in the* Drosophila *brain are critical for memory. Curr. Biol. 21 (10), 848–854.http://doi.org/10.1016/j.cub.2011.02.041.)*

These experiments utilized chronic knockdown of INX6 and INX7 in the APL and DPM neurons. The gross anatomy of APL and DPM neurons in *inx*^{RNAi} flies was not different from that in control flies, suggesting that their development was not affected by the chronic knockdown. Nevertheless, considering the critical roles of INX6 and INX7 in neural development (Ostrowski et al., 2008; Stebbings et al., 2002), we used an inducible knockdown strategy to reduce INX6 and INX7 expression specifically in adult flies. By inactivating the temperature-sensitive repressor GAL80, we selectively induced RNAi knockdown with a temperature shift 5 days before associative conditioning. Inducible knockdown of INX6 or INX7 still significantly impaired 3-h memory (Wu et al., 2011). Thus, the gap-junctional communication between APL and DPM neurons is critical for the formation of ASM but not ARM.

Although the *C316-GAL4* driver expresses in DPM neurons, it also expresses in many other cells, especially a portion of the MB KCs (Wu et al., 2011). We used *2721-GAL4* as a more specific DPM neuron driver that does not express in KCs, to show that INX6 is required in the DPM neuron but not in the APL neuron, and that INX7 is required in the APL neuron but not in the DPM neuron for the formation of ASM (Figs. 3.4C–D). Within the Vienna Tile (VT) line collection from Barry J. Dickson's group, we identified a specific APL driver, *VT43924-GAL4*, with specific expression in only the two APL neurons and later combined it with *UAS-GAL4* to enhance the expression level, as *APL-GAL4* (Wu et al., 2013). The necessity of *inx7* expression in APL neurons indeed is confirmed with *VT43924-GAL4* (Shih, 2013). By combining *GH146-GAL4* with *Cha-GAL80*, which expresses a transcriptional repressor in many neurons, we could reduce expression to undetectable levels in APL neurons, leaving visible expression in only a subset of PNs. Consistently, knockdown of INX7 in these remaining PNs did not affect 3-h memory formation (Wu et al., 2011). These results identified these gap junctions as heterotypic, a prerequisite to transmit signal asymmetrically. It is noteworthy that when we used *L0111-LexA::GAD/ GH146-GAL4, UAS-inx7*^{RNAi} (v103256)*; LexAop-GCaMP3/+* flies to knockdown the gap junctions, the odor response in the DPM neurons did not show significant changes (Jason Sih-Yu Lai, personal communication).

3. DUAL ROLE OF APL NEURON IN ITM THROUGH GAP-JUNCTIONAL AND OCTOPAMINERGIC CHEMICAL NEUROTRANSMISSION

The APL neuron innervated the whole ipsilateral MB, including the calyx and peduncle, all lobes, and contralateral β' lobes (Fig. 3.2C) (Liu and Davis, 2009). Polarity labeling with Syt::HA, which is indicative of putative axonal terminals, showed an APL neuron with more presynaptic terminals in the α'/β' lobes and heel than in other subregions in the MB (Wu et al., 2013). These putative axonal terminals include the GABAergic and octopaminergic neurotransmission (Liu and Davis, 2009; Wu et al., 2013). Putative dendrites labeled with DenMark (Nicolaï et al., 2010) were distributed evenly among all MB subregions. For the DPM neuron, both the axonal and dendritic markers were evenly distributed through the fibers (Wu et al., 2013).

Although Pitman et al. claimed that the APL neuron provides widespread inhibition for appetitive labile memory (Pitman et al., 2011), the data indicating that RNAi-mediated knockdown of gamma aminobutyric acid (GABA) in APL neurons had no effect on ITM discourage the involvement of the GABAergic APL neuron in aversive ITM (Wu et al., 2013). It is noteworthy that ITM deficit caused by blocking chemical neurotransmission in APL neuron has been categorized into ARM impairment, which is most likely due to the dysfunction of octopaminergic chemical neurotransmission (Wu et al., 2013). In contrast, GABAergic neurotransmission from APL neuron has been shown to be the MB feedback loop maintaining the sparseness of KC odor responses and enhancing the learned odor discrimination (Lin et al., 2014). A combinatorial assay with knockdowns of the ASM-required *inx7* and/or ARM-required *Tβh*, which is a key enzyme for octopamine biosynthesis, was conducted to delineate the role of APL neuron in ITM (Wu et al., 2013). Compared with control flies (Fig. 3.5(a)), 3-h memory was impaired in flies expressing either $inx7^{RNAi}$ (Fig. 3.5(b)) or $T\beta h^{RNAi}$ (Fig. 3.5(c)) in

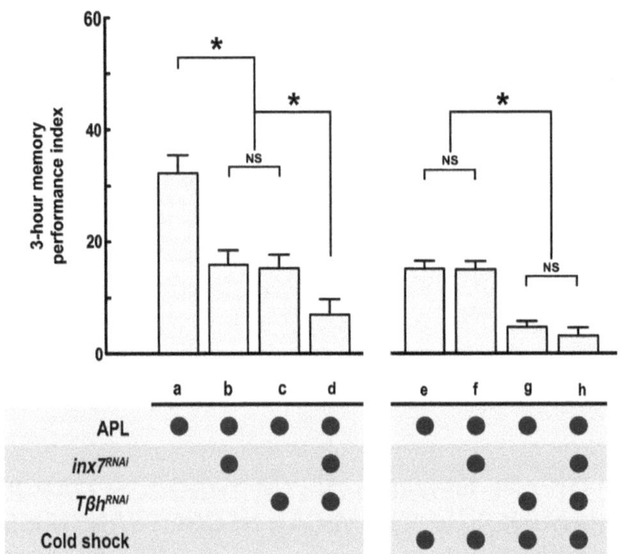

Figure 3.5 ***Additive effect of octopaminergic and gap-junctional manipulations in the anterior paired lateral neurons for anesthesia-resistant memory and anesthesia-sensitive memory.*** Each value represents mean ± S.E.M. ($n = 8$). *$p < .05$. Genotypes: (1) *+/+; APL-GAL4/+* (a, e), (2) *UAS-inx7*^{RNAi} (v103256)/+; *APL-GAL4/+* (b, f), (3) *+/UAS-Tβh*^{RNAi} (v51667); *APL-GAL4/+* (c, g), and (4) *UAS-inx7*^{RNAi} (v103256)/*UAS-Tβh*^{RNAi} (v51667); *APL-GAL4/UAS-GAL4* (d, h). *(This figure is adapted from Fig. 7 in Wu, C.-L., Shih, M.-F.M., Lee, P.-T., & Chiang, A.-S., 2013. An octopamine-mushroom body circuit modulates the formation of anesthesia-resistant memory in* Drosophila. *Curr. Biol. 23 (23), 2346–2354. http://doi.org/10.1016/j.cub.2013.09.056.)*

the APL neurons. Double *inx7* and *Tβh* knockdowns in the APL neurons further reduced 3-h memory (Fig. 3.5(d)), indicating that *inx7* and *Tβh* are independently involved in 3-h memory. Cold-induced anesthesia also impaired 3-h memory by erasing ASM (Fig. 3.5(e)), and additional *inx7* knockdown had no effect on the residual memory (Fig. 3.5(f)), confirming that our cold shock treatment thoroughly removed the ASM component from 3-h memory, and that *inx7* is exclusively involved in ASM. Importantly, the residual memory after cold shock (Fig. 3.5(e)) was further reduced in the *Tβh* knockdown flies, with or without additional *inx7* knockdown (Fig. 3.5(g and h)). Thus, APL neuron modulates ASM and ARM formations through gap-junctional and octopaminergic neurotransmissions, respectively.

To summarize, the key finding of our study is that two MB modulatory neurons, the APL and DPM neurons, form heterotypic gap junctions that are specifically required for labile ASM of aversive olfactory conditioning in *Drosophila*. This conclusion is supported by four independent lines of evidence. First, APL and DPM neurons are dye coupled (Fig. 3.3A), a diagnostic feature of cells connected via gap junctions. Second, anatomically the two neurons intersect preferentially in the α'/β' lobes and anterior peduncle, which coincides with INX6-antibody and INX7-antibody immunostaining results (Fig. 3.3C) (Wu et al., 2011). Third, 3-h ASM is impaired by inducible knockdown of either INX6 in DPM neurons or INX7 in APL neurons, but not the converse (Fig. 3.4). This suggests that the APL-DPM gap junctions are heterotypic, and raises the possibility that the connections are in some way asymmetric. Fourth, the combinatorial assay utilizing RNAi-mediated knockdowns of *inx7* and *Tβh* in APL neuron demonstrated the requirement of INX7 for 3-h ASM (Fig. 3.5). However, without a temporally precise method to manipulate gap junctions it is still infeasible to address whether the persistent gap-junctional communication throughout the memory consolidation phase is required for normal ASM expression.

Bissiere et al. showed that electrical synapses control hippocampal contributions to fear learning and memory in rats: they discovered that blocking gap junctions in the dorsal hippocampus by injecting specific gap junction blocker impaired context-dependent fear learning, memory, and extinction (Bissiere et al., 2011). Surprisingly, in flies disrupting the heterotypic gap-junctional communication between APL and DPM neurons does not perturb normal learning (Figs. 3.4A–B). Normal learning requires normal odor perception, shock response, and locomotion. The dispensability for learning suggests that the gap-junctional coupling between APL and DPM neurons is independent of the role of GABAergic transmission from APL neurons, which has been reported to sharpen the odor response in KCs (Lin et al., 2014). There are several possible explanations for this intracellular segregation. Because the APL neuron innervates the MB through two branches (Fig. 3.2C), one entering the calyx and the other entering the vertical lobe, the first possible explanation is that each APL branch plays a functionally distinct role by subcellular localization of GABA and INX7. Learning may induce the feedback inhibition for the odor response in the KCs through the calyceal branch of the APL neuron, whereas the lobe branch of the APL neuron provides the gap-junctional communication with DPM neuron to stabilize ASM. In addition, the heterotypic composition of the gap junctions may favor diffusion only from APL to DPM neurons, keeping GABA release of the APL neurons unaffected by the activity of the DPM neurons. This type of rectification gated by heterotypic gap junctions has been demonstrated in the *Drosophila* giant fiber system (Phelan et al., 2008; see also Phelan et al., 2017 (Chapter 2)). The rectified communication hypothesis is also supported by the fact that blocking neurotransmission from the DPM neuron impaired ITM without affecting learning (Keene et al., 2006; Lee et al., 2011).

The gap-junctional communication between APL and DPM neurons is not involved in 3-h ARM, either (Figs. 3.4A—B and 3.5). In contrast, 3-h ARM consolidation requires the recurrent activities of octopaminergic APL and serotonergic DPM neurons, ie, continuous *shibire*-mediated blockade in APL or DPM neurons for at least 1 h led to 3-h ARM deficit (Lee et al., 2011; Wu et al., 2013). The requirement for recurrent activity supported the idea that ARM is gradually consolidated within 2 h after training (Tully et al., 1994).

4. NONSPIKING APL AND DPM NEURAL NETWORK

APL neurons, and probably DPM neurons also (Glenn C. Turner, personal communication), function as nonspiking neurons (Papadopoulou et al., 2011), which is a type of neuron that transmits signals via graded potential instead of action potential (Roberts and Bush, 1981). Graded potential does not rely on voltage-gated sodium channels to travel through the neuronal fibers, but instead on passive propagation, which is faster but raises a concern of spatial decrement in amplitude. These features make the nonspiking neuron a good candidate to integrate multiple input signals for downstream modulation along the wide-extending fibers. Indeed, APL and DPM neurons broadly innervate the MB lobes (Fig. 3.2), most likely making synapses with every KC, and respond to various odors as well as electric shock (Liu and Davis, 2009; Wu et al., 2011; Yu et al., 2005). It has been demonstrated that a single pair of the APL neuron is the *Drosophila* equivalent of a group of the honeybee GABAergic feedback neurons, receiving odor information from the MB lobes and releasing GABA inhibition to the MB calyx (Grünewald, 1999; Lin et al., 2014). In addition, by utilizing multiple types of neurotransmissions (Table 3.1) an APL and DPM neural network offers great versatility in downstream modulation, such as ASM stabilization, ARM consolidation, decision-making, sleep regulation (Haynes et al., 2015; P.-T. Lee et al., 2011; Wu et al., 2011, 2013, 2015).

5. LABILE MEMORY CIRCUIT OF PERSISTENT ACTIVITY

ASM can be totally abolished by transient anesthetization before testing, and different timing or duration of anesthetization did not make any significant change on the effect of ASM erasure (Knapek et al., 2011; Tully et al., 1994). These findings suggest that some type of persistent neural activity arising in these neurons stabilizes ASM. What is the possible nature of the persistent activity? Persistent activity can manifest in ongoing spiking activity or subtle aspects of neural activity. For example, presynaptic residual calcium with slow clearance kinetics has been proposed as a buffer to sustain persistent activity in a recurrent circuit as working memory (Mongillo et al., 2008; Wang, 2001). However, the temporal scale for ASM, which can last for up to 6 h after one-session training, is much longer than that for working memory. Prolonging the time frame during which some transient markers of past activities are sustained may require more complex circuitry with a self-sustaining feedback loop. Moreover, during the memory consolidation phase, or the time period between training and testing, flies receive no olfactory stimulus, and no spontaneous neural activity in KCs has been noticed after classical conditioning (Akalal et al., 2010; Wang et al., 2008; Yu et al., 2006). Thus, the persistent activity for ASM might arise in the local circuits that include the presynapses of odor-responsive KCs and the nearby gap junctions between APL and DPM neurons.

DPM neurons once had been proposed to release the excitatory neurotransmitter acetylcholine (Keene et al., 2004), but a new study showed that DPM neurons are not cholinergic by negative immunoreactive signal against choline acetyltransferase in DPM cell bodies and fibers (Haynes et al., 2015). Therefore, it should be other neural mechanism(s) that work to sustain this prolonged persistent activity. For example, in the case of the dopaminergic MB-M3 neurons, or the PAM-03 neurons (Aso et al., 2014), the output of the fibers innervating the tips of the horizontal (β and β′) lobes is required during training for later ASM expression (Aso et al., 2010). PAM-03 neurons might contribute to the initiation and maintenance of the persistent activity.

How does the heterotypic gap-junctional communication between APL and DPM neurons contribute to the persistent activity for labile ASM? A theoretical study reported that persistent activity requires a fast timescale of synaptic depression, whereas the onset of a slow timescale depression terminates the persistent activity (Volman et al., 2007). The gap-junctional communication would eventually lead to the neurotransmission from the nonspiking APL and DPM neural network to the MB KCs coding for the odor identities and memory traces. All the conventional neurotransmitters released by APL and DPM neural network (Table 3.1) have been examined by *shibire* manipulation and/or specific RNAi-mediated knockdowns in terms of 3-h aversive olfactory memory performance (Lee et al., 2011; Wu et al., 2013). None of them significantly contributed to ASM. In contrast, the *amn*-encoded neuropeptide is the only known neurotransmission involved in ASM. Although the mechanism by which AMN neurotransmission mediates ASM consolidation remains unclear, several lines of evidence highlight its importance. AMN suppresses the oscillation amplitude of intracellular calcium concentration in KCs (Rosay et al., 2001). Memory traces corresponding to ITM in the DPM neurons (and very likely in $\alpha'\beta'$ KCs) and corresponding to long-term memory in $\alpha\beta$ KCs are absent in the *amn* mutant flies (Wang et al., 2008; Yu et al., 2005, 2006). It would be innovative to study whether *amn*-encoded neuropeptide can change the timescale of KC synaptic depression, and whether the gap-junctional communication between APL and DPM neurons modulates the AMN release from spatiotemporal perspectives.

Forgetting of labile ASM is regulated through Rac activity in fly (Shuai et al., 2010). Rac is a small G-protein acting as a key regulator of actin cytoskeleton dynamics by switching between GTP-bound active and GDP-bound inactive forms (Etienne-Manneville and Hall, 2002) and also contributes to forgetting in rats (Jiang et al., 2015). When the dominant-negative Rac mutant was panneuronally expressed, flies showed slower memory decay. However, the 3-h memory enhancement disappeared after an additional cold shock treatment, indicating the Rac activity did not regulate the forgetting of ARM. It has been known as a conserved cell biological pathway that Rac activity leads to the inhibition of actin depolymerization (Arber et al., 1998; Edwards et al., 1999a; Yang et al., 1998), which is critically involved in synaptic plasticity (Hanley, 2014; Luo, 2002). Interestingly, the site-specific manipulations of Rac activity identified the MB $\alpha\beta$ and γ KCs as relevant neuronal substrate (Shuai et al., 2010), suggesting that the ASM susceptible to Rac-mediated forgetting is manifested in the synaptic plasticity changes of the $\alpha\beta$ and γ KCs. The output of $\alpha\beta$ and γ KCs should be required for ASM during retrieval.

In addition to *amn*, a variety of genes are also involved in ASM. In the MB $\alpha\beta$ and γ KCs, *rutabaga*, a gene encoding Ca^{2+}-sensitive type 1 adenylyl cyclase, is critical for ASM (Isabel et al., 2004; Scheunemann et al., 2012). *DCO* is a gene encoding the catalytic subunit of PKA and is required for an early phase of ASM (Li et al., 1996; Skoulakis et al., 1993), whereas A-kinase anchoring proteins—bound pool of PKA is acting within MB KCs to support a late phase of ASM (Schwaerzel et al., 2007). Knockdown of *N*-methyl-D-aspartic acid receptors in KCs also impaired ASM (Wu et al., 2007). Interestingly, two ubiquitous axonal proteins, synapsin and bruchpilot, have respective impacts on ASM and ARM while mutated (Knapek et al., 2011, 2010; Niewalda et al., 2015). Given the fact that synapsin has preferential contribution during high-frequency stimulation and bruchpilot has preferential contribution at low-frequency stimulation (Akbergenova and Bykhovskaia, 2007; Kittel et al., 2006; Knapek et al., 2011; Sun et al., 2006), it is possible that ASM and ARM are encoded in the same population of KCs but have different modes of neurotransmission (Knapek et al., 2011). With the arsenal of split-GAL4 lines expressing in MB output neurons (Aso et al., 2014), it is possible to identify an ASM- or ARM-specific retrieval circuit.

6. SUMMARY AND IMPLICATION

It has been shown that *Drosophila* INX7 in the APL and INX6 in the DPM neurons form a gap-junctional communication within the MB, a critical brain region for olfactory

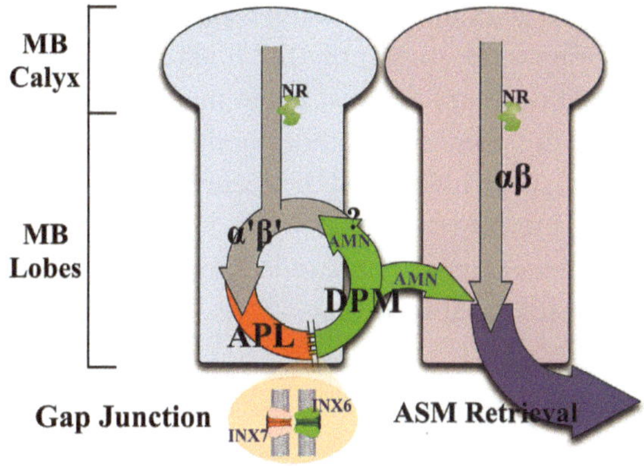

Figure 3.6 ***A model of recurrent circuits maintaining anesthesia-sensitive memory (ASM).*** We propose that recurrent circuits comprosed of, but are not limited to, α′β′ Kenyon cells (KCs) → anterior paired lateral (APL) → dorsal paired medial (DPM) neurons maintain the synaptic plasticity change coding for ASM in αβ KCs. The heterotypic gap junctions between APL and DPM neurons are a critical component of the recurrent circuits. *NR*, *N*-methyl-D-aspartic acid receptor.

associative learning and memory. The gap-junctional communication is essential for memory consolidation, rather than acquisition or retrieval. Knockdown of the gap junctions only impaired labile ASM, while keeping the other component, consolidated ARM, unaffected. Based on the vulnerability of ASM on anesthetization, a model of recurrent circuits maintaining ASM is proposed (Fig. 3.6). A recurrent reward circuit has been identified to underlie the formation and consolidation of LTM, in which an MB glutamatergic output neuron directly feedback-activates the MB upstream reinforcing dopaminergic neuron (Ichinose et al., 2015). In contrast, the source of the persistent activity in the recurrent circuits model for ASM is still missing. The seven newly identified glutamatergic MB output neuron types are ideal candidates to test (Aso et al., 2014).

ASM is a type of memory that flies gradually lose during the course of senescence (Tamura et al., 2003), an analogy to dementia in human. The global cost of dementia was over US $600 billion in 2014, and still keeps increasing (Hiscott, 2014). People are desperate to find a cure. Understanding the neural mechanisms underlying ASM in fly, for example, the gap-junctional communication in recurrent circuits, should shed light on the treatment for dementia.

OUTSTANDING ISSUES AND FUTURE RESEARCH

- To determine whether the heterotypic gap-junctional communication exhibits rectification using electrophysiological recording in *Xenopus* oocyte or cell culture system.
- To localize the gap junctions and map the microcircuit using immunoelectron microscopy.

- To develop a reversible method to interfere gap-junctional communication with spatial and temporal precision, to identify the functional role of gap junction.
- To identify the critical timing/stage that labile memory requires the gap-junctional communication using the potential method mentioned earlier.
- To decipher the neural mechanisms and the recurrent circuits stabilizing labile memory using optogenetic interrogation.
- To demonstrate the persistent activity underlying labile memory using functional imaging or electrophysiological recording.

ACKNOWLEDGMENTS

We thank Toshihide Hige for extensive comments and discussion, Glenn Turner and Jason S.Y. Lai for sharing unpublished data, and Jacob Strieb for critical reading. M.-F.M.S. is a postdoctoral researcher in the Josh Dubnau laboratory at Cold Spring Harbor Laboratory. C.-L.W. is an assistant professor at Chang Gung University. This work was supported by grants from DART NeuroScience LLC in USA (to Josh Dubnau), grants from Ministry of Science and Technology in Taiwan (to C.-L.W.), and grants from Chang Gung Memorial Hospital in Taiwan (to C.-L.W.).

REFERENCES

Adams, M.D., Celniker, S.E., Holt, R.A., Evans, C.A., Gocayne, J.D., Amanatides, P.G., Scherer, S.E., Li, P.W., Hoskins, R.A., Galle, R.F., George, R.A., Lewis, S.E., Richards, S., Ashburner, M., Henderson, S.N., Sutton, G.G., Wortman, J.R., Yandell, M.D., Zhang, Q., Chen, L.X., Brandon, R.C., Rogers, Y.H., Blazej, R.G., Champe, M., Pfeiffer, B.D., Wan, K.H., Doyle, C., Baxter, E.G., Helt, G., Nelson, C.R., Gabor, G.L., Abril, J.F., Agbayani, A., An, H.J., Andrews-Pfannkoch, C., Baldwin, D., Ballew, R.M., Basu, A., Baxendale, J., Bayraktaroglu, L., Beasley, E.M., Beeson, K.Y., Benos, P.V., Berman, B.P., Bhandari, D., Bolshakov, S., Borkova, D., Botchan, M.R., Bouck, J., Brokstein, P., Brottier, P., Burtis, K.C., Busam, D.A., Butler, H., Cadieu, E., Center, A., Chandra, I., Cherry, J.M., Cawley, S., Dahlke, C., Davenport, L.B., Davies, P., de Pablos, B., Delcher, A., Deng, Z., Mays, A.D., Dew, I., Dietz, S.M., Dodson, K., Doup, L.E., Downes, M., Dugan-Rocha, S., Dunkov, B.C., Dunn, P., Durbin, K.J., Evangelista, C.C., Ferraz, C., Ferriera, S., Fleischmann, W., Fosler, C., Gabrielian, A.E., Garg, N.S., Gelbart, W.M., Glasser, K., Glodek, A., Gong, F., Gorrell, J.H., Gu, Z., Guan, P., Harris, M., Harris, N.L., Harvey, D., Heiman, T.J., Hernandez, J.R., Houck, J., Hostin, D., Houston, K.A., Howland, T.J., Wei, M.H., Ibegwam, C., Jalali, M., Kalush, F., Karpen, G.H., Ke, Z., Kennison, J.A., Ketchum, K.A., Kimmel, B.E., Kodira, C.D., Kraft, C., Kravitz, S., Kulp, D., Lai, Z., Lasko, P., Lei, Y., Levitsky, A.A., Li, J., Li, Z., Liang, Y., Lin, X., Liu, X., Mattei, B., McIntosh, T.C., McLeod, M.P., McPherson, D., Merkulov, G., Milshina, N.V., Mobarry, C., Morris, J., Moshrefi, A., Mount, S.M., Moy, M., Murphy, B., Murphy, L., Muzny, D.M., Nelson, D.L., Nelson, D.R., Nelson, K.A., Nixon, K., Nusskern, D.R., Pacleb, J.M., Palazzolo, M., Pittman, G.S., Pan, S., Pollard, J., Puri, V., Reese, M.G., Reinert, K., Remington, K., Saunders, R.D., Scheeler, F., Shen, H., Shue, B.C., Sidén-Kiamos, I., Simpson, M., Skupski, M.P., Smith, T., Spier, E., Spradling, A.C., Stapleton, M., Strong, R., Sun, E., Svirskas, R., Tector, C., Turner, R., Venter, E., Wang, A.H., Wang, X., Wang, Z.Y., Wassarman, D.A., Weinstock, G.M., Weissenbach, J., Williams, S.M., Woodage, T., Worley, K.C., Wu, D., Yang, S., Yao, Q.A., Ye, J., Yeh, R.F., Zaveri, J.S., Zhan, M., Zhang, G., Zhao, Q., Zheng, L., Zheng, X.H., Zhong, F.N., Zhong, W., Zhou, X., Zhu, S., Zhu, X., Smith, H.O., Gibbs, R.A., Myers, E.W., Rubin, G.M., Venter, J.C., 2000. The genome sequence of *Drosophila melanogaster*. Science 287 (5461), 2185–2195.

Akalal, D.-B.G., Yu, D., Davis, R.L., 2010. A late-phase, long-term memory trace forms in the γ neurons of *Drosophila* mushroom bodies after olfactory classical conditioning. J. Neurosci. 30 (49), 16699–16708.
Akbergenova, Y., Bykhovskaia, M., 2007. Synapsin maintains the reserve vesicle pool and spatial segregation of the recycling pool in *Drosophila* presynaptic boutons. Brain Res. 1178, 52–64.
Arber, S., Barbayannis, F.A., Hanser, H., Schneider, C., Stanyon, C.A., Bernard, O., Caroni, P., 1998. Regulation of actin dynamics through phosphorylation of cofilin by LIM-kinase. Nature 393 (6687), 805–809.
Aso, Y., Siwanowicz, I., Bräcker, L., Ito, K., Kitamoto, T., Tanimoto, H., 2010. Specific dopaminergic neurons for the formation of labile aversive memory. Curr. Biol. 20 (16), 1445–1451.
Aso, Y., Hattori, D., Yu, Y., Johnston, R.M., Iyer, N.A., Ngo, T.-T.B., Dionne, H., Abbott, L.F., Axel, R., Tanimoto, H., Rubin, G.M., 2014. The neuronal architecture of the mushroom body provides a logic for associative learning. eLife 3, e04577.
Berry, J., Krause, W.C., Davis, R.L., 2008. Olfactory memory traces in *Drosophila*. Prog. Brain Res. 169, 293–304.
Bissiere, S., Zelikowsky, M., Ponnusamy, R., Jacobs, N.S., Blair, H.T., Fanselow, M.S., 2011. Electrical synapses control hippocampal contributions to fear learning and memory. Science 331 (6013), 87–91.
Condorelli, D.F., Belluardo, N., Trovato-Salinaro, A., Mudò, G., 2000. Expression of Cx36 in mammalian neurons. Brain Res. Brain Res. Rev. 32 (1), 72–85.
Davis, R.L., 2005. Olfactory memory formation in *Drosophila*: from molecular to systems neuroscience. Annu. Rev. Neurosci. 28, 275–302.
de Belle, J.S., Heisenberg, M., 1994. Associative odor learning in *Drosophila* abolished by chemical ablation of mushroom bodies. Science 263 (5147), 692–695.
Dietzl, G., Chen, D., Schnorrer, F., Su, K.-C., Barinova, Y., Fellner, M., Gasser, B., Kinsey, K., Oppel, S., Scheiblauer, S., Couto, A., Marra, V., Keleman, K., Dickson, B.J., 2007. A genome-wide transgenic RNAi library for conditional gene inactivation in *Drosophila*. Nature 448 (7150), 151–156.
Dubnau, J., Tully, T., 1998. Gene discovery in *Drosophila*: new insights for learning and memory. Annu. Rev. Neurosci. 21, 407–444.
Edwards, D.C., Sanders, L.C., Bokoch, G.M., Gill, G.N., 1999a. Activation of LIM-kinase by Pak1 couples Rac/Cdc42 GTPase signalling to actin cytoskeletal dynamics. Nat. Cell Biol. 1 (5), 253–259.
Edwards, D.H., Heitler, W.J., Krasne, F.B., 1999b. Fifty years of a command neuron: the neurobiology of escape behavior in the crayfish. Trends Neurosci. 22 (4), 153–161.
Etienne-Manneville, S., Hall, A., 2002. Rho GTPases in cell biology. Nature 420 (6916), 629–635.
Feany, M.B., Quinn, W.G., 1995. A neuropeptide gene defined by the *Drosophila* memory mutant amnesiac. Science 268 (5212), 869–873.
Folkers, E., Drain, P., Quinn, W.G., 1993. Radish, a *Drosophila* mutant deficient in consolidated memory. Proc. Natl. Acad. Sci. U.S.A. 90 (17), 8123–8127.
Frisch, C., De Souza-Silva, M.A., Sohl, G., Güldenagel, M., Willecke, K., Huston, J.P., Dere, E., 2005. Stimulus complexity dependent memory impairment and changes in motor performance after deletion of the neuronal gap junction protein connexin36 in mice. Behav. Brain Res. 157 (1), 177–185.
Glanzman, D.L., 2008. New tricks for an old slug: the critical role of postsynaptic mechanisms in learning and memory in *Aplysia*. Prog. Brain Res. 169, 277–292. Elsevier.
Grünewald, B., 1999. Physiological properties and response modulations of mushroom body feedback neurons during olfactory learning in the honeybee, *Apis mellifera*. J. Comp. Physiol. 185 (6), 565–576.
Hanley, J.G., 2014. Actin-dependent mechanisms in AMPA receptor trafficking. Front. Cell. Neurosci. 8, 381.
Haynes, P.R., Christmann, B.L., Griffith, L.C., 2015. A single pair of neurons links sleep to memory consolidation in *Drosophila melanogaster*. eLife 4.
Hiscott, R., 2014. World Alzheimer Report Identifies Preventable Risk Factors for dementia. In: Neurology Today.
Honegger, K.S., Campbell, R.A.A., Turner, G.C., 2011. Cellular-resolution population imaging reveals robust sparse coding in the *Drosophila* mushroom body. J. Neurosci. 31 (33), 11772–11785.
Ichinose, T., Aso, Y., Yamagata, N., Abe, A., Rubin, G.M., Tanimoto, H., 2015. Reward signal in a recurrent circuit drives appetitive long-term memory formation. eLife 4.

Isabel, G., Pascual, A., Preat, T., 2004. Exclusive consolidated memory phases in *Drosophila*. Science 304 (5673), 1024–1027.

Jefferis, G.S.X.E., Marin, E.C., Watts, R.J., Luo, L., 2002. Development of neuronal connectivity in *Drosophila* antennal lobes and mushroom bodies. Curr. Opin. Neurobiol. 12 (1), 80–86.

Jiang, L., Mao, R., Zhou, Q., Yang, Y., Cao, J., Ding, Y., Yang, Y., Zhang, X., Li, L., Xu, L., 2015. Inhibition of Rac1 activity in the hippocampus impairs the forgetting of contextual fear memory. Mol. Neurobiol.

Juszczak, G.R., Swiergiel, A.H., 2009. Properties of gap junction blockers and their behavioural, cognitive and electrophysiological effects: animal and human studies. Prog. Neuropsychopharmacol. Biol. Psychiatry 33 (2), 181–198.

Keene, A.C., Waddell, S., 2007. *Drosophila* olfactory memory: single genes to complex neural circuits. Nat. Rev. Neurosci. 8 (5), 341–354.

Keene, A.C., Stratmann, M., Keller, A., Perrat, P.N., Vosshall, L.B., Waddell, S., 2004. Diverse odor-conditioned memories require uniquely timed dorsal paired medial neuron output. Neuron 44 (3), 521–533.

Keene, A.C., Krashes, M.J., Leung, B., Bernard, J.A., Waddell, S., 2006. *Drosophila* dorsal paired medial neurons provide a general mechanism for memory consolidation. Curr. Biol. 16 (15), 1524–1530.

Kittel, R.J., Wichmann, C., Rasse, T.M., Fouquet, W., Schmidt, M., Schmid, A., Wagh, D.A., Pawlu, C., Kellner, R.R., Willig, K.I., Hell, S.W., Buchner, E., Heckmann, M., Sigrist, S.J., 2006. Bruchpilot promotes active zone assembly, Ca^{2+} channel clustering, and vesicle release. Science 312 (5776), 1051–1054.

Knapek, S., Gerber, B., Tanimoto, H., 2010. Synapsin is selectively required for anesthesia-sensitive memory. Learn. Mem. 17 (2), 76–79.

Knapek, S., Sigrist, S., Tanimoto, H., 2011. Bruchpilot, a synaptic active zone protein for anesthesia-resistant memory. J. Neurosci. 31 (9), 3453–3458.

Lee, P.-T., Lin, H.-W., Chang, Y.-H., Fu, T.-F., Dubnau, J., Hirsh, J., Lee, T., Chiang, A.-S., 2011. Serotonin-mushroom body circuit modulating the formation of anesthesia-resistant memory in *Drosophila*. Proc. Natl. Acad. Sci. U.S.A. 108 (33), 13794–13799.

Li, W., Tully, T., Kalderon, D., 1996. Effects of a conditional *Drosophila* PKA mutant on olfactory learning and memory. Learn. Mem. 2 (6), 320–333.

Lin, A.C., Bygrave, A.M., de Calignon, A., Lee, T., Miesenböck, G., 2014. Sparse, decorrelated odor coding in the mushroom body enhances learned odor discrimination. Nat. Neurosci. 17 (4), 559–568.

Liu, X., Davis, R.L., 2009. The GABAergic anterior paired lateral neuron suppresses and is suppressed by olfactory learning. Nat. Neurosci. 12 (1), 53–59.

Luo, L., 2002. Actin cytoskeleton regulation in neuronal morphogenesis and structural plasticity. Annu. Rev. Cell Dev. Biol. 18, 601–635.

Marder, E., 1998. Electrical synapses: beyond speed and synchrony to computation. Curr. Biol. 8 (22), R795–R797.

Margulies, C., Tully, T., Dubnau, J., 2005. Deconstructing memory in *Drosophila*. Curr. Biol. 15 (17), R700–R713.

McCracken, C.B., Roberts, D.C.S., 2006. Neuronal gap junctions: expression, function, and implications for behavior. Int. Rev. Neurobiol. 73, 125–151.

Mongillo, G., Barak, O., Tsodyks, M., 2008. Synaptic theory of working memory. Science 319 (5869), 1543–1546.

Nicolaï, L.J.J., Ramaekers, A., Raemaekers, T., Drozdzecki, A., Mauss, A.S., Yan, J., Landgraf, M., Annaert, W., Hassan, B.A., 2010. Genetically encoded dendritic marker sheds light on neuronal connectivity in *Drosophila*. Proc. Natl. Acad. Sci. U.S.A. 107 (47), 20553–20558.

Niewalda, T., Michels, B., Jungnickel, R., Diegelmann, S., Kleber, J., Kähne, T., Gerber, B., 2015. Synapsin determines memory strength after punishment- and relief-learning. J. Neurosci. 35 (19), 7487–7502.

Ostrowski, K., Bauer, R., Hoch, M., 2008. The *Drosophila* innexin 7 gap junction protein is required for development of the embryonic nervous system. Cell Commun. Adhesion 15 (1), 155–167.

Papadopoulou, M., Cassenaer, S., Nowotny, T., Laurent, G., 2011. Normalization for sparse encoding of odors by a wide-field interneuron. Science 332 (6030), 721–725.

Paul, D.L., 1986. Molecular cloning of cDNA for rat liver gap junction protein. J. Cell Biol. 103 (1), 123–134.
Phelan, P., Starich, T.A., 2001. Innexins get into the gap. Bioessays 23 (5), 388–396.
Phelan, P., Goulding, L.A., Tam, J.L.Y., Allen, M.J., Dawber, R.J., Davies, J.A., Bacon, J.P., 2008. Molecular mechanism of rectification at identified electrical synapses in the *Drosophila* giant fiber system. Curr. Biol. 18 (24), 1955–1960.
Phelan, P., Bacon, J.P., Blagburn, J.M., 2017. Neural circuits underlying escape behavior in Drosophila: focus on electrical signaling. In: Jing, J. (Ed.), Network Functions and Plasticity. Elsevier (Chapter 2).
Pitman, J.L., Huetteroth, W., Burke, C.J., Krashes, M.J., Lai, S.-L., Lee, T., Waddell, S., 2011. A pair of inhibitory neurons are required to sustain labile memory in the *Drosophila* mushroom body. Curr. Biol. 21 (10), 855–861.
Quinn, W.G., Dudai, Y., 1976. Memory phases in *Drosophila*. Nature 262 (5569), 576–577.
Quinn, W.G., Sziber, P.P., Booker, R., 1979. The *Drosophila* memory mutant amnesiac. Nature 277 (5693), 212–214.
Roberts, A., Bush, B.M.H., 1981. Neurones without Impulses. Cambridge University Press.
Rosay, P., Armstrong, J.D., Wang, Z., Kaiser, K., 2001. Synchronized neural activity in the *Drosophila* memory centers and its modulation by amnesiac. Neuron 30 (3), 759–770.
Scheunemann, L., Jost, E., Richlitzki, A., Day, J.P., Sebastian, S., Thum, A.S., Efetova, M., Davies, S.A., Schwärzel, M., 2012. Consolidated and labile odor memory are separately encoded within the *Drosophila* brain. J. Neurosci. 32 (48), 17163–17171.
Schwaerzel, M., Jaeckel, A., Mueller, U., 2007. Signaling at A-kinase anchoring proteins organizes anesthesia-sensitive memory in *Drosophila*. J. Neurosci. 27 (5), 1229–1233.
Seyedabadi, M., Fakhfouri, G., Ramezani, V., Mehr, S.E., Rahimian, R., 2014. The role of serotonin in memory: interactions with neurotransmitters and downstream signaling. Exp. Brain Res. 232 (3), 723–738.
Shih, M.-F.M., October 22, 2013. Anterior Paired Lateral Neurons in *Drosophila* Intermediate-term Memory. (A.-S. Chiang). Institute of Biotechnologies, National Tsing Hua University. Retrieved from: http://etd.lib.nctu.edu.tw/cgi-bin/gs32/hugsweb.cgi?o=dnthucdr&s=id=%22GH000934272%22.&searchmode=basic.
Shuai, Y., Lu, B., Hu, Y., Wang, L., Sun, K., Zhong, Y., 2010. Forgetting is regulated through Rac activity in *Drosophila*. Cell 140 (4), 579–589.
Skoulakis, E.M., Kalderon, D., Davis, R.L., 1993. Preferential expression in mushroom bodies of the catalytic subunit of protein kinase A and its role in learning and memory. Neuron 11 (2), 197–208.
Squire, L.R., Kandel, E.R., 2000. Memory. Macmillan.
Stebbings, L.A., Todman, M.G., Phillips, R., Greer, C.E., Tam, J., Phelan, P., Jacobs, K., Bacon, J.P., Davies, J.A., 2002. Gap junctions in *Drosophila*: developmental expression of the entire innexin gene family. Mech. Dev. 113 (2), 197–205.
Stewart, W.W., 1981. Lucifer dyes—highly fluorescent dyes for biological tracing. Nature 292 (5818), 17–21.
Stough, S., Shobe, J.L., Carew, T.J., 2006. Intermediate-term processes in memory formation. Curr. Opin. Neurobiol. 16 (6), 672–678.
Sun, J., Bronk, P., Liu, X., Han, W., Südhof, T.C., 2006. Synapsins regulate use-dependent synaptic plasticity in the calyx of Held by a Ca^{2+}/calmodulin-dependent pathway. Proc. Natl. Acad. Sci. U.S.A. 103 (8), 2880–2885.
Tamura, T., Chiang, A.-S., Ito, N., Liu, H.-P., Horiuchi, J., Tully, T., Saitoe, M., 2003. Aging specifically impairs amnesiac-dependent memory in *Drosophila*. Neuron 40 (5), 1003–1011.
Tanaka, N.K., Tanimoto, H., Ito, K., 2008. Neuronal assemblies of the *Drosophila* mushroom body. J. Comp. Neurol. 508 (5), 711–755.
Toda, A.M.A., Huganir, R.L., 2015. Regulation of AMPA receptor phosphorylation by the neuropeptide PACAP38. Proc. Natl. Acad. Sci. U.S.A. 112 (21), 6712–6717.
Tully, T., Quinn, W.G., 1985. Classical conditioning and retention in normal and mutant *Drosophila melanogaster*. J. Comp. Physiol. 157 (2), 263–277.

Tully, T., Preat, T., Boynton, S.C., Del Vecchio, M., 1994. Genetic dissection of consolidated memory in *Drosophila*. Cell 79 (1), 35–47.

Volman, V., Gerkin, R.C., Lau, P.-M., Ben-Jacob, E., Bi, G.-Q., 2007. Calcium and synaptic dynamics underlying reverberatory activity in neuronal networks. Phys. Biol. 4 (2), 91–103.

Waddell, S., Armstrong, J.D., Kitamoto, T., Kaiser, K., Quinn, W.G., 2000. The amnesiac gene product is expressed in two neurons in the *Drosophila* brain that are critical for memory. Cell 103 (5), 805–813.

Wang, Y., Mamiya, A., Chiang, A.-S., Zhong, Y., 2008. Imaging of an early memory trace in the *Drosophila* mushroom body. J. Neurosci. 28 (17), 4368–4376.

Wang, X.J., 2001. Synaptic reverberation underlying mnemonic persistent activity. Trends Neurosci. 24 (8), 455–463.

Wu, C.-L., Xia, S., Fu, T.-F., Wang, H., Chen, Y.-H., Leong, D., Chiang, A.-S., Tully, T., 2007. Specific requirement of NMDA receptors for long-term memory consolidation in *Drosophila* ellipsoid body. Nat. Neurosci. 10 (12), 1578–1586.

Wu, C.-L., Shih, M.-F.M., Lai, J.S.-Y., Yang, H.-T., Turner, G.C., Chen, L., Chiang, A.-S., 2011. Heterotypic gap junctions between two neurons in the *Drosophila* brain are critical for memory. Curr. Biol. 21 (10), 848–854.

Wu, C.-L., Shih, M.-F.M., Lee, P.-T., Chiang, A.-S., 2013. An octopamine-mushroom body circuit modulates the formation of anesthesia-resistant memory in *Drosophila*. Curr. Biol. 23 (23), 2346–2354.

Wu, C.-L., Fu, T.-F., Chou, Y.-Y., Yeh, S.-R., 2015. A single pair of neurons modulates egg-laying decisions in *Drosophila*. PLoS One 10 (3), e0121335.

Yang, N., Higuchi, O., Ohashi, K., Nagata, K., Wada, A., Kangawa, K., Nishida, E., Mizuno, K., 1998. Cofilin phosphorylation by LIM-kinase 1 and its role in Rac-mediated actin reorganization. Nature 393 (6687), 809–812.

Yu, D., Keene, A.C., Srivatsan, A., Waddell, S., Davis, R.L., 2005. *Drosophila* DPM neurons form a delayed and branch-specific memory trace after olfactory classical conditioning. Cell 123 (5), 945–957.

Yu, D., Akalal, D.-B.G., Davis, R.L., 2006. *Drosophila* alpha/beta mushroom body neurons form a branch-specific, long-term cellular memory trace after spaced olfactory conditioning. Neuron 52 (5), 845–855.

CHAPTER 4

The Role of Electrical Coupling in Rhythm Generation in Small Networks

F. Nadim, X. Li, M. Gray, J. Golowasch
New Jersey Institute of Technology and Rutgers University-Newark, Newark, NJ, United States

1. INTRODUCTION

Electrical coupling is responsible for a variety of network effects, particularly in networks that generate rhythmic activity, some of which are well established, such as regulation of phase relationships, synchrony, and pattern formation (Beierlein et al., 2000; Bucher et al., 2005; Cooke, 2002; Friesen and Hocker, 2001; Marder and Bucher, 2007; Nargeot et al., 2009; Schweighofer et al., 1999; Wenning et al., 2004b), and some are novel, such as a direct role in rhythmogenesis (Gansert et al., 2007; Lewis and Rinzel, 2000; Manor et al., 1997). Nevertheless, it is quite evident that gap junctions of various types are much more prevalent in the nervous system, including pattern generating networks, than previously acknowledged. Not only are there three distinct classes of gap junction–forming molecules (connexins, innexins, and pannexins) (Koval et al., 2014), but also combinations of the multiple subunits within each class lead to the formation of gap junctions with a wide variety of properties.

Perhaps most notable among these are rectifying versus nonrectifying gap junctions (Phelan et al., 2008). (See also the discussion of rectifying gap junctions by Phelan et al. 2017 (Chapter 2)) Rectifying gap junctions act similarly to chemical synapses in that they are asymmetric and thus pass positive charge preferentially in one direction (or negative in the opposite direction), although they function at very different time scales. Nonrectifying gap junctions, for all their apparent simplicity (indeed, they are represented as a simple passive resistor), can confer remarkably subtle and complex properties on the networks they form.

Here we focus on four aspects of electrical coupling's role in pattern generating networks: rhythmogenesis, synchronization, pattern formation, and the effects of neuromodulation of gap junctions on these properties. Within these topics, we examine how electrical coupling interacts with other components of the network, including chemical synapses and intrinsic voltage-gated ionic currents, to generate functional rhythmic activity in small networks.

Network Functions and Plasticity
ISBN 978-0-12-803471-2, http://dx.doi.org/10.1016/B978-0-12-803471-2.00004-7

2. RHYTHMOGENESIS

Electrical coupling among neurons has been shown to be involved in the generation of rhythmic activity (Deans et al., 2001; Hormuzdi et al., 2001; Maier et al., 2002), as also described by Traub et al. 2017 (Chapter 13). Here we will refer to the role of gap junctions in rhythmogenesis strictly defined as the mechanisms that engender rhythmic or oscillatory activity. Oscillations can be repetitive action potentials, periodic subthreshold oscillations or rhythmic bursts of action potentials. The role of electrical coupling in coordinating or synchronizing multiple cells is intuitively clear, given that current flow between cells through gap junctions tends to equalize their voltages. The details of the levels of coordination, synchronization, and phase locking vary greatly depending on the general properties of the coupled cells, which we discuss in a separate section. However, the role of electrical coupling in the actual generation of rhythmic activity is not evident. Such rhythmic activity will be, by definition, network-based, as opposed to pacemaker-based oscillations, in which pacemaker neurons drive the rhythmic activity of an entire network, such as in the crustacean pyloric network (Miller and Selverston, 1982).

2.1 Tonic Spiking

The role of gap junctions in regular tonic spiking activity generation has been shown theoretically in networks of varying complexities (Gansert et al., 2007; Lewis and Rinzel, 2000). In its simplest form, an action potential originates in one cell and propagates through a network of electrically coupled identical model neurons. Provided that the network coupling is strong enough for the voltage change generated by the action potential of one neuron to exceed action potential generation threshold in cells coupled to it, a new action potential will be generated in the coupled neighboring cells. The network is required to be closed, meaning that closed loops of connected cells must exist. In networks formed of simple symmetrical loops coupled via gap junctions, propagation will take place symmetrically along the two paths and the action potentials will annihilate each other at some point where the propagating waves meet. Even in more complex networks the collision of action potentials converging onto a node may block propagation. However, if the network (or the input to the network) is asymmetrical, then a unidirectional reentry loop can be formed and the action potentials can propagate (in one direction) around a close circuit of coupled neurons (in principle) indefinitely.

Asymmetry can occur in various ways (Gansert et al., 2007), for example, via one or more inhibitory chemical synapses from one neuron to one of its electrically coupled neighbors (Fig. 4.1A left). Depending on its kinetics, such a synapse can block action potential propagation in the direction of the postsynaptic neuron and only allow it to proceed in the opposite direction. Provided that conduction propagation around the loop takes longer than the refractory period of the initiating cell, regular tonic spiking may

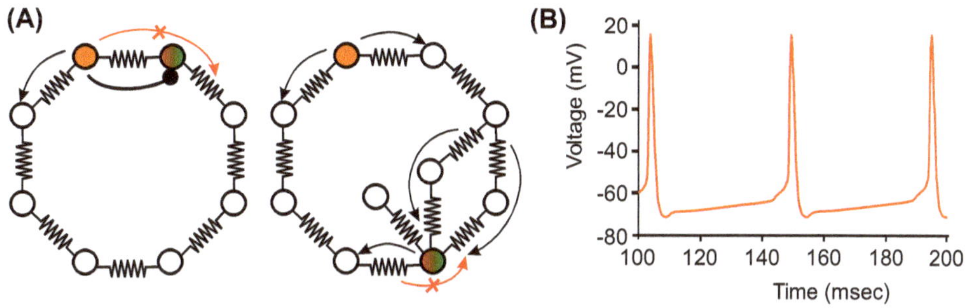

Figure 4.1 ***Generation of rhythmic activity in gap junction coupled model networks.*** (A) Asymmetric simple networks. An action potential in the starting (*orange*) neuron can trigger network activity. The network asymmetry, produced by the inhibitory synapse from this neuron to the next (clockwise) neuron can prevent clockwise propagation, depending on the kinetics of the synapse. On the right, the clockwise-propagating action potentials can effectively invade both parallel neurons and the converging action potentials onto the *red/green* cell (ie, the loaded cell) via respective electrical synapses may be enough to overcome its conductance load and thus allow clockwise propagation around the network. (B) Example of repetitive firing in a single neuron from the network shown in (A) (left).

occur (Fig. 4.1B), with the action potentials in each successive neuron in the loop lagging those of its presynaptic neighbor. Thus the number of cells in the loop, the electrical coupling strength and the refractory period determine spiking frequency.

Another source of asymmetry may be conductance load at one point along the loop (Fig. 4.1A right). Conductance load occurs, for example, when one cell receives a large number of electrical connections, making it leakier. Propagation across the loaded cell may fail if the conductance load is sufficiently high in one direction to prevent the electrical excitatory postsynaptic potential (EPSP) from reaching threshold, but may continue to a different postsynaptic cell, if the conductance load to that cell is smaller and allows a stronger input from that direction.

2.2 Network Oscillations

More complex electrically coupled model networks, built in a manner similar to the example discussed earlier but with 80 coupled neurons, produce oscillatory activity at mammalian cortical frequency bands, such as beta, gamma, and ultrafast "ripple" (Gansert et al., 2007; LeBeau et al., 2003; Lewis and Rinzel, 2000; Traub et al., 1999). The model of Gansert et al. (2007), for example, involves dendrodendritic electrical coupling and no neuron in the network is required to be spontaneously active as in other models (eg, Lewis and Rinzel, 2000). The neuron that initiates network activity may receive a transient excitatory input that triggers the oscillations. The frequency of the network oscillation (equivalent to a local field potential) will be determined by the length of the active loop(s), the electrical coupling strength, and the refractory properties of the neurons in the loops. Furthermore, these oscillations are stable because the progression of activity across the network repeats itself cycle after cycle. However, perturbations at some nodes in the network can extinguish or change the propagation wave and thus alter the frequency. Interestingly, the frequency of these oscillations as well as the architecture of the network involved sensitively depends on the diameter of the dendrites containing the electrical synapses (Gansert et al., 2007).

The model network described by Traub et al. (1999) involves axoaxonal electrical coupling between adjacent cells, whereas the network described by Lewis and Rinzel (2000) involves single compartment neurons. In both networks, spontaneous action potential activity is required. Interestingly, waves of propagation move spatially outward across the network from the points of origin, sometimes fusing and coalescing with other waves generated at different nodes (Lewis and Rinzel, 2000). Thus bursts of oscillatory activity are generated and spread through the network through electrical connections, but these bursts can only be reinitiated when spontaneous activity in neurons initiates a new wave.

2.3 Subthreshold Oscillatory Activity

Subthreshold oscillatory activity in inferior olive (IO) neurons has been linked to the ~10-Hz rhythmic activity observed in the cerebellum and its dysfunction is thought to cause the condition known as "physiological tremor" (Llinas and Sasaki, 1989). (A detailed discussion of the role of electrical coupling in the IO can be found by Welsh and Turecek 2017 (Chapter 12)) In the IO, subthreshold oscillatory activity can in principle be generated in several different ways but is independent of action potential generation. Manor et al. (1997) have presented compelling theoretical evidence that suggests that electrical coupling can be an essential element of the subthreshold oscillation generator. They show that electrical coupling can generate stable oscillations in pairs of model neurons, even when the coupled neurons are not oscillators. One compelling example is the coupling of a silent and a bistable cell, both defined by only a transient Ca^{2+} current and a leak current. Both cells are quiescent in isolation, although the bistable cell can switch between stable states in response to external input. Nevertheless, connecting the neurons with moderate to strong electrical coupling can induce oscillations in both cells (Fig. 4.2A). Strong enough electrical coupling effectively makes each cell behave as if it expressed the average of the currents of the two cells. If these average currents place the cell in an oscillatory state within the parameter space that defines these oscillations, then the entire system oscillates. This is elegantly shown using phase plane analysis (Fig. 4.2B), in which the stable fixed points of each cell (the point where the v- and h-nullclines cross) becomes unstable and generates stable oscillatory activity (Manor et al., 1997).

A more biologically realistic scenario in the IO is to consider the case of weak electrical coupling (Devor and Yarom, 2002). Subthreshold oscillations can also be produced

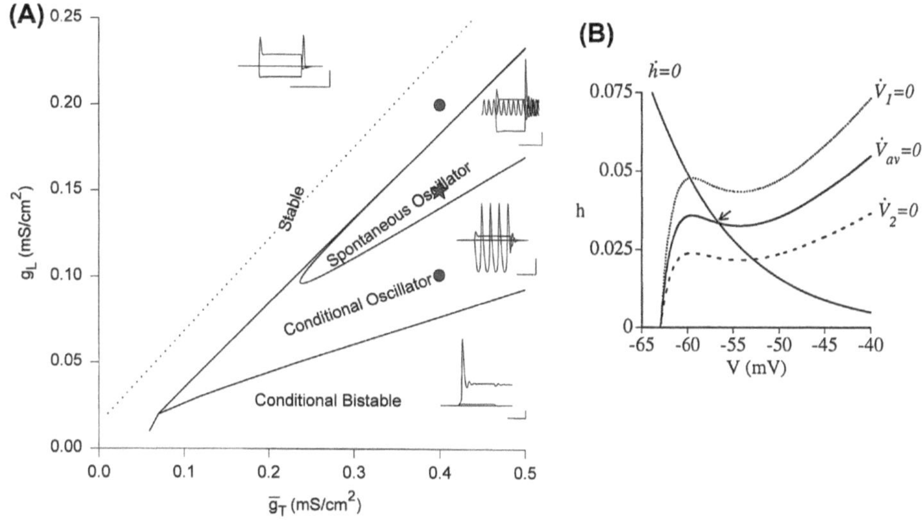

Figure 4.2 ***Subthreshold oscillatory activity in inferior olive neurons models.*** (A) Various types of neuronal activity are observed in the plane of leak-transient calcium conductance in response to constant current injection. Insets show representative activities observed in the different regions as labeled in response to both positive and negative current injection. *Red star* corresponds to the position in this parameter space of the model whose *V*-nullcline in the phase plane shown in (B) is labeled as $V_{av} = 0$ ($g_T = 0.4$ mS/cm^2, $g_L = 0.15$ mS/cm^2). *Gray circles* label model cells with $g_T = 0.4$ mS/cm^2, $g_L = 0.1$ mS/cm^2 (cell 1), $g_T = 0.4$ mS/cm^2, $g_L = 0.2$ mS/cm^2 (cell 2). Inset scale bars: 0.5 s, 5 mV. (B) Phase plane showing the voltage and h nullclines in the *V-h* plane, where *h* is the activation variable of I_h, for the three different model cells labeled with *gray circles* and a *red star* in (A). The phase plane portrays the behavior of the solution trajectories of a two-variable system in the two-dimensional variable (phase) space. The nullclines represent quasi stable states of the system on which the rate of change of one variable is 0. Thus near the nullcline, the system tends to move in the direction of the other variable, either toward or away from the curve. The overall point to which the system is attracted is the point where the two nullclines cross, referred to as the fixed point. The fixed point is stable when the slope of the *V*-nullcline (labeled $V_x = 0$, *x* being 1, av, or 2) at the fixed point is positive (eg, curves $V_1, V_2 = 0$). The fixed point is unstable, and may be the source of oscillatory behavior, if the slope of the V-nullcline is negative at that fixed point (eg, curve $V_{av} = 0$). Here, both isolated cells ($V_1, V_2 = 0$) have stable fixed points, and therefore the system does not support oscillations, but the coupled cell ($V_{av} = 0$) has an unstable fixed point and can generate oscillations. V_{av} is the average voltage of the two coupled cells. *(Reproduced with permission from Manor, Y., Rinzel, J., Segev, I., Yarom, Y., 1997. Low-amplitude oscillations in the inferior olive: a model based on electrical coupling of neurons with heterogeneous channel densities. J. Neurophysiol. 77 (5), 2736–2752.)*

when two nonoscillatory model neurons are coupled by weak or moderate electrical synapses (Manor et al., 1997). This case is not as easily amenable to phase plane analysis, yet an intuitive understanding can still be obtained. In the weakly coupled case, a hyperpolarized quiescent neuron transiently lowers the voltage of a coupled, bistable, cell that is locked in the active state, and places the system in a regime in which oscillations are possible. The flow of current into the quiescent cell generates an inward current that forces a depolarization of that cell back toward its equilibrium point. From the perspective of the second cell, the situation is reversed and thus the two-cell network can oscillate, with both cells somewhat out of phase from each other. Provided that the balance of currents in both cells is such that either or both cells transiently develop a region of negative conductance in their current—voltage relationship, the dynamics can be generated that swings the voltages of both cells back and forth across their fixed points and develop sustained oscillatory activity (Manor et al., 1997).

Interestingly, a number of studies have shown that oscillatory activity is commonly observed during early vertebrate development in the nervous system, at a time when chemical synapses are not yet established but electrical coupling through gap junctions is prevalent (Belousov and Fontes, 2013; Saint-Amant and Drapeau, 2001). Waves of oscillatory activity occur throughout the embryonic spinal cord and brain, and persist until GABAergic inhibitory synapses are established (Blankenship and Feller, 2010). It is believed that these synchronized oscillations in the embryonic nervous system are essential for the proper development of functional neuronal networks (Belousov and Fontes, 2013; Blankenship and Feller, 2010).

Similar global oscillations have been observed in the invertebrate nervous system at the embryonic stage. The crustacean stomatogastric nervous system includes several central pattern generators (CPGs) that control the activities of motor rhythms, which control distinct behaviors and operate at different frequencies (Marder and Bucher, 2007; Nusbaum and Beenhakker, 2002). The esophageal, gastric mill, and pyloric networks, for example, are distinct CPGs that, respectively, control the swallowing, chewing, and filtering of food. Yet, at the embryonic stage of the lobster *Homarus gammarus*, these three rhythms are synchronized and exhibit as a single rhythm (Casasnovas and Meyrand, 1995; Le Feuvre et al., 1999). Interestingly, the rhythms in the adult stomatogastric nervous system can be reverted back to a single embryonic-like pattern by preventing the release of gamma aminobutyric acid (GABA) (Ducret et al., 2007). These results suggest that during development electrical coupling play a significant role in the generation of oscillatory activity, and that the electrical synapses that are necessary to produce this activity are also present and active in the adult. The prevalence of electrical coupling in the adult CNS is observed in most animals. In the medicinal leech *Hirudo*, for example, removing Ca^{2+} from the bath in an in vitro CNS preparation and, therefore, blocking chemical synaptic transmission, leads to synchronized oscillations of all neurons in the CNS ganglia, indicating widespread electrical coupling among all neurons (Angstadt and Friesen, 1991). In addition, electrical coupling has been also proposed to be involved in the generation of epileptic seizures (Jin and Chen, 2011). Together, these observations strongly indicate that in both the developing and the adult nervous systems, rhythmic activity can be generated by networks of primarily electrically coupled neurons.

3. SYNCHRONIZED OSCILLATIONS

As mentioned in the previous section, intuitively it is clear that electrical coupling among neurons would result in synchronizing their activity because current flow through gap junctions tends to equalize the membrane voltages of the neurons. Although examples of synchronization of oscillatory activity through electrical connections are plentiful, in biological networks which consist of heterogeneous populations of neurons and synaptic interactions are both chemical and electrical, the output of the network can be far more complex than simple synchronous activity.

3.1 Synchronization in Electrically Coupled Networks

Examples in small oscillatory networks in invertebrates show that oscillatory activity of multiple neuron types can be synchronized by strong electrical coupling, even when the neurons are heterogeneous. An example of such synchronization through electrical coupling is observed in the crustacean cardiac ganglion, where two types of neurons (five large cells and four small cells) produce synchronized bursting activity, due to their strong electrical coupling, which controls the contractions of the heart muscle (Cooke, 2002). In the pyloric CPG (period ~1 s) of the stomatogastric ganglion (STG) of crabs and lobsters, the pacemaker group consists of the anterior burster (AB) and the two pyloric dilator (PD) neurons. All three neurons are strongly electrically coupled and burst in synchrony (Marder and Bucher, 2007) (Fig. 4.3A,B). Although the AB neuron is an intrinsic oscillator, the two PD neurons are usually not oscillatory or show very weak bursting activity when isolated (Marder, 1984).

Photoablation of the PD neurons shows that the AB neuron, in isolation, is a constant-burst-duration oscillator: when its cycle period is changed, for example, by DC current injection, the burst duration remains almost unchanged. In contrast, the AB-PD group is a constant-duty-cycle oscillator: the burst duration changes proportionally to the cycle period (Abbott et al., 1991). In addition, the coupling of the two nonintrinsically oscillatory PD neurons to the AB neuron greatly increases the range of oscillation cycle periods (Soto-Trevino et al., 2005). In the crab *Cancer borealis*, during normal pyloric activity, the two lateral posterior gastric (LPG) neurons also burst in synchrony with the AB and PD neurons. The two LPG neurons are electrically coupled to the PD neurons, but this coupling is weaker than the PD–PD coupling and is also rectifying: it passes positive current preferentially from PD to LPG. It is speculated that the weak rectifying nature of this coupling allows the LPG neuron to switch to the much slower (period ~10 s) gastric mill rhythmic activity in response to changes in sensory inputs (Shruti et al., 2014).

Synchronization mediated by electrical coupling is also widely observed in the mammalian brain circuits. Knockout studies of connexin 36 (Cx36) gap junction protein in mice show that these gap junctions are primarily expressed in inhibitory interneurons

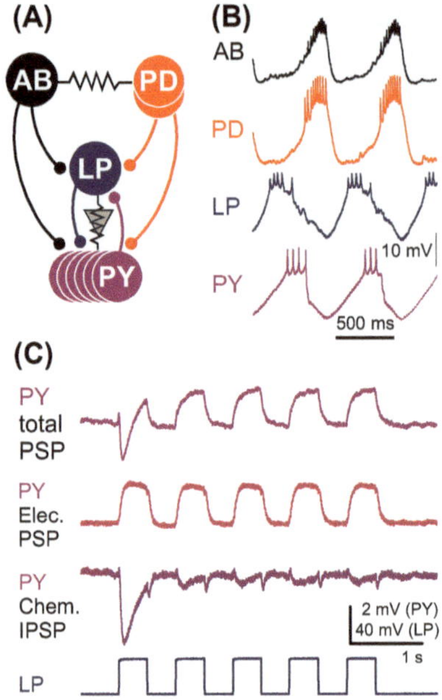

Figure 4.3 ***The pyloric central pattern generator (CPG) network of the crustacean stomatogastric nervous system.*** (A) Schematic diagram showing the connectivity of four types of pyloric neurons. Ball and stick denotes chemical inhibition, whereas resistors denote electrical coupling (with *triangle showing rectification*). (B) Triphasic oscillations of the pyloric neurons shown in simultaneous intracellular recordings. Anterior burster (AB) and pyloric dilator (PD) neurons are the electrically coupled pacemaker group. Lateral pyloric (LP) and pyloric constrictor (PY) neurons are follower neurons. (C) Compound graded postsynaptic potentials (PSPs) in the PY neuron, produced by five voltage clamp depolarizing steps in LP. The chemical inhibitory postsynaptic potential (IPSP) shows strong short-term depression and disappears after the first step. The (rectifying) electrical PSP does not change for different pulses. Consequently, the total PSP goes from mostly hyperpolarizing in the first step to depolarizing in consequent steps. *((A) and (B) reproduced with permission from Rabbah, P., Nadim, F., 2007. Distinct synaptic dynamics of heterogeneous pacemaker neurons in an oscillatory network. J. Neurophysiol. 97 (3), 2239–2253. doi:10.1152/jn.01161.2006. (C) Reproduced with permission from Mamiya, A., Manor, Y., Nadim, F., 2003. Short-term dynamics of a mixed chemical and electrical synapse in a rhythmic network. J. Neurosci. 23 (29), 9557–9564.)*

of the cerebral cortex. Low-threshold spiking (LTS) interneurons in the cortex produce correlated oscillatory activity, which persists when chemical synapses are blocked (Beierlein et al., 2000). However, the synchrony of activity among the LTS neurons is greatly reduced and more localized in the Cx36 knockout mice compared with the wild type (Deans et al., 2001). It is plausible therefore that the electrical coupling through Cx36 gap junctions is important in producing widespread coordinated oscillations among inhibitory LTS neurons in the cortex.

In the previous section, we discussed the role of electrical coupling in the generation of oscillations in the IO. Interestingly, a study using the knockout Cx36 mice has shown that the generation of oscillations in the IO does not depend on electrical coupling but the synchronous oscillations in the network are lost in the knockouts (Long et al., 2002). The role of electrical coupling in synchronizing the activity in the IO was proposed in early studies of this system (Llinas and Yarom, 1981, 1986). Following on this proposal, a two-neuron theoretical study proposed that electrical coupling in the IO can act to either synchronize or desynchronize the oscillatory activity, depending on coupling strength and modulatory states (Schweighofer et al., 1999).

Glutamatergic excitation of the neurons in the rat thalamic reticular nucleus (TRN) through metabotropic receptors results in small clusters of synchronous local rhythmic activity. These synchronized rhythms persist after blocking chemical synaptic transmission, indicating that they are coordinated through electrical connections (Long et al., 2004), which are prevalent among these neurons (Landisman et al., 2002). Such localized synchrony in the TRN is in contrast to the more widespread synchronizations observed in the cortex and may be important for the precise coordinated interaction of the TRN with different nuclei within the thalamus. Additional evidence for the role of electrical coupling in synchronization in the TRN has been provided by an in vivo study in cats, where blocking electrical coupling using halothane abolished synchronous spindle oscillations (Fuentealba et al., 2004).

3.2 Synchrony in Networks With Both Electrical Coupling and Chemical Synapses

Despite the numerous examples of the role of electrical coupling in synchronizing oscillations, it is now clear that electrical coupling does not always lead to synchronous activity. Theoretical studies have shown that electrical coupling between neurons can lead to out-of-phase or splayed activity (Chow and Kopell, 2000; Schweighofer et al., 1999; Sherman and Rinzel, 1992; Zahid and Skinner, 2009). Moreover, experimental studies have shown that synchronization in networks that involve electrical and chemical synapses may not necessarily depend on the electrical coupling.

The rhythmic activity of the mouse inspiratory system, for example, is reduced in frequency after application of gap junction blockers in vitro, but the blockers enhance the short-time-scale synchrony within each inspiratory burst. In contrast, blocking chemical inhibition increases inspiratory frequency but disrupts synchrony (Bou-Flores and Berger, 2001). The level of short-time-scale synchrony of inspiratory neurons regulates the firing rate of motor neurons, which, potentially, affects muscle activity (Murthy and Fetz, 1994).

This counterintuitive result could be explained by the differential roles of electrical coupling, depending on the intrinsic properties of the participating neurons. A computational study using electrically coupled conductance-based neurons has shown that some ionic currents promote synchrony, whereas others impede it. For example, potassium currents appear to promote synchronized neural activity, whereas the persistent sodium current promotes desynchronization (Pfeuty et al., 2003). Blocking electrical coupling enhanced synchrony in the mouse inspiratory network (Bou-Flores and Berger, 2001), which may be due to the interaction between potassium and persistent sodium currents in different neural populations, some of which have prominent levels of persistent sodium current (Powers and Binder, 2003).

Although there is ample evidence of both inhibition-mediated synchrony and electrical synapse—mediated synchrony (Connors and Long, 2004; Hu et al., 2011; Merriam et al., 2005; Rela and Szczupak, 2004), it is suggested that chemical inhibition and electrical coupling have inherently different synchronization properties and their roles in synchrony complement each other (Kopell and Ermentrout, 2004; Ostojic et al., 2009). In a model network of homogeneous interneurons coupled with both gap junction and chemical inhibition, chemical inhibition is more useful in rapid synchronization than electrical coupling. However, with heterogeneity, a small amount of electrical coupling, in addition to chemical inhibition, can achieve a similar level of synchrony as a large increase in chemical inhibition alone. With high levels of heterogeneity, chemical synapses alone fail to synchronize network (Kopell and Ermentrout, 2004). A theoretical study using integrate-and-fire neurons also shows that electrical coupling plays an important role in synchronization in the presence of heterogeneity, as well as extrinsic noise (Ostojic et al., 2009). Interestingly, once synchrony is achieved, the network produces regular oscillatory activity whose frequency is set by the intrinsic firing rate of the neurons. In contrast, synchrony can also be achieved in a network coupled by chemical synapses, but the network activity remains irregular and mostly depends on the temporal dynamics of synaptic transmission (Brunel, 2000; Brunel and Hakim, 1999). A similar point is made experimentally with cortical interneuron pairs coupled electrically using dynamic clamp (Mancilla et al., 2007). Synchrony is observed over a broad range of frequencies of the coupled neurons. More importantly, electrical coupling can synchronize pairs of neurons that have heterogeneous intrinsic firing frequencies with up to 10% difference.

3.3 Implications of Synchrony in Electrically Coupled Networks for Behavior

Theoretical and experimental studies have suggested a variety of functional roles for synchrony mediated through electrical coupling. A computational study of mitral cells in the olfactory bulb, for instance, shows that distal dendritic gap junctions, despite their small somatic coupling ratio, synchronize mitral cell action potentials within each glomerulus over a broad temporal range of inputs (Migliore et al., 2005). This suggests that electrical coupling at distal sites may play an important role in synchronizing local disparate sensory inputs within a functional unit.

In addition, studies on various systems have proposed a role for electrical coupling as a coincidence detector (Curti et al., 2012; Hestrin and Galarreta, 2005; Veruki and Hartveit, 2002). Neurons in the mouse mesencephalic trigeminal (MesV) nucleus, which is responsible for the masticatory behaviors, become electrically coupled only at a later stage of development when the chewing behaviors emerge (Curti et al., 2012). Strong correlated activity is observed and only observed in pairs or small clusters of electrically coupled neurons. Notably, when coupled neuron pairs are depolarized simultaneously, their excitability is greatly enhanced compared with when each neuron separately receives the same depolarizing input; for instance, depolarization that only evokes single spikes in each neuron can elicit strong bursts when both are coactivated. This suggests that electrically coupled MesV neurons function as coincidence detector by their enhanced synchronized response to simultaneous sensory input (see also Li and Rekling, 2017 (Chapter 11)).

Moreover, experiments using snail neuron pairs coupled with dynamic clamp have shown that, when coupled both electrically and chemically, neurons can switch between synchrony and antiphase activity in response to transient stimulation. In this system, bistability never occurs without electrical coupling (Bem et al., 2005). It is proposed that a stimulus can instantaneously reconfigure an electrically coupled network. Thus a particular signal, depending on the timing, will either be amplified or suppressed.

An interesting example of direct behavioral function of electrical coupling is the leech whole body shortening (Frank et al., 1975). Stimulation of mechanosensory neurons on the anterior end of the animal produces this shortening, which depends on two intersegmental pathways in the segmental CNS along the body of the animal. The faster of the two pathways is mediated by a single interneuron, known as the S cell, in each segmental ganglion. The S cells send axons to both adjacent ganglia through the medial connective nerve and contact the axon of the neighboring S cells through strong electrical coupling. These axons are large and have a high conduction velocity, and the electrical coupling is strong enough to allow action potentials to pass to the neighboring S cells without failure. In effect, the chain of S cells produces a rapid conduction system, with multiple spike-initiation zones, along the whole body of the animal (Baccus et al., 2001). The S cells make electrical synapses with motor neurons that cause the longitudinal muscles to contract and, in whole-body shortening, the chain of S cells is strongly activated and provides the first wave of excitation that results in the behavior (Kristan et al., 2005). The second wave appears to be due to another chain of (unidentified) interneurons with lower conduction velocity (Shaw and Kristan Jr., 1995).

The bilateral touch sensory T cells have an excitatory connection to the S cells in each ganglion and this connection is insensitive to external calcium concentrations, indicating that it is electrical in nature. Interestingly, double dye fills of T and S cells show that their dendritic arborizations do not overlap and are not even close (Muller and Scott, 1981). The electrical communication between these two neurons happens to be through a third interneuron that is electrically coupled to both the T and S cells, although the functional significance of this intermediary connection is not known (Muller and Scott, 1981).

In summary, electrical coupling often interacts both in straightforward and extremely complex ways with voltage-gated ionic currents as well as chemical synapses to regulate synchrony in a wide variety of systems. In addition, gap junction localization is likely to have significant effects on electrical coupling–mediated synchrony.

4. PATTERN GENERATION

It is now well established that neural networks underlying motor behaviors are flexible constructs. CPG networks can alter their output through neuromodulation, by functionally excluding or recruiting neurons, and through coordination with other CPG networks. For example, in feeding in the sea mollusk *Aplysia californica*, the opening and protraction of the radula (a feeding apparatus) occurs simultaneously with ingestive behavior (bites and swallowing), whereas in egestive behavior (food rejection), the radula closes during protraction (Cropper et al., 2004). The neural mechanisms underlying this switch match the motor activity and are well described (see Jing et al., 2017 (Chapter 5)). Such flexibility requires that the relative timing (phase) of the participating neurons changes to provide the appropriate modifications in the motor outputs. The locus of such flexibility, in general, is often a modification of the strengths of chemical synapses and electrical coupling. Interestingly, the combined dynamic actions of electrical coupling and chemical synapses may act either to modify phase in different behavioral contexts, as in the feeding behavior of *Aplysia* (Nargeot et al., 2009) (see Nargeot and Bédécarrats, 2017 (Chapter 6)), or to maintain phase despite changes in cycle frequency (Bucher et al., 2005; Friesen and Hocker, 2001; Wenning et al., 2004b; Zhang et al., 2014).

In CPG networks, electrical coupling is common among neurons that are active in different phases of the cycle (Mamiya et al., 2003; Rela and Szczupak, 2004; Weaver et al., 2010). Electrical coupling between out-of-phase neurons allows for appropriate phase locking of different components of the CPG but not necessarily by synchronizing these activities. In the heartbeat CPG of the medicinal leech *Hirudo medicinalis*, for example, two bilateral heart tubes produce beats with a period of ~10 s to push blood through the closed circulatory system. One of the two heart tubes is active in synchronous mode, whereas the other constricts with a rear-to-front peristalsis. Different yet homologous segmental heart interneurons (HN) comprising the pattern generating network involve electrical coupling among homologous bursting neurons in different segments (Maranto and Calabrese, 1984), yet these segmental HN neurons may be active in phase or with a phase lag, when the ipsilateral heart tube is contracting in a synchronous or a peristaltic manner, respectively (Weaver et al., 2010). As described later, electrical coupling is essential in determining whether each heart tube is active in synchronous or peristaltic mode.

These two leech heartbeats are controlled by a set of bilateral motor neurons that are found in segments 3–18 of the 21 segmental ganglia (Thompson and Stent, 1976). The motor neurons are in turn controlled by the heartbeat CPG, which primarily consists of bilateral HN cells in segments 1–7. The HN cells are connected via inhibitory synapses and rectifying electrical coupling and produce the patterned activity that ultimately controls the heartbeat behavior (Fig. 4.4) (Norris et al., 2006).

Of all bilateral HN neurons, only those in segments 3 and 4 make reciprocally inhibitory connections and, in isolated ganglia, produce out-of-phase (half-center) rhythmic bursting activity. Interestingly, only the HN neurons in segments 3, 4, 6, and 7 are premotor neurons and contact the heart motor neurons; those in segments 1, 2, and 5 are only responsible for producing the patterned activity. Two additional pairs of premotor HN cells were identified in ganglia 15 and 16, which are part of the leech heartbeat CPG (Wenning et al., 2011). These neurons do not receive chemical synaptic input from the other neurons in the CPG and are only active due to excitation they receive from electrical coupling to the axons of the other HN cells.

When one of the two heart tubes (and the corresponding motor neurons) is active in synchronous mode, the other constricts with a rear-to-front peristalsis, controlled by a corresponding rear-to-front peristaltic activity of the motor neurons. After 20–40 beats, the two sides switch (Wenning et al., 2004a). This switch is ultimately controlled by the activity of the HN(5) neurons (the switch interneurons; Fig. 4.4A). At any time only one of the two HN(5) neurons is active, whereas the other is silent (Calabrese and Peterson, 1983; Gramoll et al., 1994). The heart tube on the active HN(5) side constricts in synchronous mode because inhibition by HN(5) on the ipsilateral side synchronizes the activity of HN(6,7) with HN(3,4) (Fig. 4.4B).

A cursory examination of the leech heartbeat CPG circuit diagram (Fig. 4.4A) suggests that HN(6,7) on both sides receive inhibitory input only from the same active HN(5), which suggests that they should be coactive on the two sides. This is not the case; in fact, the two bilateral HN(6,7) are active out of phase (Fig. 4.4C). This phase difference is to a small degree because contralateral inhibition by HN(5) is weaker than ipsilateral inhibition. It was shown, through computational modeling, however, that this surprising relative phasing can be explained by the strong rectifying electrical coupling between HN(3,4) and HN(6,7), which, in combination with the inhibitory input from the active HN(5), sculpts the activities of HN(6,7) (Lamb and Calabrese, 2012; Weaver et al., 2010).

The leech swim circuit also involves extensive electrical coupling that is important for its proper function. Swimming in leech involves repeated dorsoventral undulations (cycle period 0.4–2 s) in which each segment has a phase delay of approximately 5% from its neighbor (Kristan and Calabrese, 1976). The swim activity is generated by a CPG consisting of at least 13 interneurons in each of the midbody ganglia, in which all bilaterally paired neurons are strongly electrically coupled, presumably to synchronize bilateral activity during dorsoventral undulations (Kristan and Calabrese, 1976). In addition, electrical coupling mediates some reciprocal interaction between the CPG interneurons and the dorsal and ventral inhibitory motor neurons, and a strong electrical connection

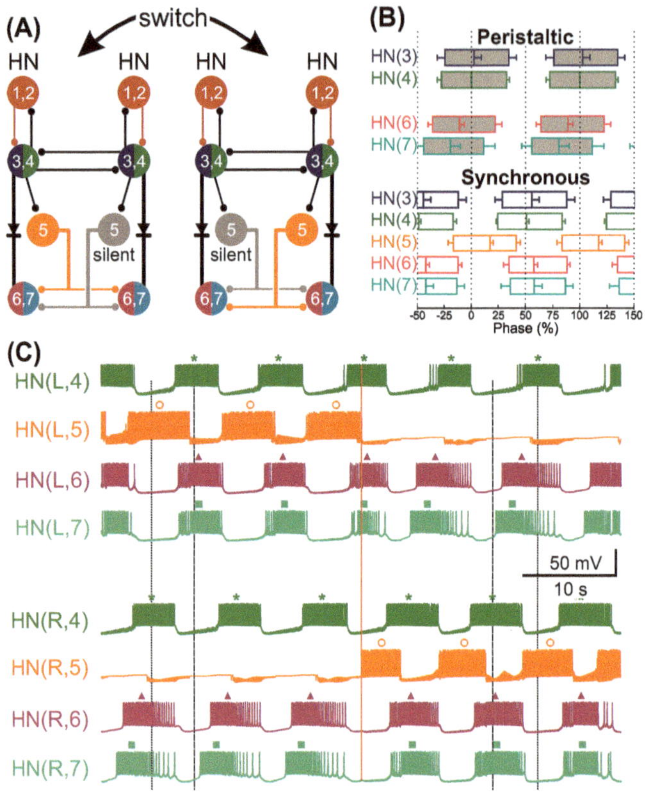

Figure 4.4 ***The leech heartbeat central pattern generator (CPG).*** (A) Schematic diagram of the leech heartbeat CPG shows bilateral pairs of heart interneurons (HN) in segments 1–7. Neurons with similar activity and connectivity on each side are combined. *Small circles* indicate chemical synaptic inhibition; diodes are rectifying electrical synapses. Switches in the coordination mode are associated with switches in the activity of HN(5) neurons (*active in color, silent in grey*; see text). (B) The phase diagram (start, mean, and end phase) of the HNs in segments 3–7 in peristaltic and synchronous modes, indicating the fraction of each cycle where the neurons were bursting. Phase 0 is the mean phase of HN(4) on the peristaltic side. (C) Computational model of the leech heartbeat CPG shows intracellular activity of the HNs in segments 4–7 during a switch (*vertical red line*) between synchronous and peristaltic mode. Note the HN(5) cells switching between active (synchronous side) and silent (peristaltic side) states. Dashed and dotted vertical lines, respectively, show the median activity of HN(4) on the synchronous and peristaltic side. *((A) and (B) reproduced with permission from Norris, B.J., Weaver, A.L., Morris, L.G., Wenning, A., Garcia, P.A., Calabrese, R.L., 2006. A central pattern generator producing alternative outputs: temporal pattern of premotor activity. J. Neurophysiol. 96 (1), 309–326. doi:10.1152/jn.00011.2006. (C) Reproduced with permission from Weaver, A.L., Roffman, R.C., Norris, B.J., Calabrese, R.L., 2010. A role for compromise: synaptic inhibition and electrical coupling interact to control phasing in the leech heartbeat CpG. Front. Behav. Neurosci. 4. doi:10.3389/fnbeh.2010.00038.)*

between the ventral stretch receptor and cell 33, another CPG interneuron, provides sensory input to the CPG (Kristan et al., 2005).

A number of mixed synapses (eg, cell 60—28 and 208) in the CPG network involve rectifying electrical coupling and chemical inhibition, an example of mixed chemical and electrical synapses that are commonly observed (Mamiya et al., 2003; Weaver et al., 2010). Such a combination can lead to complex network outputs (Rela and Szczupak, 2004). For example, the nonspiking (NS) 151 neurons in the leech ganglia are electrically coupled to all motor neurons and therefore may influence all behaviors (Wadepuhl, 1989). This is a rectifying connection that becomes stronger if the motor neurons are depolarized. Yet motor neurons also inhibit NS through a polysynaptic pathway (Rela and Szczupak, 2003). Using an elegant set of experiments, Rela and Szczupak (2003) showed that through this feedback mechanism, the NS neuron can functionally uncouple motor neurons that are otherwise electrically coupled, thus allowing these neurons to be involved in distinct behaviors.

One example of the complex interactions between chemical inhibition and electrical coupling comes from the stomatogastric pyloric network. The LP and PY neurons of the crustacean pyloric network have a rectifying electrical connection and reciprocally inhibit one another. During the triphasic pyloric rhythm, the LP and PY neurons take turns producing alternating bursting activity in between the synchronous bursts of the pacemaker AB and PD neurons (Fig. 4.3B). The out-of-phase bursting activity suggests a role for the reciprocally inhibitory synapses between LP and PY neurons. Yet, this first-glance analysis of the network output does not suggest a clear functional role for the electrical coupling between the two neurons. Two additional facts provide some insight: first, the electrical connection is strongly rectifying, passing only positive current from the LP to the PY neuron (and negative current in the opposite direction) (Graubard and Hartline, 1987). Second, although the inhibitory synapse from PY to LP is strong, the LP to PY inhibition is weak and strongly depressing (Mamiya et al., 2003). This latter point has an interesting functional consequence. If the LP neuron is depolarized periodically, the inhibition from LP to PY will depress and the resulting current from LP to PY is a depolarizing current due to the electrical coupling (Fig. 4.3C). This implies that, during ongoing pyloric activity, the effect of the electrical coupling of LP to PY is primarily to help the PY neuron recover from the strong inhibition it receives from the pacemaker neurons. Similarly, the fact that the electrical coupling is strongly rectifying implies that, when the PY neuron is inhibited by the pacemakers, it helps hyperpolarize the LP neuron. Yet, the presence of the strong inhibitory chemical synapse from PY to LP implies that depolarization of the PY neuron also leads to inhibition of the LP neuron. This allows a window of firing for the LP neuron when the PY neuron is not inhibited by the pacemakers and also not actively bursting, thus providing robustness to the triphasic activity of the pyloric rhythm (Fig. 4.3B).

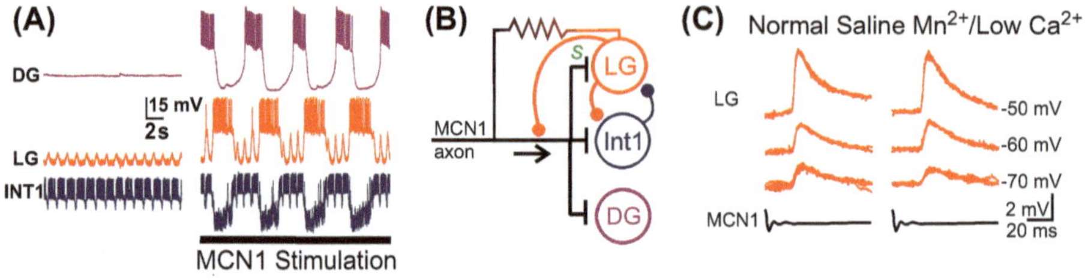

Figure 4.5 ***Activity of the gastric mill central pattern generator (CPG) network elicited by the modulatory projection neuron MCN1.*** (A) In the absence of MCN1 activity (left), the dorsal gastric (DG) and lateral gastric (LG) neurons are silent, whereas Int1 produces bursting oscillations timed with the pyloric CPG (see Fig. 4.3). MCN1 stimulation (extracellular 1-ms pulses at 15 Hz, amplitude at threshold) elicits bursting activity (period ~10 s) in these neurons with DG and Int1 active out of phase with the LG neuron. (B) Schematic diagram shows that the axon terminals of MCN1 excite LG through a slow modulatory transmitter (*s*), and also excite Int1 and DG (*bars*). LG and Int1 produce reciprocally inhibitory connections (*stick and balls*), which allows them to oscillate in antiphase when excited by MCN1. When LG is active, it presynaptically inhibits the MCN1 terminals and the excitation from MCN1 slowly decays. MCN1 axons also produce a fast excitation of LG through electrical coupling (resistor) and this fast excitation persists during the LG presynaptic inhibition of MCN1 terminals. (C) The electrical coupling from MCN1 terminals to LG is voltage dependent. Electrical EPSPs in LG from an action potential in MCN1 persist in low-Ca^{2+}, Mn^{2+} saline and increase in amplitude when LG is depolarized. Reproduced with permission from Coleman, M.J., Meyrand, P., Nusbaum, M.P., 1995. A switch between two modes of synaptic transmission mediated by presynaptic inhibition. Nature 378 (6556), 502–505. doi:10.1038/378502a0.

Another example of mixed chemical and electrical synapses, also from the stomatogastric nervous system, is in the connection between neuromodulatory projection neurons and local CPG networks. A well-studied example is the modulatory commissural neuron 1 (MCN1) in the stomatogastric nervous system of the crab *Cancer borealis*. The MCN1 soma is located in an anterior ganglion but its axon projects to the STG and releases at least three known neuromodulatory transmitters: GABA and two neuropeptides (Nusbaum et al., 2001). MCN1 release of CabTRP-Ia, one of the two neuropeptides, excites the lateral gastric (LG) neuron, which composes one half of the rhythm generating half-center oscillator at the heart of the gastric mill network (Wood et al., 2000). The excitation of the LG neuron by CabTRP-Ia allows this neuron to escape inhibition by its half-center partner, interneuron 1 (Int1), and produce a burst of action potentials (Fig. 4.5A). The LG neuron, in turn, presynaptically inhibits the MCN1 axon terminals and thereby removes the excitation by MCN1 to itself and to a number of other STG neurons (Fig. 4.5B).

However, MCN1 axon terminals also make an electrical synapse to the LG neuron, which remains active during the presynaptic inhibition of MCN1 terminals by LG, because LG is depolarized at that stage. It is this electrical synapse that allows the LG neuron to produce a strong burst of action potentials while it presynaptically inhibits

MCN1 (Coleman and Nusbaum, 1994). In an unusual twist, the electrical synapse from MCN1 axon terminals to LG is voltage dependent: as the LG neuron depolarizes, the electrical synapse becomes larger and, therefore, it is strongest when it is functionally important (Fig. 4.5C), ie, when the LG neuron produces its burst of action potentials (Coleman et al., 1995). This example not only shows a role for electrical coupling in supporting neuromodulatory actions of a projection neuron, but also provides an example of how electrical synapses could allow projection neurons to rapidly interact with local CPG circuits.

Another small neural network in which electrical coupling plays an important role in pattern generation with a direct consequence on behavioral output is the CPG that controls the crayfish swimmeret system. This system produces a phasic rhythm over a 10-fold range of frequencies in which the rhythmic movement of the four pairs of abdominal swimmerets has a rear-to-front intersegmental coordination with a next-neighbor phase difference of ~0.25 (Namba and Mulloney, 1999). This rhythm is driven by a set of four segmental CPGs in the four adjacent abdominal ganglia A2—A5.

In each of these ganglia, two sets of interneurons [inhibitory power stroke (IPS) and inhibitory return stroke (IRS)] drive the power- and return-stroke motor neurons by inhibition. The IPS and IRS neurons are bilaterally synchronized in each ganglion, but produce reciprocally inhibitory connections and produce antiphase bursting oscillations that drive the swimmerets in each segment (Smarandache-Wellmann et al., 2013). The four segments are coordinated by a set of ascending (ASC) and descending (DSC) interneurons in each ganglion that, respectively, are inhibited by the IPS or IRS rhythm-generating neurons and also, respectively, project to all anterior or posterior ganglia (Smarandache-Wellmann and Gratsch, 2014).

The coordination of activity between the CPGs of ganglia A2—A5 is through specialized commissural interneurons 1 (CI1s) that integrate inputs from the ASC and DSC neurons in each ganglion. In each ganglion, the bilateral CI1s receive electrical excitatory input from ipsilateral ASC and DSC projection neurons with a gradient of excitation that decreases with distance. The CI1s are, in turn, electrically coupled to the IRS neurons and thereby integrate and transmit the coordination information from all other ganglia to the segmental CPG within their home ganglion. This mechanism allows for the proper phasic coordination of the swimmeret CPGs in the four adjacent ganglia A2—A5, which drive the swimmerets with their 0.25 intersegmental phase lag (Smarandache-Wellmann and Gratsch, 2014; Smarandache-Wellmann et al., 2014).

A theoretical study of small oscillatory networks illustrated how the interaction between electrical coupling and chemical synapses can allow "hub" neurons to produce pattern coordination between multiple oscillatory networks with distinct oscillation frequencies or become recruited by faster or slower networks (Gutierrez et al., 2013). Finally, theoretical studies have also shown that rectification in an electrical synapse can strongly influence the interaction between electrical connections and chemical synaptic connections and greatly reshape how these synaptic interactions influence the network output (Gutierrez and Marder, 2013).

Altogether, these examples show the importance of regular and rectifying electrical coupling in producing the proper activity phase in a variety of oscillatory networks, especially in pattern generating rhythms.

5. NEUROMODULATION

Neuromodulators can influence the behavior of the electrical synapses with profound consequences that vary in scale from how a single synapse functions (Johnson et al., 1993), to how effectively a complex behavior is executed (Correa et al., 2015). Electrical synapses can be modulated directly by biochemical modifications of the channel subunits that compose gap junctions (Bao et al., 2004; Giaume et al., 1992; Herve and Sarrouilhe, 2006; Muller et al., 2010; Spray and Bennett, 1985; Stagg and Fletcher, 1990; van Rijen et al., 2000). (For a discussion of neuromodulation of electrical coupling in the retina, see Ribelayga and O'Brien, 2017 (Chapter 10)) Alternatively, their modulation can be indirect, through the modification of ionic currents that interact with the electrical signal (Haas and Landisman, 2011). This diversity in modulation of electrical coupling has only recently begun to be appreciated.

5.1 Neuromodulators Can Alter the Sign of a Synapse by Determining the Relative Weights of Electrical and Chemical Synapses

One way a neuromodulator alters the function of a neural network at the local level is illustrated in the STG of the spiny lobster *Panulirus interruptus*. As mentioned earlier, many synapses involve both a chemical and an electrical component. Interestingly, dopamine can simultaneously modulate both the electrical and the chemical component of such a synapses (Johnson et al., 1993, 1994). In the pyloric network of *P. interruptus*, the synapse from the pacemaker AB neuron to the follower ventral dilator (VD) neuron includes both a chemical inhibitory component and a rectifying electrical coupling component. In control conditions, the synapse is dominated by the electrical coupling and depolarizations of the AB neuron lead to depolarizing responses in the VD neuron. However, in the presence of dopamine, the effective sign of this synaptic connection is reversed and depolarizations in the AB neuron hyperpolarize the VD neuron. This reversal was shown to be due to the reduced strength of the electrical connection and enhancement of the chemical inhibition. Similar results were observed for other neurons (Johnson et al., 1993, 1994). Unfortunately, the functional consequences of this sign reversal are unclear given that this neuromodulator also targets several ionic conductances in the same cells (Gruhn et al., 2005; Johnson et al., 2003; Peck et al., 2006, 2001). Nevertheless, these studies suggest that, by targeting mixed electrical–chemical connections, neuromodulators can dynamically enhance the flexibility of neuronal networks and effectively switch whether a synapse is excitatory or inhibitory.

5.2 Electrical Synapses in the Mammalian Thalamic Reticular Nucleus Can Be Modulated by Ionic Currents

Many neuromodulators influence regenerative ionic currents such as the persistent sodium current (I_{NaP}) (Artinian et al., 2012; Carrillo-Reid et al., 2009; Koch et al., 2015). However, it has been shown that activation of such sodium currents appears to amplify electrical coupling in a number of systems (Curti and Pereda, 2004; Dugue et al., 2009; Haas and Landisman, 2011; Mann-Metzer and Yarom, 1999) with interesting physiological consequences. In the TRN, for example, Haas and Landisman (2011) observed that when I_{NaP} is active, an increase in synchrony of tonic spiking was observed, without affecting the synchrony of their slow bursting. Conversely, they show that the electrical coupling and the action potential synchrony decreases when I_{NaP} is blocked (Haas and Landisman, 2011). This modulatory effect on the electrical coupling can best be explained by the effect of Na^+ currents on the membrane properties of the extrajunctional membrane: in these cells, the resting membrane potential of the presynaptic neuron is near the midpoint of activation of I_{NaP} and both depolarizations and hyperpolarizations can significantly affect the current flow through these channels. Thus, if I_{NaP} is activated by depolarization from the resting state, current flow into a neighboring cell through adjacent gap junction channels leads to the depolarization of that cell, whereas hyperpolarization of the presynaptic neuron results in the partial deactivation of I_{NaP} and the consequent hyperpolarization of the coupled cell. When I_{NaP} is blocked, the amplification of current flow across the electrical synapse is eliminated and the effective electrical coupling is significantly reduced. Although this was shown for I_{NaP}, it is not the only current that can in principle have such an effect.

Indirect regulation of electrical synapses by the activation (or inactivation) of voltage-gated currents such as the one just described, which may be under neuromodulatory control, can impact the phase relationships and perhaps even rhythmogenesis in a variety of neuronal networks expressing electrical coupling.

5.3 Neuromodulators of Electrical Coupling can Modify Behavior

As mentioned earlier, disentangling the effect of a neuromodulator on electrical coupling from its effect on different conductances is not necessarily easy. This is especially true in complex networks that underlie complex behaviors such as the network that mediates spicule protractions in the nematode *Caenorhabditis elegans*. Reciprocally coupled sensory motor neurons in the male *C. elegans* tail control most

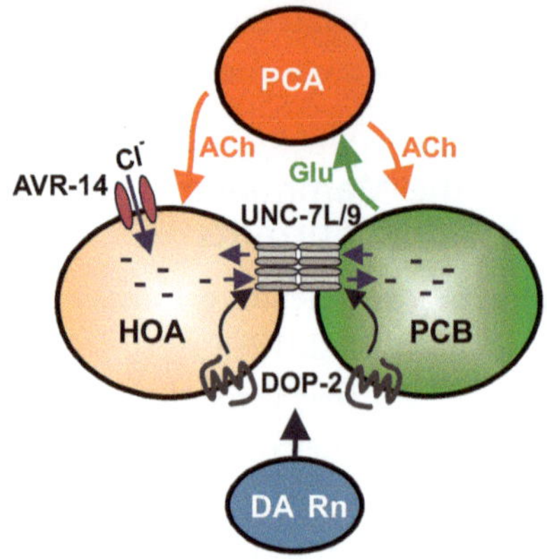

Figure 4.6 ***Spicule protraction circuit in Caenorhabditis elegans.*** Of the two post-cloaca sensilla neurons (PCA and PCB), PCA is cholinergic (ACh) and PCB glutamatergic (Glu). They are found as bilateral pairs. PCA excites PCB but inhibits the hook sensilum neuron (HOA). PCA and PCB are involved in posterior sex muscle contraction and, through electrical coupling, the rhythmic contraction of the spicule protractor muscles. The *unc-7* gene allele, *rg396*, codes for an innexin between PCB and the HOA neuron. DA upregulates the innexin-mediated electrical coupling. Thus, when PCA inhibits HOA in the presence of DA, PCB is also hyperpolarized and spicule protraction is inhibited. Modified with permission from Correa, P.A., Gruninger, T., Garcia, L.R., 2015. DOP-2 D2-Like receptor regulates UNC-7 innexins to attenuate recurrent sensory motor neurons during *C. elegans* copulation. J. Neurosci. 35 (27), 9990–10004. doi:10.1523/JNEUROSCI.0940-15.2015. *DA*, dopamine; *DOP-2*, Dopamine-2 receptor; *DA Rn*, DA ray neuron.

of the copulation behavior. These neurons include the cholinergic post-cloaca sensilla neuron A (PCA) and glutamatergic post-cloaca sensilla neuron B (PCB) bilateral pairs (Garcia et al., 2001), which are involved in posterior sex muscle contraction and, through electrical coupling, the rhythmic contraction of the spicule protractor muscles. Spicule protraction can be induced by acetylcholine (ACh) agonists. Agonist-induced protraction, however, is inhibited by simultaneous application of dopamine (Liu et al., 2011). A mutant allele of the *unc-7* gene, *rg396*, coding for an innexin, is shown to be necessary for the dopamine-induced inhibition of spicule protraction (Correa et al., 2015). The authors suggest that dopamine functionally modulates the cholinergic PCA to PCB excitation through the actions of a third neuron, the hook sensilum neuron A (HOA), which is also postsynaptic to PCA (Fig. 4.6). PCA excites PCB, whereas it uses ACh to inhibit HOA through a chloride channel receptor (AVR-14). HOA is electrically coupled to PCB via UNC-7 innexin channels. Both PCB and HOA neurons also express dopamine 2 (DOP-2) receptors, and activation of this receptor modulates the UNC-7 to increase electrical coupling between these two neurons. Therefore inhibition of HOA by PCA also provides a simultaneous inhibition of PCB through electrical coupling. These results suggest that neuromodulation of electrical synapses can have important functions not only at the synaptic level, but on high-level behaviors as well.

5.4 Expression of Identified Connexins in Model Systems Suggest That Gap Junctions Can Be Modulated Directly by Phosphorylation and Other Mechanisms

Although it is true that electrical synapses can be modulated indirectly through activation of voltage-gated ionic currents that interact with these signals, there is also evidence that the gap junction itself is the target of neuromodulators. One would expect that if gap junctions were directly targeted by neuromodulators, they would be susceptible to intracellular biochemical modifications such as phosphorylation. Indeed, many studies show that connexins are both a target of and are phosphorylated by both protein kinase C (PKC) (Bao et al., 2004; Stagg and Fletcher, 1990) and protein kinase A (PKA) (Stagg and Fletcher, 1990; TenBroek et al., 2001). An example of PKC phosphorylation reducing pore size is shown by Bao et al. (2004) who identified a phosphorylation site at serine 368 that regulates the gap junction channel permeability, and also showed that permeability and Lucifer Yellow coupling is reduced in the S368 phosphorylated state (Bao et al., 2004). In contrast to inhibition of gap junction permeability by PKC, van Rijen et al. (2000) showed that when expressing cardiac CX40 in human hepatoma cells (cell line SKHep1 cells), which are normally devoid of gap junctions, gap junctions appear and phosphorylation of CX40 by PKA led to an augmentation of the macroscopic halothane-sensitive gap junction conductance. This enhancement was abolished in the presence of the PKA antagonists. Furthermore, the authors found that Lucifer Yellow dye coupling was enhanced by cAMP and blocked by halothane (van Rijen et al., 2000).

Coupling could be mediated by the regulation of gap junction channel conductance (eg, Moreno et al., 1994), or by the regulation of expression of gap junction channels. For example, it has been shown that the increased expression of Cx36 in the mammalian central nervous system early after birth is mediated by type II mGluR-driven upregulation of the connexin via an unidentified posttranscriptional mechanism (Song et al., 2012), whereas the electrical coupling reduction that follows a few weeks thereafter is mediated by an N-methyl-D-aspartate (NMDA)-dependent activation of cAMP response element-binding protein (CREB) signaling (Arumugam et al., 2005). These results strongly suggest that phosphorylation and expression regulation of connexins in the mammalian central nervous system regulates gap junction permeability directly. Together, these studies suggest that gap junctions are the direct targets of signal transduction, which suggests a direct mechanism by which neuromodulators can influence electrical synapses.

6. SUMMARY

In this chapter, we reviewed the role of electrical coupling in small oscillatory networks. Electrical coupling is present in all networks, yet compared with chemical synapses, its role in network function is much less understood. Many advances in recent years, including the development of genetic tools, have provided the methodology to experimentally manipulate electrical coupling in large networks or whole animals to understand its function. Yet small networks, due to the limited numbers of neurons involved, still provide a more rigorous understanding of the functional role of electrical coupling, especially when these functions involve interactions with other dynamic components, such as chemical synapses and voltage-gated ionic currents.

We focused on four topics to elucidate the role of electrical coupling: rhythmogenesis, synchrony, pattern generation, and neuromodulation. It is now clear that axonal, dendritic, and somatic coupling can all be involved in the generation of rhythmic activity (see also Traub et al., 2017 (Chapter 13)). The role of electrical coupling in generating synchronized activity is intuitive. However, when electrical coupling interacts with chemical synapses or certain neuronal intrinsic properties, it may in fact lead to the reduction of synchronicity among neurons, at least in the fast time scale. We discussed examples in which these effects, at the level of small networks, are often understood at the behavioral level, but these same mechanisms often scale up to larger networks such as those in the mammalian brain.

Although the synchronizing role of electrical coupling implies that it should produce in-phase activity in rhythmic networks, it often plays an important role in producing out-of-phase oscillations, as described in the section on pattern generation. We provide a number of examples that demonstrate this unexpected role for electrical coupling, including some in which the out-of-phase activity crucially depends on rectifying gap junctions.

Finally, we provide a few examples of how neuromodulators can modify electrical coupling and, through these modifications, lead to novel behaviors. Understanding the direct and indirect neuromodulation of electrical coupling and how this could reshape network activity is an expanding area of research, especially in mammalian systems where little is known. We predict that studies of gap junction modulation in the near future will produce substantial advances in this area.

OUTSTANDING ISSUES AND FUTURE DIRECTIONS

We believe that the following directions are important ways in which small networks may provide important contributions in the near future.

- Our understanding of the role of electrical coupling in rhythmogenesis is still very basic. Coupling location, strength, and rectification are all topics barely studied to date and how the interaction between electrical coupling and chemical synapses may contribute to rhythm generation is lacking.
- How different voltage-gated ionic currents may influence the role of electrical coupling in network function is an important subject that deserves much further scrutiny.

- Direct neuromodulation of gap junctions has been shown in few cases and the mechanisms are not understood. Furthermore, understanding how modulation of gap junctions affects electrical coupling, and therefore its functional influence on network activity, has been barely examined.

ACKNOWLEDGMENTS

The authors thank Ronald L. Calabrese, Eve Marder, and Michael P. Nusbaum for helpful discussions. This work was supported by NIH grants MH060605, NS083319, and NS085330.

REFERENCES

Abbott, L.F., Marder, E., Hooper, S.L., 1991. Oscillating networks: control of burst duration by electrically coupled neurons. Neural Comput. 3, 487–497.

Angstadt, J.D., Friesen, W.O., 1991. Synchronized oscillatory activity in leech neurons induced by calcium channel blockers. J. Neurophysiol. 66 (6), 1858–1873.

Artinian, L., Zhong, L., Yang, H., Rehder, V., 2012. Nitric oxide as intracellular modulator: internal production of NO increases neuronal excitability via modulation of several ionic conductances. Eur. J. Neurosci. 36 (10), 3333–3343.

Arumugam, H., Liu, X., Colombo, P.J., Corriveau, R.A., Belousov, A.B., 2005. NMDA receptors regulate developmental gap junction uncoupling via CREB signaling. Nat. Neurosci. 8 (12), 1720–1726.

Baccus, S.A., Sahley, C.L., Muller, K.J., 2001. Multiple sites of action potential initiation increase neuronal firing rate. J. Neurophysiol. 86 (3), 1226–1236.

Bao, X., Reuss, L., Altenberg, G.A., 2004. Regulation of purified and reconstituted connexin 43 hemichannels by protein kinase C-mediated phosphorylation of Serine 368. J. Biol. Chem. 279 (19), 20058–20066.

Beierlein, M., Gibson, J.R., Connors, B.W., 2000. A network of electrically coupled interneurons drives synchronized inhibition in neocortex. Nat. Neurosci. 3 (9), 904–910.

Belousov, A.B., Fontes, J.D., 2013. Neuronal gap junctions: making and breaking connections during development and injury. Trends Neurosci. 36 (4), 227–236.

Bem, T., Le Feuvre, Y., Rinzel, J., Meyrand, P., 2005. Electrical coupling induces bistability of rhythms in networks of inhibitory spiking neurons. Eur. J. Neurosci. 22 (10), 2661–2668.

Blankenship, A.G., Feller, M.B., 2010. Mechanisms underlying spontaneous patterned activity in developing neural circuits. Nat. Rev. Neurosci. 11 (1), 18–29.

Bou-Flores, C., Berger, A.J., 2001. Gap junctions and inhibitory synapses modulate inspiratory motoneuron synchronization. J. Neurophysiol. 85 (4), 1543–1551.

Brunel, N., 2000. Dynamics of sparsely connected networks of excitatory and inhibitory spiking neurons. J. Comput. Neurosci. 8 (3), 183–208.

Brunel, N., Hakim, V., 1999. Fast global oscillations in networks of integrate-and-fire neurons with low firing rates. Neural Comput. 11 (7), 1621–1671.

Bucher, D., Prinz, A.A., Marder, E., 2005. Animal-to-animal variability in motor pattern production in adults and during growth. J. Neurosci. 25 (7), 1611–1619.

Calabrese, R.L., Peterson, E., 1983. Neural control of heartbeat in the leech, *Hirudo medicinalis*. Symp. Soc. Exp. Biol. 37, 195–221.

Carrillo-Reid, L., Tecuapetla, F., Vautrelle, N., Hernandez, A., Vergara, R., Galarraga, E., Bargas, J., 2009. Muscarinic enhancement of persistent sodium current synchronizes striatal medium spiny neurons. J. Neurophysiol. 102 (2), 682–690.

Casasnovas, B., Meyrand, P., 1995. Functional differentiation of adult neural circuits from a single embryonic network. J. Neurosci. 15 (8), 5703–5718.

Chow, C., Kopell, N., 2000. Dynamics of spiking neurons with electrical coupling. Neur. Comput. 12, 1643–1678.

Coleman, M.J., Meyrand, P., Nusbaum, M.P., 1995. A switch between two modes of synaptic transmission mediated by presynaptic inhibition. Nature 378 (6556), 502–505.

Coleman, M.J., Nusbaum, M.P., 1994. Functional consequences of compartmentalization of synaptic input. J. Neurosci. 14 (11 Pt. 1), 6544–6552.

Connors, B.W., Long, M.A., 2004. Electrical synapses in the mammalian brain. Annu. Rev. Neurosci. 27, 393–418.

Cooke, I.M., 2002. Reliable, responsive pacemaking and pattern generation with minimal cell numbers: the crustacean cardiac ganglion. Biol. Bull. 202 (2), 108–136.

Correa, P.A., Gruninger, T., Garcia, L.R., 2015. DOP-2 D2-Like receptor regulates UNC-7 innexins to attenuate recurrent sensory motor neurons during *C. elegans* copulation. J. Neurosci. 35 (27), 9990–10004.

Cropper, E.C., Evans, C.G., Hurwitz, I., Jing, J., Proekt, A., Romero, A., Rosen, S.C., 2004. Feeding neural networks in the mollusc *Aplysia*. Neurosignals 13 (1–2), 70–86.

Curti, S., Hoge, G., Nagy, J.I., Pereda, A.E., 2012. Synergy between electrical coupling and membrane properties promotes strong synchronization of neurons of the mesencephalic trigeminal nucleus. J. Neurosci. 32 (13), 4341–4359.

Curti, S., Pereda, A.E., 2004. Voltage-dependent enhancement of electrical coupling by a subthreshold sodium current. J. Neurosci. 24 (16), 3999–4010.

Deans, M.R., Gibson, J.R., Sellitto, C., Connors, B.W., Paul, D.L., 2001. Synchronous activity of inhibitory networks in neocortex requires electrical synapses containing connexin36. Neuron 31 (3), 477–485.

Devor, A., Yarom, Y., 2002. Electrotonic coupling in the inferior olivary nucleus revealed by simultaneous double patch recordings. J. Neurophysiol. 87 (6), 3048–3058.

Ducret, E., Le Feuvre, Y., Meyrand, P., Fenelon, V.S., 2007. Removal of GABA within adult modulatory systems alters electrical coupling and allows expression of an embryonic-like network. J. Neurosci. 27 (14), 3626–3638.

Dugue, G.P., Brunel, N., Hakim, V., Schwartz, E., Chat, M., Levesque, M., Dieudonne, S., 2009. Electrical coupling mediates tunable low-frequency oscillations and resonance in the cerebellar Golgi cell network. Neuron 61 (1), 126–139.

Frank, E., Jansen, J.K., Rinvik, E., 1975. A multisomatic axon in the central nervous system of the leech. J. Comp. Neurol. 159 (1), 1–13.

Friesen, W.O., Hocker, C.G., 2001. Functional analyses of the leech swim oscillator. J. Neurophysiol. 86 (2), 824–835.

Fuentealba, P., Crochet, S., Timofeev, I., Bazhenov, M., Sejnowski, T.J., Steriade, M., 2004. Experimental evidence and modeling studies support a synchronizing role for electrical coupling in the cat thalamic reticular neurons in vivo. Eur. J. Neurosci. 20 (1), 111–119.

Gansert, J., Golowasch, J., Nadim, F., 2007. Sustained rhythmic activity in gap-junctionally coupled networks of model neurons depends on the diameter of coupled dendrites. J. Neurophysiol. 98 (6), 3450–3460.

Garcia, L.R., Mehta, P., Sternberg, P.W., 2001. Regulation of distinct muscle behaviors controls the *C. elegans* male's copulatory spicules during mating. Cell 107 (6), 777–788.

Giaume, C., Cordier, J., Glowinski, J., 1992. Endothelins inhibit junctional permeability in cultured mouse astrocytes. Eur. J. Neurosci. 4 (9), 877–881.

Gramoll, S., Schmidt, J., Calabrese, R.L., 1994. Switching in the activity state of an interneuron that controls coordination of the hearts in the medicinal leech (*Hirudo medicinalis*). J. Exp. Biol. 186, 157–171.

Graubard, K., Hartline, D.K., 1987. Full-wave rectification from a mixed electrical-chemical synapse. Science 237 (4814), 535–537.

Gruhn, M., Guckenheimer, J., Land, B., Harris-Warrick, R.M., 2005. Dopamine modulation of two delayed rectifier potassium currents in a small neural network. J. Neurophysiol. 94 (4), 2888–2900.

Gutierrez, G.J., Marder, E., 2013. Rectifying electrical synapses can affect the influence of synaptic modulation on output pattern robustness. J. Neurosci. 33 (32), 13238–13248.

Gutierrez, G.J., O'Leary, T., Marder, E., 2013. Multiple mechanisms switch an electrically coupled, synaptically inhibited neuron between competing rhythmic oscillators. Neuron 77 (5), 845–858.

Haas, J.S., Landisman, C.E., 2011. State-dependent modulation of gap junction signaling by the persistent sodium current. Front. Cell Neurosci. 5, 31.

Herve, J.C., Sarrouilhe, D., 2006. Protein phosphatase modulation of the intercellular junctional communication: importance in cardiac myocytes. Prog. Biophys. Mol. Biol. 90 (1–3), 225–248.

Hestrin, S., Galarreta, M., 2005. Electrical synapses define networks of neocortical GABAergic neurons. Trends Neurosci. 28 (6), 304–309.

Hormuzdi, S.G., Pais, I., LeBeau, F.E., Towers, S.K., Rozov, A., Buhl, E.H., Monyer, H., 2001. Impaired electrical signaling disrupts gamma frequency oscillations in connexin 36-deficient mice. Neuron 31 (3), 487–495.

Hu, H., Ma, Y., Agmon, A., 2011. Submillisecond firing synchrony between different subtypes of cortical interneurons connected chemically but not electrically. J. Neurosci. 31 (9), 3351–3361.

Jin, M.M., Chen, Z., 2011. Role of gap junctions in epilepsy. Neurosci. Bull. 27 (6), 389–406.

Jing, J., Cropper, E.C., Weiss, K.R., 2017. Network functions of electrical Coupling Present in Multiple and specific sites in behavior-generating circuits. In: Jing, J. (Ed.), Network Functions and Plasticity. Elsevier (Chapter 5).

Johnson, B.R., Kloppenburg, P., Harris-Warrick, R.M., 2003. Dopamine modulation of calcium currents in pyloric neurons of the lobster stomatogastric ganglion. J. Neurophysiol. 90 (2), 631–643.

Johnson, B.R., Peck, J.H., Harris-Warrick, R.M., 1993. Amine modulation of electrical coupling in the pyloric network of the lobster stomatogastric ganglion. J. Comp. Physiol. A 172 (6), 715–732.

Johnson, B.R., Peck, J.H., Harris-Warrick, R.M., 1994. Differential modulation of chemical and electrical components of mixed synapses in the lobster stomatogastric ganglion. J. Comp. Physiol. A 175 (2), 233–249.

Koch, H., Caughie, C., Elsen, F.P., Doi, A., Garcia 3rd, A.J., Zanella, S., Ramirez, J.M., 2015. Prostaglandin E2 differentially modulates the central control of eupnoea, sighs and gasping in mice. J. Physiol. 593 (1), 305–319.

Kopell, N., Ermentrout, B., 2004. Chemical and electrical synapses perform complementary roles in the synchronization of interneuronal networks. Proc. Natl. Acad. Sci. U. S. A. 101 (43), 15482–15487.

Koval, M., Isakson, B.E., Gourdie, R.G., 2014. Connexins, pannexins and innexins: protein cousins with overlapping functions. FEBS Lett. 588 (8), 1185.

Kristan Jr., W.B., Calabrese, R.L., 1976. Rhythmic swimming activity in neurones of the isolated nerve cord of the leech. J. Exp. Biol. 65 (3), 643–668.

Kristan Jr., W.B., Calabrese, R.L., Friesen, W.O., 2005. Neuronal control of leech behavior. Prog. Neurobiol. 76 (5), 279–327.

Lamb, D.G., Calabrese, R.L., 2012. Small is beautiful: models of small neuronal networks. Curr. Opin. Neurobiol. 22 (4), 670–675.

Landisman, C.E., Long, M.A., Beierlein, M., Deans, M.R., Paul, D.L., Connors, B.W., 2002. Electrical synapses in the thalamic reticular nucleus. J. Neurosci. 22 (3), 1002–1009.

Le Feuvre, Y., Fenelon, V.S., Meyrand, P., 1999. Central inputs mask multiple adult neural networks within a single embryonic network. Nature 402 (6762), 660–664.

LeBeau, F.E., Traub, R.D., Monyer, H., Whittington, M.A., Buhl, E.H., 2003. The role of electrical signaling via gap junctions in the generation of fast network oscillations. Brain Res. Bull. 62 (1), 3–13.

Lewis, T.J., Rinzel, J., 2000. Self-organized synchronous oscillations in a network of excitable cells coupled by gap junctions. Network 11 (4), 299–320.

Li, W.-C., Rekling, J.C., 2017. Electrical Coupling in the Generation of Vertebrate Motor Rhythms. In: Jing, J. (Ed.), Network Functions and Plasticity. Elsevier (Chapter 11).

Liu, Y., LeBeouf, B., Guo, X., Correa, P.A., Gualberto, D.G., Lints, R., Garcia, L.R., 2011. A cholinergic-regulated circuit coordinates the maintenance and bi-stable states of a sensory-motor behavior during *Caenorhabditis elegans* male copulation. PLoS Genet. 7 (3), e1001326.

Llinas, R., Sasaki, K., 1989. The functional organization of the olivo-cerebellar system as examined by multiple Purkinje cell recordings. Eur. J. Neurosci. 1 (6), 587–602.

Llinas, R., Yarom, Y., 1981. Electrophysiology of mammalian inferior olivary neurones in vitro. Different types of voltage-dependent ionic conductances. J. Physiol. 315, 549–567.

Llinas, R., Yarom, Y., 1986. Oscillatory properties of guinea-pig inferior olivary neurones and their pharmacological modulation: an in vitro study. J. Physiol. 376, 163–182.

Long, M.A., Deans, M.R., Paul, D.L., Connors, B.W., 2002. Rhythmicity without synchrony in the electrically uncoupled inferior olive. J. Neurosci. 22 (24), 10898–10905.

Long, M.A., Landisman, C.E., Connors, B.W., 2004. Small clusters of electrically coupled neurons generate synchronous rhythms in the thalamic reticular nucleus. J. Neurosci. 24 (2), 341–349.

Maier, N., Guldenagel, M., Sohl, G., Siegmund, H., Willecke, K., Draguhn, A., 2002. Reduction of high-frequency network oscillations (ripples) and pathological network discharges in hippocampal slices from connexin 36-deficient mice. J. Physiol. 541 (Pt. 2), 521–528.

Mamiya, A., Manor, Y., Nadim, F., 2003. Short-term dynamics of a mixed chemical and electrical synapse in a rhythmic network. J. Neurosci. 23 (29), 9557–9564.

Mancilla, J.G., Lewis, T.J., Pinto, D.J., Rinzel, J., Connors, B.W., 2007. Synchronization of electrically coupled pairs of inhibitory interneurons in neocortex. J. Neurosci. 27 (8), 2058–2073.

Mann-Metzer, P., Yarom, Y., 1999. Electrotonic coupling interacts with intrinsic properties to generate synchronized activity in cerebellar networks of inhibitory interneurons. J. Neurosci. 19 (9), 3298–3306.

Manor, Y., Rinzel, J., Segev, I., Yarom, Y., 1997. Low-amplitude oscillations in the inferior olive: a model based on electrical coupling of neurons with heterogeneous channel densities. J. Neurophysiol. 77 (5), 2736–2752.

Maranto, A.R., Calabrese, R.L., 1984. Neural control of the hearts in the leech, *Hirudo medicinalis*. J. Comp. Physiol. A 154 (3), 367–380.

Marder, E., 1984. Roles for electrical coupling in neural circuits as revealed by selective neuronal deletions. J. Exp. Biol. 112, 147–167.

Marder, E., Bucher, D., 2007. Understanding circuit dynamics using the stomatogastric nervous system of lobsters and crabs. Annu. Rev. Physiol. 69, 291–316.

Merriam, E.B., Netoff, T.I., Banks, M.I., 2005. Bistable network behavior of layer I interneurons in auditory cortex. J. Neurosci. 25 (26), 6175–6186.

Migliore, M., Hines, M.L., Shepherd, G.M., 2005. The role of distal dendritic gap junctions in synchronization of mitral cell axonal output. J. Comput. Neurosci. 18 (2), 151–161.

Miller, J.P., Selverston, A.I., 1982. Mechanisms underlying pattern generation in lobster stomatogastric ganglion as determined by selective inactivation of identified neurons. II. Oscillatory properties of pyloric neurons. J. Neurophysiol. 48 (6), 1378–1391.

Moreno, A.P., Saez, J.C., Fishman, G.I., Spray, D.C., 1994. Human connexin43 gap junction channels. Regulation of unitary conductances by phosphorylation. Circ. Res. 74 (6), 1050–1057.

Muller, K., Scott, S., 1981. Transmission at a 'direct' electrical connexion mediated by an interneurone in the leech. J. Physiol. 311, 565.

Muller, L.P., Dedek, K., Janssen-Bienhold, U., Meyer, A., Kreuzberg, M.M., Lorenz, S., Weiler, R., 2010. Expression and modulation of connexin 30.2, a novel gap junction protein in the mouse retina. Vis. Neurosci. 27 (3–4), 91–101.

Murthy, V.N., Fetz, E.E., 1994. Effects of input synchrony on the firing rate of a three-conductance cortical neuron model. Neural Comput. 6 (6), 1111–1126.

Namba, H., Mulloney, B., 1999. Coordination of limb movements: three types of intersegmental interneurons in the swimmeret system and their responses to changes in excitation. J. Neurophysiol. 81 (5), 2437–2450.

Nargeot, R., Bédécarrats, A., 2017. Electrical synapses and learning–induced plasticity in motor rhythmogenesis. In: Jing, J. (Ed.), Network Functions and Plasticity. Elsevier (Chapter 6).

Nargeot, R., Le Bon-Jego, M., Simmers, J., 2009. Cellular and network mechanisms of operant learning-induced compulsive behavior in *Aplysia*. Curr. Biol. 19 (12), 975–984.

Norris, B.J., Weaver, A.L., Morris, L.G., Wenning, A., Garcia, P.A., Calabrese, R.L., 2006. A central pattern generator producing alternative outputs: temporal pattern of premotor activity. J. Neurophysiol. 96 (1), 309–326.

Nusbaum, M.P., Beenhakker, M.P., 2002. A small-systems approach to motor pattern generation. Nature 417 (6886), 343–350.

Nusbaum, M.P., Blitz, D.M., Swenson, A.M., Wood, D., Marder, E., 2001. The roles of co-transmission in neural network modulation. Trends Neurosci. 24 (3), 146–154.

Ostojic, S., Brunel, N., Hakim, V., 2009. Synchronization properties of networks of electrically coupled neurons in the presence of noise and heterogeneities. J. Comput. Neurosci. 26 (3), 369–392.

Peck, J.H., Gaier, E., Stevens, E., Repicky, S., Harris-Warrick, R.M., 2006. Amine modulation of Ih in a small neural network. J. Neurophysiol. 96 (6), 2931–2940.

Peck, J.H., Nakanishi, S.T., Yaple, R., Harris-Warrick, R.M., 2001. Amine modulation of the transient potassium current in identified cells of the lobster stomatogastric ganglion. J. Neurophysiol. 86 (6), 2957–2965.

Pfeuty, B., Mato, G., Golomb, D., Hansel, D., 2003. Electrical synapses and synchrony: the role of intrinsic currents. J. Neurosci. 23 (15), 6280–6294.

Phelan, P., Bacon, J.P., Blagburn, J.M., 2017. Neural circuits underlying escape behavior in Drosophila: focus on electrical signaling. In: Jing, J. (Ed.), Network Functions and Plasticity. Elsevier (Chapter 2).

Phelan, P., Goulding, L.A., Tam, J.L., Allen, M.J., Dawber, R.J., Davies, J.A., Bacon, J.P., 2008. Molecular mechanism of rectification at identified electrical synapses in the *Drosophila* giant fiber system. Curr. Biol. 18 (24), 1955–1960.

Powers, R.K., Binder, M.D., 2003. Persistent sodium and calcium currents in rat hypoglossal motoneurons. J. Neurophysiol. 89 (1), 615–624.

Rabbah, P., Nadim, F., 2007. Distinct synaptic dynamics of heterogeneous pacemaker neurons in an oscillatory network. J. Neurophysiol. 97 (3), 2239–2253.

Rela, L., Szczupak, L., 2003. Coactivation of motoneurons regulated by a network combining electrical and chemical synapses. J. Neurosci. 23 (2), 682–692.

Rela, L., Szczupak, L., 2004. Gap junctions: their importance for the dynamics of neural circuits. Mol. Neurobiol. 30 (3), 341–357.

Ribelayga, C.P., O'Brien, J., 2017. Circadian and light-adaptive control of electrical synaptic plasticity in the vertebrate retina. In: Jing, J. (Ed.), Network Functions and Plasticity. Elsevier (Chapter 10).

Saint-Amant, L., Drapeau, P., 2001. Synchronization of an embryonic network of identified spinal interneurons solely by electrical coupling. Neuron 31 (6), 1035–1046.

Schweighofer, N., Doya, K., Kawato, M., 1999. Electrophysiological properties of inferior olive neurons: a compartmental model. J. Neurophysiol. 82 (2), 804–817.

Shaw, B., Kristan Jr., W., 1995. The whole-body shortening reflex of the medicinal leech: motor pattern, sensory basis, and interneuronal pathways. J. Comp. Physiol. A 177 (6), 667–681.

Sherman, A., Rinzel, J., 1992. Rhythmogenic effects of weak electrotonic coupling in neuronal models. Proc. Natl. Acad. Sci. U. S. A. 89 (6), 2471–2474.

Shruti, S., Schulz, D.J., Lett, K.M., Marder, E., 2014. Electrical coupling and innexin expression in the stomatogastric ganglion of the crab *Cancer borealis*. J. Neurophysiol. 112 (11), 2946–2958.

Smarandache-Wellmann, C., Gratsch, S., 2014. Mechanisms of coordination in distributed neural circuits: encoding coordinating information. J. Neurosci. 34 (16), 5627–5639.

Smarandache-Wellmann, C., Weller, C., Mulloney, B., 2014. Mechanisms of coordination in distributed neural circuits: decoding and integration of coordinating information. J. Neurosci. 34 (3), 793–803.

Smarandache-Wellmann, C., Weller, C., Wright Jr., T.M., Mulloney, B., 2013. Five types of nonspiking interneurons in local pattern-generating circuits of the crayfish swimmeret system. J. Neurophysiol. 110 (2), 344–357.

Song, J.H., Wang, Y., Fontes, J.D., Belousov, A.B., 2012. Regulation of connexin 36 expression during development. Neurosci. Lett. 513 (1), 17–19.

Soto-Trevino, C., Rabbah, P., Marder, E., Nadim, F., 2005. Computational model of electrically coupled, intrinsically distinct pacemaker neurons. J. Neurophysiol. 94 (1), 590–604.

Spray, D.C., Bennett, M.V., 1985. Physiology and pharmacology of gap junctions. Annu. Rev. Physiol. 47, 281–303.

Stagg, R.B., Fletcher, W.H., 1990. The hormone-induced regulation of contact-dependent cell-cell communication by phosphorylation. Endocr. Rev. 11 (2), 302–325.

TenBroek, E.M., Lampe, P.D., Solan, J.L., Reynhout, J.K., Johnson, R.G., 2001. Ser364 of connexin43 and the upregulation of gap junction assembly by cAMP. J. Cell Biol. 155 (7), 1307–1318.

Thompson, W.J., Stent, G.S., 1976. Neuronal control of heartbeat in the medicinal leech. II. Intersegmental coordination of heart motor neuron activity by heart interneurons. J. Comp. Physiol. 111, 287–307.
Traub, R.D., Schmitz, D., Jefferys, J.G., Draguhn, A., 1999. High-frequency population oscillations are predicted to occur in hippocampal pyramidal neuronal networks interconnected by axoaxonal gap junctions. Neuroscience 92 (2), 407–426.
Traub, R.D., Whittington, M.A., Draguhn, A., 2017. Gap junctions between pyramidal cells account for a variety of very fast network oscillations (>80 Hz) in cortical structures. In: Jing, J. (Ed.), Network Functions and Plasticity. Elsevier (Chapter 13).
van Rijen, H.V., van Veen, T.A., Hermans, M.M., Jongsma, H.J., 2000. Human connexin40 gap junction channels are modulated by cAMP. Cardiovasc. Res. 45 (4), 941–951.
Veruki, M.L., Hartveit, E., 2002. AII (Rod) amacrine cells form a network of electrically coupled interneurons in the mammalian retina. Neuron 33 (6), 935–946.
Wadepuhl, M., 1989. Depression of excitatory motoneurones by a single neurone in the leech central nervous system. J. Exp. Biol. 143, 509–527.
Weaver, A.L., Roffman, R.C., Norris, B.J., Calabrese, R.L., 2010. A role for compromise: synaptic inhibition and electrical coupling interact to control phasing in the leech heartbeat CpG. Front. Behav. Neurosci. 4.
Welsh, J.P., Turecek, J., 2017. Implications of electrical synapse plasticity in the inferior olive. In: Jing, J. (Ed.), Network Functions and Plasticity. Elsevier (Chapter 12).
Wenning, A., Cymbalyuk, G.S., Calabrese, R.L., 2004a. Heartbeat control in leeches. I. Constriction pattern and neural modulation of blood pressure in intact animals. J. Neurophysiol. 91 (1), 382–396.
Wenning, A., Hill, A.A., Calabrese, R.L., 2004b. Heartbeat control in leeches. II. Fictive motor pattern. J. Neurophysiol. 91 (1), 397–409.
Wenning, A., Norris, B.J., Doloc-Mihu, A., Calabrese, R.L., 2011. Bringing up the rear: new premotor interneurons add regional complexity to a segmentally distributed motor pattern. J. Neurophysiol. 106 (5), 2201–2215.
Wood, D.E., Stein, W., Nusbaum, M.P., 2000. Projection neurons with shared cotransmitters elicit different motor patterns from the same neural circuit. J. Neurosci. 20 (23), 8943–8953 [pii].
Zahid, T., Skinner, F.K., 2009. Predicting synchronous and asynchronous network groupings of hippocampal interneurons coupled with dendritic gap junctions. Brain Res. 1262, 115–129.
Zhang, C., Guy, R.D., Mulloney, B., Zhang, Q., Lewis, T.J., 2014. Neural mechanism of optimal limb coordination in crustacean swimming. Proc. Natl. Acad. Sci. U. S. A. 111 (38), 13840–13845.

CHAPTER 5

Network Functions of Electrical Coupling Present in Multiple and Specific Sites in Behavior-Generating Circuits

J. Jing[1,2], E.C. Cropper[2], K.R. Weiss[2]
[1]Nanjing University, Nanjing, Jiangsu, China; [2]Icahn School of Medicine at Mount Sinai, New York, NY, United States

1. INTRODUCTION

Neural networks generate activity that ultimately mediates diverse animal behaviors. How much we understand a behavior largely depends on our knowledge of the underlying neural circuit. Thus scientists who strive to understand behavior expend significant effort in attempts to analyze neural circuits. Indeed, circuitry studies are experiencing a renaissance of a sort as new approaches, eg, optogenetics and connectomics, are being developed and put in use (Deisseroth, 2015; Jarrell et al., 2012; Sporns and Bullmore, 2014). In this chapter, we describe in-depth circuitry studies primarily in experimentally advantageous molluscan model systems (Arshavsky et al., 1998; Benjamin, 2012; Jing and Weiss, 2005; Nargeot and Simmers, 2012; Newcomb et al., 2012; Sasaki et al., 2013). In particular, we describe recent research that provides novel insights into network functions by revealing the diverse locations and specific roles of electrical coupling in multifunctional rhythm-generating circuits that mediate motivated behaviors, eg, feeding and locomotion.

We highlight a generic circuit structure consisting of a local rhythm-generating network and remote circuit elements (Fig. 5.1). In our opinion, we think most, if not all, brain circuits, no matter how simple or complex, can be considered in these general terms. Rhythmic activity is generated by a local circuit (or a microcircuit) through synaptic interactions between local circuit elements (see Chapters 4 (Nadim et al., 2017), 6 (Nargeot and Bédécarrats, 2017), 1 (Li and Rekling, 2017), and 13 (Traub et al., 2017)). Here, we focus on the rhythmic activity that results from the activity of a central pattern generator (CPG) (Cropper et al., 2004; Grillner, 2006; Marder and Calabrese, 1996; Marder et al., 2014; Yuste et al., 2005) (Chapters 4 (Nadim et al., 2017), 6 (Nargeot and Bédécarrats, 2017), and 1 (Li and Rekling, 2017)). CPG-mediated behaviors have been described in both invertebrates and vertebrates. In this situation, the underlying networks are often multifunctional (Briggman and Kristan, 2008; Jing et al., 2009), ie, the local circuit generates multiple outputs. The activity generated by

Network Functions and Plasticity
ISBN 978-0-12-803471-2, http://dx.doi.org/10.1016/B978-0-12-803471-2.00005-9

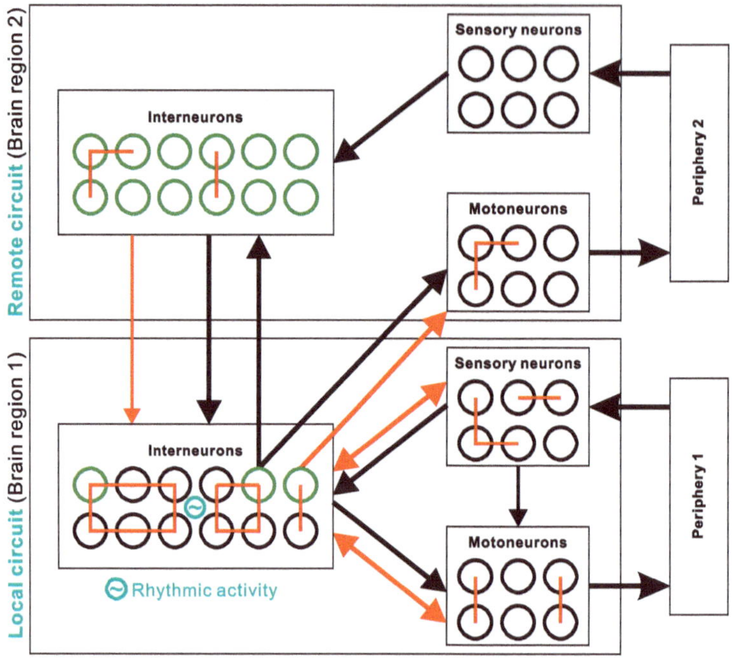

Figure 5.1 ***Schematic diagram of a circuit structure consisting of a local circuit and remote circuit elements.*** This circuit is generalized from a multifunctional rhythm-generating circuit in *Aplysia* feeding system which controls coordinated activity of two peripheral structures. The diagram illustrates synaptic connections between the local circuit (bottom half) and remote circuit elements (top half) involved in controlling two sensory and motor structures. In particular, electrical coupling among different classes of neurons (*red arrows*) and within same classes of neurons (*red lines* connecting the neurons) are distributed in a broad, but specific manner. Chemical connections are shown in *black arrows*. Shown in green are projection neurons with their axons projecting to the other brain region. The *blue* ~ indicates the main rhythmic activity generated by interneurons in the local circuit. Although some of connection patterns may be specific to *Aplysia* feeding network, the general functional and structural schemes, with modifications, can be applied to many other circuits.

the local circuit is propagated to and from at least one remote brain region. In part, the latter consists of input to the CPG from projection neurons that trigger activity. In addition, it consists of output to remote elements, eg, motor neurons. Thus one basic question that arises is what is the nature of functional connections (including chemical and electrical synapses) between local circuit elements, and between elements of the local circuit and remote elements, and how do these connections contribute to (1) generation of rhythmic network activity and multifunctionality and (2) propagation of activity.

The work reviewed in this chapter suggests that electrical coupling appears to be broadly distributed in the circuit. However, electrical coupling is present in a specific, nonrandom manner (Fig. 5.1). Most commonly, electrical coupling is present in neurons of the same type, or homogenous cells, eg, similar cells that are bilaterally located, or ipsilateral cells that have identical intrinsic properties. For cells that are not of the same type, ie, heterogeneous cells, coupling is often present between elements that are active in a single phase. Within the same phase, electrical coupling may further organize these elements into functional subgroups, eg, building blocks or modules, thereby helping to define functional classes in a heterogeneous population of neurons. Overall, electrical coupling, wherever present, makes a major contribution to promoting synchronous or similar activity of neurons firing at the same phase of rhythmic activity, to promoting generation of multiple outputs, or to transmitting activity to and from remote circuit elements.

2. THE FEEDING NEURAL CIRCUIT IN *APLYSIA*

The *Aplysia* feeding network is typical of a multifunctional rhythm-generating network in that it consists of a local circuit in the buccal ganglion that generates cyclic activity, and remote circuit elements in the cerebral ganglion that interact with the local circuit (Fig. 5.2B). The feeding system controls coordinated movements of two major structures: the mouth (lips) and the radula (Kupfermann, 1974; Rosen et al., 1982) (Fig. 5.2A). The lips are controlled by neural elements in the cerebral ganglion and the radula is controlled by neural elements in the buccal ganglion. The lips are an external structure that is part of the mouth; whereas the radula is an internal structure that is located inside the buccal mass. The radula is a hand-like structure that grasps and manipulates food (or an object). The radula generates two sets of movements; protraction—retraction and opening—closing. The lips open during radula protraction to allow *Aplysia* access to food or related objects.

There are three aspects to network function. First, the local circuit (the CPG) in the buccal ganglion generates biphasic rhythmic radula movements. Each cycle of feeding-related activity consists of a radula protraction—retraction sequence (Fig. 5.2C). One major question is how buccal interneurons and motoneurons are organized synaptically to generate this biphasic pattern.

The second, and perhaps more important question stems from the multifunctionality of the *Aplysia* feeding network, ie, the network generates multiple types of motor patterns (Fig. 5.2C). For example, there are at least two major types of feeding behavior, ingestion and egestion (Jing et al., 2004; Lu et al., 2015; Morton and Chiel, 1993a,b). In all cases, rhythmic movements of the radula begin with protraction and end with retraction. However, phase relationships between opening—closing and protraction—retraction differ and depend on the nature of the behavior expressed. In ingestion, the radula is open during protraction, and closed during retraction. With this combination of movements food is pulled into the buccal cavity. In egestion, the radula is closed during protraction, and open during retraction. This combination of movements pushes an object out. In addition, ingestive behaviors can be divided into two subtypes; biting and swallowing. During biting, the protraction duration is extended to allow time for the

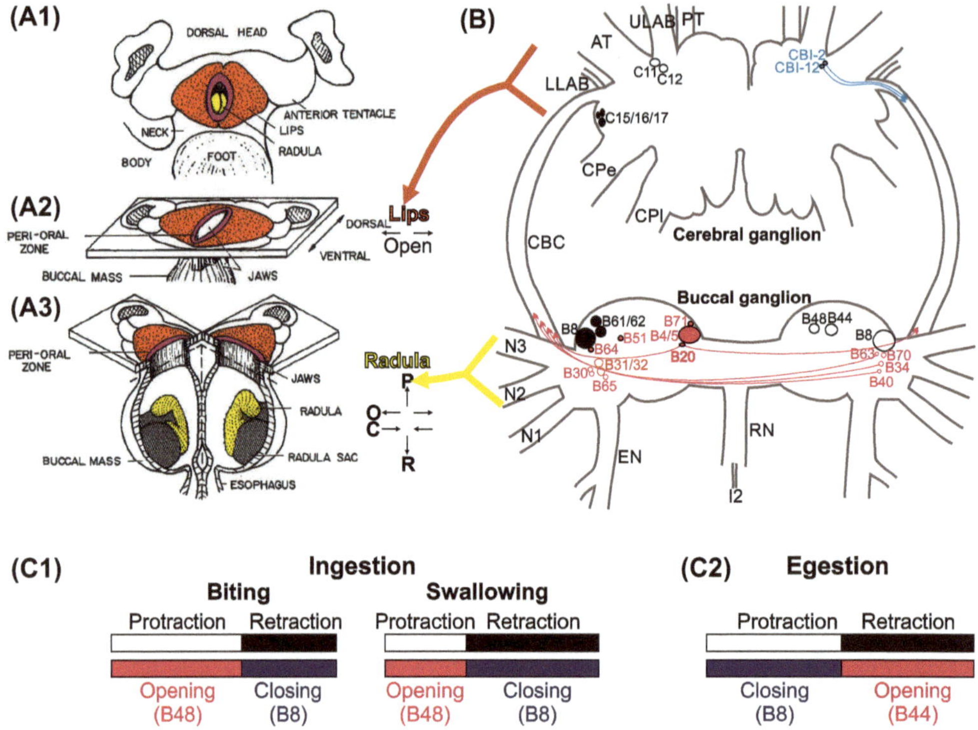

Figure 5.2 ***The structure of the periphery, distribution of the major central circuit elements, and multiple types of motor patterns generated by the CPG in the* Aplysia *feeding system.*** (A) Illustration of peripheral sensory motor structures: the lips (red) and the radula (yellow). (A1) intact head with mouth partially open; (A2) dissected head exposing buccal mass; (A3) dissected buccal mass exposing radula. *(Adapted from Rosen, S.C., Weiss, K.R., Cohen, J.L., Kupfermann, I., 1982. Interganglionic cerebral-buccal mechanoafferents of Aplysia: receptive fields and synaptic connections to different classes of neurons involved in feeding behavior. J. Neurophysiol. 48, 271–288.).* (B) Major central elements of the *Aplysia* feeding circuit in the cerebral and buccal ganglia that control the lips and radula, respectively. In the cerebral ganglion (ventral surface), shown are two CBIs, CBI-2 and CBI-12 (axons to the buccal ganglion), and motoneurons C11, C12, C15/16/17 (axons out to the cerebral nerves). In the buccal ganglion, neurons that are on the caudal surface are shown in open circles, whereas neurons that are on the rostral surface are shown in filled circles. CPG neurons are shown in red. Shown are BCIs, B63, B34, B40, B20, B71 (axons to the cerebral ganglion), and buccal interneurons, B65, B30, B70, B64 (axons to the contralateral buccal ganglion, but not out of the buccal ganglion). Other buccal neurons, B8, B61/62, B51, B4/5, B48, B44, B31/32, have axons in the buccal nerves. Note that almost all *Aplysia* neurons are bilaterally paired. Thus, although only one neuron is shown, all neurons illustrated have contralateral homologues. Nerve abbreviations: *AT*, Anterior tentacular; *EN*, esophageal nerve; *LLAB*, lower labial; *PT*, posterior tentacular; *RN*, radular nerve; *ULAB*, upper labial. Connectives: *CPe*, Cerebral-pedal; *CPl*, Cerebral-pleural; *CBC*, cerebral-buccal connective. (C) Phase relationship of protraction–retraction versus opening–closing movements in ingestive and egestive motor patterns. Ingestive patterns are also divided into two subtypes, biting, with a long protraction duration, and swallowing, with a short protraction duration. Note that in ingestive versus egestive programs, both opening and closing change their phases relative to protraction and retraction. However, phase shifting is implemented differently at the level of motoneurons. Two different opening motoneurons are recruited differentially during protraction (B48) versus retraction (B44). A single motoneuron that phase shifts (B8) mediates closing.

radula to move outside the mouth and reach/grasp food. During swallowing, the protraction duration is short because food is already inside the mouth, whereas the retraction duration is long when food grasp is successful and food is pulled deeper into the buccal cavity toward the esophagus. Thus the second important question is how these multiple motor patterns are generated by the CPG.

Third, the local circuit interacts with remote circuit elements in the cerebral ganglion. This involves at least two types of interactions. Cerebral-buccal interneurons (CBIs) provide input to the buccal ganglion, and the buccal CPG drives lip-opening motoneurons. The CBIs receive sensory input from the cerebral nerves that innervate the head, and provide command or modulatory input to the buccal CPG to drive or modulate the motor patterns (Hurwitz et al., 2003; Jing et al., 2004; Jing and Weiss, 2001; Rosen et al., 1991; Wu et al., 2014). Here, the CBIs that can evoke a motor program when stimulated are considered to be command-like, whereas the other CBIs are considered to be modulatory in that, although they cannot directly evoke a motor program, they may change specific characteristics of a motor program. The lips must be open during protraction (Perrins and Weiss, 1996) and protraction is generated by the buccal CPG. Consequently, buccal/cerebral activity should be coordinated by the buccal-cerebral interneurons (BCIs) that are active during protraction (Jing et al., 2011).

In the following sections, we discuss how the above-mentioned functions are implemented by the *Aplysia* feeding circuit, and describe the role of electrical coupling.

2.1 Local Circuit—Rhythm Generation in a Multifunctional Network

2.1.1 Generation of Biphasic Rhythmic Activity (Protraction—Retraction)

Rhythmic radula movements, and more specifically, the alternating protraction—retraction sequence, are the primary motor pattern that is present in the feeding network. Most, if not all, activity in the feeding system is associated with activity patterns that correspond to protraction—retraction (Fig. 5.3A—C). In the buccal ganglion, the CPG consists of a number of interneurons that contribute to the generation of this biphasic activity. Indeed, regardless of the type of motor program expressed, most interneurons in the buccal ganglion are active during either protraction or retraction. However, activity levels of interneurons vary and depend on the version of the program that is expressed. The protraction group (P group) of interneurons consists of at least six members: including B63 and B34 (Hurwitz et al., 1997), B20 (Jing and Weiss, 2001; Teyke et al., 1993), B65 (Kabotyanski et al., 1998), B30A and B30B (Jing et al., 2004), and B40 (Jing and Weiss, 2002) (Figs. 5.2B and 5.3A,D). The retraction group (R group) of interneurons consists of at least five members, including B64 (Hurwitz and Susswein, 1996), B4/5 (Gardner, 1971), B70 (Sasaki et al., 2009), B51 (Evans and Cropper, 1998; Jing et al., 2004; Nargeot et al., 1999; Plummer and Kirk, 1990), and B71 (Sasaki et al., 2013) (Figs. 5.2B and 5.3B,E).

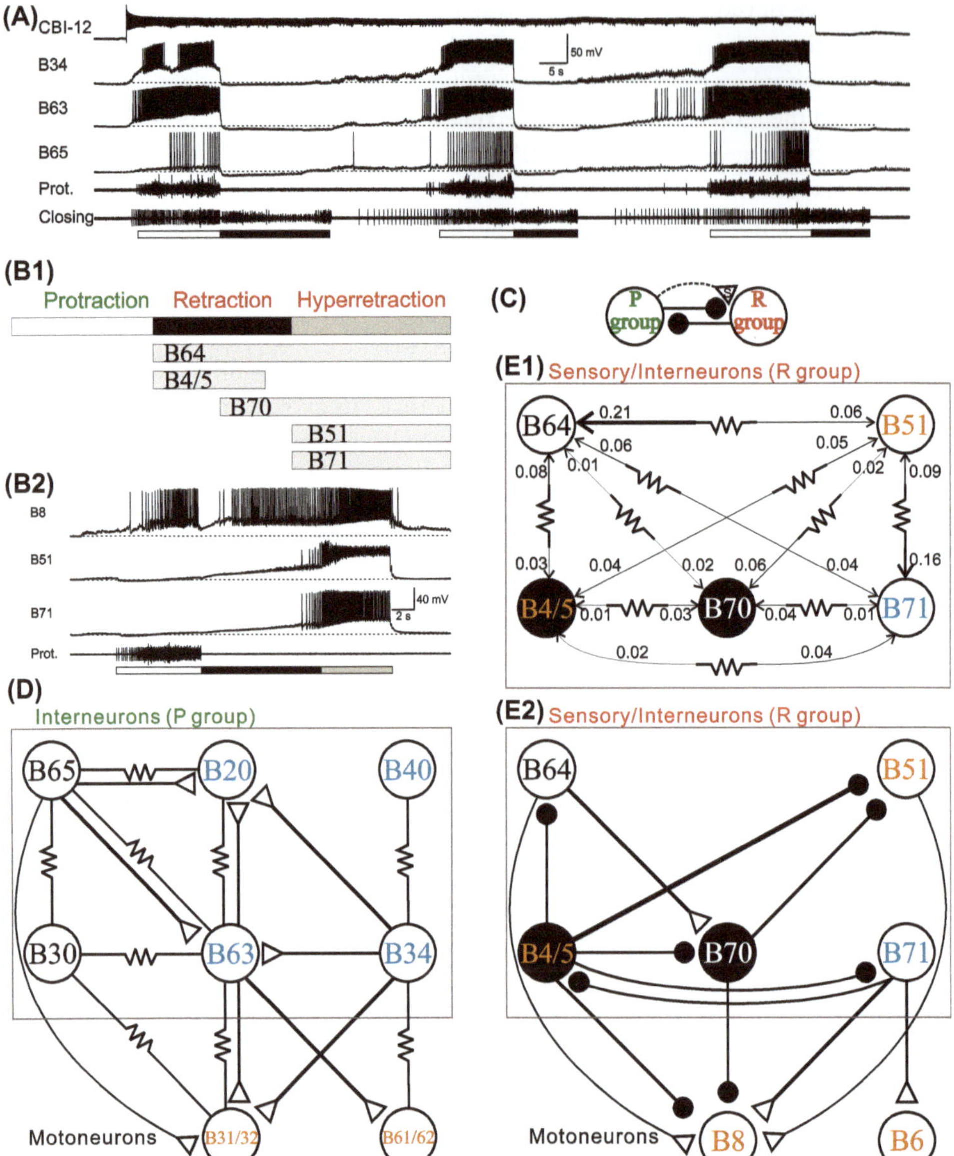

Figure 5.3 ***Rhythmic activity generated by the local circuit and synaptic connectivity in the*** **Aplysia** ***buccal ganglion.*** (A) Simultaneous activity of three protraction interneurons B34, B63, and B65 during rhythmic motor programs elicited by CBI-12 (three cycles). For the protraction interneurons, note the slow depolarization during the early phase of protraction and the hyperpolarization during retraction. Dotted lines: resting membrane potentials. Prot., protraction (I2). (B) Illustration of protraction–retraction–hyperretraction sequence (B1) and activity of retraction sensory and interneurons (B51, B71) during one cycle of motor program (B2). Protraction (I2), open bar; retraction, filled bar; hyper-retraction, gray bar. B8, closing motoneuron. (C) Connections between P group neurons and R group

Because of the presence of multiple elements in both the protraction and retraction groups, two requirements must be satisfied. One requirement is that these multiple protraction or retraction neurons need to maintain their activity within each phase. The second requirement is that the network needs to prevent the spillover of interneuronal firing to inappropriate phases. For the first requirement, electrical coupling appears to play a major role as evidenced by the fact that electrical coupling is extensively observed between neurons that are coactive in the same phase, ie, fire together during either protraction (Fig. 5.3D) or retraction (Fig. 5.3E1), but is largely absent between neurons that are active in different phases. For the protraction phase, chemical excitation also plays a role, as it is observed between several protraction interneurons. In contrast, electrical coupling is likely to be the main player during retraction. This is suggested by the fact that chemical excitation is observed only between two neurons (B64 and B70) (Fig. 5.3E2).

For the second requirement, ie, prevention of the spillover of firing, two factors are important. On the one hand, there is no electrical coupling between protraction and retraction neurons. On the other hand, chemical inhibition plays a critical role because protraction and retraction neurons mutually inhibit each other. Specifically, many protraction neurons (B63, B34, B30, B40) inhibit retraction neurons B64, B70, and B71; in turn, B64 and B71 inhibit all major protraction neurons (Hurwitz and Susswein, 1996; Sasaki et al., 2009, 2013) (Fig. 5.3C, see also Fig. 5.5B). Protraction phase neurons slowly depolarize and fire when a sufficient depolarization is reached during protraction. During the retraction phase, they receive strong synaptic inhibition that terminates their activity and hyperpolarizes them well below their resting membrane potentials (Fig. 5.3A). This inhibition originates mainly from retraction interneurons, eg, B64. The reverse is true for retraction neurons, ie, retraction neurons receive inhibition during protraction, whereas they depolarize during retraction (Fig. 5.3B2). Thus electrical coupling appears to

neurons. Mutual inhibition between P group neurons (eg, B63, B34, B30, B40) and R group neurons (eg, B64, CBI-5/6, B71) mediates the alternating protraction–retraction sequence. Slow excitation from some P group neurons (eg, B30, B65) to some R group neurons (eg, B64, CBI-5/6) enables the phase switching from protraction to retraction. (D) Protraction circuit. The P group neurons are organized into two functional subgroups based on electrical coupling. Note that for clarity, not all electrical coupling, eg, between B65 and B31, is shown. (E) Retraction circuit. (E1) Retraction neurons are electrically coupled, with varying coupling ratios as indicated. (E2) Illustration of chemical connections that fine-tune the activity patterns of retraction neurons. Projection neurons are shown in blue, whereas neurons with peripheral axons (with either sensory or motor functions) are shown in orange. *(B) and (E) are from Sasaki, K., Cropper, E.C., Weiss, K.R., Jing, J., 2013. Functional differentiation of a population of electrically coupled heterogeneous elements in a microcircuit. J. Neurosci. 33(1), 93–105.* Symbols in this and the rest of the figures: open triangle, chemical excitation; filled circle, chemical inhibition; resistance symbol, electrical coupling; s, slow synaptic connection.

contribute to the depolarization of protraction neurons during protraction, and of retraction neurons during retraction.

We can therefore suggest that one major function of electrical coupling is to promote coordinated activity between neurons active in the same phase. This is consistent with results obtained in other model systems (see Chapters 4 (Nadim et al., 2017), 6 (Nargeot and Bédécarrats, 2017), 1 (Li and Rekling, 2017), and 13 (Traub et al., 2017)). However, as we describe later, the contribution of electrical coupling is not limited to promoting coordinated firing of neurons active in the same phase. Specifically, examination of activity and connectivity of multiple interneurons that are active during either protraction or retraction shows that electrical coupling between specific neurons within a single phase contributes to the multifunctionality of the network.

2.1.2 Protraction: Electrical Coupling—Based Separation Into Two Functional Subgroups/Modules for Neurons Active in One Phase

A further question is whether all the neurons that fire in a single phase are coupled together. Examination of protraction neurons shows that this is not necessarily the case. If we only examine the intracellular recording of the protraction interneurons (Fig. 5.3A), there is no obvious basis to separate them into functional groups, because these neurons similarly depolarize as the protraction phase initiates and spike when depolarization reaches threshold. However, based on electrical coupling between interneurons (Fig. 5.3D), protraction interneurons appear to be separated into two electrically coupled subgroups: subgroup 1 that includes B63—B20—B65—B30 and subgroup 2 that includes B34—B40.

What is the functional significance of this electrical coupling—based segregation? Perhaps we can understand this from at least two perspectives. First, one important function of the protraction interneurons is to drive the protraction motoneurons, B31/32 and B61/62 (Fig. 5.3D). Indeed, the separation of coupling between interneurons can be extended to protraction motoneurons B31/32 and B61/62. Specifically, B31 is coupled to B32, and B61 is coupled to B62, but B31/32 are not coupled to B61/62. In addition, B31/32 are coupled to subgroup 1 protraction neurons, whereas B61/62 are coupled to subgroup 2. Thus subgroup 1 interneurons use electrical excitation to drive B31/32, whereas subgroup 2 interneurons similarly drive B61/62. It is notable that electrical excitation is supplemented by chemical excitation, as subgroup 1 or 2 interneurons do provide chemical excitation to B61/62 or B31/32, respectively (Hurwitz et al., 1997).

Second, the two subgroups play opposite roles in mediating protraction duration: subgroup 1 interneurons B65 and B30 tend to promote motor programs of shorter protraction, whereas subgroup 2 interneurons B34 and B40 promote motor programs of longer duration (Jing et al., 2004; Jing and Weiss, 2002, 2005). Indeed, the higher-order interneuron CBI-2 acts preferentially on subgroup 2 to promote biting-like programs with a long protraction duration, whereas CBI-4 acts preferentially on subgroup 1 neurons (B30 and B65) to promote swallowing-like programs, which have a short protraction duration (Jing et al., 2004) (see Fig. 5.5A). Furthermore, we showed that CBI-2 and CBI-12 use specific synaptic connections (both excitation and inhibition) of various strengths to act on these two subgroups of neurons to generate biting-like programs with various protraction durations (Jing and Weiss, 2005). More generally, this work demonstrated that the two subgroups of protraction neurons function as two building blocks or modules that can be combined flexibly to generate variants of motor programs. The aforementioned studies therefore support the modular concept that has originally been proposed by investigators (Bizzi and Mussa-Ivaldi, 2004; Bizzi et al., 1991; Grillner, 1985; Hart and Giszter, 2004; Stein and Smith, 1997; Tresch et al., 2002) who work on vertebrates. However, note that the composition of vertebrate modules has yet to be clearly demonstrated at the level of interneurons.

2.1.3 Retraction: Generation of Variable Phasic Activity Within One Phase and Roles of Asymmetric Coupling

Similar to the coupling between the protraction interneurons, there is extensive electrical coupling between retraction interneurons. However, retraction differs from protraction in that it has multiple components, ie, a hyperretraction phase can be added, which extends retraction duration in a subset of motor programs (Fig. 5.2C1). This hyperretraction is believed to be important for swallowing, which occurs when animals successfully grasp food, and move it deeper into the buccal cavity (Evans and Cropper, 1998; Jing et al., 2004; Morton and Chiel, 1993a,b; Nargeot et al., 1999). The variability of retraction durations is reflected in the variable activity patterns of the five retraction neurons. At least four patterns can be distinguished (Fig. 5.3B) in which: (1) interneurons are active throughout retraction including hyperretraction, ie, B64; (2) interneurons are active during the early phase of retraction, ie, B4/5; (3) interneurons are active during the late phase of retraction, including hyperretraction, B70; and (4) interneurons are only active during hyperretraction, ie, B51 and B71.

Despite the diversity of their activity patterns, retraction neurons are all electrically coupled (Sasaki et al., 2013). The question then is why do their activity patterns differ? Three important factors can account for this variability. (1) The coupling strengths between these neurons differ (Fig. 5.3E1). In fact, the difference in coupling strength can provide at least a partial explanation for the difference in firing patterns of these neurons. For example, the B51 firing pattern most closely resembles the B71 firing pattern (Fig. 5.3B2) and these two neurons are the most strongly coupled. In contrast, the B70 firing pattern is relatively unique and B70 is not strongly coupled to other retraction neurons. In effect, strong coupling can define a functional class. (2) There is extensive synaptic inhibition between some retraction neurons (Fig. 5.3E2). In particular, B4/5 fires early during retraction, and its inhibitory input to B70 contributes to the delayed activation of this neuron (Sasaki et al., 2009). (3) Intrinsic properties also play an important role. Although B64 is also inhibited by B4/5, B64 has a plateau potential, which sustains its activity throughout retraction (despite the inhibition). In contrast, although B51 also has a plateau potential, its threshold is higher than that of B64 (Fig. 5.4E), so that B51 initiates plateau activity late during retraction, ie, hyperretraction. Thus the lesson here is that even among electrically coupled neurons active in a single phase, coupling strengths and chemical synapses between them can differ and further define functional subgroups.

In addition to our suggestion that different coupling strengths serve to generate distinct functional subgroups, there are additional, although not necessarily mutually exclusive, explanations for the presence of weak coupling in the network. (1) Weak coupling may make a significant contribution to steady-state changes in membrane potentials, eg, during the buildup of the retraction phase. (2) Coupling strengths may be underestimated if sites of coupling are further away from the soma. (3) Weak coupling

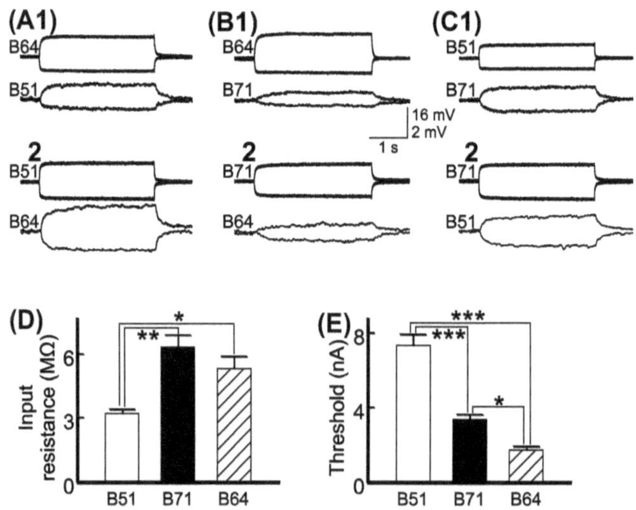

Figure 5.4 ***Asymmetric electrical coupling (A–C) and intrinsic properties (D–E) of retraction neurons (B64, B51, and B71).*** (A) From B64 (both depolarizing and hyperpolarizing currents) to B51 (A1), and from B51 to B64 (A2). (B) From B64 to B71 (B1), and from B71 to B64 (B2). (C) From B51 to B71 (C1), and from B71 to B51 (C2). For each panel, the top recordings show the presynaptic cells that were depolarized and hyperpolarized for 3 s, and the lower recordings show corresponding responses in the postsynaptic cells. Coupling ratios measured by depolarizing and hyperpolarizing pulses were similar: from B64 to B51, 0.086 ± 0.005 versus 0.098 ± 0.006, and from B51 to B64, 0.171 ± 0.013 versus 0.168 ± 0.014, ($n = 4$); from B64 to B71, 0.048 ± 0.004 versus 0.045 ± 0.004, and from B71 to B64, 0.074 ± 0.010 versus 0.079 ± 0.012, ($n = 3$); from B51 to B71, 0.155 ± 0.009 versus 0.152 ± 0.008, and from B71 to B51, 0.104 ± 0.007 versus 0.102 ± 0.003, ($n = 4$). The data suggested that these electrical synapses are nonrectifying. (D) Apparent input resistances. (E) Spike thresholds: minimum depolarizing currents needed to induce spiking. Bonferroni post hoc tests: $*p < .05$; $**p < .01$; $***p < .001$. Error bars indicate SEM. *(From Sasaki, K., Cropper, E.C., Weiss, K.R., Jing, J., 2013. Functional differentiation of a population of electrically coupled heterogeneous elements in a microcircuit. J. Neurosci. 33(1), 93–105.)*

may represent some part of a development program, and could be important or stronger early during development, with chemical connections becoming more important later (see He and Shi, 2017 (Chapter 14)).

Most notably, the coupling between heterogeneous neurons is often asymmetrical, meaning the coupling ratio between a pair of neurons A and B is different when measured from A to B versus from B to A (Fig. 5.3E1). Asymmetrical coupling can result from differences in intrinsic properties of the coupled neurons or rectifying electrical coupling between them (Bennett, 1966; Furshpan and Potter, 1959; Johnson et al., 1993). Because rectifying electrical coupling passes current better in one direction, one of its hallmarks is that for two coupled neurons A and B, depolarizing current injected into A will produce larger responses in B than hyperpolarizing current injected into A. Furthermore, hyperpolarizing current injected into B will produce larger responses in A than depolarizing current in B (Furshpan and Potter, 1959; Marder, 2009). Notably, rectifying electrical coupling likely results from heterotypic gap junctions between the two coupled neurons (Marder, 2009; Phelan et al., 2008; Rash et al., 2013; Phelan et al., 2017 (Chapter 2)). Our experiments with three pairs of neurons, B64−B51, B64−B71, and B51−B71 (Fig. 5.4A−C) (Sasaki et al., 2013) showed that coupling ratios measured with both depolarization and hyperpolarization currents are similar, suggesting that the asymmetrical coupling between these neurons is likely not due to rectifying electrical coupling. Rather, it is, at least partly, due to the observed differences in input resistance of the three neurons (Fig. 5.4D).

Similar to protraction interneurons, retraction interneurons, eg, B64 and B71, also play a role in driving activity in retraction motoneurons. Here, electrical coupling also plays a role as retraction motoneurons (eg, B6) are electrically coupled to retraction neurons (eg, B71) (Hurwitz and Susswein, 1996; Sasaki et al., 2013).

2.1.4 Opening/Closing: Different Roles of Electrical Coupling in the Control of Motoneurons Showing Different Phasic Activity in Different Programs

As noted earlier, opening and closing movements are similar in that their relative timing differs during ingestive versus egestive behavior (Fig. 5.2C). Specifically, during ingestion, opening occurs during protraction, and closing occurs during retraction. During egestion, the reverse is true.

At the circuit level, this phase shifting is implemented using different mechanisms. For opening, there are two different motoneurons, B48 and B44, that are active during protraction and retraction, respectively (Fig. 5.2C) (Evans et al., 1996; Friedman et al., 2009). During ingestive programs, B48 is strongly active, whereas B44 is not, thus opening occurs during protraction. During egestive programs, B48 is not active, whereas B44 is strongly active, thus opening occurs during retraction. In effect, phase shifting of opening is implemented by the differential recruitment of two motor neurons, ie, motoneurons do not change the phase in which they are active. In this case, electrical coupling is an effective means of driving motor neuron activity: ie, B48 is electrically coupled to protraction neurons B34-B40, whereas B44 is electrically coupled to retraction neurons. The activity levels of B48/B44 in different motor programs are determined by a combination of network inputs and states that promote the expression of different types of programs (Friedman et al., 2009).

In contrast to opening, phase shifting of closing is implemented at the level of motoneurons. The closing motoneurons, B8, unlike opening motoneurons, do not restrict their activity to either protraction or retraction. Rather, in ingestive programs, B8 is active during retraction, whereas during egestive programs, it is active during protraction (Fig. 5.2C). In theory, this phase shifting could be implemented by phase shifting interneurons, which has been postulated in vertebrate motor networks (Grillner, 1981, 1985). However, no such interneuron has been identified in *Aplysia*. Rather, the *Aplysia* feeding network utilizes two different interneurons, B40 and B20, that are normally active during protraction to drive activity in closing motoneurons (Jing and Weiss, 2001, 2002). B40 is preferentially active during ingestive programs, and uses fast inhibition and slow chemical excitation to drive closing motoneurons during retraction. In contrast, B20 is preferentially active during egestive programs, and uses fast chemical excitation to drive closing motoneurons during protraction. As might be expected, electrical connections between protraction interneurons or most retraction interneurons and closing motoneurons have not been observed. The inflexibility of electrical coupling likely makes it an ineffective candidate for providing excitation to phase shifting neurons. For example, in ingestive programs, B40 and B8 are active during different phases of the motor program. Consequently, electrical synapses from B40 to B8 would not produce the desired pattern of B8 activity.

2.2 Interaction With Remote Circuit Elements

2.2.1 Input Activation of the Local Circuit by CBIs

Input activation of the buccal CPG is mediated by the CBIs in the cerebral ganglion, higher order interneurons that receive sensory input from the head and project their axons to the buccal ganglion. There are about 13 CBIs. Except CBI-7, all other CBIs, including CBI-1, CBI-2, CBI-3, CBI-4 (Hurwitz et al., 2003; Jing et al., 2004; Jing and Weiss, 2001; Rosen et al., 1991), CBI-5/6 (Perrins and Weiss, 1998; Sasaki et al., 2007, 2008), CBI-8/9 (Xin et al., 1999), CBI-10 (Chen et al., 2015), CBI-11 (Wu et al., 2014), and CBI-12 (Hurwitz et al., 1999; Jing and Weiss, 2005) have been studied. Interestingly, the only monosynaptic connections that have been described between the CBIs are electrical (Fig. 5.5A). For example, CBI-5 and CBI-6 are identical, and are electrically coupled. The same applies for CBI-8 and CBI-9. Based on current findings, there are at least two groups of electrically coupled CBIs: one group consists of CBI-2, CBI-3, and CBI-11 (Morgan et al., 2002; Wu et al., 2014); the second group consists of CBI-12 and CBI-8/9 (Xin et al., 1999). Thus electrical coupling apparently plays a role in

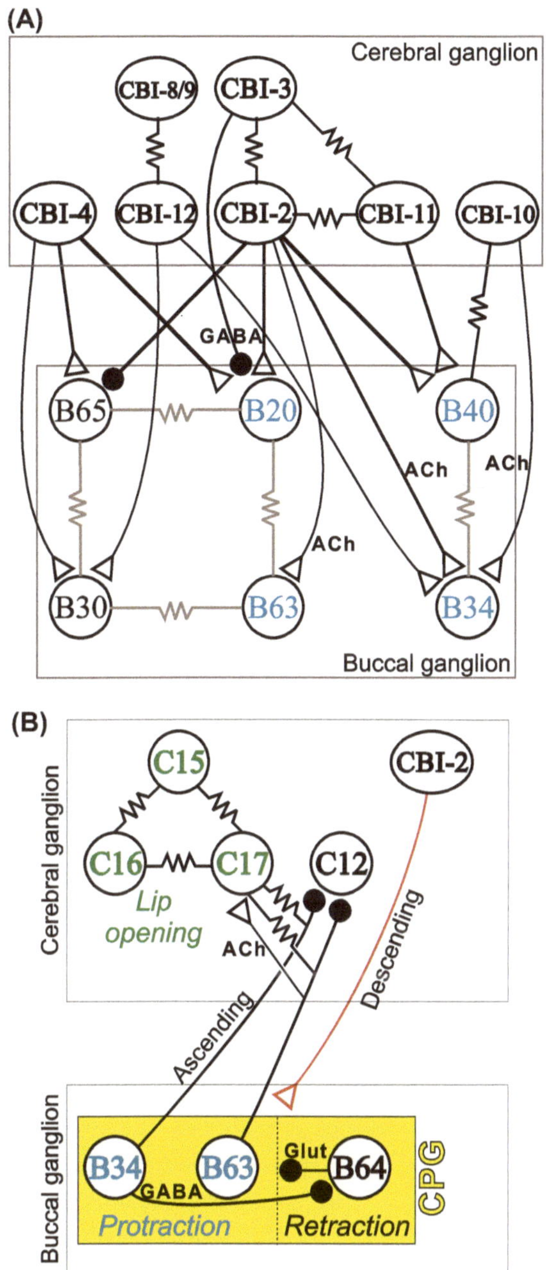

Figure 5.5 ***Electrical coupling among remote circuit elements and between remote circuit elements and local circuit elements in the* Aplysia *feeding network.*** (A) Connections between CBIs are only electrical, whereas connections from CBIs to buccal protraction interneurons are largely chemical, with one (from CBI-10 to B40) identified as electrical. CBI-5/6 are not illustrated because they act primarily during retraction (Sasaki et al., 2007, 2008), and do not provide excitatory inputs to protraction interneurons. (B) Connections from BCIs (B34 and B63) to cerebral motoneurons C15/16/17. Motor programs in the buccal CPG are elicited by descending activity in CBI-2 from the cerebral ganglion. In turn, protraction interneurons B34 and B63 send ascending information back to cerebral motoneurons. The connections from B34 to C15/16/17 are purely electrical, whereas those from B63 are mixed, ie, both chemical and electrical. Also shown are the inhibitory connections from B34 and B63 to another cerebral motoneuron C12. Transmitters: acetylcholine (ACh); GABA (Jing et al., 2003); glutamate (Glut) (Jing et al., 2015). *((B) is modified from Jing, J., Sasaki, K., Perkins, M.H., Siniscalchi, M.J., Ludwar, B.C., Cropper, E.C., et al., 2011. Coordination of distinct motor structures through remote axonal coupling of projection interneurons. J. Neurosci. 31(43), 15438–15449. Symbols as in Fig. 5.3.)*

separating these CBIs into distinct functional groups. It is conceivable that electrical coupling promotes similar activity patterns between the coupled CBIs when sensory input and/or feedback input from the buccal CPG activate these neurons (eg, Wu et al., 2014). Notably, although the CBIs do not have monosynaptic chemical connections with each other, they do receive chemical connections from other sources, including cerebral and buccal neurons (Hurwitz et al., 1999; Jing et al., 2008; Sasaki et al., 2007).

One of the most important functions of the CBIs is to provide synaptic input to specific buccal local-circuit elements (including buccal interneurons and motoneurons). These inputs drive and/or modulate motor programs. These CBIs include command like neurons: CBI-2, CBI-4, CBI-12, CBI-11, CBI-10, and CBI-1, and modulatory neurons: CBI-3 and CBI-8/9. So far, the most prominent synaptic connections from the CBIs to buccal neurons are chemical connections (Hurwitz et al., 2003; Jing and Weiss, 2001, 2002) (Fig. 5.5A). Chemical connections include cholinergic excitation from CBI-2, CBI-11, and CBI-10 (Chen et al., 2015; Dacks and Weiss, 2013; Hurwitz et al., 2003; Wu et al., 2014), and GABAergic inhibition from CBI-3 (Jing et al., 2003). However, Chen et al. (2015) described electrical coupling between at least one CBI, ie, CBI-10, and buccal interneuron B40. This coupling makes a major contribution to the activation of B40 during CBI-10-elicited motor programs. Although the role of electrical coupling in mediating synaptic inputs from CBIs to buccal neurons has not been fully characterized, it is clear that electrical coupling may contribute importantly to the excitation of buccal interneurons by at least a subset of CBIs.

2.2.2 Output From the Local Circuit: Coordination With Remote Elements

One of the major functions of the feeding system is to coordinate the activity of the cerebral lip motoneurons with radula protraction. There are three lip motoneurons (Perrins and Weiss, 1996), C15/16/17, and all are strongly electrically coupled to each other (Figs. 5.2B and 5.5B). Initially, it was hypothesized that these lip motoneurons are controlled by the CBIs (Perrins and Weiss, 1996). However, despite significant efforts, no CBIs with such a function have been identified. In our study (Jing et al., 2011), we considered an alternative explanation that BCIs may actually mediate excitation of lip motoneurons. Specifically, protraction is mediated by protraction interneurons in the buccal ganglion, and four of these protraction interneurons, including B63, B20, B34, and B40, function as BCIs, ie, their action potentials propagate from the buccal ganglion to the cerebral ganglion. Because they are active during protraction, they are well suited to transmit information about the protraction phase from the buccal ganglion to the cerebral ganglion. Indeed, our study showed that B63, B34, and B40 all excite lip motoneurons C15/16/17, with B63 and B34 providing the strongest input (Jing et al., 2011) (Fig. 5.5B). Interestingly, further evidence from axonal recordings and pharmacological experiments indicates that the excitatory input from B63 is both electrical and chemical, and input from B34 is purely electrical. This suggests that axodendritic electrical coupling can act alone (B34) or together with chemical excitation (B63) to drive remote output elements to fire in the same phase of the buccal motor program.

2.2.3 Experimental Procedures Used to Establish Electrical Coupling Between Remote Circuit Elements

It is useful to discuss here the experimental proofs for electrical coupling between remote circuit elements, ie, axodendritic coupling. In gastropod molluscs, by and large, chemical synaptic connections between neurons are not graded. Thus for a pair of neurons A and B within the same ganglion, if we apply a hyperpolarization current in A, and observe hyperpolarization in B, this is usually sufficient to demonstrate electrical coupling between A and B.

However, the situation may be more complex for a projection neuron A and a remote neuron B when their somata are located in the two distinct ganglia (particularly when the two ganglia are separated by a long connective). In this situation, if we hyperpolarize the soma of neuron A, and do not observe a hyperpolarization in the soma of B, we cannot conclude that there is no electrical coupling. The site of electrical synaptic contact may be distant, so that hyperpolarization in A may be negligible when it reaches the site of synaptic contact in the remote ganglion. In this case, other evidence can be sought. The most direct, convincing demonstration of electrical coupling between the two cells would be to record from the remote axon of A and the soma of B in the same ganglion, and perform the hyperpolarization experiment. This was indeed how electrical synapses from B34 to C15/16/17 were demonstrated (Jing et al., 2011). If axonal recording proves to be too difficult or impossible, other experimental proof can be obtained. We can perfuse Ca^{2+}-free saline, which would eliminate a chemical excitatory synaptic connection, but would not block an electrical synapse. In addition, if the fast excitatory neurotransmitter of the neuron A is known, we can use known pharmacological antagonists of the receptor. Using 0 Ca^{2+} saline and receptor antagonists, if the synaptic potential is blocked completely, this would suggest the presence of a chemical synapse. If the block is partial or there is no effect at all, this would suggest the presence of an electrical component or a pure electrical synapse. Indeed, we have realized that even in a gastropod molluscan system such as *Aplysia*, electrical synapses from projection interneurons to remote circuit elements might be overlooked. Thus we suggest that at a minimum, all excitatory connections from a projection interneuron to its remote targets should be tested with Ca^{2+}-free saline to determine whether the connections are chemical or electrical.

2.3 Summary of Electrical Coupling in the *Aplysia* Feeding Circuit

Electrical coupling is involved in all aspects of communication between circuit elements in the *Aplysia* feeding circuit. Within this circuit, electrical coupling fulfills various functions, often acting in conjunction with chemical connections and intrinsic properties of circuit elements. (1) In rhythm generation, electrical coupling plays a major role in promoting synchronous or similar patterns of activity. Among homologous neurons, which fire similarly, electrical coupling is often the only functional connection observed. This is true for ipsilateral motoneurons (eg, between B31 and B32), and interneurons (between B30A and B30B, CBI-8 and CBI-9). For neurons with contralateral axons that connect bilaterally located homologous cells, most connections are electrical (eg, interneuron B34 and contralateral B34) (Fig. 5.2B). Notably, the contralateral coupling could promote bilateral synchronization. Among heterogeneous neurons, electrical coupling is prominently observed for those neurons that fire in the same phase, and is largely absent between neurons that fire in different phases. (2) For heterogeneous neurons that fire in a single phase, electrical coupling can define interneuronal subgroups with each subgroup serving a similar function that is distinct from that of other subgroups. This helps the network to achieve multifunctionality. This is accomplished in two ways. First, electrical coupling is restricted to specific subgroups, with each subgroup serving a specific function. Importantly, the coupled neurons often serve as a building block or a module that controls specific motor parameters for a behavior. Second, coupling strengths vary, and neurons with more similar firing patterns and functions are coupled more strongly. Electrical coupling also acts in conjunction with chemical inhibition to generate variable firing patterns within a single phase. Notably, electrical coupling is effective for interneurons to drive motoneurons that remain active in one phase during different programs, but was not effective in driving motoneurons that shift their phases in which they are active during different programs. (3) For neurons that are remotely located, electrical coupling serves an important role in providing excitation for neurons that fire in a similar phase and the neurons are connected through axodendritic coupling. In this case, electrical coupling can work alone or together with chemical excitation.

3. OTHER NEURAL CIRCUITS IN GASTROPOD MOLLUSCS

How relevant or general are the findings concerning the role of electrical coupling in the multifunctional feeding rhythm-generating network of *Aplysia*? A proper first step would be to compare the *Aplysia* feeding network with neural circuits in other molluscs, more specifically, gastropod molluscs. We focus on a few selected rhythm-generating neural circuits where critical circuit elements, including the major CPG elements and/or higher order command/modulatory neurons, have been identified.

3.1 The Feeding Circuits in *Lymnaea* and *Pleurobranchaea*

The feeding networks of *Lymnaea* and *Pleurobranchaea* appear to be most similar to the *Aplysia* feeding network. The network organizations in *Lymnaea* and *Pleurobranchaea* are similar to that of *Aplysia* in that the feeding CPG elements are located in the buccal ganglion, whereas higher order interneurons (CBIs) are located in the cerebral ganglion. The feeding networks of *Helisoma*, *Planorbis*, and other gastropod molluscs (Murphy, 2001) also have a similar organization, but they are not discussed further because connectivity between interneurons in these circuits, especially electrical coupling, is not well described.

In *Lymnaea*, the feeding motor pattern is triphasic, with each cycle consisting of protraction—rasp—swallowing phases (Fig. 5.6A) (Benjamin, 2012). Although they have different names in *Lymnaea*, rasp corresponds to retraction in *Aplysia*, and swallowing corresponds to hyperretraction. Thus overall motor patterns in the two animals are similar with one difference: in *Lymnaea*, hyperretraction always occurs, hence a triphasic pattern, whereas in *Aplysia*, hyperretraction occurs only in a subset of programs. This suggests that *Lymnaea* motor patterns appear to display less variability. In fact, the lesser degree of variability in *Lymnaea* is also manifested in another important difference between the two animals, ie, *Lymnaea* do not appear to produce multiple types of motor patterns, whereas *Aplysia* do.

In *Lymnaea*, there are three major groups of buccal CPG interneurons mediating each phase; for protraction: N1M and N1L; for retraction (rasp): N2d and N2v; and for hyperretraction (swallowing): N3p, N3t (Fig. 5.6). Note that consistent with a triphasic pattern in *Lymnaea* versus a biphasic pattern in *Aplysia*, retraction interneuron N2d in *Lymnaea* is only active during rasp, not during swallowing phase; whereas retraction interneuron B64 in *Aplysia* is active during both retraction and hyperretraction (if it occurs). Regardless, the overall localizations and role of electrical coupling are similar to *Aplysia*. For example, there is electrical coupling between two retraction (rasp) interneurons N2d and N2v. In addition, many interneurons are separately electrically coupled to motoneurons active in their respective phases. However, compared with *Aplysia*, *Lymnaea* appear to have fewer interneurons. In particular, there are fewer interneurons that are active during the protraction phase. Finally, there do not appear to be any electrical coupling-based subgroups for protraction interneurons. The absence of this network feature may be a major reason for which *Lymnaea* feeding network does not produce multiple types of motor patterns.

In contrast, similar to *Aplysia*, *Pleurobranchaea* produce biphasic motor programs with a protraction—retraction sequence, and do generate multiple types of motor patterns (Croll et al., 1985a). However, little information is available regarding the composition of the buccal CPG in *Pleurobranchaea* (Croll et al., 1985b; Gillette et al., 1982), precluding a useful comparison with *Aplysia*.

In terms of the higher order interneurons in the cerebral ganglion, similar to the CBIs in *Aplysia*, there are also multiple CBIs in *Lymnaea*, and moreover, a subset of them (CV1a, CV1b, CA1) are electrically coupled. Similarly, in *Pleurobranchaea*, PC_P interneurons (Gillette et al., 1978), *Aplysia* CBI-2 homologs, are coupled with two other CBIs: ET_{II} (Kovac et al., 1983), which might be homologs of CBI-11 and/or CBI-3 in *Aplysia*.

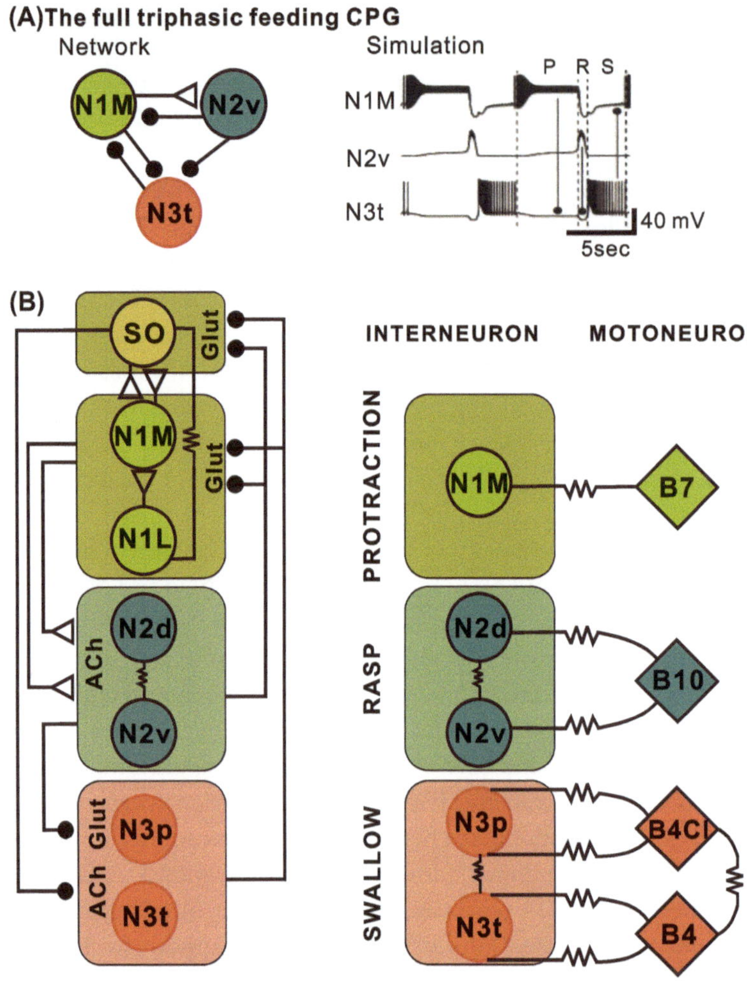

Figure 5.6 Lymnaea *feeding network.* (A) Illustration of connections of three major CPG interneurons, Protraction (P) neuron, N1M, Rasp (R) neuron, N2v, and swallowing (S) neuron (N3t) (left panel), and their activity during the triphasic motor pattern in a computer simulation (right panel). (B) A complete diagram of the connectivity of the feeding CPG in the buccal ganglion. The cells that fire in the same phase are shown in the same color. Note electrical coupling between CPG interneurons (eg, N2d and N2v), and extensive electrical coupling between CPG interneurons with the motoneurons they control. Known transmitters for chemical synapses are indicated: glutamate (Glut), acetylcholine (ACh). *(Adapted from Benjamin, P.R., 2012. Distributed network organization underlying feeding behavior in the mollusk Lymnaea. Neural Syst. Circuits 2, 4. Symbols as in Fig. 5.3.)*

As in *Aplysia*, some CBIs in *Lymnaea* provide direct chemical excitatory inputs to N1M, the buccal protraction CPG interneuron, to drive/modulate motor patterns; but so far, no electrical connection has been reported.

3.2 Locomotor Networks in Gastropod Molluscs

We consider locomotor behaviors in *Tritonia*, *Pleurobranchaea*, *Aplysia*, *Clione*, *Hermissenda*, *Melibe*, and *Dendronotus*, because some information about their locomotor networks is available. The primary form of locomotion in benthic gastropod molluscs is crawling (Newcomb et al., 2012), eg, mucociliary locomotion in *Tritonia*, *Pleurobranchaea*, and *Hermissenda*; muscular crawling (pedal wave) in *Aplysia californica*. Pelagic species, like *Clione*, locomote by wing flapping (swimming). The locomotor motor patterns are rhythmic (with the possible exception of mucociliary crawling in *Tritonia*, *Pleurobranchaea*, and *Hermissenda*). Another characteristic of locomotor motor patterns is that, although the motor patterns do vary in speed, usually, there do not appear to be multiple types of motor patterns.

Despite the various forms of locomotion in different animals, the underlying networks for *Tritonia*, *Pleurobranchaea*, *Aplysia*, and *Hermissenda* crawling and *Clione* swimming share a similar organization with the *Aplysia* feeding network: ie, the locomotor CPG is located in the pedal ganglion that innervates locomotor structures, the foot and body wall, whereas higher order neurons are located in the cerebral ganglion. The *Clione* swimming CPG in the pedal ganglion (Arshavsky et al., 1998; Orlovsky et al., 1999) is the most extensively characterized. Each swimming cycle consists of two phases: a dorsal wing flexion (D-phase) and a ventral wing flexion (V-phase). These movements are produced by alternating contractions of antagonistic wing muscles. Electrical coupling plays an important role in the CPG (Fig. 5.7A). D-phase and V-phase are controlled by two separate groups of motoneurons, ie, group 1/3 versus group 2/4, respectively, with each group electrically coupled. In turn, these motoneurons are excited, partly through electrical coupling, by interneurons group 7 or group 8d/e, respectively. Group 7 and group 8d/e neurons are themselves electrically coupled within their groups. Alternating D-phase and V-phase activity is mediated by reciprocal inhibition between group 7 and group 8d/e neurons (Arshavsky et al., 1998; Satterlie, 1985). Little information is available about the composition of the other locomotor CPGs, but work in *Aplysia* (Yu et al., 2015) suggests that electrical coupling in the pedal ganglion may play a prominent role in generating pedal waves, ie, neurons that are active in a similar phase tend to be electrically coupled, whereas those that are active in different phases are not coupled.

One type of higher order neurons, ie, serotonergic neurons in the cerebral ganglion that project their axons to the pedal ganglion where the locomotor CPG is located have been identified in all five species, including DSI-A/B/C *Tritonia* (Getting, 1981; Popescu

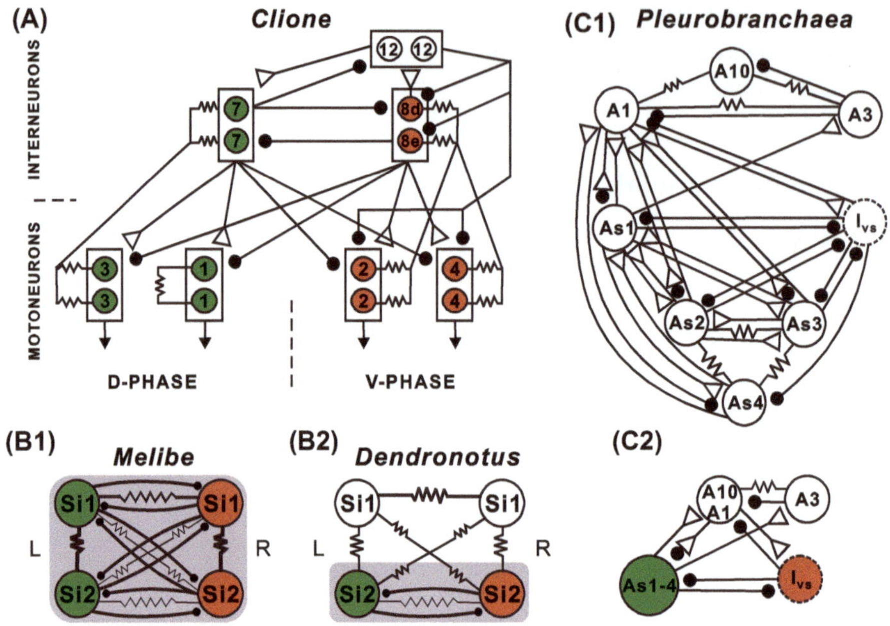

Figure 5.7 ***CPGs for locomotion and swimming in gastropod molluscs.*** All these rhythmic behaviors are biphasic (shown in green vs. red). (A) The locomotor CPG in *Clione* that mediates dorsal (D-phase) and ventral (V-phase) wing flapping. Alternating activity in D-phase (green) and V-phase (red) neurons are mediated by reciprocal inhibition between interneurons 7 and interneurons 8d/e. Note extensive electrical coupling within CPG interneurons, motoneurons, and between interneurons and motoneurons. *(Adapted from Arshavsky, Y.I., Deliagina, T.G., Orlovsky, G.N., Panchin, Y.V., Popova, L.B., Sadreyev, R.I., 1998. Analysis of the central pattern generator for swimming in the mollusk Clione. Ann. N. Y. Acad. Sci. 860, 51–69.).* (B) The swimming CPGs in *Melibe* (B1) and *Dendronotus* (B2) that mediate left (green) and right (red) body flexion swimming. CPG interneurons are shaded in light blue. Left and right flexions in *Melibe* are mediated by reciprocal inhibition between bilateral Si1/Si2 interneurons, whereas left and right flexions in *Dendronotus* are mediated by reciprocal inhibition only between bilateral Si2 interneurons. Note extensive electrical coupling. Although there exists electrical coupling between bilateral Si1 and Si2 in *Melibe* and bilateral Si2 in *Dendronotus*, the coupling is weak. Line thickness of resistance symbols indicates strength of electrical coupling. *(Adapted from Newcomb, J.M., Sakurai, A., Lillvis, J.L., Gunaratne, C.A., Katz, P.S., 2012. Homology and homoplasy of swimming behaviors and neural circuits in the Nudipleura (Mollusca, Gastropoda, Opisthobranchia). Proc. Natl. Acad. Sci. U.S.A., 109 (Suppl. 1), 10669–10676.).* (C) The swimming CPG in *Pleurobranchaea* that mediates dorsal and ventral body flexion escape swimming. (C1) Ipsilateral connections of the CPG interneurons. (C2) A simplified diagram of the CPG circuit. Dorsal (green) and ventral (red) body flexions are mediated by reciprocal inhibition between dorsal flexion interneurons As1–4 and ventral flexion interneuron I_{VS}. Ventral flexion interneuron A3 serves to terminate activity of A10/A1. Note electrical coupling between A1 and A10; and between As2-4. *(Adapted from Jing, J., Gillette, R., 1999. Central pattern generator for escape swimming in the notaspid sea slug* Pleurobranchaea californica. *J. Neurophysiol. 81, 654–667. Symbols as in Fig. 5.3.)*

and Frost, 2002), CPB1 and CPA1 neurons in *Clione* (Panchin et al., 1995; Satterlie and Norekian, 1995), As1/2/3/4 in *Pleurobranchaea* (Jing and Gillette, 1999, 2000, 2003), CC9/10 in *Aplysia* (Jing et al., 2008), and three CPTs in *Hermissenda* (Tian et al., 2006). Except that CPB1 and CPA1 in *Clione* are two groups of neurons (each with more than four members), DSI-B/C, As2/3/4, CC9/10, and CPTs consist of two or three similar single neurons, and are electrically coupled in each species. Most notably, As2-4 in *Pleurobanchaea* are also electrically coupled to A4, an interneuron commanding avoidance turns (Jing and Gillette, 2003), indicating that electrical connections are not restricted to single behavioral networks. It is unknown if there is electrical coupling within or between CPB1 and CPA1 interneurons. Regardless, perhaps with the exception of the CPTs in *Hermissenda*, all these serotonergic neurons provide primarily excitatory input to motoneurons, modulatory neurons, and possibly CPG interneurons in the pedal ganglion to drive and/or modulate locomotor patterns.

3.3 Swimming Networks in Benthic Gastropod Molluscs

A limited number of benthic gastropod molluscs, eg, *Tritonia*, *Pleurobranchaea*, *Melibe*, and *Dendronotus*, occasionally swim (Newcomb et al., 2012). Swimming motor patterns in these animals are all biphasic, but details of swimming differ. *Melibe* and *Dendronotus* swim through left and right body flexion (L-R swimming), whereas *Tritonia* and *Pleurobranchaea* swim through dorsal and ventral flexion (D-R swimming).

The network organization of *Melibe* and *Dendronotus* swimming CPGs differs somewhat from the *Aplysia* feeding network and the locomotor networks discussed earlier (Section 3.2) in that these swimming CPG interneurons are not restricted to the pedal ganglia. For *Melibe* (Sakurai et al., 2011; Thompson and Watson, 2005), there are at least two CPG interneurons, Si1 in the pleural ganglion, and Si2 in the pedal ganglion (Fig. 5.7B). Left and right Si1/Si2 mutually inhibit each other, thereby generating alternating L-R flexion. Electrical coupling is critical for promoting activity within a single phase, as ipsilateral Si1 and Si2, which fire in the same phase, are strongly coupled (ratio: ~0.18). Interestingly, however, Si1/Si2 and their contralateral homologs, which fire in opposite phases, are also electrically coupled, but the coupling is quite weak (<0.03). In the case of *Dendronotus*, this animal appears to use only mutually inhibited left and right Si2 to generate L-R flexion, despite the fact that Si1 is present in *Dendronotus*. There is also electrical coupling between contralateral Si2, but again this coupling is quite weak (<0.02). In contrast, coupling between the nonrhythmic contralateral Si1 is higher (~0.06).

The network organization of *Tritonia* and *Pleurobranchaea* swimming networks differs even more dramatically from *Aplysia* feeding network and the locomotor networks of gastropod molluscs in that these swimming CPGs are not located in the pedal ganglia, but are located in the cerebral ganglion. In fact, the swimming CPGs in *Tritonia* and *Pleurobranchaea* utilize cerebral serotonergic interneurons DSI-A/B/C and As1/2/3/4 that drive crawling CPG in the pedal ganglion and, therefore, function as higher order interneurons in crawling (Fig. 5.7C). Thus DSI-A/B/C and As1/2/3/4 represent a dramatic example of multifunctional neurons in that these neurons contribute to multiple behaviors, but their contributions are different in different behaviors (Jing et al., 2009; Popescu and Frost, 2002), ie, they are higher order interneurons for crawling, but act as CPG interneurons for swimming. Other swimming CPG interneurons, like C2, VSIs in *Tritonia*, and A1, A3, and A10 in *Pleurobranchaea*, are also cerebral interneurons that project to the pedal ganglion. In fact, rather than functioning as a mode of locomotion as in *Melibe* and *Dendronotus*, swimming in *Tritonia* and *Pleurobranchaea* are primarily used to escape from predators or noxious stimuli. Thus it was suggested (Katz et al., 2001) that the escape swimming CPG in *Tritonia* and *Pleurobranchaea* arose, at least in part, from a nonrhythmic neuromodulatory arousal system that involves higher order neurons in the cerebral ganglion (Jing and Gillette, 2000; Jing et al., 2009). Notably, higher order neurons also mediate escape behaviors in other invertebrate animals, including crayfish (Edwards et al., 1999), *Drosophila* (Phelan et al. 2017 (Chapter 2)).

Regardless of the evolutionary origin of these CPG interneurons, electrical coupling plays an important role in the operation of *Tritonia* and *Pleurobranchaea* swimming CPGs in that electrical coupling is observed most commonly in cells active in one phase, eg, DSI-B/C in *Tritonia* and As2–4 in *Pleurobranchaea*, as mentioned earlier, and A1 and A10 in *Pleurobranchaea* (Fig. 5.7C) (Getting, 1981; Jing and Gillette, 1999).

3.4 Variations of Electrical Coupling Strength Between Homologous Neurons in Different Species: Implications for Evolution

One important conclusion based on molluscan work is that one needs to measure coupling ratios whenever electrical coupling is present. If the coupling between two neurons is asymmetric, coupling ratio should also be measured using both hyperpolarization and depolarization currents in presynaptic neurons to determine whether the asymmetric coupling is due to rectifying synapses or intrinsic properties of the neurons (see Section 2.1.3) (Fig. 5.4A). Coupling ratios are important for understanding circuit mechanisms for a specific neural circuit in a single animal (Getting, 1981; Jing and Gillette, 1999; Sakurai et al., 2011; Sasaki et al., 2013). In addition, they are helpful to study plasticity of electrical coupling that may play an important role in learning (Sieling et al., 2014).

Another important reason for quantifying the coupling ratio is that it is a quantitative parameter that can be conveniently used to compare roles of electrical coupling between homologous neurons in different species. Such a comparison has been made between *Tritonia* and *Pleurobranchaea* swimming CPGs (Jing and Gillette, 1999), as well as the swimming CPGs in *Melibe* and *Dendronotus* (Sakurai et al., 2011). Importantly, differences in coupling ratios could explain, at least partially, some difference in network activity in these animals, eg, the coupling between DSI-B/C (0.19) in *Tritonia* is stronger than that between As2–3 (0.09) in *Pleurobranchaea*, which could account for the higher frequency firing of DSI-B/C than As2–3 during swimming. Overall, the work provided important clues as to how the same or different behaviors can be generated by homologous neurons, or how the same behaviors can be generated by nonhomologous neurons (Newcomb et al., 2012). Ultimately, more extensive studies along these lines may shed more light on how circuits and behaviors have evolved.

4. SUMMARY

In this chapter, we reviewed the role of electrical coupling in rhythm-generating networks responsible for feeding and locomotion in selected molluscan model systems. The evidence in *Aplysia* and other gastropod molluscs support the general conclusion that electrical coupling is distributed in diverse but specific sites in neural circuits.

(1) For rhythm generation, electrical coupling acts to promote coordinated activity for neurons that are active in the same phase. A similar pattern is observed in all molluscan circuits for electrical coupling between ipsilateral neurons of the same types (eg, cerebral serotonergic interneurons in *Tritonia*, *Pleurobranchaea*, and *Aplysia*) or contralateral homologues. Perhaps the only paradox is the electrical coupling between contralateral Si1 and/or Si2 homologues in *Melibe* and *Dendronotus*, despite the fact that the contralateral homologues actually fire in antiphase. In this case, however, the coupling is quite weak, and presumably does not overcome mutual inhibition between the contralateral homologues that enables their alternating firing. In the case of heterogeneous cells, electrical coupling is most prominently observed in cells that fire in the same phase. This includes coupling within interneurons (eg, N2v and N2d in *Lymnaea*, group 7 or group 8 interneurons in *Clione*), or within motoneurons, and between interneurons and motoneurons in *Aplysia* and *Lymnaea* feeding, and *Clione* swimming networks.

(2) There is no evidence yet in the feeding networks of *Lymnaea* or *Plerobranchaea* of the existence of electrically coupled functional subgroups of neurons that fire in the same phase of motor programs, which in *Aplysia* function as modules that are responsible for the multifunctionality of the feeding network. We believe that there are explanations for both cases: in *Lymnaea*, it is likely because *Lymnaea* feeding circuit apparently does not produce multiple types of motor patterns; whereas in *Pleurobranchaea*, it is because the feeding CPG has not been well described. We believe that the multifunctional CPGs in *Pleurobranchaea* and other related species will be a good testing ground and we do expect a similar organization like *Aplysia* feeding will be discovered.

(3) For the role of electrical coupling in coordinating activity between remote circuit elements, there are also few other cases in the selected molluscan circuits that can add to the *Aplysia* cases. However, projection neurons that utilize pure electrical synapses or mixed chemical and electrical synapses to drive their remote targets have been well described in *Drosophila* (Phelan et al., 2017 (Chapter 2)), crayfish (Edwards et al., 1999), crab (Coleman et al., 1995), and reticulospinal neurons in lamprey

(Rovainen, 1979) and fish (Mathner cells) (Korn and Faber, 2005; Pereda, 2014). Thus we do expect that similar cases will be found in more molluscan networks as well as in other model systems.

In summary, circuitry work in gastropod molluscan model systems as well as other invertebrate model systems including *Caenorhabditis elegans* (Chatzigeorgiou and Schafer, 2011; Kawano et al., 2011; Rabinowitch et al., 2014), leech (Kristan et al., 2005), crustacean (Edwards et al., 1999; Furshpan and Potter, 1959; Marder and Bucher, 2007; Nadim and Bucher, 2014), and *Drosophila* (Phelan et al., 2008; Wu et al., 2011), some of which have been described in other chapters of this book, has provided novel insights and greatly advanced our knowledge of the role of electrical coupling in multifunctional rhythm-generating networks. Understanding the specific roles of electrical coupling also helps define functional roles of other factors, ie, chemical connections and intrinsic properties of circuit elements. We expect that future work in these model systems will continue to provide essential data that will help formulate general principles for understanding the synaptic basis of circuit operations.

FUTURE DIRECTIONS

- Although important progress has been made, more work is needed to gain a complete understanding of the role of electrical coupling in circuit function. One important lesson from this review is that it is important to measure electrical coupling ratio and define the nature of asymmetric electrical synapses. Thus whenever electrical coupling is present, it is critical to document electrical coupling ratios as much as possible, as they provide a quantitative basis for understanding circuit and evolutionary mechanisms.
- In particular, more extended work in *Aplysia* should clarify roles of electrical coupling used by higher order interneurons (CBIs) to drive/modulate the buccal CPG, as well as feedback from BCIs to cerebral neurons. Similar circuit work in *Lymnaea* and *Pleurobranchaea* feeding networks is also needed, especially considering that the *Pleurobranchaea* feeding network is a good example of a multifunctional rhythm-generating network. The findings of electrical coupling–based separation of functional subgroups for neurons (ie, modules) that fire within a single phase should be verified in other neural circuits, eg, in *Pleurobranchaea.*
- The role of electrical coupling in the locomotor CPGs of the pedal ganglia for *Aplysia* and other molluscs, eg, *Tritonia*, *Pleurobranchaea*, and *Lymnaea*, should be studied more extensively.
- Because many neural networks in other invertebrates, including *C. elegans*, *Drosophila*, crustacean, and leech, and in vertebrates, including zebra fish, turtle, and rodents, are also multifunctional rhythm-generating networks, it would be of interest to determine whether and how electrical coupling may define functional subtypes (ie, modules) in

these animal model systems as well. Given the growing interest in defining functional cell types in contemporary neuroscience research, we expect that future studies of electrical coupling in model systems shall help further advance this line of research.

ACKNOWLEDGMENTS

A number of mentors, colleagues, and students of the first author, eg, Dr. Rhanor Gillette, Dr. Kosei Sasaki, Mr. Song-an Chen, Mr. Ke Yu, Ms. Ting-ting Chen, and Ms. Dan-dan Liu, contributed significantly to the research described here. Mr. Song-an Chen and Mr. Ke Yu also helped in making illustrations. We also thank Drs. Paul Benjamin and Ildiko Kemenes for providing original figures adapted as Fig. 5.6. Our research was supported by the grants from the National Natural Science Foundation of China (31371104), and the National Institutes of Health (NS066587, NS070583, and MH051393) of the United States of America.

REFERENCES

Arshavsky, Y.I., Deliagina, T.G., Orlovsky, G.N., Panchin, Y.V., Popova, L.B., Sadreyev, R.I., 1998. Analysis of the central pattern generator for swimming in the mollusk *Clione*. Ann. N. Y. Acad. Sci. 860, 51–69.

Benjamin, P.R., 2012. Distributed network organization underlying feeding behavior in the mollusk *Lymnaea*. Neural Syst. Circuits 2, 4.

Bennett, M.V., 1966. Physiology of electrotonic junctions. Ann. N. Y. Acad. Sci. 137 (2), 509–539.

Bizzi, E., Mussa-Ivaldi, F.A., 2004. Toward a neurobiology of coordinate transformation. In: Gazzaniga, M.S. (Ed.), The Cognitive Neuroscience III. MIT Press, Cambridge, MA, pp. 413–425.

Bizzi, E., Mussa-Ivaldi, F.A., Giszter, S., 1991. Computations underlying the execution of movement: a biological perspective. Science 253 (5017), 287–291.

Briggman, K.L., Kristan, W.B., 2008. Multifunctional pattern-generating circuits. Annu. Rev. Neurosci. 31, 271–294.

Chatzigeorgiou, M., Schafer, W.R., 2011. Lateral facilitation between primary mechanosensory neurons controls nose touch perception in *C. elegans*. Neuron 70 (2), 299–309.

Chen, S.-A., Chen, T.-T., Yu, K., Liu, D.-D., Vilim, F.S., Cropper, E.C., et al., 2015. Identification of specific targets of a command-like interneuron reveals novel roles of pattern-generating interneurons in *Aplysia* feeding network. In: Paper Presented at the Program No. 420.29, 2015 Neuroscience Meeting Planner.

Coleman, M.J., Meyrand, P., Nusbaum, M.P., 1995. A switch between two modes of synaptic transmission mediated by presynaptic inhibition. Nature 378 (6556), 502–505.

Croll, R.P., Davis, W.J., Kovac, M.P., 1985a. Neural mechanisms of motor program switching in the mollusc *Pleurobranchaea*. I. Central motor programs underlying ingestion, egestion, and the "neutral" rhythm(s). J. Neurosci. 5, 48–55.

Croll, R.P., Davis, W.J., Kovac, M.P., 1985b. Neural mechanisms of motor program switching in the mollusc *Pleurobranchaea*. II. Role of the ventral white cell, anterior ventral, and B3 buccal neurons. J. Neurosci. 5, 56–63.

Cropper, E.C., Evans, C.G., Hurwitz, I., Jing, J., Proekt, A., Romero, A., et al., 2004. Feeding neural networks in the mollusc *Aplysia*. Neurosignals 13 (1–2), 70–86.

Dacks, A.M., Weiss, K.R., 2013. Release of a single neurotransmitter from an identified interneuron coherently affects motor output on multiple time scales. J. Neurophysiol. 109 (9), 2327–2334.

Deisseroth, K., 2015. Optogenetics: 10 years of microbial opsins in neuroscience. Nat. Neurosci. 18 (9), 1213–1225.

Edwards, D.H., Heitler, W.J., Krasne, F.B., 1999. Fifty years of a command neuron: the neurobiology of escape behavior in the crayfish. Trends Neurosci. 22 (4), 153–161.

Evans, C.G., Cropper, E.C., 1998. Proprioceptive input to feeding motor programs in *Aplysia*. J. Neurosci. 18 (19), 8016–8031.

Evans, C.G., Rosen, S., Kupfermann, I., Weiss, K.R., Cropper, E.C., 1996. Characterization of a radula opener neuromuscular system in *Aplysia*. J. Neurophysiol. 76 (2), 1267–1281.

Friedman, A., Zhurov, Y., Ludwar, B., Weiss, K.R., 2009. Motor outputs in a multi-tasking network: relative contributions of inputs and experience-dependent states. J. Neurophysiol. 102, 3711–3727.

Furshpan, E.J., Potter, D.D., 1959. Transmission at the giant motor synapses of the crayfish. J. Physiol. 145 (2), 289–325.

Gardner, D., 1971. Bilateral symmetry and interneuronal organization in the buccal ganglia of *Aplysia*. Science 173 (996), 550–553.

Getting, P.A., 1981. Mechanisms of pattern generation underlying swimming in *Tritonia*. I. Neuronal network formed by monosynaptic connections. J. Neurophysiol. 46, 65–79.

Gillette, R., Kovac, M.P., Davis, W.J., 1978. Command neurons in *Pleurobranchaea* receive synaptic feedback from the motor network they excite. Science 199, 798–801.

Gillette, R., Gillette, M.U., Davis, W.J., 1982. Substrates of command ability in a buccal neuron of *Pleurobranchaea*. I. Mechanisms of action potential broadening. J. Comp. Physiol. 146, 449–459.

Grillner, S., 1981. Control of locomotion in bipeds, tetrapods, and fish. In: Brooks, V. (Ed.), Handbook of Physiology, Sec. 1, vol. 2. American Physiological Society, Bethesda, MD, pp. 1179–1236.

Grillner, S., 1985. Neurobiological bases of rhythmic motor acts in vertebrates. Science 228 (4696), 143–149.

Grillner, S., 2006. Biological pattern generation: the cellular and computational logic of networks in motion. Neuron 52 (5), 751–766.

Hart, C.B., Giszter, S.F., 2004. Modular premotor drives and unit bursts as primitives for frog motor behaviors. J. Neurosci. 24 (22), 5269–5282.

He, S., Shi, S.-H., 2017. Lineage-dependent electrical synapse formation in the mammalian neocortex. In: Jing, J. (Ed.), Network Functions and Plasticity. Elsevier (Chapter 14).

Hurwitz, I., Susswein, A.J., 1996. B64, a newly identified central pattern generator element producing a phase switch from protraction to retraction in buccal motor programs of *Aplysia californica*. J. Neurophysiol. 75 (4), 1327–1344.

Hurwitz, I., Kupfermann, I., Susswein, A.J., 1997. Different roles of neurons B63 and B34 that are active during the protraction phase of buccal motor programs in *Aplysia californica*. J. Neurophysiol. 78 (3), 1305–1319.

Hurwitz, I., Perrins, R., Xin, Y., Weiss, K.R., Kupfermann, I., 1999. C-PR neuron of *Aplysia* has differential effects on "Feeding" cerebral interneurons, including myomodulin-positive CBI-12. J. Neurophysiol. 81 (2), 521–534.

Hurwitz, I., Kupfermann, I., Weiss, K.R., 2003. Fast synaptic connections from CBIs to pattern-generating neurons in *Aplysia*: initiation and modification of motor programs. J. Neurophysiol. 89 (4), 2120–2136.

Jarrell, T.A., Wang, Y., Bloniarz, A.E., Brittin, C.A., Xu, M., Thomson, J.N., et al., 2012. The connectome of a decision-making neural network. Science 337 (6093), 437–444.

Jing, J., Gillette, R., 1999. Central pattern generator for escape swimming in the notaspid sea slug *Pleurobranchaea californica*. J. Neurophysiol. 81, 654–667.

Jing, J., Gillette, R., 2000. Escape swim network interneurons have diverse roles in behavioral switching and putative arousal in *Pleurobranchaea*. J. Neurophysiol. 83, 1346–1355.

Jing, J., Gillette, R., 2003. Directional avoidance turns encoded by single interneurons and sustained by multifunctional serotonergic cells. J. Neurosci. 23 (7), 3039–3051.

Jing, J., Weiss, K.R., 2001. Neural mechanisms of motor program switching in *Aplysia*. J. Neurosci. 21, 7349–7362.

Jing, J., Weiss, K.R., 2002. Interneuronal basis of the generation of related but distinct motor programs in *Aplysia*: implications for current neuronal models of vertebrate intralimb coordination. J. Neurosci. 22, 6228–6238.

Jing, J., Weiss, K.R., 2005. Generation of variants of a motor act in a modular and hierarchical motor network. Curr. Biol. 15, 1712–1721.

Jing, J., Vilim, F.S., Wu, J.S., Park, J.H., Weiss, K.R., 2003. Concerted GABAergic actions of *Aplysia* feeding interneurons in motor program specification. J. Neurosci. 23 (12), 5283–5294.

Jing, J., Cropper, E.C., Hurwitz, I., Weiss, K.R., 2004. The construction of movement with behavior-specific and behavior-independent modules. J. Neurosci. 24 (12), 6315–6325.

Jing, J., Vilim, F.S., Cropper, E.C., Weiss, K.R., 2008. Neural analog of arousal: persistent conditional activation of a feeding modulator by serotonergic initiators of locomotion. J. Neurosci. 28 (47), 12349–12361.
Jing, J., Gillette, R., Weiss, K.R., 2009. Evolving concepts of arousal: insights from simple model systems. Rev. Neurosci. 20, 405–427.
Jing, J., Sasaki, K., Perkins, M.H., Siniscalchi, M.J., Ludwar, B.C., Cropper, E.C., et al., 2011. Coordination of distinct motor structures through remote axonal coupling of projection interneurons. J. Neurosci. 31 (43), 15438–15449.
Jing, J., Alexeeva, V., Chen, S.A., Yu, K., Due, M.R., Tan, L.N., et al., 2015. Functional characterization of a vesicular glutamate transporter in an interneuron that makes excitatory and inhibitory synaptic connections in a molluscan neural circuit. J. Neurosci. 35 (24), 9137–9149.
Johnson, B.R., Peck, J.H., Harris-Warrick, R.M., 1993. Amine modulation of electrical coupling in the pyloric network of the lobster stomatogastric ganglion. J. Comp. Physiol. A 172 (6), 715–732.
Kabotyanski, E.A., Baxter, D.A., Byrne, J.H., 1998. Identification and characterization of catecholaminergic neuron B65, which initiates and modifies patterned activity in the buccal ganglia of *Aplysia*. J. Neurophysiol. 79 (2), 605–621.
Katz, P.S., Fickbohm, D.J., Lynn-Bullock, C.P., 2001. Evidence that the central pattern generator for swimming in *Tritonia* arose from a non-rhythmic neuromodulatory arousal system: implications for the evolution of specialized behavior. Am. Zool. 41 (4), 962–975.
Kawano, T., Po, M.D., Gao, S., Leung, G., Ryu, W.S., Zhen, M., 2011. An imbalancing act: gap junctions reduce the backward motor circuit activity to bias *C. elegans* for forward locomotion. Neuron 72 (4), 572–586.
Korn, H., Faber, D.S., 2005. The Mauthner cell half a century later: a neurobiological model for decision-making? Neuron 47 (1), 13–28.
Kovac, M.P., Davis, W.J., Matera, E.M., Croll, R.P., 1983. Organization of synaptic inputs to paracerebral feeding command interneurons of *Pleurobranchaea californica*. I. Excitatory inputs. J. Neurophysiol. 49, 1517–1538.
Kristan Jr., W.B., Calabrese, R.L., Friesen, W.O., 2005. Neuronal control of leech behavior. Prog. Neurobiol. 76 (5), 279–327.
Kupfermann, I., 1974. Feeding behavior in *Aplysia*: a simple system for the study of motivation. Behav. Biol. 10, 1–26.
Li, W.-C., Rekling, J.C., 2017. Electrical coupling in the generation of vertebrate motor rhythms. In: Jing, J. (Ed.), Network Functions and Plasticity. Elsevier (Chapter 11).
Lu, H., McManus, J.M., Cullins, M.J., Chiel, H.J., 2015. Preparing the periphery for a subsequent behavior: motor neuronal activity during biting generates little force but prepares a retractor muscle to generate larger forces during swallowing in *Aplysia*. J. Neurosci. 35 (12), 5051–5066.
Marder, E., Bucher, D., 2007. Understanding circuit dynamics using the stomatogastric nervous system of lobsters and crabs. Annu. Rev. Physiol. 69, 291–316.
Marder, E., Calabrese, R.L., 1996. Principles of rhythmic motor pattern generation. Physiol. Rev. 76 (3), 687–717.
Marder, E., O'Leary, T., Shruti, S., 2014. Neuromodulation of circuits with variable parameters: single neurons and small circuits reveal principles of state-dependent and robust neuromodulation. Annu. Rev. Neurosci. 37, 329–346.
Marder, E., 2009. Electrical synapses: rectification demystified. Curr. Biol. 19 (1), R34–R35.
Morgan, P.T., Jing, J., Vilim, F.S., Weiss, K.R., 2002. Interneuronal and peptidergic control of motor pattern switching in *Aplysia*. J. Neurophysiol. 87, 49–61.
Morton, D.W., Chiel, H.J., 1993a. In vivo buccal nerve activity that distinguishes ingestion from rejection can be used to predict behavioral transitions in *Aplysia*. J. Comp. Physiol. A 172, 17–32.
Morton, D.W., Chiel, H.J., 1993b. The timing of activity in motor neurons that produce radula movements distinguishes ingestion from rejection in *Aplysia*. J. Comp. Physiol. A 173, 519–536.
Murphy, A.D., 2001. The neuronal basis of feeding in the snail, *Helisoma*, with comparisons to selected gastropods. Prog. Neurobiol. 63 (4), 383–408.
Nadim, F., Bucher, D., 2014. Neuromodulation of neurons and synapses. Curr. Opin. Neurobiol. 29C, 48–56.
Nadim, F., Li, X., Gray, M., Golowasch, J., 2017. The role of electrical coupling in rhythm generation in small networks. In: Jing, J. (Ed.), Network Functions and Plasticity. Elsevier (Chapter 4).

Nargeot, R., Bédécarrats, A., 2017. Electrical synapses and learning–induced plasticity in motor rhythmogenesis. In: Jing, J. (Ed.), Network Functions and Plasticity. Elsevier (Chapter 6).

Nargeot, R., Simmers, J., 2012. Functional organization and adaptability of a decision-making network in *Aplysia*. Front. Neurosci. 6, 113.

Nargeot, R., Baxter, D.A., Byrne, J.H., 1999. In vitro analog of operant conditioning in *Aplysia*. I. Contingent reinforcement modifies the functional dynamics of an identified neuron. J. Neurosci. 19 (6), 2247–2260.

Newcomb, J.M., Sakurai, A., Lillvis, J.L., Gunaratne, C.A., Katz, P.S., 2012. Homology and homoplasy of swimming behaviors and neural circuits in the Nudipleura (Mollusca, Gastropoda, Opisthobranchia). Proc. Natl. Acad. Sci. U.S.A. 109 (Suppl. 1), 10669–10676.

Orlovsky, G.N., Deliagina, T.G., Grillner, S., 1999. Neural Control of Locomotion. Oxford University Press, New York, NY.

Panchin, Y.V., Popova, L.B., Deliagina, T.G., Orlovsky, G.N., Arshavsky, Y.I., 1995. Control of locomotion in marine mollusk *Clione limacina*. VIII. Cerebropedal neurons. J. Neurophysiol. 73 (5), 1912–1923.

Pereda, A.E., 2014. Electrical synapses and their functional interactions with chemical synapses. Nat. Rev. Neurosci. 15 (4), 250–263.

Perrins, R., Weiss, K.R., 1996. A cerebral central pattern generator in *Aplysia* and its connections with buccal feeding circuitry. J. Neurosci. 16 (21), 7030–7045.

Perrins, R., Weiss, K.R., 1998. Compartmentalization of information processing in an *Aplysia* feeding circuit interneuron through membrane properties and synaptic interactions. J. Neurosci. 18 (10), 3977–3989.

Phelan, P., Bacon, J.P., Blagburn, J.M., 2017. Neural circuits underlying escape behavior in Drosophila: focus on electrical signaling. In: Jing, J. (Ed.), Network Functions and Plasticity. Elsevier (Chapter 2).

Phelan, P., Goulding, L.A., Tam, J.L., Allen, M.J., Dawber, R.J., Davies, J.A., et al., 2008. Molecular mechanism of rectification at identified electrical synapses in the *Drosophila* giant fiber system. Curr. Biol. 18 (24), 1955–1960.

Plummer, M.R., Kirk, M.D., 1990. Premotor neurons B51 and B52 in the buccal ganglia of *Aplysia californica*: synaptic connections, effects on ongoing motor rhythms, and peptide modulation. J. Neurophysiol. 63 (3), 539–558.

Popescu, I.R., Frost, W.N., 2002. Highly dissimilar behaviors mediated by a multifunctional network in the marine mollusk *Tritonia diomedea*. J. Neurosci. 22 (5), 1985–1993.

Rabinowitch, I., Chatzigeorgiou, M., Zhao, B., Treinin, M., Schafer, W.R., 2014. Rewiring neural circuits by the insertion of ectopic electrical synapses in transgenic *C. elegans*. Nat. Commun. 5, 4442.

Rash, J.E., Curti, S., Vanderpool, K.G., Kamasawa, N., Nannapaneni, S., Palacios-Prado, N., et al., 2013. Molecular and functional asymmetry at a vertebrate electrical synapse. Neuron 79 (5), 957–969.

Rosen, S.C., Weiss, K.R., Cohen, J.L., Kupfermann, I., 1982. Interganglionic cerebral-buccal mechanoafferents of *Aplysia*: receptive fields and synaptic connections to different classes of neurons involved in feeding behavior. J. Neurophysiol. 48, 271–288.

Rosen, S.C., Teyke, T., Miller, M.W., Weiss, K.R., Kupfermann, I., 1991. Identification and characterization of cerebral-to-buccal interneurons implicated in the control of motor programs associated with feeding in *Aplysia*. J. Neurosci. 11 (11), 3630–3655.

Rovainen, C.M., 1979. Neurobiology of lampreys. Physiol. Rev. 59 (4), 1007–1077.

Sakurai, A., Newcomb, J.M., Lillvis, J.L., Katz, P.S., 2011. Different roles for homologous interneurons in species exhibiting similar rhythmic behaviors. Curr. Biol. 21 (12), 1036–1043.

Sasaki, K., Due, M.R., Jing, J., Weiss, K.R., 2007. Feeding CPG in *Aplysia* directly controls two distinct outputs of a compartmentalized interneuron that functions as a CPG element. J. Neurophysiol. 98 (6), 3796–3801.

Sasaki, K., Jing, J., Due, M.R., Weiss, K.R., 2008. An input-representing interneuron regulates spike timing and thereby phase switching in a motor network. J. Neurosci. 28 (8), 1916–1928.

Sasaki, K., Brezina, V., Weiss, K.R., Jing, J., 2009. Distinct inhibitory neurons exert temporally specific control over activity of a motoneuron receiving concurrent excitation and inhibition. J. Neurosci. 29, 11732–11744.

Sasaki, K., Cropper, E.C., Weiss, K.R., Jing, J., 2013. Functional differentiation of a population of electrically coupled heterogeneous elements in a microcircuit. J. Neurosci. 33 (1), 93–105.

Satterlie, R.A., Norekian, T.P., 1995. Serotonergic modulation of swimming speed in the pteropod mollusc *Clione limacina*. III. Cerebral neurons. J. Exp. Biol. 198 (Pt 4), 917–930.

Satterlie, R.A., 1985. Reciprocal inhibition and postinhibitory rebound produce reverberation in a locomotor pattern generator. Science 229, 402–404.

Sieling, F., Bedecarrats, A., Simmers, J., Prinz, A.A., Nargeot, R., 2014. Differential roles of nonsynaptic and synaptic plasticity in operant reward learning-induced compulsive behavior. Curr. Biol. 24 (9), 941–950.

Sporns, O., Bullmore, E.T., 2014. From connections to function: the mouse brain connectome atlas. Cell 157 (4), 773–775.

Stein, P.S., Smith, J.L., 1997. Neural and biomechanical control strategies for different forms of vertebrate hindlimb motor tasks. In: Stein, P.S.G., Grillner, S., Selverston, A.I., Stuart, D.G. (Eds.), Neurons, Networks, and Motor Behavior. MIT Press, Cambridge, MA, pp. 61–73.

Teyke, T., Rosen, S.C., Weiss, K.R., Kupfermann, I., 1993. Dopaminergic neuron-B20 generates rhythmic neuronal activity in the feeding motor circuitry of *Aplysia*. Brain Res. 630, 226–237.

Thompson, S., Watson 3rd, W.H., 2005. Central pattern generator for swimming in *Melibe*. J. Exp. Biol. 208 (Pt 7), 1347–1361.

Tian, L.M., Kawai, R., Crow, T., 2006. Serotonin-immunoreactive CPT interneurons in *Hermissenda*: identification of sensory input and motor projections. J. Neurophysiol. 96 (1), 327–335.

Traub, R.D., Whittington, M.A., Draguhn, A., 2017. Gap junctions between pyramidal cells account for a variety of very fast network oscillations (>80 Hz) in cortical structures. In: Jing, J. (Ed.), Network Functions and Plasticity. Elsevier (Chapter 13).

Tresch, M.C., Saltiel, P., d'Avella, A., Bizzi, E., 2002. Coordination and localization in spinal motor systems. Brain Res. Brain Res. Rev. 40 (1–3), 66–79.

Wu, C.L., Shih, M.F., Lai, J.S., Yang, H.T., Turner, G.C., Chen, L., et al., 2011. Heterotypic gap junctions between two neurons in the *Drosophila* brain are critical for memory. Curr. Biol. 21 (10), 848–854.

Wu, J.S., Wang, N., Siniscalchi, M.J., Perkins, M.H., Zheng, Y.T., Yu, W., et al., 2014. Complementary interactions between command-like interneurons that function to activate and specify motor programs. J. Neurosci. 34 (19), 6510–6521.

Xin, Y., Hurwitz, I., Perrins, R., Evans, C.G., Alexeeva, V., Weiss, K.R., et al., 1999. Actions of a pair of identified cerebral-buccal interneurons (CBI-8/9) in *Aplysia* that contain the peptide myomodulin. J. Neurophysiol. 81 (2), 507–520.

Yu, K., Liu, D.D., Jia, R.N., Zheng, Y.T., Chen, S.A., Su, Y.N., et al., 2015. Contributions of electrical coupling to the generation of different activity phases in *Aplysia* locomotor network. In: Paper Presented at the Program No. 420.30, 2015 Neuroscience Meeting Planner.

Yuste, R., MacLean, J.N., Smith, J., Lansner, A., 2005. The cortex as a central pattern generator. Nat. Rev. Neurosci. 6 (6), 477–483.

CHAPTER 6

Electrical Synapses and Learning-Induced Plasticity in Motor Rhythmogenesis

R. Nargeot, A. Bédécarrats
Université de Bordeaux, INCIA, CNRS UMR 5287, Bordeaux, France

1. INTRODUCTION

Learning procedures induce long-term changes in the synaptic and membrane properties of central network neurons. By changing the strength of chemical synaptic connections via the amount of chemical transmitter released, and the density and sensitivity of postsynaptic receptors, learning reorganizes in both a short- and long-lasting manner the functional operation of central circuits (Kandel, 2001). Similarly, an increasing amount of data indicates that the intrinsic membrane properties that govern the excitability of network neurons and their capability to generate specific patterns of electrical activity can be modified by learning protocols (Martin and Morris, 2002; Mozzachiodi and Byrne, 2010). Such plasticity resulting from alterations in membrane ionic conductances modifies the spontaneous firing modes of neurons and/or the postsynaptic responses to presynaptic inputs, thereby drastically changing the functional properties of neuronal networks as a whole.

In addition to chemical synapses and intrinsic membrane properties, electrical synapses contribute to the functional organization of central neuronal networks (Bennett, 1997). These synapses, which are formed by the gap junctions, constitute a fundamental mechanism of neuronal communication and are present in most circuits involved in learning, memory, and the genesis of behaviors (Connors and Long, 2004; Dere and Zlomuzica, 2012). Surprisingly, however, much less is known about their putative contribution to the induction and/or expression of neuronal changes that underlie learning and memory-related adaptation of behaviors.

In this chapter, we show that electrical synapses and their plasticity contribute to the organization of behaviors and their adaptation by learning in vertebrates and invertebrates. We illustrate that these synapses, alone or in association with intrinsic neuronal membrane properties, play fundamental roles in the production of oscillatory activity in both homogenous and heterogeneous populations of neurons by regulating the synchronization, frequency, and/or regularity of network oscillations. By modifying electrical synapse strength, modulatory transmitters, sensory inputs and

Network Functions and Plasticity
ISBN 978-0-12-803471-2, http://dx.doi.org/10.1016/B978-0-12-803471-2.00006-0

learning procedures are able to produce long-term plasticity in network oscillatory properties and the genesis of behaviors.

We then describe the contribution of learning-induced plasticity in electrical synapses to changes in the dynamic properties of an identified decision-making network for feeding behavior in the marine mollusk *Aplysia*. In this animal model, an operant-reward learning procedure has been found to transform the irregular and infrequent emission of a buccal motor pattern to a more frequent and periodic, compulsive-like expression (Nargeot et al., 2007, 2009). This behavioral transition is associated with a dopamine (DA)-dependent strengthening of buccal circuit electrical synapses and an increase in the intrinsic membrane excitability of identified decision neurons (Bédécarrats et al., 2013). Experimental manipulations have been used to analyze the respective contributions of the plasticity in electrical synapses and intrinsic excitability to the changes in motor network output (Sieling et al., 2014). Specifically, learning-induced plasticity in electrical synaptic coupling has been found to play a critical role in the synchrony and regularity of central neuron rhythmic bursting and motor pattern genesis, whereas plasticity in intrinsic excitability regulates the frequency of cell bursting and resultant motor output. Thus, the membrane and synaptic plasticity, which, respectively, intensifies and rigidifies the operation of a decision-making network, underlies two different aspects of the learning-induced transition from a flexible motivated act, to an automatic and rhythmic, compulsive behavior.

2. ELECTRICAL SYNAPSES IN THE ORGANIZATION OF BEHAVIORAL ACTIONS

2.1 Electrical Synapses in Rhythmic Motor Pattern Genesis

Electrical synapses are present in many structures of the central nervous system (CNS) that underlie the genesis of behaviors in vertebrates and invertebrates. In vertebrates, for example, they are widespread in the premotor and motor cortices (Gibson et al., 1999), the striatum (Koos and Tepper, 1999), the olivary—cerebellar complex (Sotelo and Llinas, 1972; De Zeeuw et al., 1998), and the spinal cord (Fulton et al., 1980; Tresch and Kiehn, 2000).

The functional implication of electrical synapses in motor rhythmogenesis has been investigated in the spinal cord where rhythmic locomotor movements of the fore- and hind limbs are generated. In specific modulatory conditions, a periodic locomotor activity pattern, so-called fictive locomotion, can be recorded from the ventral roots of isolated spinal cords from the neonatal rat. A periodic activity can still be recorded from these spinal ventral roots in the absence of calcium ions which blocks chemical synapses, or in the presence of tetrodotoxin, a sodium channel blocker that abolishes action potential—dependent chemical synaptic transmission (Tresch and Kiehn, 2000). This persistent rhythmic motor activity, as well as normal fictive locomotion, was found to be suppressed by carbenoxolone, a pharmacological blocker of gap junctions. Thus, electrical synapses appear to be essential, although not necessarily exclusively responsible, for the periodic motor pattern genesis in the spinal cord of newborn animals (Tresch and Kiehn, 2000; Li and Rekling, 2017(Chapter 11). Although the neuronal origin of this oscillatory activity has yet to be identified, electrical synapses have been found between different spinal neurons including excitatory interneurons and motoneurons that contribute to the motor pattern genesis (Fulton et al., 1980; Rash et al., 1996; Kiehn and Tresch, 2002; Hinckley and Ziskind-Conhaim, 2006; Song et al., 2016).

Theoretical studies have indicated that electrical synapses among a population of nonoscillating neurons are sufficient to produce a regular oscillatory activity (Sherman and Rinzel, 1992; Smolen et al., 1993; Traub et al., 2017 (Chapter 13)). However, in biological networks, these synapses often operate in conjunction with oscillatory intrinsic membrane properties of the electrically coupled neurons (Tresch and Kiehn, 2000; Rekling et al., 2000; Blatow et al., 2003; Leznik and Llinas, 2005). Thus, in addition to their participation in the actual genesis of oscillatory activity, electrical synapses can regulate the synchronization, frequency, and variability of an oscillatory activity produced by endogenous pacemaker or bursting neurons.

Such regulatory roles have been investigated in the pyloric network of the crustacean stomatogastric nervous system (Nadim et al., 2017 (Chapter 4) where this network generates the cyclic motor output pattern that drives alternate constriction/dilatation movements of the foregut pyloric chamber. This highly regular repetitive motor activity is driven by a kernel of electrically coupled oscillatory cells composed of the AB interneuron and two PD motoneurons (Selverston and Miller, 1980). In experimental conditions of in situ synaptic isolation from their partners in the pyloric network, each of these neurons can individually generate spontaneous oscillatory bursting activity (Miller and Selverston, 1982; Bal et al., 1988). This activity differs according to the cell type as a result of their differing intrinsic membrane properties. The AB neuron is the only pacemaker cell that spontaneously generates a regular oscillation at a frequency similar to that of the normal pyloric motor rhythm. In contrast, the two PD neurons separately generate repeated bursts of action potentials with irregular burst durations and interburst intervals and at a mean cycle frequency that is lower than the intact pyloric network rhythm. Thus, an important role of the strong electrical synapses between the AB and the PD neurons is to synchronize their inherent bursting activities with the more regular and rapid pacemaker activity of the AB neuron (Bal et al., 1988; Abbott et al., 1991).

In addition to such a role in the synchronization of neuronal intrinsic oscillatory activities, experimental and modeling studies have shown that electrical synapses can regulate the frequency, burst duration, and variability of oscillations in ensembles of electrically coupled neurons (Kepler et al., 1990; Sharp et al., 1992; Szucs et al., 2000). The regulatory control of oscillatory activity in a heterogeneous population of neurons has been investigated in a hybrid system in which an electronic neuron was artificially electrically coupled to in situ isolated PD neurons (Szucs et al., 2000). The electronic neuron was then configured to generate either a periodic or an irregular bursting activity. In either case, depending on the coupling strength, the addition of an electrical synapse decreased the variability of the individual oscillatory activity and led to periodic oscillations in the neuronal ensemble. Concordant results were also obtained by using a computational model of two heterogeneous oscillatory neurons or in an experimental approach in which two isolated and oscillating biological neurons were coupled via an artificial electrical synapse (Sharp et al., 1992; Soto-Trevino et al., 2005). Again in both cases, changes in strength of the electrical synapse controlled the regularity and frequency of rhythmic bursting in the coupled oscillators.

Therefore, electrical synapses not only simply synchronize activity within neuronal networks, but also exert subtle control over the variability and frequency of endogenous bursting or oscillatory properties of coupled neurons. Consequently, depending on electrical synaptic strength and the heterogeneity of the intrinsic membrane properties of coupled neurons, a given network can express complex and variable patterns of activity.

2.2 Electrical Synapses in the Organization of Motivated Behaviors

Action patterns in motivated or goal-directed actions, including feeding and sexual and social behaviors, are repeated with a huge variability. Although directed toward attainment of a specific goal, these patterns are flexible and their emissions change depending on the dynamic properties of the underlying central networks and their regulation by rewarding or punishing outcomes through various learning processes.

In mammals, the expression of motivated and goal-directed behaviors depends to a large extent on the functional operations of the basal ganglia. These central structures integrate information from the cortex and the limbic structures, including the hippocampus, and redistribute this information to the premotor and motor cortical regions. This major decision-making network is modulated by dopaminergic input pathways, including those from the ventral tegmental area, which mediate responses to rewarding stimuli (Vandecasteele et al., 2007; Schultz, 1998).

Electrical synapses have been found in all these neuronal structures between local interneurons and/or projection neurons (Gibson et al., 1999; Vandecasteele et al., 2007). Their role has been investigated in genetically engineered animals deficient in connexin-36, a major neuronal gap junction protein, or by the use of pharmacological gap junction blockers. Both experimental conditions drastically weakened or even suppressed synchronized spiking activity and the different bioelectrical oscillations that are normally expressed by these circuits (Draguhn et al., 1998; Onn and Grace, 2000; Hormuzdi et al., 2001; Deans et al., 2001). However, although in some cases these rhythmic activities, such as hippocampal theta rhythms, have been associated with memory processes and/or behavioral states (Buzsáki, 2005), it remains unclear whether in the basal ganglia they contribute critically to the normal decision-making processes that underpin the expression and variability of motivated behaviors (Berke, 2009).

The implication of electrical synapses in the expression of a motivated behavior can be more readily investigated in the simpler and identified decision-making networks of invertebrates. Such analyses have been successfully made on the central neuronal network that generates aspects of feeding behavior in *Aplysia* (Nargeot and Simmers, 2012; Jing et al., 2017(Chapter 5)). During food-seeking behavior, while randomly exploring its environment, the animal performs several motor acts that are all directed toward the goal of obtaining appropriate nutriments. Among these actions, previous studies have focused on the repeated movements of the tongue-like organ, the radula, as they are easily quantifiable (Kupfermann, 1974). A striking feature of this behavior is the variability with which the radula movement cycles are emitted (Fig. 6.1A). Each cycle is basically composed of sequential protraction and retraction phases, with a closure of the two radula halves occurring during one or other of these two main phases. However, successive cycles are produced spontaneously at highly irregular time intervals and with different phase relationships, individual pattern durations, and strength (Horn et al., 2004; Lum et al., 2005; Brezina et al., 2006; Nargeot et al., 2007; Cullins et al., 2015).

These radula movement cycles are generated by a decision-making neuronal network located in the bilateral buccal ganglia. This network is composed of a population of bilaterally symmetrical pairs of neurons that belong to either protractor or retractor groups of cells (Fig. 6.1B). The neurons of the protractor group organize the autonomous and variable drive that randomly triggers the initial protractor phase of motor activity. By virtue of their electrical and/or chemical synapses with buccal network partners, the protractor neurons also instigate activity in the retractor neuron group and closure motoneurons, thereby triggering expression of a complete radula motor pattern (Sasaki et al., 2013; Jing et al., 2017(Chapter 5)).

The protractor group of the buccal network thus consists of a kernel of heterogeneous and electrically coupled neurons that is essential to the decision-making processes and that autonomously and variably drives the motor output for radula behavior (Fig. 6.1B–D) (Nargeot and Simmers, 2012). These neurons, the so-called B30, B63, B65 cells, fire coincident bursts of action potentials that precede and sustain the initial protraction phase of the radula motor pattern (Fig. 6.1D1). Impulse bursts in the bilateral B63 decision neurons are both necessary and sufficient to drive the protraction phase and hence expression of a complete radula motor pattern (Fig. 6.1D2). Although bursting in either B30 or B65 is not critical for pattern genesis, these neurons are able to initiate radula motor patterns indirectly by activating the B63 decision neurons through their electrical synapses (Fig. 6.1D3–D4) (Nargeot et al., 2009).

In addition to synaptic connections with other elements in the buccal network, the B30/B63/B65 neurons possess specific intrinsic membrane properties that support bursting (Hurwitz et al., 1997; Kabotyanski et al., 1998; Jing et al., 2004). This

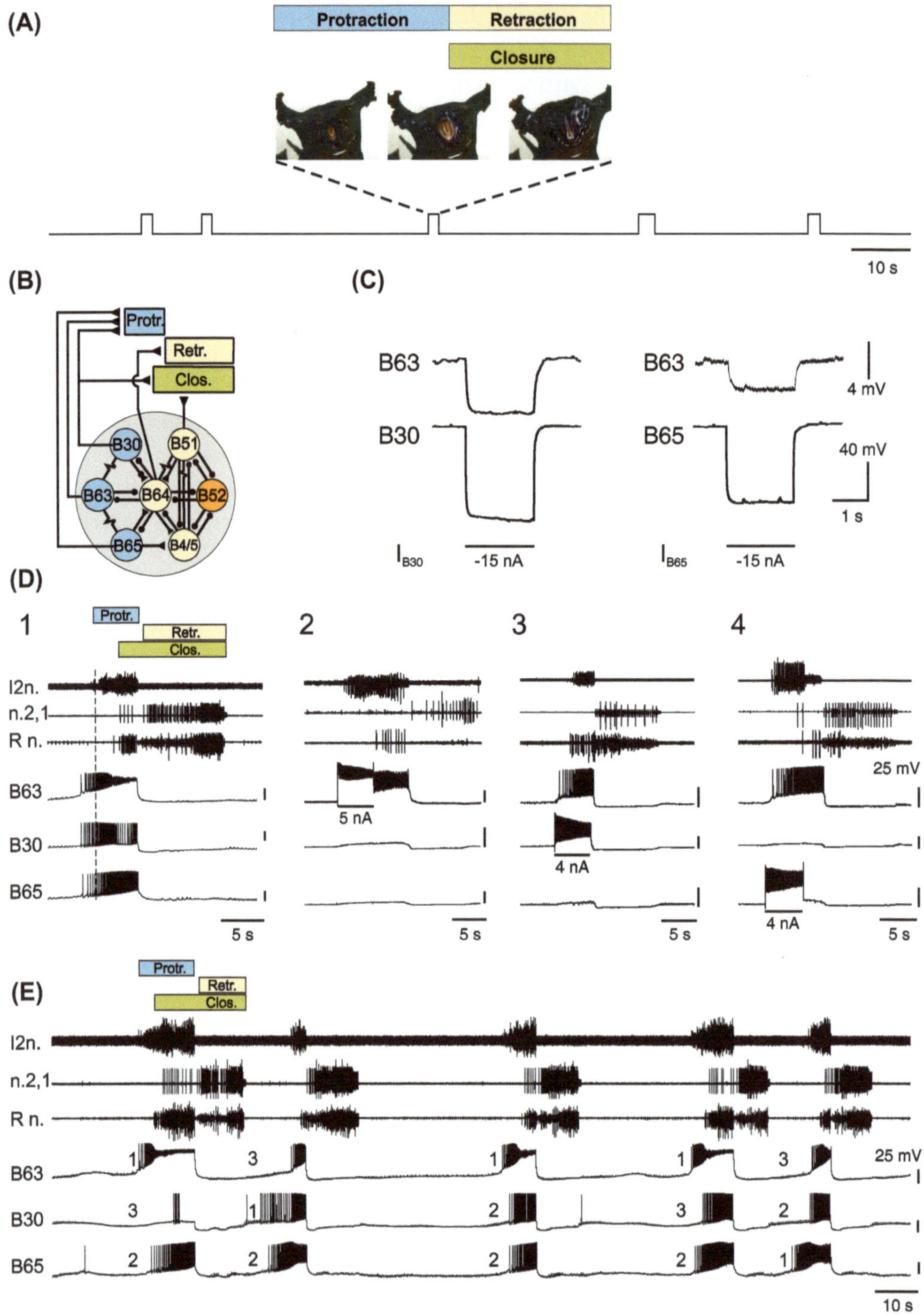

Figure 6.1 ***Feeding behavior and in vitro radula motor pattern genesis in*** **Aplysia.** (A) Sample recording of repeated cycles of movements of the tongue-like radula during spontaneous food-seeking behavior. Each movement cycle (vertical deflection) is composed of a protraction phase

endogenous property is evident in experimental conditions where neurons are functionally isolated from the rest of the network: each individual neuron continues to generate repetitive bursts of action potentials that vary spontaneously in duration and timing from one cell to another. These irregular and unpredictable bursting properties associated with weak electrical synapses are in turn responsible for the irregularity of radula motor pattern genesis. Because the key B63 decision neuron can be activated intrinsically or extrinsically by the electrically coupled B30 or B65 neurons, the relative order in which bursting commences in this heterogeneous subpopulation as well as burst duration and interburst period can vary considerably from one cycle to another (Fig. 6.1E). It is important to remember that this irregular bursting activity could underlie the variability of the radula

followed by a retraction phase. Closure of the two radula halves may overlap with either phase. The occurrence and duration of movements are highly variable when the animal is randomly sampling its environment. (B) Simplified schematic representation of the radula motor pattern generating network (*large circle*) and its synaptic connections. This network located in the buccal ganglia is basically composed of a protractor group (*blue circles*) and a retractor group of interneurons (*pale yellow circles*). They activate the protractor (Protr.) and retractor (Retr.) motoneurons, respectively. Some of these neurons also activate the closure (Clos.) motoneurons. The *small filled triangles and circles* represent excitatory and inhibitory chemical synapses, respectively. The resistance symbols represent electrical synapses *(Adapted with permission from Sieling, F., Bédécarrats, A., Simmers, J., Prinz, A.A., Nargeot, R., 2014. Differential role of nonsynaptic and synaptic plasticity in operant conditioning-induced compulsive behavior. Curr. Biol. 24, 941–950).* (C) Two-electrode current clamp recording in isolated buccal ganglia of the electrical coupling between neuronal pairs B63/B30 (*left*; mean coupling coefficient, 5.7%) and B63/B65 (*right*; mean coupling coefficient, 3.4%) in response to a brief hyperpolarizing current injection (I, horizontal bars) into B30 or B65, respectively *(Adapted with permission from Nargeot, R., Le Bon-Jego, M., Simmers, J., 2009. Cellular and network mechanisms of operant learning-induced compulsive behavior in Aplysia. Curr. Biol. 19, 975–984).* (D) Radula motor pattern genesis in isolated buccal ganglia. The pattern consists of protraction (Protr.), retraction (Retr.), and closure (Clos.) motor activity (*top three traces* corresponding to extracellular recordings of motor nerves to the intrinsic muscle 2, I2 n.; the buccal nerve 2, branch 1, n.2,1; the radula nerve, R n., respectively). (D1) Expression of an individual motor pattern by in-phase bursts of action potentials in the decision B63, B30, and B65 neurons (*three lower traces*), the onsets of which precede expression of the motor pattern (*dashed line*). An impulse burst in the B63 neuron is necessary to initiate a motor pattern, which can be triggered by depolarizing current (*horizontal lines*) injected directly into B63 (D2) or into B30 (D3) and B65 (D4), which in turn excite B63 via their electrical synapse *(Adapted with permission from Nargeot, R., Le Bon-Jego, M., Simmers, J., 2009. Cellular and network mechanisms of operant learning-induced compulsive behavior in Aplysia. Curr. Biol. 19, 975–984).* (E) Sequence of repeated spontaneous cycles of radula motor pattern genesis (*three top traces*) and associated impulse bursts in the decision neurons B63, B30, B65 (*three lower traces*) in an isolated buccal ganglia preparation. The irregular distribution in onset of these cycles (compare with the radula movement cycles in A) was correlated with uncoordinated bursting and cycle-to-cycle variability in the order of burst onsets (indicated by numbering) in the B63, B65, B30 neurons. Note that in this experiment and those illustrated in subsequent figures radula motor patterns and neuronal burst cycling were emitted during tonic (2 Hz) stimulation of the input nerve 2,3.

movement cycles that are expressed when the animal is randomly sampling its environment in search for food (Nargeot et al., 2007; Nargeot and Simmers, 2012).

Thus, spontaneous repetitive bursting activity in the buccal decision-making circuit derives from a combination of electrical synapses and the intrinsic oscillatory membrane properties of coupled neurons. However, in contrast to networks responsible for regular rhythmic behaviors, in decision-making networks, the heterogeneous neuronal population with irregular intrinsic bursting properties and weak electrical synapses provides the flexible substrate necessary for variable motivated action expression.

3. PLASTICITY OF ELECTRICAL SYNAPSES

3.1 Modulatory and Sensory-Induced Plasticity

Various CNS transmitters, including DA, serotonin, gamma-aminobutyric acid, and glutamate, have been found to induce long-term modifications in the electrical coupling between neurons (Lasater and Dowling, 1985; Colombaioni and Brunelli, 1988; Pereda et al., 1992; Landisman and Connors, 2005; Best and Regehr, 2009; Lefler et al., 2014; Mathy et al., 2014; Turecek et al., 2014). Electrical coupling strength is quantified by the ratio of postjunctional over prejunctional voltage responses to prejunctional current injection (Bennett and Zukin, 2004). A modification in this coefficient can result from a change in extrajunctional conductances and/or the junctional conductance itself. Transmitters can modulate extrajunctional ionic membrane channels, and, consequently, can modify electrical coupling between neurons with no effect on actual synaptic strength. However, there is evidence that transmitters also modulate the intercellular channels at gap junctions and thus change the strength of the electrical synapse in question (Lasater and Dowling, 1985; Bloomfield and Völgyi, 2009; D'hondt et al., 2013).

Evidence for modulation-induced plasticity at CNS electrical synapses was initially found in the escape reflex circuit of goldfish (Yang et al., 1990; Pereda et al., 1992, 2004; Korn and Faber, 2005). In this circuit, auditory eighth nerve endings monosynaptically excite the ipsilateral Mauthner cell (M-cell) via a dual electrical and chemical synapse, which subsequently activates spinal motoneurons and elicits a behavioral escape reaction (Nakajima, 1974; Lin and Faber, 1988). Varicosities containing DA have also been found in the M-cell synaptic bed and application of exogenous DA in the vicinity of this dual synapse with the M-cell causes a long-term enhancement of its strength. Although the junctional conductance was not directly measured, other evidence indicated that the plasticity results partly from a modification in the electrical synapse itself (Pereda et al., 1994). First, the bioamine did not modify the functional properties of the presynaptic afferent fibers. Second, DA did not change the efficacy of the chemical and electrical transmission to the same proportion. Third, although DA caused an increase in the input conductance of the M-cell, this effect would decrease rather than increase the electrical coupling.

The subcellular mechanisms underlying the plasticity of this electrical synapse were also investigated. The coupling plasticity was prevented by bath-perfusion of a D1-receptor antagonist or by postsynaptic intracellular injection of a cyclic adenosine monophosphate (cAMP)-dependent protein kinase (PKA) inhibitor, suggesting the involvement of a D1 receptor-mediated activation of the cAMP/PKA cascade (Pereda et al., 1992, 1994). Independent studies indicated that such a metabotropic pathway can phosphorylate gap-junction proteins and modify the permeability and/or opening probability of gap junction channels (D'hondt et al., 2013). Together these findings indicate that a modulatory transmitter, in addition to potential effects on extrajunctional membrane conductances, can evoke a long-lasting modification in the strength of electrical synapses via a direct action on their junctional conductances.

3.2 Activity-Dependent Plasticity

Coupling plasticity can also arise from the conjoint actions of a modulatory transmitter and ongoing electrical activity in electrically connected cells. Again in goldfish, repetitive high-frequency stimulation of the eighth afferent nerve produces a long-term potentiation of the synaptic transmission to the M-cell (Yang et al., 1990; Pereda and Faber, 1996). This potentiation results from an increase in both the electrical and chemical components of the synapse and persists for several tens of minutes, involving a conjoint presynaptic release of glutamate and an activity-dependent activation of postsynaptic NMDA (N-methyl-D-aspartate) receptors. This cellular mechanism, which is similar to the long-term potentiation of chemical synapses found in mammals, increases the intracellular calcium concentration and activates a calcium/calmodulin kinase II cascade in the postsynaptic M-cell (Pereda et al., 1998; Smith and Pereda, 2003). Likely targets of the kinase are the phosphorylation sites of gap-junction proteins (Pereda et al., 2003).

These results indicate that a long-term increase in efficacy of electrical synapses may depend on both the modulatory environment and ongoing neuronal activity. Alternatively, modulatory influences can also cause a long-lasting decrease in electrical synaptic efficacy (Lasater and Dowling, 1985; Yang and Faber, 1991; Haas et al., 2011). Such changes not only modify the flow of information through sensory-motor circuits, but can also regulate rhythmic activity in neuronal networks.

3.3 Dopamine and Activity-Dependent Regulation of Rhythmogenesis in *Aplysia*

The implication of electrical synaptic plasticity in regulating rhythmogenesis has been investigated in the buccal ganglia of *Aplysia*. During feeding behavior, in vivo nerve recordings have shown that each radula movement cycle leading to food ingestion elicits a high-frequency burst of action potentials in the esophageal sensory input nerves (Brembs et al., 2002), which in turn causes the frequency of ongoing radula movement cycles to increase. Moreover, pulsed direct electrical stimulation of the esophageal nerve in vivo in conjunction with each spontaneous (in absence of food) radula movement cycle, similarly and persistently increased the frequency of subsequent cycles. Significantly, such a change was not induced when the timing of the stimulation was uncorrelated with spontaneous cycle emissions.

The cellular mechanisms underlying this plasticity have been analyzed in the buccal ganglia in vitro. In correspondence with the aforementioned in vivo findings, following a period of transient esophageal nerve stimulation delivered contingently with the expression of radula motor patterns (Fig. 6.2A) or underlying bursting activity in B30/B63/B65 (Fig. 6.2C), these output patterns and cellular activities become emitted repetitively at high frequency and at regular time interval (Nargeot et al., 1999; Bédécarrats et al., 2013). Again, these changes strictly depend on the association of the afferent nerve stimulation with the ongoing spontaneous activity: when the nerve stimulation was delivered at fixed time intervals regardless of radula motor pattern expression or bursting activity in the B30/B63/B65 neurons (ie, noncontingent stimulation), the radula motor patterns and neuronal bursting continued to be generated repetitively at a low frequency and irregular time interval.

Histo- and immunofluorescence studies indicated that the esophageal nerves convey dopaminergic fibers to the buccal ganglia (Kabotyanski et al., 1998; Martinez-Rubio et al., 2009) where they make monosynaptic excitatory connections with the B30/B63/B65 neurons (Nargeot et al., 1999; Bédécarrats et al., 2013). In a high divalent solution that blocks polysynaptic pathways, transient nerve stimulation causes a prolonged depolarization of the B30/B63/B65 neurons, a postsynaptic response that is blocked by bath application of flupenthixol, a DA receptor antagonist (Fig. 6.2B). Thus, the changes in frequency and regularity of rhythmic motor pattern genesis with contingent esophageal nerve stimulation (Fig. 6.2A) are thought to result from the association of DA release by the esophageal fibers with the ongoing impulse bursts in the B30/B63/B65 neurons, which again is blocked by flupenthixol (Fig. 6.2C) (Bédécarrats et al., 2013).

The DA-dependent modulation of radula motor pattern genesis and the correlated increase in regularity and frequency of rhythmic bursting in B30/B63/B65 (Fig. 6.2C) result from specific long-term changes in these cells' membrane properties and their electrical interconnections. This dual plasticity is evidenced by a decrease in threshold for action potential generation by these cells after contingent esophageal nerve stimulation (Fig. 6.3A) and a concomitant increase in the strength of electrical coupling between the bilateral B63/B63 homologues and the ipsilateral B30/63 cell pairs (Fig. 6.3B). However, whether these effects are mediated by DA's influence on junctional conductances was not established (but see later discussion).

These findings, therefore, show that electrical synapses are not an invariable form of communication between neurons, but rather, they may express different forms of

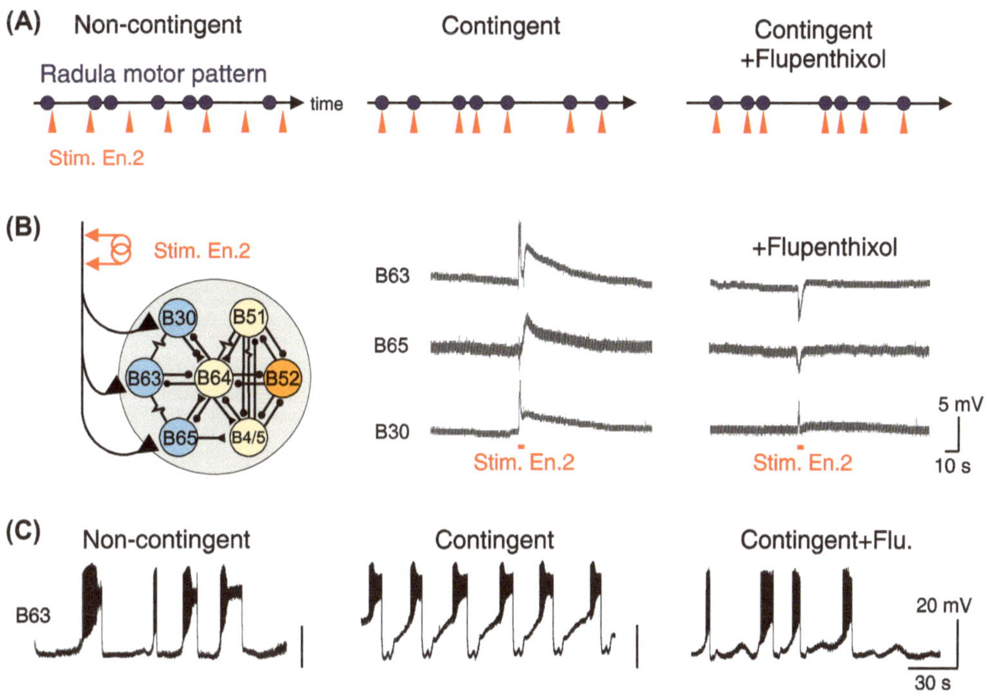

Figure 6.2 ***Dopamine and activity-dependent modulation of buccal rhythmic activity.*** (A) Protocols for stimulation of the esophageal input nerve (En.2) in isolated buccal ganglia. The stimulation (10 Hz, 5 s) was used to mimic activation of this afferent nerve during an in vivo radula movement cycle that leads to food ingestion. In a Noncontingent group (*left*), repetitive stimulations (*arrowheads*) were applied at constant intervals, independently of radula motor pattern occurrences (*filled spots*). In a Contingent group exposed either to control saline (*middle*) or to a saline containing a dopamine receptor antagonist, flupenthixol (*right*), each esophageal afferent nerve stimulation was applied in conjunction with the spontaneous expression of a radula motor pattern. (B) Simplified representation of the monosynaptic connection of the esophageal input fibers (En.2) to the decision B63/B65/B30 neurons (*left*). In a solution rich in high divalent ions to block polysynaptic pathways (*middle*), the esophageal nerve stimulation (horizontal lines, 10 Hz, 5 s) caused a long-lasting depolarization of the B63/B65/B30 neurons. These responses were blocked in the same solution containing 10^{-5} M flupenthixol (*right*). (C) Intracellular recording of B63 activity during a 45-min test period after the above mentioned protocols of En.2 stimulation. The frequency and regularity of bursting activity was durably increased after contingent En.2 stimulation compared with noncontingent stimulation. This plasticity was blocked in the presence of flupenthixol (Flu.). *(Adapted with permission from Bédécarrats, A., Cornet, C., Simmers, J., Nargeot, R., 2013. Implication of dopaminergic modulation in operant reward learning and the induction of compulsive-like feeding behavior in Aplysia. Learn. Mem. 20, 318–327).*

long-term changes in efficacy. Such plasticity can be induced by sensory input and/or modulatory transmitters, eventually in association with ongoing neuronal activity. In this way, therefore, electrical synapses express an adaptive capability similar to chemical synapses and as such they may similarly contribute to learning and memory processes.

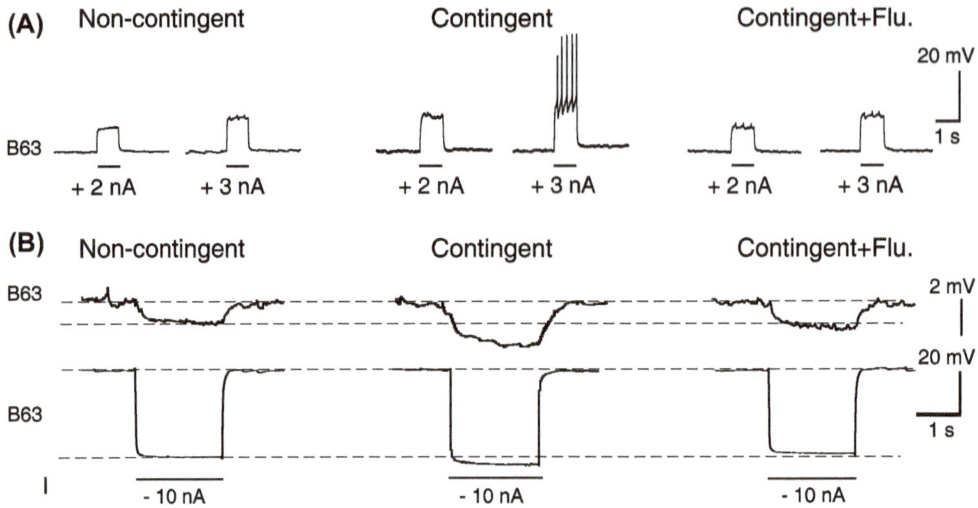

Figure 6.3 ***Activity-dependent, dopamine-mediated modulation of intrinsic excitability and electrical coupling of buccal decision neurons.*** In a test period 45 min after noncontingent and contingent protocols of En.2 stimulation (see Fig. 6.2A), the excitability of B63, defined by its spike threshold (A), and the electrical coupling between bilateral B63 neurons (B) were increased in Contingent preparations compared with either Non-contingent preparations or Contingent preparations under flupenthixol (Flu.). Recordings were made using a two-electrode current-clamp procedure *(Adapted with permission from Bédécarrats, A., Cornet, C., Simmers, J., Nargeot, R., 2013. Implication of dopaminergic modulation in operant reward learning and the induction of compulsive-like feeding behavior in Aplysia. Learn. Mem. 20, 318–327).*

4. IMPLICATION OF ELECTRICAL SYNAPSES IN LEARNING, MEMORY, AND MOTOR RHYTHMOGENESIS IN MAMMALS

4.1 Learning and Memory

The contribution of electrical synapses to learning and memory has been addressed by analyzing the loss of cognitive and/or motor performances in models of connexin-deficient animals, or following injection of gap junction blockers. Pharmacological treatments can be used in designated structures but have a weak or nonselective effect on the different types of connexins. On the other hand, genetic manipulation can be used to suppress expression of a specific type of connexin, but this occurs in the whole organism. Different genetically engineered animals deficient in different connexins have been studied (Dere and Zlomuzica, 2012). For example, the loss of a learning and memory capability was found in connexin-36 knockout mice (but see Zlomuzica et al., 2010). Although these animals display no obvious spontaneous motor and emotional deficit (Kistler et al., 2002; Frisch et al., 2005), they express weakened or no oscillatory activities in different neuronal structures including the inferior olive, hippocampus, and the corticostriatal circuits.

Connexin-36-deficient mice show deficits in olivary-dependent motor learning capability, including an alteration in Pavlovian eye-blink conditioning (Frisch et al., 2005; Van Der Giessen et al., 2008), and similar observations have been made in humans following intake of the gap junction blocker mefloquine (van Essen et al., 2010). In this associative conditioning process, the subject modifies its response to a conditioned stimulus (eg, a tone) after making a temporal association with an unconditioned stimulus (eg, an air puff). The unconditioned stimulus is partly conveyed by the inferior olive to the cerebellum. In genetically modified animals, the alteration of oscillatory activity in the inferior olive desynchronizes spiking activity, and the resulting loss of precise temporal firing in this unconditioned pathway is thought to alter the association encoding required for learning.

Connexin-36-deficient animals also express a deficiency in spatial coding and short-term spatial memory (Allen et al., 2011). This alteration is associated with a modification of spatial selectivity in hippocampal place cells, which may result from altered temporal firing. Such a mechanism may also impair context-dependent fear learning and memory (Bissiere et al., 2011). In this conditioning process, the animal learns from the association of two stimuli, a tone and an aversive stimulus, in a specific environmental context. Intraperitoneal or systemic injection of the gap junction blocker, carbenoxolone, or the more selective connexin-36 blocker mefloquine, alters this context-dependent learning and memory. Intracerebroventricular infusion of carbenoxolone reduces the power of hippocampal theta rhythms and contextual memories. Moreover, microinfusions of the blockers into the dorsal hippocampus, which contributes to context coding, similarly affects learning. Thus, given that hippocampal place cell activity is correlated to theta rhythm, these observations suggest that the blockade of gap junctions disrupts hippocampal cell firing and place coding processes required for learning and memory.

The synchronized or oscillatory activity mediated by electrical coupling in neuronal ensembles may also contribute to rewarding processes in reward-dependent learning. Natural rewards, such as food activate the lateral hypothalamus and dopaminergic neurons in the ventral tegmental area, which mainly project to the nucleus accumbens (Schultz, 1998; Wise, 2013). Electrical stimulation of the lateral hypothalamus can be substituted for natural rewards and electrode-implanted animals can be trained to self-stimulate their hypothalamus by pressing a lever that triggers this stimulation. This training protocol modifies the animal's behavior in a long-term manner, leading to continuous emissions, at an elevated rate, of the lever pressing behavior. Infusion of carbenoxolone into the nucleus accumbens decreases the rate of level pressing (Kokarovtseva et al., 2009). Similar results were obtained by pharmacological treatments that decrease excitability in this structure. Considering that electrical coupling can synchronize accumbens cell firing (Onn and Grace, 2000), it was suggested that the resulting in-phase activity in this nucleus might provide an effective excitation for activating neuronal chains ultimately leading to the motor activity. The threshold for self-stimulation behavior was also raised by pharmacological blockade, which alters gap junctions in the ventral tegmental area (Allison et al., 2006; Lassen et al., 2007). Although the cellular mechanisms remain elusive, it was suggested that electrical coupling may contribute to the in-phase activity of the dopaminergic neurons that is needed to produce strong rewarding effects in the target networks.

Therefore, these results suggest that electrical synapses, via their contribution to the synchronization and/or oscillatory expression of network neuronal activity, may induce long-lasting plasticity that underlies learning and memory.

4.2 Plasticity of Electrical Synapses and the Expression of Habitual or Compulsive Behaviors

Further data indicate that through learning-induced plasticity, electrical synapses can shape the overt expression of behavioral patterns. Similarly to natural rewards, repeated exposure to addictive drugs, through reward-dependent learning, can lead to a transition from goal-directed to habitual and compulsive behaviors (Wise, 2013; Robinson and Berridge, 2003; Hyman et al., 2006; Everitt and Robbins, 2016). These latter behaviors, in contrast to the former, are stereotyped and automatically repeated regardless of the incentive value of the action outcome.

Psychostimulant drugs, such as amphetamine and cocaine, through a sensitization process, can lead to the expression of frequent, repetitive, and stereotyped actions, which result from increased extracellular DA concentrations in the corticostriatal decision-making circuits (Delfs and Kelley, 1990; Goto and Grace, 2008). DA induces long-lasting modifications in intrinsic membrane properties and chemical synapses as well as electrical synapses that ensure synchronized and oscillatory neuronal activities (O'Donnell and Grace, 1993; Berke, 2009; Tritsch and Sabatini, 2012; Schultz, 2013). Such plasticities are believed to modify the functional operation of corticostriatal circuits and lead to the emission of habitual behavioral sequences (Berke et al., 2004; Balleine and O'Doherty, 2010).

Experimental evidence indicates that the expression of stereotyped and rhythmic behaviors, such as hyperlocomotion, biting, licking, sniffing induced by psychostimulant drugs, and DA agonists could modulate the strength of electrical synapses and expression of oscillatory activity at least in the nucleus accumbens (Moore and Grace, 2002; Onn and Grace, 2000). This nucleus, a major target of ventral tegmental dopaminergic neurons, contributes to the expression of stereotyped and compulsive behaviors (Howe et al., 2011; Burguière et al., 2015). The repeated injection of psychostimulant drugs transiently increases the level of connexin-36 mRNA in both the prefrontal cortex and ventral striatum, including the nucleus accumbens (McCracken et al., 2005a,b). This mRNA increase is not associated with a significant increase in protein level. However, separate experiments revealed that repeated systemic injections of the DA receptor agonist apomorphine induce a persistent increase in dye coupling between the projection neurons in the prefrontal cortex and the nucleus accumbens (Onn and Grace, 2000). This long-lasting change in dye coupling was found to be associated with an increase in the proportion of cortical and striatal neurons that generate membrane potential oscillations. Finally, intrastriatal infusions of the gap junction blockers carbenoxolone or anandamine prevented the apomorphine-induced expression of rhythmic and stereotyped behaviors (Moore and Grace, 2002).

Thus, DA-induced strengthening of electrical synapses and associated oscillatory activity could be a cellular correlate of addictive drug-induced rhythmic and stereotyped compulsive behaviors. Such synaptic plasticity could decrease the functional flexibility in the decision-making circuits needed for the completion of behavioral sequences and thereby lead to recurrent actions. Nonetheless, the functional relationship between changes in electrical synapses in decision-making networks and the expression of rhythmic behaviors remains unclear. However, as described later, direct evidence for such a role in the transition from motivated to rhythmic, compulsive-like, behavior has been gained from studies of *Aplysia* feeding.

5. ROLE OF ELECTRICAL SYNAPSES IN THE INDUCTION OF COMPULSIVE-LIKE BEHAVIOR IN *APLYSIA*

5.1 Learning-Induced Plasticity in Motor Pattern Genesis

In a form of associative learning, the delivery of food reward in an operant conditioning procedure was found to transform the infrequent and irregular emissions of radula movement cycles to frequent and rhythmic stereotyped actions (Nargeot et al., 2007; Sieling et al., 2014). Three groups of animals were compared in this learning procedure (Fig. 6.4A). In a contingent group, animals received a food reward in association with each spontaneous emission of a radula movement cycle to reproduce the action-reward contingency characteristic of operant conditioning. In a noncontingent group, animals received the same number of rewards as each contingently trained animal, but the reward was delivered periodically and regardless of movement cycle emission. By comparison with contingently trained animals, this group allowed establishing the essential importance of action–reward association to any resultant behavioral and neuronal plasticities. In a third control group, naïve animals received no food reward.

The radula movement cycles emitted by animals from the three experimental groups were then assessed during a posttraining test period. The naïve control animals generated movement cycles at relatively low frequency and irregular time intervals (Fig. 6.4B). In contrast, in contingently trained animals, movement cycles were emitted repetitively at a high frequency and at regular intervals (Fig. 6.4B). This behavioral transition depended on the action–reward association as it was not induced by the noncontingent training procedure. It also depended on the appetitive value of the

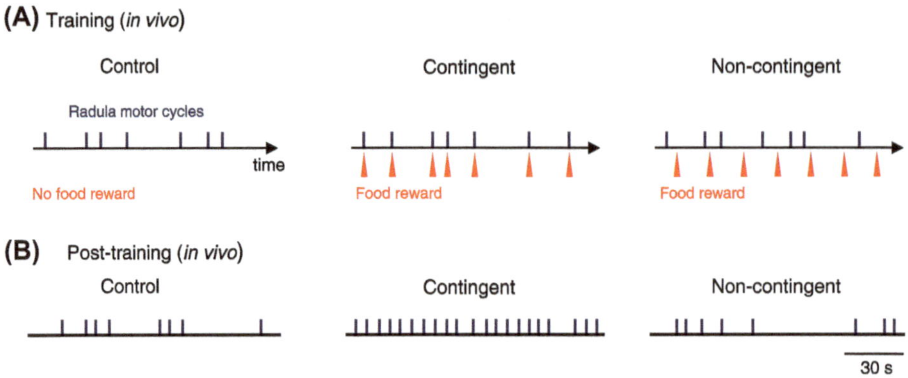

Figure 6.4 ***Operant-reward learning in*** **Aplysia** ***feeding behavior.*** (A) In vivo training protocol. In a control group (*left*), animals were not allowed to ingest any food during the spontaneous expression of radula movement cycles (*vertical bars*). In a contingent group (*middle*), animals received a food reward (a seaweed extract, at *arrowheads*) in strict association with each spontaneous movement cycle. In a Noncontingent group, the food reward was delivered at regular intervals (6 s) independently of the occurrences of radula movements. (B) In vivo posttraining test period. The frequency and regularity of spontaneous radula movement cycles (*vertical bars*) were strongly increased in contingently trained animals (Contingent) compared with either nonrewarded naïve (Control) or noncontingently rewarded animals (Noncontingent).

reward as it was not induced when the contingent reward was replaced by its nonpalatable nutriment (Nargeot et al., 2007).

This reward- and contingency-dependent increase in frequency and regularity of the radula movement cycles persevered for several hours after training. Although ineffective in producing actual ingestion, this modified behavioral activity after termination of food stimulation was expressed at the exclusion of other food seeking-related actions, such as locomotion and head waving. The accelerated and inflexible, rhythmically repeating motor action of *Aplysia* thus shared features with reward learning- or drug-induced compulsive behaviors in mammals (Wise and Koob, 2014).

This behavioral transition from a motivated to compulsive-like feeding action of *Aplysia* after operant conditioning was found to be associated with neuronal changes in the decision-making buccal network (Fig. 6.1B). In isolated ganglia from naïve and noncontingently trained animals, radula motor patterns are generated at low frequency and irregular intervals (Fig. 6.5A), as for radula movements in vivo (Fig. 6.4B). In contrast, in isolated ganglia from contingently trained animals, radula motor patterns are rhythmically generated at a higher frequency and with a regular period, again as for actual radula behavior (Fig. 6.5A, c.f., 6.4B). These modifications in buccal network motor output are associated with an increased excitability and synchrony of impulse bursts in the electrically coupled decision neurons B30/B63/B65 (Fig. 6.5B) (Nargeot et al., 2007; Sieling et al., 2014). In isolated buccal ganglia from naïve and noncontingently trained animals, spike bursts are not only emitted at a low, irregular frequency as for the corresponding motor pattern expression, but also occurred with variable delays and the order in which these cells started to fire in each cycle. In contrast, in ganglia from contingently trained animals, also in correspondence with the network motor output, bursts in the decision neurons are generated at a higher frequency and at regular intervals (Fig. 6.5B). Furthermore, the relative timing of neuronal activity becomes regularized: burst onsets in the three cells are now synchronized and occur in a fixed order with the essential decision neuron B63, probably due to its higher excitability, invariably leading the other two cells in each cycle.

This cell-wide plasticity is associated with long-term changes in the intrinsic membrane properties of, and the electrical synapses between, the buccal network decision neurons (Nargeot et al., 2009). Specifically, the use of two electrode current-clamp established that operant-reward learning increases the excitability of individual

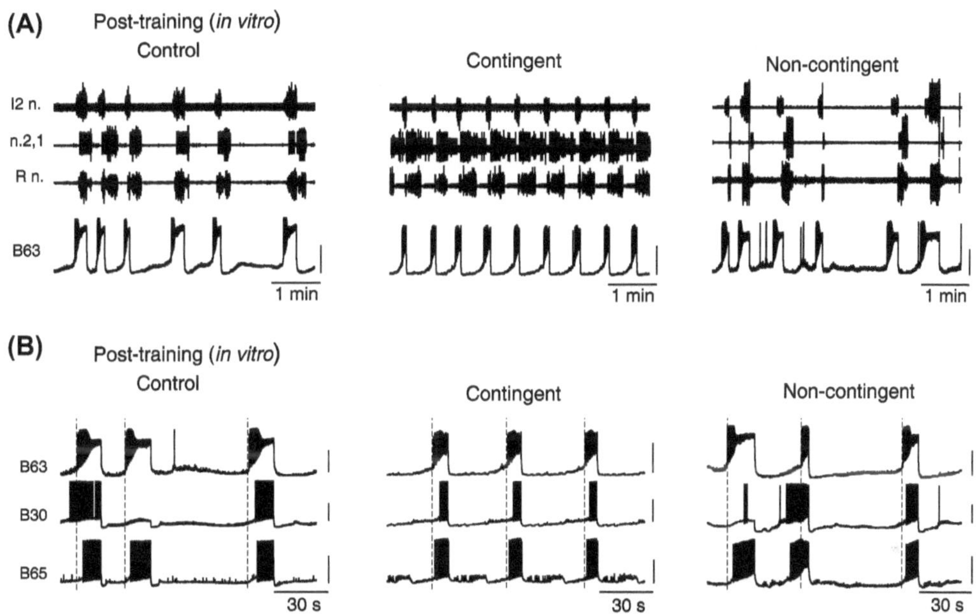

Figure 6.5 ***In vitro neuronal correlates of operant-reward learning.*** (A) After training the frequency and regularity of radula motor pattern expression and associated impulse bursts in the B63 neuron were strongly increased in buccal ganglia isolated from contingently trained animals (Contingent) compared with either naïve nonrewarded (Control) or noncontingently rewarded animals (Noncontingent). (B) Burst onsets in B30 and B65 relative to burst onsets in B63 (indicated by *vertical dotted lines*) that trigger radula motor pattern genesis (see A) are highly variable in Control and Noncontingent buccal ganglia preparations. This variability was replaced by strictly coordinated and stereotyped bursting in the three decision neurons after learning in Contingent preparations. *(Adapted with permission from Nargeot, R., Le Bon-Jego, M., Simmers, J., 2009. Cellular and network mechanisms of operant learning-induced compulsive behavior in* Aplysia. *Curr. Biol. 19, 975–984.)*

B30/B63/B65 neurons (Fig. 6.6A). This change is expressed as a decrease in spike threshold potentials and an increase in membrane input resistance (ie, corresponding to a decrease in input conductance). In addition, the electrical coupling between the B63/B30 and B63/B65 cell pairs increase (Fig. 6.6B and C), with measured synaptic conductances being higher in ganglia from trained animals (B63/B30: 14.8 nS; B63/B65: 4.0 nS) than from naïve animals (B63/B30: 10.6 nS; B63/B65: 2.6 nS). Thus, the operant learning modifies both extrajunctional and the junctional conductances.

The membrane and synaptic plasticity are thought to be partly induced by DA and an activity-dependent modulation of the decision-making neurons. As previously mentioned (see Section 2.3), food delivery in vivo in conjunction with the expression of cyclic radula movements activates esophageal nerve afferents (Brembs et al., 2002). In in vitro preparations, direct and contingent electrical stimulation (and resultant DA release) of the esophageal nerve with ongoing radula motor patterns and impulse bursts

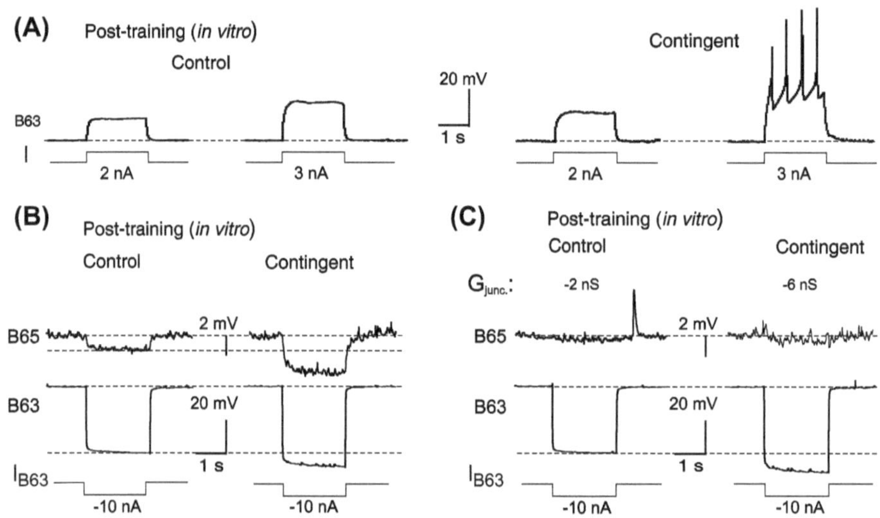

Figure 6.6 ***Learning-induced plasticity of intrinsic excitability and electrical synapses.*** (A, B) After in vivo training, the intrinsic excitability of a B63 neuron (A) and the electrical coupling of the B63/B65 pair (B) were increased in isolated buccal ganglia from a contingently rewarded *Aplysia* (Contingent) compared with a naïve (Control) animal. Membrane excitability was measured by the action potential threshold in response to intracellular injection of depolarizing current pulses (I). The electrical coupling was measured by the voltage deflection of B65 in response to a hyperpolarizing current injected into B63 (I_{B63}). In both cases, two-electrode current clamp was employed. (C) Correlated with the increase in electrical coupling (B), the junctional conductance between the B63 and B65 neurons was increased in a preparation from a Contingent, but not a Control animal. The junctional conductance was measured as the absolute value of an artificial synaptic conductance ($G_{junc.}$) introduced by dynamic clamp required to nullify the natural electrical coupling. *(Adapted with permission from Sieling, F., Bédécarrats, A., Simmers, J., Prinz, A.A., Nargeot, R., 2014. Differential role of nonsynaptic and synaptic plasticity in operant conditioning-induced compulsive behavior. Curr. Biol. 24, 941–950.)*

in the individual B30/B63/B65 cells, thereby replicating the input nerve's activation during actual behavior, similarly increases the intrinsic excitability of these neurons and their electrical couplings (Figs. 6.3 and 6.6) (Bédécarrats et al., 2013). Although additional mechanisms may be implicated in vivo, these findings strongly suggest that DA and DA-dependent synaptic and cellular plasticity make at least a major contribution to the operant conditioning-induced motor changes.

5.2 Respective Roles of Electrical Synapses and Neuronal Excitability in the Plasticity of Motor Rhythmogenesis

To determine the relative contributions of the plasticity of decision neuron electrical synapses and membrane properties to the learning-induced changes in radula motor rhythmogenesis, the dynamic clamp technique was employed with buccal ganglia isolated from naïve and trained *Aplysia* (Sieling et al., 2014). This technique allows an identified neuron(s) to be injected in real time with an artificial synaptic and/or membrane current computed from the recorded cell(s) membrane potential and a user-defined conductance (Sharp et al., 1992, 1993a,b; Prinz et al., 2004). In this way, an artificial membrane and/or synaptic conductance and the resultant current can be used to replicate or suppress learning-induced plasticity in specific electrical synapses and/or membrane properties.

Fig. 6.7A and B illustrates the effects of introducing artificial junctional conductances between the B63/B30 and B63/B65 pairs in buccal ganglia from naïve animals to selectively replicate the learning-induced strengthening of their electrical synapses (Sieling et al., 2014). In these preparations, which initially generated radula motor patterns and underlying B30/B63/B65 impulse bursts at low frequency and irregular intervals, the artificial synaptic strengthening (Fig. 6.7A) enhanced the regularity, but not the frequency, of the decision neuron bursting and motor pattern expression (Fig. 6.7B). In the same naïve preparations, a selective experimental decrease in the membrane leak conductance of the B30/B63/B65 neurons was also used to replicate the learning-induced increase in their excitability (Fig. 6.7C). This membrane property manipulation in turn enhanced the frequency, but not the regularity, of ongoing decision cell bursting activity and motor pattern genesis (Fig. 6.7D). These findings thus indicated that plasticity in electrical synaptic coupling and in membrane excitability govern different, yet complementary, aspects of radula network rhythmogenesis and its response to learning. This conclusion was further supported by applying a conjoint experimental increase in synaptic strength and neuronal excitability in naïve preparations to reproduce both aspects of the operant learning-induced plasticity in the B30/B63/B65 neurons (Fig. 6.8A). This combined conductance introduction immediately transformed their infrequent/irregular bursting and resultant motor patterns to the accelerated/regularized cell bursting and motor pattern phenotypes observed in operantly trained preparations (Sieling et al., 2014).

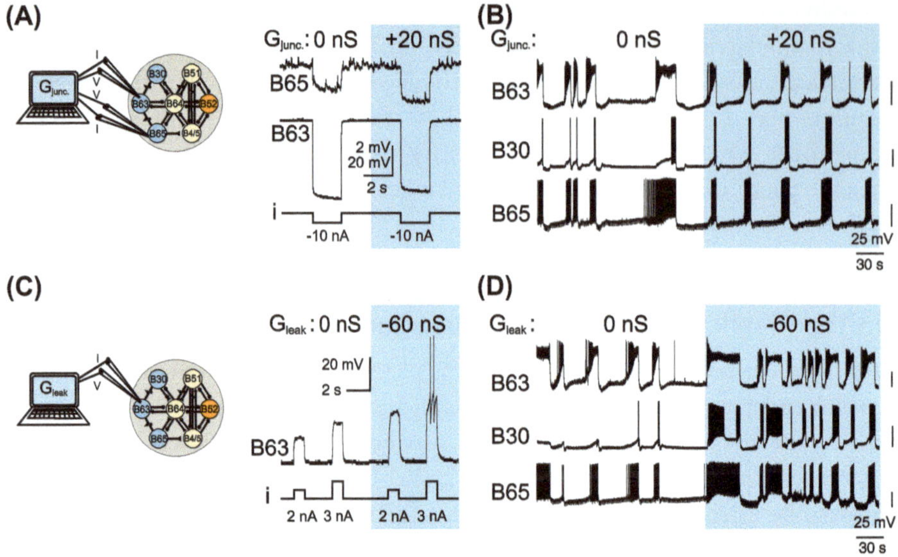

Figure 6.7 ***Plasticity in intrinsic excitability and electrical synapses of decision neurons contribute to different aspects of motor rhythmogenesis.*** (A) Experimental protocol (*left*) for the addition of a dynamic-clamp-defined junctional conductance ($G_{junc.}$: 20 nS, *blue panel*) in in vitro buccal ganglia from a naïve control animal. The artificial conductance was used to reproduce the learning-induced increase in the electrical coupling between the B63/B65 neurons (*right*, compare with Fig. 6.6B). (B) An artificial $G_{junc.}$ (+20 nS, *blue panel*) introduced simultaneously into the B63/B30 and B63/B65 cell pairs of a control preparation increased the regularity, but not the frequency, of motor pattern genesis and associated spike bursts. (C) Experimental protocol (*left*) for addition of a dynamic-clamp-defined leak conductance (G_{leak}, −60 nS, *blue panel*). In in vitro buccal ganglia from a naïve control animal, this artificial conductance introduction reproduces the learning-induced increase in intrinsic excitability of B63 (*right*, compare with Fig. 6.6A). (D) In a control preparation, a dynamic-clamp-defined G_{leak} of −60 nS (*blue panel*) introduced simultaneously into the B63, B30, B65 neurons, increased the frequency, but not the regularity, of spontaneous motor patterns and associated impulse bursts. *(Adapted with permission from Sieling, F., Bédécarrats, A., Simmers, J., Prinz, A.A., Nargeot, R., 2014. Differential role of nonsynaptic and synaptic plasticity in operant conditioning-induced compulsive behavior. Curr. Biol. 24, 941–950.)*

In a reciprocal approach with in vitro preparations from previously trained animals, an artificial and selective decrease in the strength of the electrical synapses between the decision neurons was now used to nullify the learning-induced plasticity in these synapses. This manipulation desynchronized the spontaneous activity in the B30/B63/B65 neurons and produced irregular bursting and motor pattern genesis, but with no change in cycle frequency. In contrast, when the learning-induced increase in decision neuron membrane excitability was experimentally canceled, the frequency, but not the regularity, of bursting and motor patterns was decreased. Accordingly, a dynamic clamp-induced suppression of both synaptic and membrane plasticities in trained preparations immediately transformed ongoing regular/frequent bursting and motor pattern production to a slow irregular state as recorded in naïve preparations (Fig. 6.8B) (Sieling et al., 2014).

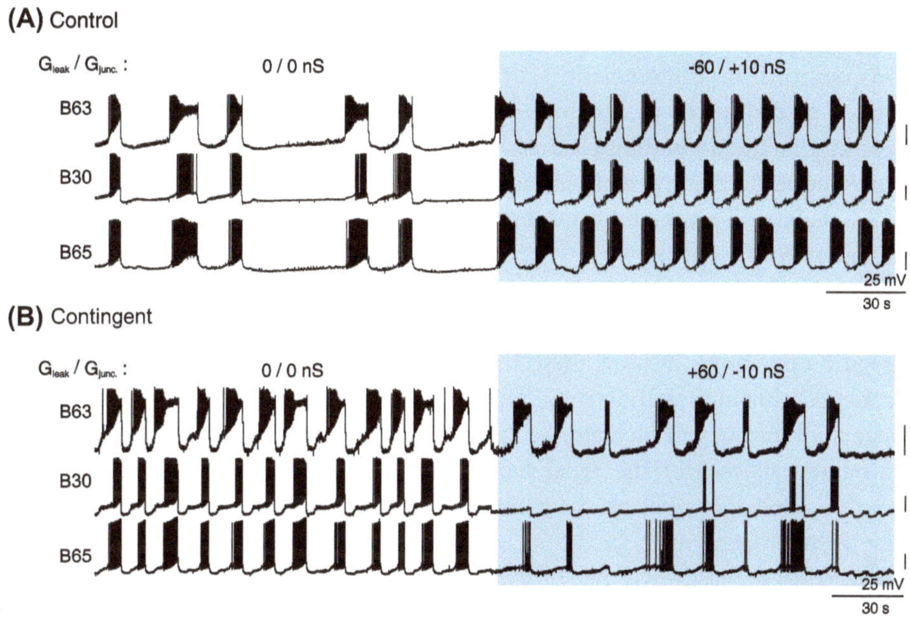

Figure 6.8 ***Plasticity in intrinsic excitability and electrical synapses of decision neurons contribute to different aspects of the motor rhythmogenesis.*** (A) In a control preparation with the conjoint addition of a dynamic-clamp-defined G_{leak} of −60 nS to the B63, B30, B65 cells and a $G_{junc.}$ of +10 nS to the B63/B30 and B63/B65 neuron pairs (*blue panel*), buccal motor patterns and underlying B63/B30/B65 bursting, which were otherwise generated at low frequency and irregular intervals, switched immediately to a rapid, regular expression. (B) In a preparation from a contingently trained animal, the frequent and regular bursting of B63/B30/B65 and motor pattern genesis were transformed to slower, irregular activity during introduction of a dynamic clamp-defined G_{leak} of +60 nS and a $G_{junc.}$ of −10 nS to B63/B30 and B63/B65 pairs (*blue panel*). These artificial conductances (G_{leak}, G_{junc}), respectively, returned intrinsic excitability and synaptic strengths to naïve levels. *(Adapted with permission from Sieling, F., Bédécarrats, A., Simmers, J., Prinz, A.A., Nargeot, R., 2014. Differential role of nonsynaptic and synaptic plasticity in operant conditioning-induced compulsive behavior. Curr. Biol. 24, 941–950.)*

Altogether, these data established a causal relationship between the strength of electrical synapses and the regularity of synchronized periodic bursting in the decision neurons that drive radula motor output. A separate mechanism, involving plasticity in the membrane excitability of these same neurons controls the frequency of their burst activities and resultant motor output. In a more general perspective, these results clearly indicate that a functional plasticity at electrical synapses, such as plasticity of intrinsic membrane properties and of chemical synapses, can provide a memory trace that participates in long-term motor adaptations involving motor rhythmogenesis and a transition to the expression of compulsive behavior.

6. CONCLUSION

The data reviewed in this chapter illustrate the important contribution of electrical synapses to neuronal rhythmogenesis, learning and memory, and learning-induced transitions from goal-directed to compulsive behaviors. Along with chemical synapses and intrinsic membrane properties, electrical synapses constitute the essential building blocks of neural networks, being present in most, if not all, neuronal structures that generate and control behavior. This includes sensory-motor reflex pathways, decision-making networks underlying the expression of goal-directed behaviors and central pattern generators that drive rhythmic behaviors. Within these circuits, electrical synapses can interconnect both homogenous and heterogeneous populations of neurons responsible for specific operations. Within these functional neuron groups, electrical connections may act in combination with intrinsic membrane properties of individual cells to generate and regulate bursting and oscillatory activity of the neuronal network ensemble. Electrical coupling can control the synchronization, frequency, and regularity of neuronal bursting or oscillatory activity. Such regulatory roles may also be critical to the functioning of heterogeneous populations of electrically coupled neurons. In such populations, electrical synapses may be sufficiently weak to be unable to prevent a temporal diversity of spiking, bursting, or oscillatory activity in the different network neurons. As a consequence, network functioning and thus motor output could be expressed with a wide flexibility and variability of activity patterns. In contrast, strong electrical interconnections would constrain the disparate activities of individual neurons into producing unified and thereby regularized oscillatory network functioning and resultant motor output. Evidence now supports the conclusion that electrical synapses, like chemical synapses and intrinsic membrane properties, can be durably modified by modulatory transmitters, sensory inputs, and learning protocols. Thus, such electrical coupling plasticity can be seen to be capable of reducing the variability in network operation and motor output and lead to the production of stereotyped behavior.

Learning and modulatory transmitters can induce plasticity in the electrical synapses and intrinsic membrane properties of the same neuron, making it difficult to clearly distinguish the respective contributions of these plasticities to any long-term modifications in a network's functioning. Although in theory, either form of plasticity could modify the electrical coupling between neurons, each has different functional consequences for network computations and motor adaptability. Taking advantage of the simple decision-making network for feeding behavior in *Aplysia*, which is amenable to cellular analysis, the causal and respective roles of learning-induced plasticity in electrical synapses and intrinsic membrane properties have been investigated in identified neurons.

The data indicate that learning and DA-induced changes in intrinsic excitability of a heterogeneous population of decision neurons alter the frequency, but not the variability of neuronal impulse bursts and resultant motor pattern genesis. In contrast, learning and DA-induced strengthening of these cells' electrical synapses play a fundamental role in the synchronization and regularity of neuronal bursting and patterned network output. This strengthening of the electrical synapses among the heterogeneous decision neuron subpopulation effectively "rigidifies" the buccal circuit's operation and leads to regularized rhythmic motor output. Such plasticity in a decision-making network, which can be induced by reward-dependent operant learning, could underlie transitions of flexible, goal-directed behaviors to the rhythmic and stereotyped actions observed in compulsive and addictive behaviors of other animals, including humans.

OUTSTANDING QUESTIONS

- Learning in vertebrates and invertebrates induces changes in electrical synapse strength, but little is known about the causal role of this plasticity in behavioral adaptations. Is plasticity at electrical synapses essential for the expression of learning-induced behavioral changes?
- Sensory and modulatory inputs can cause long-lasting modifications in the strength of chemical and electrical synapses. Do these different synaptic plasticities have distinct or equivalent functional consequences for network operations and the genesis of behaviors?
- Electrical synaptic plasticity can decrease the variability of behavioral expression and contribute to discharge synchronization and oscillatory activity in network neurons. Is such a role also implicated in the expression of pathological behaviors such as the stereotyped repetitive motor actions found in obsessive and addictive disorders?

ACKNOWLEDGMENTS

We thank Dr. John Simmers and Victoria Normand for helpful discussions and corrections. The preparation of the manuscript was supported by the grant BRAIN *ANR*-10-LABX-43 and *ANR*-10-IDEX-03-02.

REFERENCES

Abbott, L.F., Marder, E., Hooper, S.L., 1991. Oscillating networks: control of burst duration by electrically coupled neurons. Neural Comput. 3, 487–497.

Allen, K., Fuchs, E.C., Jaschonek, H., Bannerman, D.M., Monyer, H., 2011. Gap junctions between interneurons are required for normal spatial coding in the hippocampus and short-term spatial memory. J. Neurosci. 31, 6542–6552.

Allison, D.W., Ohran, A.J., Stobbs, S.H., Mameli, M., Valenzuela, C.F., Sudweeks, S.N., Ray, A.P., Henriksen, S.J., Steffensen, S.C., 2006. Connexin-36 gap junctions mediate electrical coupling between ventral tegmental area GABA neurons. Synapse 60, 20–31.

Bal, T., Nagy, F., Moulins, M., 1988. The pyloric central pattern generator in crustacea: a set of conditional neuronal oscillators. J. Comp. Physiol. 163, 715–727.

Balleine, B.W., O'Doherty, J.P., 2010. Human and rodent homologies in action control: corticostriatal determinants of goal-directed and habitual action. Neuropsychopharmacology 35, 48–69.

Bédécarrats, A., Cornet, C., Simmers, J., Nargeot, R., 2013. Implication of dopaminergic modulation in operant reward learning and the induction of compulsive-like feeding behavior in *Aplysia*. Learn. Mem. 20, 318–327.

Bennett, M.V.L., Zukin, R.S., 2004. Electrical coupling and the neuronal synchronization in the mammalian brain. Neuron 41, 495–511.

Bennett, M.V.L., 1997. Gap junctions as electrical synapses. J. Neurocytol. 26, 349–366.

Berke, J.D., Okatan, M., Skurski, J., Eichenbaum, H.B., 2004. Oscillatory entrainment of striatal neurons in freely moving rats. Neuron 43, 883–896.

Berke, J.D., 2009. Fast oscillations in cortical-striatal networks switch frequency following rewarding events and stimulant drugs. Eur. J. Neurosci. 30, 848–859.

Best, A.R., Regehr, W.G., 2009. Inhibitory regulation of electrically coupled neurons in the inferior olive is mediated by asynchronous release of GABA. Neuron 62, 555–565.

Bissiere, S., Zelikowsky, M., Ponnusamy, R., Jacobs, N.S., Blair, H.T., Fanselow, M.S., 2011. Electrical synapses control hippocampal contributions to fear learning and memory. Science 331, 87–91.

Blatow, M., Rozov, A., Katona, I., Hormuzdi, S.G., Meyer, A.H., Whittington, M.A., Caputi, A., Monyer, H., 2003. A novel network of multipolar bursting interneurons generates theta frequency oscillations in neocortex. Neuron 38, 805–817.

Bloomfield, S.A., Völgyi, B., 2009. The diverse functional roles and regulation of the neuronal gap junctions in the retina. Nat. Rev. Neurosci. 10, 495–506.

Brembs, B., Lorenzetti, F.D., Reyes, F.D., Baxter, D.A., Byrne, J.H., 2002. Operant reward learning in *Aplysia*: neuronal correlates and mechanisms. Science 296, 1706–1709.

Brezina, V., Proekt, A., Weiss, K.R., 2006. Cycle-to-cycle variability as an optimal behavioral strategy. Neurocomputing 69, 1120–1124.

Burguière, E., Monteiro, P., Mallet, L., Feng, G., Graybiel, A.M., 2015. Striatal circuits, habits, and implications for obsessive-compulsive disorder. Curr. Opin. Neurobiol. 30, 59–65.

Buzsáki, G., 2005. Theta rhythm of navigation: link between path integration and landmark navigation, episodic and semantic memory. Hippocampus 15, 827–840.

Colombaioni, L., Brunelli, M., 1988. Neurotransmitter-induced modulation of an electrotonic synapse in the CNS of *Hirudo medicinalis*. Exp. Biol. 47, 139–144.

Connors, B.W., Long, M.A., 2004. Electrical synapses in the mammalian brain. Annu. Rev. Neurosci. 27, 393–418.

Cullins, M.J., Shaw, K.M., Gill, J.P., Chiel, H.J., 2015. Motor neuronal activity varies least among individuals when it matters most for behavior. J. Neurophysiol. 113, 981–1000.

De Zeeuw, C.I., Simpson, J.I., Hoogenraad, C.C., Galjaart, N., Koekkoek, S.K., Ruigrok, T.J., 1998. Microcircuitry and function of the inferior olive. Trends Neurosci. 21, 391–400.

Deans, M.R., Gibson, J.R., Sellitto, C., Connors, B.W., Paul, D.L., 2001. Synchronous activity of inhibitory networks in neocortex requires electrical synapses containing connexin36. Neuron 31, 477–485.

Delfs, J.M., Kelley, A.E., 1990. The role of D1 and D2 dopamine receptors in oral stereotypy induced by dopaminergic stimulation of the ventrolateral striatum. Neuroscience 39, 59–67.

Dere, E., Zlomuzica, A., 2012. The role of gap junctions in the brain in health and disease. Neurosci. Biobehav. Rev. 36, 206–217.

D'hondt, C., Iyyathurai, J., Vinken, M., Rogiers, V., Leybaert, L., Himpens, B., Bultynck, G., 2013. Regulation of connexin- and pannexin-based channels by post-translational modifications. Biol. Cell. 105, 373–398.

Draguhn, A., Traub, R.D., Schmitz, D., Jefferys, J.G., 1998. Electrical coupling underlies high-frequency oscillations in the hippocampus in vitro. Nature 394, 189–192.

Everitt, B.J., Robbins, T.W., 2016. Drug addiction: updating actions to habits to compulsions ten years on. Annu. Rev. Psychol. 67, 23–50.

Frisch, C., De Souza-Silva, M.A., Söhl, G., Güldenagel, M., Willecke, K., Huston, J.P., Dere, E., 2005. Stimulus complexity dependent memory impairment and changes in motor performance after deletion of the neuronal gap junction protein connexin36 in mice. Behav. Brain Res. 157, 177–185.

Fulton, B.P., Miledi, R., Takahashi, T., 1980. Electrical synapses between motoneurons in the spinal cord of the newborn rat. Proc. R. Soc. Lond. B. Biol. Sci. 208, 115–120.

Gibson, J.R., Beierlein, M., Connors, B.W., 1999. Two networks of electrically coupled inhibitory neurons in neocortex. Nature 402, 75–79.

Goto, Y., Grace, A.A., 2008. Limbic and cortical information processing in the nucleus accumbens. Trends Neurosci. 31, 552–558.

Haas, J.S., Zavala, B., Landisman, C.E., 2011. Activity-dependent long-term depression of the electrical synapses. Science 334, 389–393.

Hinckley, C.A., Ziskind-Conhaim, L., 2006. Electrical coupling between locomotor-related excitatory interneurons in the mammalian spinal cord. J. Neurosci. 26, 8477–8483.

Hormuzdi, S.G., Pais, I., LeBeau, F.E., Towers, S.K., Rozov, A., Buhl, E.H., Whittington, M.A., Monyer, H., 2001. Impaired electrical signaling disrupts gamma frequency oscillations in connexin 36-deficient mice. Neuron 31, 487–495.

Horn, C.C., Zhurov, Y., Orekhova, I.V., Proekt, A., Kupfermann, I., Weiss, K.R., Brezina, V., 2004. Cycle-to-cycle variability of neuromuscular activity in *Aplysia* feeding behavior. J. Neurophysiol. 92, 157–180.

Howe, M.W., Atallah, H.E., McCool, A., Gibson, D.J., Graybiel, A.M., 2011. Habit learning is associated with major shifts in frequencies of oscillatory activity and synchronized spike firing in striatum. Proc. Natl. Acad. Sci. U.S.A. 108, 16801–16806.

Hurwitz, I., Kupfermann, I., Susswein, A.J., 1997. Different roles of neurons B63 and B34 that are active during the protraction phase of the buccal motor programs in *Aplysia californica*. J. Neurophysiol. 78, 1305–1319.

Hyman, S.E., Malenka, R.C., Nestler, E.J., 2006. Neural mechanisms of addiction: role of reward-related learning and memory. Ann. Rev. Neurosci. 29, 565–598.

Jing, J., Cropper, E.C., Hurwitz, I., Weiss, K.R., 2004. The construction of movement with behavior-specific and behavior-independent modules. J. Neurosci. 24, 6315–6325.

Jing, J., Cropper, E.C., Weiss, K.R., 2017. Network functions of electrical coupling present in multiple and specific sites in behavior-generating circuits. In: Jing, J. (Ed.), Network Functions and Plasticity: Perspectives from Studying Neuronal Electrical Coupling in Microcircuits. Elsevier (Chapter 5).

Kabotyanski, E.A., Baxter, D.A., Byrne, J.H., 1998. Identification and characterization of catecholaminergic neuron B65, which initiates and modifies patterned activity in the buccal ganglia of *Aplysia*. J. Neurophysiol. 79, 605–621.

Kandel, E.R., 2001. The molecular biology of memory storage: a dialogue between genes and synapses. Science 294, 1030–1038.

Kepler, T.B., Marder, E., Abbott, L.F., 1990. The effect of electrical coupling on the frequency of model neuronal oscillators. Science 248, 83–85.

Kiehn, O., Tresch, M.C., 2002. Gap junctions and motor behavior. Trends Neurosci. 25, 108–115.

Kistler, W.M., De Jeu, M.T., Elgersma, Y., Van Der Giessen, R.S., Hensbroek, R., Luo, C., Koekkoek, S.K., Hoogenraad, C.C., Hamers, F.P., Gueldenagel, M., Sohl, G., Willecke, K., De Zeeuw, C.I., 2002. Analysis of Cx36 knockout does not support tenet that olivary gap junctions are required for complex spike synchronization and normal motor performance. Ann. N.Y. Acad. Sci. 978, 391–404.

Kokarovtseva, L., Jaciw-Zurakiwsky, T., Mendizabal Arbocco, R., Frantseva, M.V., Perez Velazquez, J.L., 2009. Excitability and gap junction-mediated mechanisms in nucleus accumbens regulate self-stimulation reward in rats. Neuroscience 159, 1257–1263.

Koós, T., Tepper, J.M., 1999. Inhibitory control of neostriatal projection neurons by GABAergic interneurons. Nat. Neurosci. 5, 467–472.

Korn, H., Faber, D.S., 2005. The Mauthner cell half a century later: a neurobiological model for decision-making? Neuron 47, 13–28.

Kupfermann, I., 1974. Feeding behavior in *Aplysia*: a simple system for the study of motivation. Behav. Biol. 10, 1–26.
Landisman, C.E., Connors, B.W., 2005. Long-term modulation of electrical synapses in the mammalian thalamus. Science 310, 1809–1813.
Lasater, E.M., Dowling, J.E., 1985. Dopamine decreases conductance of the electrical junctions between cultured retinal horizontal cells. Proc. Natl. Acad. Sci. U.S.A. 82, 3025–3029.
Lassen, M.B., Brown, J.E., Stobbs, S.H., Gunderson, S.H., Maes, L., Valenzuela, C.F., Ray, A.P., Henriksen, S.J., Steffensen, S.C., 2007. Brain stimulation reward is integrated by a network of electrically coupled GABA neurons. Brain Res. 1156, 46–58.
Lefler, Y., Yarom, Y., Uusisaari, M.Y., 2014. Cerebellar inhibitory input to the inferior olive decreases electrical coupling and blocks subthreshold oscillations. Neuron 81, 1389–1400.
Leznik, E., Llinas, R., 2005. Role of gap junctions in synchronized neuronal oscillations in the inferior olive. J. Neurophysiol. 94, 2447–2456.
Li, W.-C., Rekling, J.C., 2017. Electrical coupling in the generation of vertebrate motor rhythms. In: Jing, J. (Ed.), Network Functions and Plasticity: Perspectives from Studying Neuronal Electrical Coupling in Microcircuits. Elsevier (Chapter 11).
Lin, J.-W., Faber, D.S., 1988. Synaptic transmission mediated by single club endings on the goldfish Mauthner cell. I. Characteristics of electrotonic and chemical postsynaptic potentials. J. Neurosci. 8, 1302–1312.
Lum, C.S., Zhurov, Y., Cropper, E.C., Weiss, K.R., Brezina, V., 2005. Variability of swallowing performance in intact, freely feeding *Aplysia*. J. Neurophysiol. 94, 2427–2446.
Martin, S.J., Morris, R.G., 2002. New life in an old idea: the synaptic plasticity and memory hypothesis revisited. Hippocampus 12, 609–636.
Martínez-Rubio, C., Serrano, G.E., Miller, M.W., 2009. Localization of biogenic amines in the foregut of *Aplysia californica*: catecholaminergic and serotoninergic innervation. J. Comp. Neurol. 514, 329–342.
Mathy, A., Clark, B.A., Häusser, M., 2014. Synaptically induced long-term modulation of electrical coupling in the inferior olive. Neuron 81, 1290–1296.
McCracken, C.B., Patel, K.M., Vrana, K.E., Paul, D.L., Roberts, D.C., 2005a. Amphetamine withdrawal produces region-specific and time-dependent changes in connexin36 expression in rat brain. Synapse 56, 39–44.
McCracken, C.B., Hamby, S.M., Patel, K.M., Morgan, D., Vrana, K.E., Roberts, D.C., 2005b. Extended cocaine self-administration and deprivation produces region-specific and time-dependent changes in connexin36 expression in rat brain. Synapse 58, 141–150.
Miller, J.P., Selverston, A.I., 1982. Mechanisms underlying pattern generation in lobster stomatogastric ganglion as determined by selective inactivation of identified neurons. IV. Network properties of pyloric system. J. Neurophysiol. 48, 1416–1432.
Moore, H., Grace, A.A., 2002. A role for electrotonic coupling in the striatum in the expression of dopamine receptor-mediated stereotypies. Neuropsychopharmacology 27, 980–992.
Mozzachiodi, R., Byrne, J.H., 2010. More than synaptic plasticity: role of nonsynaptic plasticity in learning and memory. Trends Neurosci. 33, 17–26.
Nadim, F., Li, X., Gray, M., Golowasch, J., 2017. The role of electrical coupling in rhythm generation in small networks. In: Jing, J. (Ed.), Network Functions and Plasticity: Perspectives from Studying Neuronal Electrical Coupling in Microcircuits. Elsevier (Chapter 4).
Nakajima, Y., 1974. Fine structure of the synaptic endings on the Mauthner cell of the goldfish. J. Comp. Neurol. 156, 375–402.
Nargeot, R., Simmers, J., 2012. Functional organization and adaptability of a decision-making network in *Aplysia*. Front. Neurosci. 6, 113.
Nargeot, R., Baxter, D.A., Patterson, G.W., Byrne, J.H., 1999. Dopaminergic synapses mediate neuronal changes in an analogue of operant conditioning. J. Neurophysiol. 81, 1983–1987.
Nargeot, R., Petrissans, C., Simmers, J., 2007. Behavioral and *in vitro* correlates of compulsive-like food-seeking induced by operant conditioning in *Aplysia*. J. Neurosci. 27, 8059–8070.

Nargeot, R., Le Bon-Jego, M., Simmers, J., 2009. Cellular and network mechanisms of operant learning-induced compulsive behavior in *Aplysia*. Curr. Biol. 19, 975–984.

O'Donnell, P., Grace, A.A., 1993. Dopaminergic modulation of dye coupling between neurons in the core and shell regions of the nucleus accumbens. J. Neurosci. 13, 3456–3471.

Onn, S.P., Grace, A.A., 2000. Amphetamine withdrawal alters bistable states and cellular coupling in rat prefrontal cortex and nucleus accumbens neurons recorded in vivo. J. Neurosci. 20, 2332–2345.

Pereda, A.E., Faber, D.S., 1996. Activity-dependent short-term enhancement of intercellular coupling. J. Neurosci. 16, 983–992.

Pereda, A., Triller, A., Korn, H., Faber, D.S., 1992. Dopamine enhances both electrotonic coupling and chemical excitatory postsynaptic potentials at mixed synapses. Proc. Natl. Acad. Sci. U.S.A. 89, 12088–12092.

Pereda, A.E., Nairn, A.C., Wolszon, L.R., Faber, D.S., 1994. Postsynaptic modulation of synaptic efficacy at mixed synapses on the Mauthner cell. J. Neurosci. 14, 3704–3712.

Pereda, A.E., Bell, T.D., Chang, B.H., Czernik, A.J., Nairn, A.C., Soderling, T.R., Faber, D.S., 1998. Ca^{2+}/calmodulin-dependent kinase II mediates simultaneous enhancement of gap-junctional conductance and glutamatergic transmission. Proc. Natl. Acad. Sci. U.S.A. 95, 13272–13277.

Pereda, A., O'Brien, J., Nagy, J.I., Bukauskas, F., Davidson, K.G.V., Kamasawa, N., Yasumura, T., Rash, J.E., 2003. Connexin35 mediates electrical transmission at mixed synapses on Mauthner cells. J. Neurosci. 23, 7489–7503.

Pereda, A.E., Rash, J.E., Nagy, J.I., Bennett, M.V.L., 2004. Dynamics of electrical transmission at club endings on the Mauthner cells. Brain Res. Rev. 47, 227–244.

Prinz, A.A., Abbott, L.F., Marder, E., 2004. The dynamic clamp comes of age. Trends Neurosci. 27, 218–224.

Rash, J.E., Dillman, R.K., Bilhartz, B.L., Duffy, H.S., Whalen, L.R., Yasumura, T., 1996. Mixed synapses discovered and mapped throughout mammalian spinal cord. Proc. Natl. Acad. Sci. U.S.A. 93, 4235–4239.

Rekling, J.C., Shao, X.M., Feldman, J.L., 2000. Electrical coupling and excitatory synaptic transmission between rhythmogenic respiratory neurons in the preBotzinger complex. J. Neurosci. 20 pp. RC113(1–5).

Robinson, T.E., Berridge, K.C., 2003. Addiction. Annu. Rev. Psychol. 54, 25–53.

Sasaki, K., Cropper, E.C., Weiss, K.R., Jing, J., 2013. Functional differentiation of a population of electrically coupled heterogeneous elements in a microcircuit. J. Neurosci. 33, 93–105.

Schultz, W., 1998. Predictive reward signal of dopamine neurons. J. Neurophysiol. 80, 1–27.

Schultz, W., 2013. Updating dopamine reward signals. Curr. Opin. Neurobiol. 23, 229–238.

Selverston, A.I., Miller, J.P., 1980. Mechanisms underlying pattern generation in lobster stomatogastric ganglion as determined by selective inactivation of identified neurons. I. Pyloric system. J. Neurophysiol. 44, 1102–1121.

Sharp, A.A., Abbott, L.F., Marder, E., 1992. Artificial Electrical synapses in oscillatory networks. J. Neurophysiol. 67, 1691–1694.

Sharp, A.A., O'Neil, M.B., Abbott, L.F., Marder, E., 1993a. The dynamic clamp: artificial conductances in biological neurons. Trends Neurosci. 16, 389–394.

Sharp, A.A., O'Neil, M.B., Abbott, L.F., Marder, E., 1993b. Dynamic clamp: computer-generated conductances in real neurons. J. Neurophysiol. 69, 992–995.

Sherman, A., Rinzel, J., 1992. Rhythmogenic effects of weak electrotonic coupling in neuronal models. Proc. Natl. Acad. Sci. U.S.A. 89, 2471–2474.

Sieling, F., Bédécarrats, A., Simmers, J., Prinz, A.A., Nargeot, R., 2014. Differential role of nonsynaptic and synaptic plasticity in operant conditioning-induced compulsive behavior. Curr. Biol. 24, 941–950.

Smith, M., Pereda, A.E., 2003. Chemical synaptic activity modulates nearby electrical synapses. Proc. Natl. Acad. Sci. U.S.A. 100, 4849–4854.

Smolen, P., Rinzel, J., Sherman, A., 1993. Why pancreatic islets burst but single beta cells do not. The heterogeneity hypothesis. Biophys. J. 64, 1668–1680.

Song, J., Ampatzis, K., Björnfors, E.R., El Manira, A., 2016. Motor neurons control locomotor circuit function retrogradely via gap junctions. Nature 529, 399–402.
Sotelo, C., Llinas, R., 1972. Specialized membrane junctions between neurons in the vertebrate cerebellar cortex. J. Cell. Biol. 53, 271–289.
Soto-Treviño, C., Rabbah, P., Marder, E., Nadim, F., 2005. Computational model of electrically coupled, intrinsically distinct pacemaker neurons. J. Neurophysiol. 94, 590–604.
Szucs, A., Varona, P., Volkovskii, A.R., Abarbanel, H.D., Rabinovich, M.I., Selverston, A.I., 2000. Interacting biological and electronic neurons generate realistic oscillatory rhythms. Neuroreport 11, 63–69.
Traub, D., Whittington, M.A., Draguhn, A., 2017. Gap junctions between pyramidal cells account for a variety of very fast network oscillations (> 80 Hz) in cortical structures. In: Jing, J. (Ed.), Network Functions and Plasticity: Perspectives from Studying Neuronal Electrical Coupling in Microcircuits. Elesevier (Chapter 13).
Tresch, C., Kiehn, O., 2000. Motor coordination without action potentials in the mammalian spinal cord. Nat. Neurosci. 3, 593–599.
Tritsch, N.X., Sabatini, B.L., 2012. Dopaminergic modulation of synaptic transmission in cortex and striatum. Neuron 76, 33–50.
Turecek, J., Yuen, G.S., Han, V.Z., Zeng, X.H., Bayer, K.U., Welsh, J.P., 2014. NMDA receptor activation strengthens weak electrical coupling in mammalian brain. Neuron 81, 1375–1388.
Van Der Giessen, R.S., Koekkoek, S.K., van Dorp, S., De Gruijl, J.R., Cupido, A., Khosrovani, S., Dortland, B., Wellershaus, K., Degen, J., Deuchars, J., Fuchs, E.C., Monyer, H., Willecke, K., De Jeu, M.T., De Zeeuw, C.I., 2008. Role of olivary electrical coupling in cerebellar motor learning. Neuron 58, 599–612.
van Essen, T.A., van der Giessen, R.S., Koekkoek, S.K., Vanderwerf, F., Zeeuw, C.I., van Genderen, P.J., Overbosch, D., de Jeu, M.T., 2010. Anti-malaria drug mefloquine induces motor learning deficits in humans. Front. Neurosci. 4, 191.
Vandecasteele, M., Deniau, J.M., Glowinski, J., Venance, L., 2007. Electrical synapses in basal ganglia. Rev. Neurosci. 18, 15–35.
Wise, R.A., Koob, G.F., 2014. The development and maintenance of drug addiction. Neuropsychopharmacology 39, 254–262.
Wise, R.A., 2013. Dual roles of dopamine in food and drug seeking: the drive-reward paradox. Biol. Psychiatry 73, 819–826.
Yang, X.D., Faber, D.S., 1991. Initial synaptic efficacy influences induction and expression of long-term changes in transmission. Proc. Natl. Acad. Sci. U.S.A. 88, 4299–4303.
Yang, X.D., Korn, H., Faber, D.S., 1990. Long-term potentiation of electrotonic coupling at mixed synapses. Proc. Natl. Acad. Sci. U.S.A. 348, 542–545.
Zlomuzica, A., Reichinnek, S., Maxeiner, S., Both, M., May, E., Wörsdörfer, P., Draguhn, A., Willecke, K., Dere, E., 2010. Deletion of connexin45 in mouse neurons disrupts one-trial object recognition and alters kainate-induced gamma-oscillations in the hippocampus. Physiol. Behav. 101, 245–253.

CHAPTER 7

Electrical Synapses and Neuroendocrine Cell Function

N.S. Magoski
Queen's University, Kingston, ON, Canada

In a diversity of species, many fundamental behaviors and essential processes are controlled or strongly influenced by neuroendocrine and neurosecretory cells. This includes flight-or-fight responses, fluid balance, growth, lactation, metabolism, ovulation, parturition, phenotypic changes, and stress (Conn and Kaczmarek, 1989; Veelaert et al., 1998; Toivonen and Partridge, 2009; Koene, 2010; Colomer et al., 2011; Rorsman et al., 2012; Katayama et al., 2013). Such an impact over a broad range of events arises from the ability of these cells to release biologically potent hormones in a coordinated manner (Bargmann and Scharrer, 1951; Mason and Bern, 1977; Hartenstein, 2006; Brown et al., 2013). This chapter is concerned with how electrical coupling shapes the activity and output of neuroendocrine cells.

1. NEUROENDOCRINE CELLS

Neuroendocrine or neurosecretory cells have the capacity to secrete hormones directly into either the blood stream or the local extracellular space in an activity-dependent manner (Bargmann and Scharrer, 1951; Hartenstein, 2006; Brown et al., 2013). Using histological techniques, neurosecretory cells were discovered or proposed as a distinct type of neuron by Dahlgren (1914) and Speidel (1919) in the spinal cord of skates, a cartilaginous fish. Subsequently, Scharrer (1928) noted similar glandular cells in the hypothalamus of the minnow, *Phoxinus*. This was followed by descriptions of neurosecretory cells in central nervous system (CNS) of the gastropod molluscs, *Aplysia* and *Pleurobranchaea* (Scharrer, 1935), the honey bee, *Apis* (Weyer, 1935), and several decapod crustaceans (Hanstrom, 1931, 1935). Early intracellular recordings (Graham and Gerard, 1946; Ling and Gerard, 1949) from neurosecretory cells in *Aplysia* (Kupfermann, 1967; Kupfermann and Kandel, 1970), the crayfish, *Procambarus* (Iwasaki and Satow, 1971), the caudal spinal cord of both the eel, *Anguilla* (Morita et al., 1961), and skate (Bennett and Fox, 1962), as well as in the hypothalamus of the gold fish, *Carassius* (Kandel, 1964), and cat or dog (Koizumi and Yamashita, 1972) revealed these cells had negative resting membrane potentials, fired bona fide action potentials that overshot zero millivolts, and received synaptic input.

Network Functions and Plasticity
ISBN 978-0-12-803471-2, http://dx.doi.org/10.1016/B978-0-12-803471-2.00007-2

With the exception of cells in certain glands (eg, adrenal medulla or pancreas), the morphology of neuroendocrine cells is like that of regular neurons, but frequently more minimal, with a single, sometimes branching axon and a limited number of shorter dendrites extending from the soma (Fig. 7.1A) (Mason and Bern, 1977). The truly distinguishing feature of neuroendocrine cells is the presence of electron-dense vesicles (also referred to as dense-core vesicles) that are 100–300 nm in diameter and contain peptides or other hormones, such as catecholamines (Mason and Bern, 1977; Hartenstein, 2006;

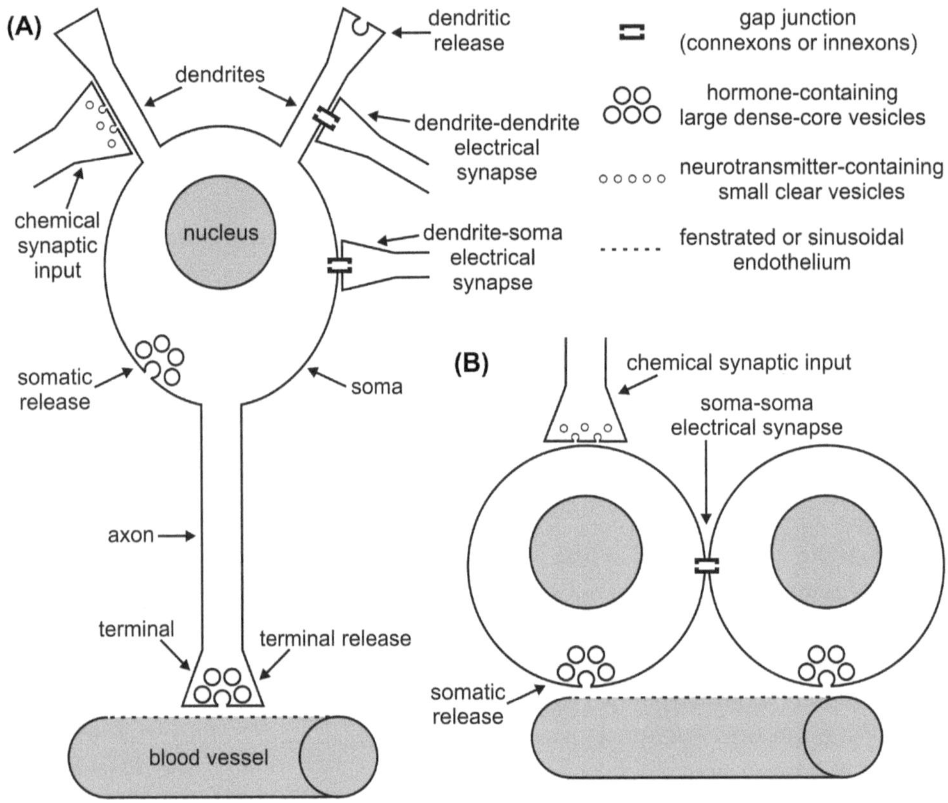

Figure 7.1 ***Neuroendocrine cells can receive both chemical and electrical synaptic input as well as release hormone at different cellular regions.*** (A) An idealized neuroendocrine cell is the recipient of both chemical transmission, mainly onto a small number of dendrites, and electrical transmission, where gap junctions are formed with adjacent neuroendocrine cells at dendrite–dendrite as well as dendrite–soma contacts. A single axon extends from the soma to a neurohemal area, which has blood vessels with an endothelial layer that is fenestrated or sinusoidal, and allows hormones released from terminals to enter the circulation. Some neuroendocrine cells also secrete hormone directly from the soma or dendrites into the local extracellular space. (B) Certain neuroendocrine cells lack any form of morphological process or extension, and both the electrical synapses with neighboring cells and hormone exocytosis occur at the soma. In some cases, these somata also receive chemical inputs from neurons elsewhere in the nervous system.

Brown et al., 2013). These vesicles are found in the soma and axon terminals as well as in dendrites. Hormone release into the circulation occurs primarily from the terminals, at a region of close apposition with blood vessels termed the neurohemal organ (Knowles and Carlisle, 1956; Mason and Bern, 1977). That stated, secretion can also take place from the soma or even dendrites (Pow and Morris, 1989; Ludwig et al., 2002); indeed, for pancreatic beta cells and chromaffin cells of the adrenal medulla, which lack processes, somatic secretion is the only option (Fig. 7.1B) (Houy et al., 2013).

Dense-core vesicle release appears to involve much of the same protein machinery required for Ca^{2+}-dependent secretion of classical neurotransmitters, like acetylcholine or glutamate, including the SNARE (soluble *N*-ethylmaleimide-sensitive factor attachment protein receptor) complex (Rettig and Neher, 2002; Toescu and Dayanithi, 2012). However, unlike the majority of regular neurons, which are usually involved in moment-to-moment communication on the order of milliseconds to seconds (Bennett, 1977), neuroendocrine cells are commonly required to signal *en masse* for more lengthy periods lasting seconds to minutes (Mason and Bern, 1977). When acting over prolonged time frames as a group, in a glandular or gland-like fashion, it may be favorable for neuroendocrine cells to synchronize both the synaptic input and resulting action potential firing. This would allow for a substantial and temporally appropriate degree of hormone release. One manner in which such synchronization could be achieved is through electrical coupling between adjacent neuroendocrine cells.

2. GAP JUNCTIONS AND ELECTRICAL COUPLING IN NEUROENDOCRINE CELLS

In the majority of cases, physiological-, anatomical-, and/or molecular-based lines of inquiry find a given group of neuroendocrine cells to be coupled by gap junctions (Michon et al., 2005; Bosco et al., 2011). However, there are some instances where the evidence for coupling is less conclusive or varies between species, such as gonadotropin-releasing hormone neurons of the vertebrate hypothalamus (Vazquez-Martinez et al., 2001; Campbell et al., 2011; Ma et al., 2015). Nevertheless, for a nucleus or cluster of neuroendocrine cells that are established as being coupled, the electrical synapses are most likely to be dendrite—dendrite or dendrite—soma; however, for nonbranching cell types, the coupling is soma—soma (Fig. 7.1) (Orci et al.,1973; Kaczmarek et al., 1979; Grynszpan-Wynograd and Nicolas, 1980; Cobbett and Hatton, 1984). The molecular basis for these gap junctions is the presence of cell-to-cell pores consisting of paired hemichannels, assembled from connexin (for vertebrates) or innexin (for invertebrates) proteins (Fig. 7.2) (Sosinsky and Nicholson, 2005; Phelan et al., 2008). Each connexin/innexin is a protein with four transmembrane domains, with both the amino- and carboxy-terminus residing intracellularly; in turn, six connexins/innexins make up a connexon/innexon, which is also referred to as a hemichannel. Two hemichannels, one from each cell, form a gap junction channel that links

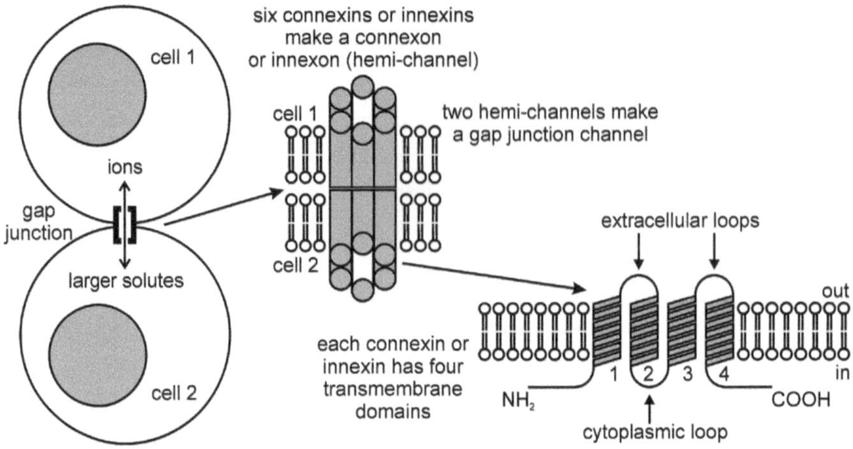

Figure 7.2 ***Gap junctions are intercellular channels formed by water-lined proteinaceous pores.*** (Left) A gap junction channel passes current, in the form of small intracellular ions, as well as larger solutes, like glucose or second messengers, from one cell to another. A grouping of gap junction channels makes an array or plaque, which is considered an electrical synapse, particularly when the coupled cells are excitable. (Middle) To make a gap junction, each cell contributes one hemichannel, which is also known as a connexon (in vertebrate) or innexon (in invertebrates). The two hemichannels not only cross the plasma membrane of their respective cells, but also bridge the gap between the cells to meet in the extracellular space and make an intercellular pathway. A given hemichannel comprises of six protein subunits, called connexins or innexins. Thus, 12 individual proteins make up a complete gap junction channel. (Right) A single connexin or innexin is a protein with both the amino and carboxy ends inside the cell, as well as four transmembrane domains joined by two extracellular loops, between domains 1 and 2 and domains 3 and 4, along with a cytoplasmic loop linking domains 2 and 3.

the cytoplasm of adjacent neurons (for further details see Pereda and Bennett, 2017 (Chapter 8); Phelan et al., 2017 (Chapter 2)).

Gap junctions pass current, in the form of intracellular ions, principally K^+ and Cl^-, under physiological conditions, and also Ca^{2+}, larger anions, second messengers, and small metabolites (Verselis et al., 1986; Bukauskas and Verselis, 2004; Krysko et al., 2005). Some gap junctions are opened or closed by changes in membrane potential (both the transjunctional voltage and the difference in voltage between the inside and outside of the neuron), whereas intracellular Ca^{2+} closes most gap junctions (Bennett, 1977; Bukauskas and Verselis, 2004), a phenomenon first noted in the exocrine salivary gland cells of the midge, *Chironomus* (Loewenstein et al., 1967). In terms of action potential firing synchrony, the flow of current is key; hence, when referring to coupling between excitable cells, "electrical synapse" is often used interchangeably with "gap junction." The first dual recordings demonstrating authentic electrical coupling between neuroendocrine cells involved pairs of pancreatic beta cells from rat (Meissner, 1976) and bag cell neurons from *Aplysia* (Blankenship and Haskins, 1979). Fig. 7.3 displays an

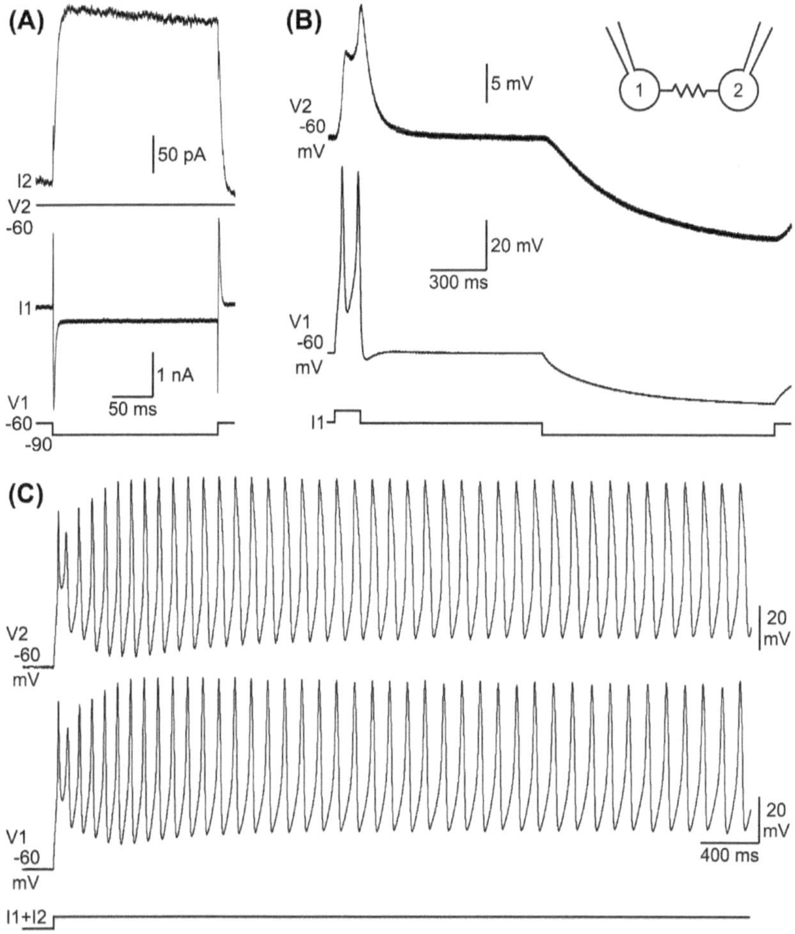

Figure 7.3 ***Electrical coupling between neuroendocrine cells.*** (A) Whole-cell voltage-clamp recordings from a pair of cultured *Aplysia* bag cell neurons that have made physical contact after being in vitro for 2 days. Both neuroendocrine cells are initially held at −60 mV, but when the voltage of cell 1 (V1) is stepped to −90 mV while continuing to hold cell 2 (V2) at −60 mV, junctional current is apparent in cell 2 (I2). This current is what the voltage-clamp amplifier must pass to keep cell 2 at −60 mV, despite the 30 mV difference imposed across the gap junction by pulsing cell 1 to −90 mV. The current in cell 1 (I1) is simply the current flowing across the plasma membrane, whereas the fast inward and outward current deflections at the start and end of the step reflect the charging and discharging of cell 1 membrane capacitance. (B) Whole-cell current-clamp recordings from the same pair in panel (A). Inset reflects the recording configuration with the two bag cell neurons electrically coupled and monitored with separate whole-cell electrodes. At first, the membrane potential of both neuroendocrine cells (V1 and V2) is set to −60 mV with constant bias current. Then a brief step of +1 nA into cell 1 (I1) evokes both action potentials in cell 1 and electrotonic potentials in cell 2. Note how the rising phase of the electrotonic potential begins before the peak of the action potential. Furthermore, a longer injection of −100 pA into cell 1 results in hyperpolarization of cell 2. The neurons are strongly coupled, with a coupling coefficient, calculated as the change in V2 over the change in V1, of around 0.65. (C) Synchronous firing in a different pair of cultured bag cell neurons that have made physical contact and formed an electrical synapse after 1 day in vitro. The neuroendocrine cells are again whole-cell current clamped at −60 mV (V1 and V2); subsequent injection of +1 nA into both neurons (I1 + I2) elicits repetitive action potential firing, where very early in the response the spikes phase lock with one another. All data are unpublished observations of N.S. Magoski, but see Dargaei et al. (2014) for similar results.

example of electrical coupling between bag cell neurons, a type of neuroendocrine cell, in culture (see also Section 8). Under voltage clamp, creating a voltage difference across the gap junction, by hyperpolarizing the membrane potential of one neuron, caused junctional current to flow and appear as outward current in the second neuron (Fig. 7.3A). Under current clamp, the electrical synapse enabled action potentials stimulated in one cell to evoke electrotonic potentials in the second cell, while hyperpolarizing one neuron lead to hyperpolarization of the other (Fig. 7.3B). Finally, when both neurons were made to spike simultaneously, their action potentials quickly synchronized (Fig. 7.3C). Unlike the pioneering finding in crayfish (Furshpan and Potter, 1959), where giant motor axons are connected by a rectifying electrical synapse (current flows one way), the coupling between neuroendocrine cells appears to be nonrectifying (current flows equally well in both directions) and, overall, is voltage independent. That neuroendocrine cells are linked by nonrectifying electrical synapses may again reflect the need for these neurons to act together, and fire in a synchronous manner to secrete hormone. The subsequent sections provide some detail on electrical synapses between neuroendocrine cells from various species (Table 7.1), while attempting to point out common mechanisms.

3. THE X-ORGAN-SINUS GLAND COMPLEX OF CRUSTACEA

The eyestalk of crustaceans, such as crab, crayfish, lobster, or shrimp, contains a group of neurosecretory cells referred to as the X-organ (Carlisle and Passano, 1953; García and Aréchiga, 1998). These neurons project axons to the sinus gland, so named because of its proximity to a blood sinus near the optic ganglia situated below the retina. Via the sinus gland, the X-organ secretes a number of peptides into the hemolymph, including hyperglycemic, molt-inhibiting, vitellogenesis-inhibiting, red-pigment-concentrating, and pigment-dispersing hormones (García and Aréchiga, 1998; Katayama et al., 2013). The designations of these peptides reflect their role in metabolism, reproduction (yolk formation), and both short- and long-term phenotypic changes (pigment distribution and molting).

Intracellular recordings from both the crayfish, *Procambarus* (Iwasaki and Satow, 1969, 1971), and the crabs, *Cardisoma* and *Podophthahus* (Nagano and Cooke, 1987), showed that X-organ neuroendocrine cells fire prolonged bursts of action potentials, lasting several minutes. Moreover, staining of neurons in crayfish X-organs with the fluorescent dye Lucifer yellow (Stewart, 1978) revealed some but not all of the cells to be dye coupled to one or several other cells (Aréchiga et al., 1985; Alvarado Alvárez et al., 1993). Electron microscopy demonstrated gap junction arrays (an assembly or plaque of gap junction channels) at dendrite—soma and soma—soma contacts (Aréchiga et al., 1985). Paired recordings, again in the crayfish, found about 30% of X-organ neuroendocrine cells to be coupled by nonrectifying electrical synapses (Alvarado Alvárez et al., 1993).

Table 7.1 Select Electrically Coupled Neuroendocrine Cells (Alphabetical by Taxonomic Rank)[a]

Taxonomic Rank	Cell	Structure	Location	Hormones	Function
Crustacea[b]	Neurosecretory cells	X-organ	Peripheral	Hyperglycemic hormone	Metabolism
				Molt-inhibiting hormone	Molting
				Vitellogenesis-oocyte inhibiting hormone	Maturation
				Red pigment-concentrating hormone	Chromatophore control
				Pigment-dispersing hormone	Chromatophore control
Insecta[c]	Intrinsic neurosecretory cells	Corpus cardiacum	Peripheral	Adipokinectic hormone I and II	Lipid metabolism
Insecta[c]	Gland cells	Prothoracic gland	Peripheral	Ecdysone	Molting
Mollusca[d]	Bag cell neurons	Abdominal ganglion[e]	Central	Egg-laying hormone	Reproduction
Mollusca[d]	Caudodorsal cells	Cerebral ganglion[f]	Central	Caudodorsal cell hormone	Reproduction
Vertebrata[g]	β-cells	Pancreas	Peripheral	Insulin	Metabolism
Vertebrata[g]	Chromaffin cells	Adrenal medula	Peripheral	Epinephrine or norepinephrine	Flight or fight
Vertebrata[g]	Magnocellular neurons	Hypothalamus[h]	Central	Oxytocin or vasopressin	Lactation, parturition, antidiuresis

[a]Phylum, subphylum, or class.
[b]Subphylum of phylum Arthropoda.
[c]Class of phylum Arthropoda.
[d]Phylum.
[e]Bag cell neuron clusters in the abdominal ganglion of the marine snail, *Aplysia californica*.
[f]Caudodorsal cell clusters in the cerebral ganglia of the pond snail, *Lymnaea stagnalis*.
[g]Subphylum of phylum Chordata.
[h]In the mammalian hypothalamus, the paraventricular and supraoptic nuclei.

This was apparent both as synchronous spiking and the response to hyperpolarizing current injection, the latter revealing unusually large coupling coefficients of near 1. The coupling coefficient reflects the extent of voltage transfer from a neuron to its coupled partner; typically, this ratio is far less than 1. As not all cells are coupled, it would seem the X-organ is not an electrical syncytium, which may reflect heterogeneity, where different hormones are expressed and released by distinct populations of neuroendocrine cells. In the isopod, *Oniscus*, simultaneous extracellular recordings of axons entering the left and right sinus gland showed that, although bursts from the two X-organs were largely synchronous, there were differences in frequency and duration, supporting the notion that specific groups of cells may be coupled and fire at select rates (Chiang and Steel, 1985).

4. THE PROTHORACIC GLAND AND INTRINSIC NEUROSECRETORY CELLS OF THE CORPORA CARDIACA FROM INSECTA

Molting in insects is controlled by the paired or single prothoracic gland, a neuroendocrine structure that is located dorsally just after the head, in the prothorax (Fukuda, 1940; Hartenstein, 2006; Toivonen and Partridge, 2009). The cells of the prothoracic gland synthesize and secrete ecdysteroid molting hormones, eg, ecdysone, in response to prothoracicotropic hormone, a peptide originating elsewhere in the brain (Hartenstein, 2006; Toivonen and Partridge, 2009). Ecdysone release correlates well with tonic action potential firing in the gland of the cockroach, *Periplaneta* (Richter and Gersch, 1983); also, for the tobacco hornworm, *Manduca*, surges in ecdysone synthesis require Ca^{2+} influx (Smith et al., 1985). Hornworm prothoracic neuroendocrine cells are round or oval in shape, and present soma—soma gap junctions at the level of electron microscopy (Hanton et al., 1993). Accordingly, Eusebio and Moody (1986) found that not only did these cells have long-duration, Ca^{2+}-dependent action potentials, but they also were electrically coupled. Furthermore, recording from cells as far as 20 somata apart showed both passive spread of voltage as well as propagation of action potentials over many cells. Spiking may be transmitted by electrical synapses across much or all of the gland, and the ensuing Ca^{2+} entry could play a role in the synthesis and/or release of ecdysone.

Another critical neuroendocrine center in the insect is the singular corpus cardiacum or the paired corpora cardiaca (Thomsen, 1943, 1952; Veelaert et al., 1998; Hartenstein, 2006). In a manner analogous to the hypothalamic—pituitary gland arrangement of vertebrates, neurosecretory cells from higher centers in the insect brain project axons to a neurohemal area in the corpus cardiacum, designated the storage lobe, where terminals deliver numerous peptide hormones into the blood (Nässel, 2002). In addition, the corpus cardiacum also has a glandular lobe, which contains intrinsic neurosecretory cells that secrete adipokinetic hormones into the blood so as to increase hemolymph lipid levels during flight (Mayer and Candy, 1969). Lucifer yellow staining of the intrinsic neuroendocrine cells of locust revealed monopolar neurons with moderately branching axons that are dye coupled to two or three proximal cells about one-third of the time (Orchard and Shivers, 1986). Electron microscopy and electrophysiology found gap junction plaques and modest, nonrectifying electrical coupling between these neurons (Orchard and Shivers, 1986). Such electrical synapses could function to synchronize intrinsic neuroendocrine cell activity following stimulation by a common excitatory chemical synaptic input, possibly octopamine, from higher brain areas (Orchard and Loughton, 1981; Orchard et al., 1983).

5. THE BETA CELLS OF THE VERTEBRATE PANCREAS

Beta cells, along with alpha and delta cells, form the islets of Langerhans in the pancreas of Vertebrata (Rorsman et al., 2012). This conglomeration of endocrine cells surround or are juxtaposed to the vasculature, allowing for the secretion of hormone into the blood. Glucose exposure depolarizes beta cells, resulting in action potential firing and the Ca^{2+}-influx-mediated release of insulin (Dean and Matthews, 1968; Rorsman et al., 2012). Beta cells in the rat, which are oval to polygonal shaped, were shown with the aid of electron microscopy to be connected by soma—soma gap junction arrays (Orci et al., 1973). The first dual recordings, in this instance from mouse, demonstrated that the regular bursting pattern evoked by glucose was synchronous between the majority of beta cells near each another in a given islet (Meissner, 1976). Using both dissected islets and dissociated beta-cell pairs from mouse, nonrectifying, voltage-independent electrical synapses, with coupling coefficients of around 0.5 at physiological glucose concentrations, were observed (Eddlestone et al., 1984; Pérez-Armendariz et al., 1991). Coupling was only detected if the cells were 35 μm or less apart (Eddlestone et al., 1984).

Insulin secretion from isolated rat islets was reduced following block of gap junction channels with the long-chain alcohol heptanol (Johnston et al., 1980; Meda et al., 1990). In addition, when individual beta cells were stained with Lucifer yellow, the antagonist lowered the average number of dye-coupled cells from around three to close to one-and-a-half (Meda et al., 1990). These data, along with the 35-μm coupling distance limit of Eddlestone et al. (1984), intimated that the electrical synapses within an islet were limited and the unit is not a syncytium. In situ hybridization and immunostaining confirmed that the molecular composition of rat beta-cell gap junctions was the result of connexin 36 expression (named as such because of the predicted molecular weight, in kilodaltons, for the protein) (Serre-Beinier et al., 2000). Not surprisingly, the frequency of beta-cell dye coupling, as well as the amplitude and synchrony of glucose-induced intracellular Ca^{2+} waves, as measured by imaging of the Ca^{2+}-sensitive fluorescent dye fura (Vorndran et al., 1995), was less in islets harvested from connexin 36 knockout mice (Ravier et al., 2005). Moreover, although the beta cells from the mutant mice still released insulin to elevated glucose, they did not do so in a pulsatile fashion, as was the case in the wild type (Ravier et al., 2005). Thus, the secretory output of beta cells appears to strongly depend on electrical synapses.

6. THE CHROMAFFIN CELLS OF THE VERTEBRATE ADRENAL MEDULLA

The flight-or-fight response is mediated by chromaffin cells of the adrenal medulla in Vertebrata; following sympathetic cholinergic input, the activation of mainly nicotinic receptors provokes action potentials and the neurohemal release of adrenaline or noradrenaline, leading to a state of whole-animal arousal in the face of potential danger (Cannon, 1915; Feldberg et al., 1934; Douglas and Rubin, 1963; Brandt et al., 1976; Colomer et al., 2011). Chromaffin cells range from oval, to polygonal, to columnar in shape, and are grouped together primarily near medullary veins (Colomer et al., 2011). Grynszpan-Wynograd and Nicolas (1980) employed electron microscopy to establish that chromaffin cells from guinea pigs, hamsters, and rats have bands or even loops of gap junctions rather than more dense plaques. Immunostaining, Northern blot analysis, and/or polymerase chain reaction then indicated that the rat adrenal medulla expressed connexin 43 (Meda et al., 1993) as well as connexin 36 (Martin et al., 2003), whereas the most abundant gap junction protein in human chromaffin cells was connexin 50 (Willenberg et al., 2006).

Dual recordings from rat adrenal medullary slices initially showed that adjacent chromaffin cells were coupled by nonrectifying, voltage-independent electrical synapses (Martin et al., 2001, 2003). However, a large variance in the coupling coefficient (from less than 0.1 to approaching 1) was also observed, which in itself could have been due to multiple cell coupling (Martin et al., 2001). Lucifer yellow fills of single chromaffin cells lead to dye spread to one or two neighboring cells about 30% of the time, and this was lessened by treatment with the gap junction blocker, carbenoxolone (Davidson and Baumgarten, 1988; Martin et al., 2001). Imaging these slices uncovered either spontaneous or nicotine-induced synchronous Ca^{2+} oscillations; moreover, action potential firing in one chromaffin cell resulted in catecholamine release, as detected by amperometry, from an adjacent cell, but only if that cell was coupled (Martin et al., 2001). In whole mice, carbenoxolone lowered splanchnic nerve-evoked catecholamine secretion from intact adrenal medulla, whereas connexin 36 knockdown animals had a paucity of secretion to similar stimulation (Desarménien et al., 2013). This indicates how electrical synapses can enlist partnered neuroendocrine cells and bring about concerted hormone exocytosis. Thus, despite likely not being a syncytium, the presence of gap junctions, in combination with an essentially simultaneous and widespread cholinergic input, may allow for multichromaffin cell recruitment.

Subsequent work suggested that chromaffin cell electrical coupling is regulated by presynaptically released acetylcholine. Specifically, short-term (minutes) incubation of rat adrenal medullary slices in the general noncompetitive nicotinic antagonist, hexamethonium (Paton and Zaimis, 1948; Tauc and Gerschenfeld, 1961), increased the frequency of Lucifer yellow dye spread from one chromaffin cell to another, elevated the conductance of the electrical synapse between paired cells, and enhanced the degree of Ca^{2+} oscillation propagating from cell to cell (Martin et al., 2003). Furthermore, removal of cholinergic afferents, through splanchnicectomy, resulted in greater dye spread (Martin et al., 2003). Adrenal medullary coupling was also regulated by cold stress, which led to more connexin 36 and 43 being expressed, as well as stronger chromaffin cell electrical synapses and Ca^{2+} oscillations (Colomer et al., 2008; Desarménien et al., 2013). This plasticity may be used by chromaffin cells to fortify coupling and maintain hormone release in the face of diminished chemical input or adverse conditions.

7. THE MAGNOCELLULAR NEUROENDOCRINE CELLS OF THE MAMMALIAN HYPOTHALAMUS

The hypothalamic paraventricular and supraoptic nuclei contain magnocellular neuroendocrine cells, which usually have one to three main dendrites and a single axon that extends to the posterior pituitary (Bargmann and Scharrer, 1951; Leng and Brown, 1997). At a neurohemal area in the posterior pituitary, these neurons secrete oxytocin or vasopressin into the blood so as to elicit lactation, parturition, or antidiuresis. The release of these peptide hormones involves peripheral synaptic input arising during suckling or delivery, as well as elevated plasma osmolarity, causing a characteristic discharge of action potentials and Ca^{2+} entry (Douglas and Poisner, 1964; Dreifuss et al., 1971; Arnauld et al., 1974; Lincoln and Wakerley, 1974; Summerlee, 1981). The supraoptic nucleus has been shown to express connexin 32, as assayed by in situ hybridization, and the levels of this RNA rose both during bouts of lactation and prior to the onset of parturition (Micevych et al., 1996). Immunopositive staining for connexin 43 was also demonstrated in the same region, but in this instance it was enhanced by water deprivation rather than lactation (Eisner and Colombo, 2002).

Andrew et al. (1981) were the first to establish the presence of electrical synapses between rat magnocellular neuroendocrine cells in slices from both paraventricular and supraoptic nuclei, where approximately half of Lucifer yellow–stained neurons dye coupled to a second cell. By electron microscopy, gap junction arrays were found at dendrite–soma contacts, although dendrite–dendrite coupling was also inferred (Andrew et al., 1981). Dual intracellular recordings in supraoptic nucleus slice then showed nonrectifying electrical synapses between magnocellular neuroendocrine cells, with coupling coefficients between 0.05 and 0.2 (Yang and Hatton, 1988). Immunostaining for oxytocin versus vasopressin in rat paraventricular neurons already filled with Lucifer yellow revealed that dye coupling occurred only between cells expressing the same peptide (Cobbett et al., 1985). Thus, like the X-organ, gap junctions may form between a particular variety of hypothalamic neuroendocrine cell, such that they secrete specific hormones in an integrated fashion. As a case in point, paired extracellular recordings of oxytocin neurons from nursing rats presented synchronous bursts of spikes that lasted several seconds and correlated with subsequent milk ejection (Belin and Moos, 1986). Simultaneous action potential firing of greater than 10 min could also be induced by intracerebroventricular injection of acetylcholine (Belin and Moos, 1986).

Similar to the chromaffin cells of the adrenal medulla, the extent of magnocellular neuroendocrine cell electrical coupling was regulated. For example, several minutes of histaminergic input from the tuberomammillary nucleus decreased the incidence of Lucifer yellow dye coupling among supraoptic oxytocin neurons from male rats, and this required postsynaptic activation of adenylate cyclase (Hatton and Yang, 2001). Dye transfer between cells in the paraventricular nucleus of male rats was also subject to modulation, where castration caused dye coupling to go down, and this could be prevented by reintroducing testosterone (Cobbett et al., 1987). Finally, dehydration of rats attenuated an index of paraventricular neuroendocrine cell dye coupling from near 30% to around 12% (Cobbett and Hatton, 1984). Interestingly, that work reported almost three-quarters of the dye coupling to be mediated by soma—soma gap junctions, based on the very close apposition of the cell bodies. Thus, both relatively rapid- (histamine) and slow- (testosterone) acting first messengers, as well as osmotic signals, can impact gap junction expression to potentially alter the coordination of firing and peptide secretion.

8. THE BAG CELL NEURONS OF *APLYSIA* AND CAUDODORSAL CELLS OF *LYMNAEA*

Neurons in the CNS of the gastropod molluscs *Aplysia californica* (a marine anaspidean snail) and *Lymnaea stagnalis* (a freshwater pulmonate snail) have been used for decades to study ion channel function, learning, memory, neural control of behavior, and neuroendocrine function (Kandel, 1979; Kemenes and Benjamin, 2009). Egg laying in these animals is governed by the activity of a homologous group of neuroendocrine cells known as the bag cell neurons in *Aplysia* and the caudodorsal cells in *Lymnaea* (Kupfermann, 1967; Pinsker and Dudek, 1977; Geraerts and Bohlken, 1976; ter Maat et al., 1986). The bag cell neurons are situated in two paired clusters of 200—400 cells each, located left and right in the rostral region of the abdominal ganglion; they receive descending chemical synaptic input from the head ganglia via the pleuroabdominal connective, and also send a projection a short distance back up the connective to a neurohemal area (Coggeshall, 1967; Frazier et al., 1967; Haskins et al., 1981). The caudodorsal cells are also found in two paired clusters of about 30 (left) and 70 (right) neurons in the ventromedial area of the cerebral ganglia; they too are innervated by chemical synapses, and project axons to a neurohemal area in the intercerebral commissure (Joosse, 1964; Wendelaar Bonga, 1971). These neuroendocrine cells secrete several peptides into the hemolymph, but the principle hormones that induce egg-laying behavior are egg-laying hormone for *Aplysia* and caudodorsal hormone for *Lymnaea* (Arch, 1972; Chiu et al., 1979; Geraerts et al., 1983a,b; Hatcher and Sweedler, 2008). These peptides are each 36 amino acids long, share around 40% homology, and act on both neuronal and peripheral targets to bring about a series of behaviors that finish with egg deposition and, ostensibly, propagation of the species (Vreugdenhil et al., 1988; Conn and Kaczmarek, 1989; Zhang and Kaczmarek, 2008; Koene, 2010).

Intracellular and extracellular recordings from either the bag cell neurons or the caudodorsal cells showed that stimulation of excitatory synaptic input triggered a prolonged afterdischarge, ie, a burst of action potentials lasting 30–60 min (Kupfermann and Kandel, 1970; De Vlieger et al., 1980; Kaczmarek et al., 1982; ter Maat et al., 1986). Both types of neuroendocrine cells were depolarized by exogenous acetylcholine (ter Maat et al., 1983; White and Magoski, 2012); furthermore, bag cell neurons have been established as expressing functional acetylcholinesterase, the enzyme that breaks down acetylcholine (Roepke, 1937; Giller and Schwartz, 1971), whereas for the caudodorsal cells, histochemical detection of acetylcholine esterase uncovered putative cholinergic inputs (ter Maat et al., 1983). Moreover, acetylcholine elicited bona fide afterdischarges in bag cell neurons, and synaptically provoked afterdischarges were prevented by the cholinergic blockers, α-conotoxin-ImI and mecamylamine (Stone et al., 1956; McIntosh et al., 1994; White and Magoski, 2012). Therefore, like the analogous magnocellular neurons (Belin and Moos, 1986; Hatton and Yang, 2002), acetylcholine may be a critical input transmitter to set off spiking and exocytosis.

Lucifer yellow staining revealed that bag cell neurons were either mono- or multipolar, had multiple branching of secondary processes, and dye coupled to their neighbors about one-third of the time (Kaczmarek et al., 1979; Haskins et al., 1981). When investigated by electron microscopy, gap junction plaques were apparent between processes in the cluster (Kaczmarek et al., 1979). No dye-coupling data have been published for caudodorsal cells, although horseradish peroxidase injections (Snow et al., 1976) demonstrated that the neurons were exclusively monopolar, with finely branching secondary axons (de Vlieger et al., 1980). Ultrastructural analysis of caudodorsal cells using electron microscopy provided evidence for both process–process and soma–soma gap junction arrays (Roubos et al., 1985). Furthermore, axons of individual caudodorsal cells were traced from one cluster to another, which provided a morphological basis for potential intercluster electrical coupling.

Dual or triple intracellular recordings gave credence to the notion that the bag cell neuron or caudodorsal cell afterdischarge constituted the synchronized spiking of all neurons in both clusters (Kupfermann and Kandel, 1970; Blankenship and Haskins, 1979; de Vlieger et al., 1980; Brown and Mayeri, 1989). Nonrectifying electrical synapses were observed between neurons in the same cluster, but the coupling coefficients were weak, ranging from 0.0075 to 0.015 in *Aplysia* (Blankenship and Haskins, 1979; Dargaei et al., 2014). Although cross-cluster coupling was also noted, it was even more diminutive, particularly for *Aplysia* compared with *Lymnaea* (Blankenship and Haskins, 1979; de Vlieger et al., 1980). Nevertheless, all attempts at detecting electrical coupling between two given caudodorsal cells were successful (de Vlieger et al., 1980), whereas for bag cell neurons most pairs were linked, although some cells failed to show coupling (Blankenship and Haskins, 1979; Dargaei et al., 2014). Overall, these data are consistent with some degree of an electrical syncytium.

When bag cell neurons were dissociated from the cluster and paired in culture, electrical synapses formed (Dargaei et al., 2014). Dual whole-cell current- or voltage-clamp recording (Hamill et al., 1981) affirmed the neurons were connected by nonrectifying and voltage-independent gap junctions; however, the strength of the synapse was higher, with coupling coefficients of around 0.6 and a junctional conductance of about 10 nS (Dargaei et al., 2014). This is in contrast to the weak electrical synapses of the intact cluster, and may have been because only two neurons were partnered, whereas in the cluster, coupling to multiple partners would act as a current sink and dampen the coupling coefficient. Alternatively, in vitro gap junctions may have been stronger because the contact site is physically closer to or even at the soma (the location of recording); within the cluster, coupling could be on remote processes and result in voltage attenuation. Regarding pharmacology, the cultured bag cell neuron electrical synapse was virtually eliminated by the gap junction blockers, niflumic acid (Harks et al., 2001) and 5-nitro-2-(3-phenylpropylamino)benzoic acid (Srinivas and Spray, 2003; Dargaei et al., 2014). Furthermore, when these reagents were applied to the intact cluster, not only was coupling suppressed, but also afterdischarges could no longer be triggered (Dargaei et al., 2014). Thus, gap junctions may be necessary for propagation of action potentials throughout the bag cell neuron network or to extend synaptic input if only a small number of neurons actually receive excitatory cholinergic synapses.

9. THE INFLUENCE OF THE EXTENT AND STRENGTH OF ELECTRICAL COUPLING ON NEUROENDOCRINE CELL FUNCTION

Peptide or catecholamine exocytosis often requires Ca^{2+} influx that is greater or more prolonged than that for classical neurotransmitter release (Neher, 1998; Soldo et al., 2004; Hickey et al., 2013). As such, neuroendocrine cells commonly fire bursts of action potentials, presumably to elevate Ca^{2+} to high levels (Iwasaki and Satow, 1969; Kupfermann and Kandel, 1970; Lincoln and Wakerley, 1974; Meissner, 1976; Mason and Bern, 1977; De Vlieger et al., 1980). Electrical coupling may facilitate the distribution of synaptic excitation or other sources of depolarizing drive, like nonselective cation currents or persistent Ca^{2+} current (Orchard et al., 1983; ter Maat et al., 1983; Leech and Habener, 1998; Magoski et al., 2002; Hatton and Yang, 2002; Tam et al., 2009;

White and Magoski, 2012). Some systems appear to contain cells that make extensive electrical synapses throughout the particular nucleus, cluster, or gland, eg, bag cell neurons, caudodorsal cells, corpus cardiacum, or prothoracic gland, whereas others are seemingly less thoroughly coupled and perhaps consist of subdivisions of cells linked together in smaller groups that do not constitute a syncytium, eg, adrenal medulla, hypothalamus, pancreas, or X-organ. The latter may reflect a need to segregate specific neuroendocrine cells on the basis of which hormones they secrete, so to avoid the potentially adverse consequences of coincidental activation.

Nevertheless, if circumstances require a more collective activation of those neuroendocrine cells that are apparently less coupled, an intensified or better coordinated excitatory chemical synaptic input could bring about higher levels of overall firing. In the rat supraoptic nucleus, the concentration of glutamate rose just prior to parturition (Herbison et al., 1997), whereas histochemical staining for acetylcholinesterase revealed that cholinergic axons formed a nexus around bovine chromaffin cells, in keeping with the majority of cells being innervated (de Diego, 2010). That stated, the extent of coupling may be higher, but not fully elucidated because the electrical synapses are possibly remote from the recording electrode or the number of gap junction channels is low. The flow of current or the passage of dye would be impeded, especially if either must travel through several cells. Finally, dye coupling itself may underreport the extent of linkage if the gap junctions involved simply do not pass the dye, as is the case for some *Aplysia* (Bodmer et al., 1988; Dargaei et al., 2014) and *Lymnaea* (Audesirk et al., 1982; Ewadinger et al., 1994) neurons, as well as pancreatic beta cells (Pérez-Armendariz et al., 1991).

In addition to the degree to which neuroendocrine cells are coupled, there is also the matter of the strength of the electrical synapse. Specifically, coupling strength is determined by the junctional conductance, which arises from the number, open probability, and single-channel conductance of the gap junction channels, as well as the membrane conductance (input resistance) of the two connected cells, and intersomatic distance, which would influence the leak of current as it flows between neurons (Bennett, 1966, 1977; Bennett and Zukin, 2004). At first glance, a parsimonious strategy would be to evolve electrical synapses that are very strong and would, on the face of it, ensure a given cell receives depolarizing current from as many adjacent cells as possible. This may be the situation for X-organ neuroendocrine cells, which displayed coupling coefficients near 1 (Alvarado Alvárez et al., 1993). However, very strong coupling is not necessarily the best manner to encourage a group of cells to burst simultaneously, and synchrony can be achieved even when electrical synapses are modest (Bennett and Zukin, 2004). For example, in a mathematical model of a pancreatic beta-cell cluster, compared to when cells were somatically coupled with a very high junctional conductance, the burst period was greater and the Ca^{2+} change was more prominent when the junctional conductance was actually far smaller and more like that measured physiologically (Sherman and Rizel, 1991). Similarly, a substantial portion of the resting

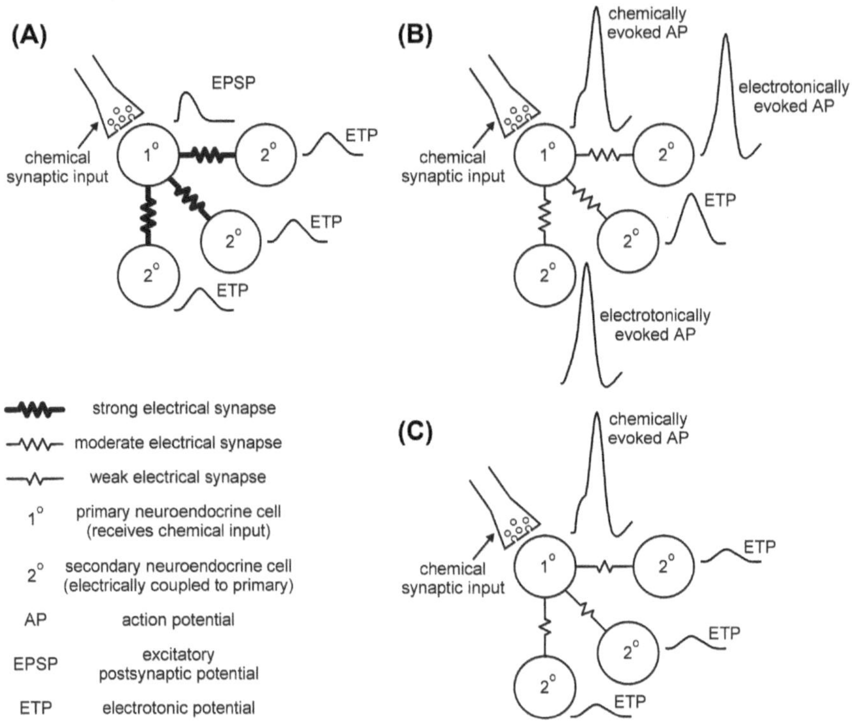

Figure 7.4 ***Strength of electrical coupling in a neuroendocrine cell network influences the spread of excitation.*** Schematic illustration of how the strength of electrical coupling may impact the flow of chemical synaptic input and activity in a neuroendocrine cell network connected by gap junctions. (A) A primary (1°) neuroendocrine cell is coupled by strong electrical synapses (bold resistor symbols) to multiple secondary (2°) neuroendocrine cells. The primary cell receives excitatory chemical synaptic input, which provokes an excitatory postsynaptic potential (EPSP) that in turn elicits electrotonic potentials (ETP) in the secondary cells. On account of the strong coupling, much of the chemically derived synaptic current leaks to the secondary cells. As a result, the EPSP does not take the primary cell above the threshold for firing, nor do any of the ETPs cause spiking in the secondary cells. (B) When the cells are coupled by modest electrical synapses (light resistor symbols) the chemically derived synaptic current from the EPSP does not leak so readily from the primary to the secondary cells. Consequently, the EPSP depolarizes the primary cell to threshold and leads to a chemically evoked action potential (AP). Because the current from the action potential is substantial, it spreads to the secondary cells and induces other action potentials or relatively large ETPs, despite the weaker gap junction. Thus, under certain circumstances, such as restricted chemical excitation, lesser coupling may be more effective at promoting firing in groups of electrically coupled neuroendocrine cells. (C) If the coupling is mediated by weak electrical synapses (light and less-substantial resistor symbols), the chemical input readily stimulates an action potential in the primary neuron. However, this fails to propagate as a spike; rather, only small ETPs are seen in the secondary cells. Because little synaptic current leaks to the secondary neurons, the majority of the chemical input goes toward evoking the action potential in the primary neuron. This spike is transmitted to the secondary cells through the weak gap junctions as a mere small ETP.

conductance of rat somatosensory cortex interneurons, coupled between dendrites or dendrite to soma, was derived from electrical connections to many other neurons (Amitai et al., 2002). The net result being that the amount of current required to evoke action potentials was quite high; thus, strong coupling may be effective for synchrony, but less conducive for spike propagation if the input to a network is limited.

In *Aplysia* bag cell neurons, the action potential firing during the afterdischarge resulted in substantial Ca^{2+} entry (Fisher et al., 1994; Michel and Wayne, 2002), which in turn reduced the junctional conductance between these neurons by about 25%, in a calmodulin-kinase-dependent manner (Dargaei et al., 2015). A straightforward mathematical model showed that, following synaptic input to a subset of cells, the likelihood of all the bag cell neurons bursting synchronously was increased when coupling was lowered modestly; too much junctional conductance resulted in the synaptic current failing to bring the membrane potential to the threshold because of leak to neighboring neurons (Dargaei et al., 2015). Of course, when the junctional conductance was made very low, as if gap junction blockers had been added, the model also failed to afterdischarge, but in that instance because of a lack of spread of excitation altogether (Dargaei et al., 2015). Thus, for effective *en masse* firing there may be an optimal junctional conductance (Sherman and Rizel, 1991); Fig. 7.4 provides a simplified illustration of how this phenomenon may operate.

10. CONCLUDING REMARKS

Electrical synapses are common features of neuroendocrine cell systems. A given group of neuroendocrine cells is often responsible for controlling key behaviors or events, including aspects of reproduction, growth, and metabolism. Considering that prompt and appreciable delivery of hormone into the blood stream is imperative for this to occur, electrical synapses are well suited to ensure that specific cells fire and secrete together.

REFERENCES

Alvarado Alvárez, R., García Hernández, U., Aréchigá, H., 1993. Electrotonic coupling between neurosecretory cells in the crayfish eyestalk. Brain Res. 613, 43–48.

Amitai, Y., Gibson, J.R., Beierlein, M., Patrick, S.L., Ho, A.M., Connors, B.W., Golomb, D., 2002. The spatial dimensions of electrically coupled networks of interneurons in the neocortex. J. Neurosci. 22, 4142–4152.

Andrew, R.D., MacVicar, B.A., Dudek, F.E., Hatton, G.I., 1981. Dye transfer through gap junctions between neuroendocrine cells of rat hypothalamus. Science 211, 1187–1189.

Arch, S., 1972. Polypeptide secretion from the isolated parietovisceral ganglion of *Aplysia californica*. J. Gen. Physiol. 59, 47–59.

Aréchiga, H., Chávez, B., Glantz, R.M., 1985. Dye coupling and gap junctions between crustacean neurosecretory cells. Brain Res. 326, 183–187.

Arnauld, E., Vincent, J.D., Dreifuss, J.J., 1974. Firing patterns of hypothalamic supraoptic neurons during water deprivation in monkeys. Science 185, 535–537.

Audesirk, G., Audesirk, T., Bowsher, P., 1982. Variability and frequent failure of Lucifer Yellow neurones to pass between two electrically coupled neurons in *Lymnaea stagnalis*. J. Neurobiol. 13, 369–375.

Bargmann, W., Scharrer, E., 1951. The site of origin of the hormones of the posterior pituitary. Am. Sci. 39, 255–259.

Belin, V., Moos, F., 1986. Paired recordings from supraoptic and paraventricular oxytocin cells in suckled rats: recruitment and synchronization. J. Physiol. 377, 369–390.

Bennett, M.V.L., Fox, S., 1962. Electrophysiology of caudal neurosecretory cells in the skate and fluke. Gen. Comp. Endocrinol. 2, 77–95.

Bennett, M.V.L., Zukin, R.S., 2004. Electrical coupling and neuronal synchronization in the mammalian brain. Neuron 41, 495–511.

Bennett, M.V.L., 1966. Physiology of electrotonic junctions. Ann. N.Y. Acad. Sci. 137, 509–539.

Bennett, M.V.L., 1977. Electrical transmission: a functional analysis and comparison with chemical transmission. Cellular biology of the neurosecretory neuron. In: Kandel, E.R. (Ed.), Cellular Biology of Neurons, Handbook of Physiology. The Nervous System. Williams and Wilkins, , Baltimore, pp. 357–416.

Blankenship, J., Haskins, J., 1979. Electrotonic coupling among neuro-endocrine cells in *Aplysia*. J. Neurophysiol. 42, 347–355.

Bodmer, R., Verselis, V., Levitan, I.B., Spray, D.C., 1988. Electrotonic synapses between *Aplysia* neurons in situ and in culture: aspects of regulation and measurements of permeability. J. Neurosci. 8, 1656–1670.

Bosco, D., Haefliger, J.A., Meda, P., 2011. Connexins: key mediators of endocrine function. Physiol. Rev. 91, 1393–1445.

Brandt, B.L., Hagiwara, S., Kidokoro, Y., Miyazaki, S., 1976. Action potentials in the rat chromaffin cell and effects of acetylcholine. J. Physiol. 263, 417–439.

Brown, R.O., Mayeri, E., 1989. Positive feedback by autoexcitory neuropeptides in neuroendocrine bag cells of *Aplysia*. J. Neurosci. 9, 1443–1451.

Brown, C.H., Bains, J.S., Ludwig, M., Stern, J.E., 2013. Physiological regulation of magnocellular neurosecretory cell activity: integration of intrinsic, local and afferent mechanisms. J. Neuroendocrinol. 25, 678–710.

Bukauskas, F.F., Verselis, V.K., 2004. Gap junction channel gating. Biochim. Biophys. Acta 1662, 42–60.

Campbell, R.E., Ducret, E., Porteous, R., Liu, X., Herde, M.K., Wellerhaus, K., Sonntag, S., Willecke, K., Herbison, A.E., 2011. Gap junctions between neuronal inputs but not gonadotropin-releasing hormone neurons control estrous cycles in the mouse. Endocrinology 152, 2290–2301.

Cannon, W.B., 1915. Bodily Changes in Pain, Hunger, Fear and Rage. D. Appleton and Company, New York.

Carlisle, D.B., Passano, L.M., 1953. The X-organ of crustacea. Nature 171, 1070–1071.

Chiang, R.G., Steel, C.G., 1985. Coupling of electrical activity from contralateral sinus glands. Brain Res. 331, 142–144.

Chiu, A.Y., Hunkapiller, M.W., Heller, E., Stuart, D.K., Hood, L.E., Strumwasser, F., 1979. Purification and primary structure of the neuropeptide egg-laying hormone of *Aplysia californica*. Proc. Natl. Acad. Sci. 76, 6656–6660.

Cobbett, P., Hatton, G.I., 1984. Dye coupling in hypothalamic slices: dependence on in vivo hydration state and osmolality of incubation medium. J. Neurosci. 4, 3034–3038.

Cobbett, P., Smithson, K.G., Hatton, G.I., 1985. Dye-coupled magnocellular peptidergic neurons of the rat paraventricular nucleus show homotypic immunoreactivity. Neuroscience 16, 885–895.

Cobbett, P., Yang, Q.Z., Hatton, G.I., 1987. Incidence of dye coupling among magnocellular paraventricular nucleus neurons in male rats is testosterone dependent. Brain Res. Bull. 18, 365–370.

Coggeshall, R.E., 1967. A light and electron microscope study of the abdominal ganglion of *Aplysia californica*. J. Neurophysiol. 30, 1263–1287.

Colomer, C., Olivos Ore, L.A., Coutry, N., Mathieu, M.N., Arthaud, S., Fontanaud, P., Iankova, I., Macari, F., Thouënnon, E., Yon, L., Anouar, Y., Guérineau, N.C., 2008. Functional remodeling of gap junction-mediated electrical communication between adrenal chromaffin cells in stressed rats. J. Neurosci. 28, 6616–6626.

Colomer, C., Martin, A.O., Desarménien, M.G., Guérineau, N.C., 2011. Gap junction-mediated intercellular communication in the adrenal medulla: an additional ingredient of stimulus-secretion coupling regulation. Biochim. Biophys. Acta 1818, 1937–1951.

Conn, P.J., Kaczmarek, L.K., 1989. The bag cell neurons of *Aplysia*. Mol. Neurobiol. 3, 237–273.

Dahlgren, U., 1914. The electric motor nerve centers in the skates (Rajidae). Science 40, 862–863.

Dargaei, Z., Colmers, P.L.W., Hodgson, H.M., Magoski, N.S., 2014. Electrical coupling between *Aplysia* bag cell neurons: characterization and role in synchronous firing. J. Neurophysiol. 112, 2680–2696.

Dargaei, Z., Standage, D., Groten, C.J., Blohm, G., Magoski, N.S., 2015. Ca^{2+}-induced uncoupling of *Aplysia* bag cell neurons. J. Neurophysiol. 113, 808–821.

Davidson, J.S., Baumgarten, I.M., 1988. Glycyrrhetinic acid derivatives: a novel class of inhibitors of gap-junctional intercellular communication. Structure-activity relationships. J. Pharmacol. Exp. Ther. 246, 1104–1107.

de Diego, A.M., 2010. Electrophysiological and morphological features underlying neurotransmission efficacy at the splanchnic nerve-chromaffin cell synapse of bovine adrenal medulla. Am. J. Physiol. Cell Physiol. 298, C397–C405.

De Vlieger, T.A., Kits, K.S., Ter Maat, A., Lodder, J.C., 1980. Morphology and electrophysiology of the ovulation hormone producing neuro-endocrine cells of the freshwater snail *Lymnaea stagnalis* (L.). J. Exp. Biol. 84, 259–271.

Dean, P.M., Matthews, E.K., 1968. Electrical activity in pancreatic islet cells. Nature 219, 389–390.

Desarménien, M.G., Jourdan, C., Toutain, B., Vessières, E., Hormuzdi, S.G., Guérineau, N.C., 2013. Gap junction signalling is a stress-regulated component of adrenal neuroendocrine stimulus-secretion coupling *in vivo*. Nat. Commun. 4, 2938.

Douglas, W.W., Poisner, A.M., 1964. Stimulus-secretion coupling in a neurosecretory organ: the role of calcium in the release of vasopressin from the neurohypophysis. J. Physiol. 172, 1–18.

Douglas, W.W., Rubin, R.P., 1963. The mechanism of catecholamine release from the adrenal medulla and the role of calcium in stimulus-secretion coupling. J. Physiol. 167, 288–310.

Dreifuss, J.J., Kalnins, I., Kelly, J.S., Ruf, K.B., 1971. Action potentials and release of neurohypophysial hormones *in vitro*. J. Physiol. 215, 805–817.

Eddlestone, G.T., Gonçalves, A., Bangham, J.A., Rojas, E., 1984. Electrical coupling between cells in islets of Langerhans from mouse. J. Membr. Biol. 77, 1–14.

Eisner, I., Colombo, J.A., 2002. Detection of a novel pattern of connexin 43 immunoreactivity responsive to dehydration in rat hypothalamic magnocellular nuclei. Exp. Neurol. 177, 321–325.

Eusebio, E.J., Moody, W.J., 1986. Calcium-dependent action potentials in the prothoracic gland of Manduca sexta. J. Exp. Biol. 126, 531–536.

Ewadinger, N.M., Syed, N.I., Lukowiak, K., Bulloch, A.G.M., 1994. Differential tracer coupling between pairs of identified neurones of the mollusc, *Lymnaea stagnalis*. J. Exp. Biol. 192, 291–297.

Feldberg, W., Minz, B., Tsudzimura, H., 1934. The mechanism of the nervous discharge of adrenaline. J. Physiol. 81, 286–304.

Fisher, T.E., Levy, S., Kaczmarek, L.K., 1994. Transient changes in intracellular calcium associated with a prolonged increase in excitability in neurons of *Aplysia californica*. J. Neurophysiol. 71, 1254–1257.

Frazier, W.T., Kandel, E.R., Kupfermann, I., Waziri, R., Coggeshall, R.E., 1967. Morphological and functional properties of identified neurons in the abdominal ganglion of *Aplysia californica*. J. Neurophysiol. 30, 1288–1351.

Fukuda, S., 1940. Induction of pupation in silkworm by transplanting the prothoracic gland. Proc. Imp. Acad. 16, 414–416.

Furshpan, E.J., Potter, D.D., 1959. Transmission at the giant motor synapses of the crayfish. J. Physiol. 145, 289–325.

García, U., Aréchiga, H., 1998. Regulation of crustacean neurosecretory cell activity. Cell Mol. Neurobiol. 18, 81–99.

Geraerts, W.P., Bohlken, S., 1976. The control of ovulation in the hermaphroditic freshwater snail *Lymnaea stagnalis* by the neurohormone of the caudodorsal cells. Gen. Comp. Endocrinol. 28, 350–357.

Geraerts, W.P., Cheeseman, P., Ebberink, R.H., Nuyt, K., Hogenes, T.M., 1983a. Partial purification and characterization of the ovulation hormone of the freshwater pulmonate snail *Lymnaea stagnalis*. Gen. Comp. Endocrinol. 51, 471–476.

Geraerts, W.P.M., Tensen, C.P., Hogenes, ThM., 1983b. Multiple release of peptides by electrically active neurosecretory caudodorsal cells of *Lymnaea stagnalis*. Neurosci. Lett. 41, 151–155.

Giller, E., Schwartz, J.H., 1971. Acetylcholinesterase in identified neurons of abdominal ganglion of *Aplysia californica*. J. Neurophysiol. 34, 108–123.

Graham, J., Gerard, R.W., 1946. Membrane potentials and excitation of impaled single muscle fibers. J. Cell. Comp. Physiol. 28, 99–117.

Grynszpan-Wynograd, O., Nicolas, G., 1980. Intercellular junctions in the adrenal medulla: a comparative freeze-fracture study. Tissue Cell 12, 661–672.

Hamill, O.P., Marty, A., Neher, E., Sakmann, B., Sigworth, F.J., 1981. Improved patch-clamp techniques for high-resolution current recording from cells and cell-free membrane patches. Pflugers Arch. 391, 85–100.

Hanstrom, B., 1931. Neue untersuchungen fiber sinnesorgane und nervensystem der crustaceen.I. Z. Morphol. Oekol. Tiere. 23, 80–236.

Hanstrom, B., 1935. Preliminary report on the probable connection between the blood gland and the chromatophore activator in decapod crustaceans. Proc. Natl. Acad. Sci. 21, 584–585.

Hanton, W.K., Watson, R.D., Bollenbacher, W.E., 1993. Ultrastructure of prothoracic glands during larval-pupal development of the tobacco hornworm, *Manduca sexta*: a reappraisal. J. Morphol. 216, 95–112.

Harks, E.G., de Roos, A.D., Peters, P.H., de Haan, L.H., Brouwer, A., Ypey, D.L., van Zoelen, E.J., Theuvenet, A.P., 2001. Fenamates: a novel class of reversible gap junction blockers. J. Pharmacol. Exp. Ther. 298, 1033–1041.

Hartenstein, V., 2006. The neuroendocrine system of invertebrates: a developmental and evolutionary perspective. J. Endocrinol. 190, 555–570.

Haskins, J.T., Price, C.H., Blankenship, J.E., 1981. A light and electron microscopic investigation of the neurosecretory bag cells of *Aplysia*. J. Neurocytol. 10, 729–747.

Hatcher, N.G., Sweedler, J.V., 2008. *Aplysia* bag cells function as a distributed neurosecretory network. J. Neurophysiol. 99, 333–343.

Hatton, G.I., Yang, Q.Z., 2001. Ionotropic histamine receptors and H_2 receptors modulate supraoptic oxytocin neuronal excitability and dye coupling. J. Neurosci. 21, 2974–2982.

Hatton, G.I., Yang, Q.Z., 2002. Synaptic potentials mediated by $\alpha 7$ nicotinic acetylcholine receptors in supraoptic nucleus. J. Neurosci. 22, 29–37.

Herbison, A.E., Voisin, D.L., Douglas, A.J., Chapman, C., 1997. Profile of monoamine and excitatory amino acid release in rat supraoptic nucleus over parturition. Endocrinology 138, 33–40.

Hickey, C.M., Groten, C.J., Sham, L., Carter, C.J., Magoski, N.S., 2013. Voltage-gated Ca^{2+} influx and mitochondrial Ca^{2+} initiate secretion from *Aplysia* neuroendocrine cells. Neuroscience 250, 755–772.

Houy, S., Croisé, P., Gubar, O., Chasserot-Golaz, S., Tryoen-Tóth, P., Bailly, Y., Ory, S., Bader, M.F., Gasman, S., 2013. Exocytosis and endocytosis in neuroendocrine cells: inseparable membranes! Front. Endocrinol. 4, 135.

Iwasaki, S., Satow, Y., 1969. Spontaneous grouped discharge of secretory neuron soma in X-organ of crayfish. *Procambarus clarki*. J. Physiol. Soc. Jap. 31, 629–630.

Iwasaki, S., Satow, Y., 1971. Sodium- and calcium-dependent spike potentials in the secretory neuron soma of the X-organ of the crayfish. J. Gen. Physiol. 57, 216–236.

Johnston, M.F., Simon, S.A., Ramon, R., 1980. Interaction of anaesthetics with electrical synapses. Nature 286, 498–500.

Joosse, J., 1964. Dorsal bodies and dorsal neurosecretory cells of the cerebral ganglia of *Lymnaea stagnalis* L. Arch. Neer. Zoo. 15, 1–103.

Kaczmarek, L.K., Finbow, M., Revel, J.R., Strumwasser, F., 1979. The morphology and coupling of *Aplysia* bag cells within the abdominal ganglion and in cell culture. J. Neurobiol. 10, 535–550.

Kaczmarek, L.K., Jennings, K., Strumwasser, F., 1982. An early sodium and a late calcium phase in the after-discharge of peptide-secreting neurons of *Aplysia*. Brain Res. 238, 105–115.

Kandel, E.R., 1964. Electrical properties of hypothalamic neuroendocrine cells. J. Gen. Physiol. 47, 691–717.

Kandel, E.R., 1979. Behavioral Biology of *Aplysia*: A Contribution to the Comparative Study of Opisthorbranch Molluscs. W.H. Freeman, New York.

Katayama, H., Ohira, T., Nagasawa, H., 2013. Crustacean peptide hormones: structure, gene expression and function. Aqua-BioSci. Monogr. 6, 49–90.

Kemenes, G., Benjamin, P.R., 2009. Lymnaea. Curr. Biol. 19, R9–R11.

Knowles, S.F.G.W., Carlisle, D.B., 1956. Endocrine control in the Crustacea. Biol. Rev. 31, 396–473.

Koene, J.M., 2010. Neuro-endocrine control of reproduction in hermaphroditic freshwater snails: mechanisms and evolution. Front. Behav. Neurosci. 4, 167.

Koizumi, K., Yamashita, H., 1972. Studies of antidromically identified neurosecretory cells of the hypothalamus by intracellular and extracellular recordings. J. Physiol. 221, 683–705.

Krysko, D., Leybaert, L., Vandenabeele, P., D'Herde, K., 2005. Gap junctions and the propagation of cell survival and cell death signals. Apoptosis 10, 459–469.

Kupfermann, I., Kandel, E.R., 1970. Electrophysiological properties and functional interconnections of two symmetrical neurosecretory clusters (bag cells) in abdominal ganglion of *Aplysia*. J. Neurophysiol. 33, 865–876.

Kupfermann, I., 1967. Stimulation of egg laying: possible neuroendocrine function of bag cells of abdominal ganglion of *Aplysia californica*. Nature 216, 814–815.

Leech, C.A., Habener, J.F., 1998. A role for Ca^{2+}-sensitive nonselective cation channels in regulating the membrane potential of pancreatic beta-cells. Diabetes 47, 1066–1073.

Leng, G., Brown, D., 1997. The origins and significance of pulsatility in hormone secretion from the pituitary. J. Neuroendocrinol. 9, 493–513.

Lincoln, D.W., Wakerley, J.B., 1974. Electrophysiological evidence for the activation of supraoptic neurones during the release of oxytocin. J. Physiol. 242, 533–554.

Ling, G., Gerard, R.W., 1949. The normal membrane potential of frog sartorius fibers. J. Cell. Comp. Physiol. 34, 383–396.

Loewenstein, W.R., Nakas, M., Socolar, S.J., 1967. Junctional membrane uncoupling. Permeability transformations at a cell membrane junction. J. Gen. Physiol. 50, 1865–1891.

Ludwig, M., Sabatier, N., Bull, P.M., Landgraf, R., Dayanithi, G., Leng, G., 2002. Intracellular calcium stores regulate activity-dependent neuropeptide release from dendrites. Nature 418, 85–89.

Ma, Y., Juntti, S.A., Hu, C.K., Huguenard, J.R., Fernald, R.D., 2015. Electrical synapses connect a network of gonadotropin releasing hormone neurons in a cichlid fish. Proc. Natl. Acad. Sci. 112, 3805–3810.

Magoski, N.S., Wilson, G.F., Kaczmarek, L.K., 2002. Protein kinase modulation of a neuronal cation channel requires protein-protein interactions mediated by an Src homology 3 domain. J. Neurosci. 22, 1–9.

Martin, A.O., Mathieu, M.-N., Chevillard, C., Guerineau, N.C., 2001. Gap junctions mediate electrical signalling and ensuing cytosolic Ca^{2+} increases between chromaffin cells in adrenal slices: a role in catecholamine release. J. Neurosci. 21, 5397–5405.

Martin, A.O., Mathieu, M.-N., Guerineau, N.C., 2003. Evidence for long-lasting cholinergic control of gap junctional communication between adrenal chromaffin cells. J. Neurosci. 23, 3669–3678.

Mason, C.A., Bern, H.A., 1977. Cellular biology of the neurosecretory neuron. In: Kandel, E.R. (Ed.), Cellular Biology of Neurons, Handbook of Physiology. The Nervous System. Williams and Wilkins, , Baltimore, pp. 651–689.

Mayer, R.J., Candy, D.J., 1969. Control of haemolymph lipid concentration during locust flight. An adipokinetic hormone from the corpora cardiaca. J. Insect Physiol. 15, 598–611.

McIntosh, J.M., Yoshikami, D., Mahe, E., Nielsen, D.B., Rivier, J.E., Gray, W.R., Olivera, B.M., 1994. A nicotinic acetylcholine-receptor ligand of unique specificity, alpha-conotoxin-ImI. J. Biol. Chem. 269, 16733–16739.

Meda, P., Bosco, D., Chanson, M., Giordano, E., Vallar, L., Wollheim, C., Orci, L., 1990. Rapid and reversible secretion changes during uncoupling of rat insulin-producing cells. J. Clin. Invest. 86, 759–768.

Meda, P., Pepper, M.S., Traub, O., Willecke, K., Gros, D., Beyer, E., Nicholson, B., Paul, D., Orci, L., 1993. Differential expression of gap junction connexins in endocrine and exocrine glands. Endocrinology 133, 2371–2378.

Meissner, H.P., 1976. Electrophysiological evidence for coupling between beta cells of pancreatic islets. Nature 262, 502–504.

Micevych, P.E., Popper, P., Hatton, G.I., 1996. Connexin 32 mRNA levels in the rat supraoptic nucleus: up-regulation prior to parturition and during lactation. Neuroendocrinology 3, 39–45.

Michel, S., Wayne, N.L., 2002. Neurohormone secretion persists after post-afterdischarge membrane depolarization and cytosolic calcium elevation in peptidergic neurons in intact nervous tissue. J. Neurosci. 22, 9063–9069.

Michon, L., Nlend, R., Bavamian, S., Bischoff, L., Boucard, N., Caille, D., Cancela, J., Charollais, A., Charpantier, E., Klee, P., Peyrou, M., Populaire, C., Zulianello, L., Meda, P., 2005. Involvement of gap junctional communication in secretion. Biochim. Biophys. Acta 1719, 82–101.

Morita, H., Ishibashi, T., Yamashita, S., 1961. Synaptic transmission in neurosecretory cells. Nature 191, 183.

Nagano, M., Cooke, I.M., 1987. Comparison of electrical responses of terminals, axons, and somata of a peptidergic neurosecretory system. J. Neurosci. 7, 634–648.

Nässel, D.R., 2002. Neuropeptides in the nervous system of *Drosophila* and other insects: multiple roles as neuromodulators and neurohormones. Prog. Neurobiol. 68, 1–84.

Neher, E., 1998. Vesicle pools and Ca^{2+} microdomains: new tools for understanding their roles in neurotransmitter release. Neuron 20, 389–399.

Orchard, I., Loughton, B.G., 1981. The neural control of release of hyperlipaemic hormone from the corpus cardiacum of *Locusta migratoria*. Comp. Biochem. Physiol. 68A, 25–30.

Orchard, I., Shivers, R., 1986. Electrical coupling and gap junctions between neurosecretory cells in an insect. Brain Res. 397, 359–364.

Orchard, I., Loughton, B.G., Gole, J.W., Downer, R.G., 1983. Synaptic transmission elevates adenosine 3′, 5′-monophosphate (cyclic AMP) in locust neurosecretory cells. Brain Res. 258, 152–155.

Orci, L., Unger, R.H., Renold, A.E., 1973. Structural coupling between pancreatic islet cells. Experientia 29, 1015–1018.

Paton, W.D.M., Zaimis, E.J., 1948. Clinical potentialities of certain bisquaternary salts causing neuromuscular and ganglionic block. Nature 162, 810.

Pereda, A.E., Bennett, M.V.L., 2017. Electrical synapses in fishes: their relevance to synaptic transmission. In: Jing, J. (Ed.), Network Functions and Plasticity. Elsevier (Chapter 8).

Pérez-Armendariz, M., Roy, C., Spray, D.C., Bennett, M.V., 1991. Biophysical properties of gap junctions between freshly dispersed pairs of mouse pancreatic beta cells. Biophys. J. 59, 76–92.

Phelan, P., Bacon, J.P., Blagburn, J.M., 2017. Neural circuits underlying escape behavior in Drosophila: focus on electrical signaling. In: Jing, J. (Ed.), Network Functions and Plasticity. Elsevier (Chapter 2).

Phelan, P., Goulding, L.A., Tam, J.L., Allen, M.J., Dawber, R.J., Davies, J.A., Bacon, J.P., 2008. Molecular mechanism of rectification at identified electrical synapses in the *Drosophila* giant fiber system. Curr. Biol. 18, 1955–1960.

Pinsker, H.M., Dudek, F.E., 1977. Bag cell control of egg laying in freely behaving *Aplysia*. Science 197, 490–493.

Pow, D.V., Morris, J.F., 1989. Dendrites of hypothalamic magnocellular neurons release neurohypophysial peptides by exocytosis. Neuroscience 32, 435–439.

Ravier, M.A., Güldenagel, M., Charollais, A., Gjinovci, A., Caille, D., Söhl, G., Wollheim, C.B., Willecke, K., Henquin, J.C., Meda, P., 2005. Loss of connexin36 channels alters beta-cell coupling, islet synchronization of glucose-induced Ca^{2+} and insulin oscillations, and basal insulin release. Diabetes 54, 1798–1807.

Rettig, J., Neher, E., 2002. Emerging roles of presynaptic proteins in Ca^{2+}-triggered exocytosis. Science 298, 781–785.

Richter, K., Gersch, M., 1983. Electrophysiological evidence of nervous involvement in the control of the prothoracic gland in *Periplaneta americana*. Experientia 39, 917–918.

Roepke, M.H., 1937. A study of choline esterase. J. Pharmacol. Exp. Ther. 59, 264–276.

Rorsman, P., Braun, M., Zhang, Q., 2012. Regulation of calcium in pancreatic α- and β-cells in health and disease. Cell Calcium 51, 300–308.

Roubos, E.W., van Leeuwen, J.P., Maijers, A., 1985. Ultrastructure of gap junctions in the central nervous system of *Lymnaea stagnalis* with particular reference to electrotonic coupling between the neuroendocrine caudodorsal cells. Neuroscience 14, 711–721.

Scharrer, E., 1928. Die lichtempfindlichkeit blinder elritzen. (Untersuchungen uber das zwischenhirn der fische. I.). Z. Vergl. Pkytiol. 7, 1–38.

Scharrer, B., 1935. Uber class hanstromsche organ X sei opithobranchiery. Publ. Stat. Zool. Napoli. 15, 132–142.
Serre-Beinier, V., Le Gurun, S., Belluardo, N., Trovato-Salinaro, A., Charollais, A., Haefliger, J.A., Condorelli, D.F., Meda, P., 2000. Cx36 preferentially connects beta-cells within pancreatic islets. Diabetes 49, 727–734.
Sherman, A., Rinzel, J., 1991. Model for synchronization of pancreatic beta-cells by gap junction coupling. Biophys. J. 59, 547–559.
Smith, W.A., Gilbert, L.I., Bollenbacher, W.E., 1985. Calcium-cyclic AMP interactions in prothoracicotropic hormone stimulation of ecdysone synthesis. Mol. Cell. Endocrinol. 39, 71–78.
Snow, P.J., Rose, P.K., Brown, A.G., 1976. Tracing axons and axon collaterals of spinal neurons using intracellular injection of horseradish peroxidase. Science 191, 312–313.
Soldo, B.I., Giovannucci, D.R., Stuenkel, E.L., Moises, H.C., 2004. Ca^{2+} and frequency dependence of exocytosis in isolated somata of magnocellular supraoptic neurones of the rat hypothalamus. J. Physiol. 555, 699–711.
Sosinsky, G.E., Nicholson, B.J., 2005. Structural organization of gap junction channels. Biochim. Biophys. Acta 1711, 99–125.
Speidel, C.C., 1919. Gland-cells of internal secretion in the spinal cord of the skates. Publ. Carneg. Inst. 13, 1–31.
Srinivas, M., Spray, D.C., 2003. Closure of gap junction channels by arylaminobenzoates. Mol. Pharmacol. 63, 1389–1397.
Stewart, W.W., 1978. Functional connections between cells as revealed by dye-coupling with a highly fluorescent naphthalimide tracer. Cell 14, 741–759.
Stone, C.A., Torchiana, M.L., Navarro, A., Beyer, K.H., 1956. Ganglionic blocking properties of 3-methylamino-isocamphane hydrochloride (mecamylamine) – a secondary amine. J. Pharmacol. Exp. Ther. 117, 169–183.
Summerlee, A., 1981. Extracellular recordings from oxytocin neurones during the expulsive phase of birth in unaesthetized rats. J. Physiol. 321, 1–9.
Tam, A.K.H., Geiger, J.E., Hung, A.Y., Groten, C.J., Magoski, N.S., 2009. Persistent Ca^{2+} current contributes to a prolonged depolarization in *Aplysia* bag cell neurons. J. Neurophysiol. 102, 3753–3765.
Tauc, L., Gerschenfeld, H.M., 1961. Cholinergic transmission mechanisms for both excitation and inhibition in molluscan central synapses. Nature 192, 366–367.
ter Maat, A., Roubos, E.W., Lodder, J.C., Buma, P., 1983. Integration of biphasic synaptic input by electrotonically coupled neuroendocrine caudodorsal cells in the pond snail. J. Neurophysiol. 49, 1392–1409.
ter Maat, A., Dijcks, F.A., Bos, N.P.A., 1986. *In vivo* recordings of neuroendocrine cells (caudo-dorsal cells) in the pond snail. J. Comp. Physiol. 158A, 853–859.
Thomsen, M., 1943. Effect of corpus cardiacum and other insect organs on the colour change of the shrimp, *Leander adtpersus*. Damke Selsk. Biol. Medd. 19, 3–38.
Thomsen, E., 1952. Functional significance of the neurosecretory brain cells and the corpus cardiacum in the female blow-fly, *Calliphora erythrocephala* Meig. J. Exp. Biol. 29, 137–172.
Toescu, E.C., Dayanithi, G., 2012. Neuroendocrine signalling: natural variations on a Ca^{2+} theme. Cell Calcium 51, 207–211.
Toivonen, J.M., Partridge, L., 2009. Endocrine regulation of aging and reproduction in *Drosophila*. Mol. Cell. Endocrinol. 299, 39–50.
Vazquez-Martinez, R., Shorte, S.L., Boockfor, F.R., Frawley, L.S., 2001. Synchronized exocytotic bursts from gonadotropin-releasing hormone-expressing cells: dual control by intrinsic cellular pulsatility and gap junctional communication. Endocrinology 142, 2095–2101.
Veelaert, D., Schoofs, L., De Loof, A., 1998. Peptidergic control of the corpus cardiacum-corpora allata complex of locusts. Int. Rev. Cytol. 182, 249–302.
Verselis, V., White, R.L., Spray, D.C., Bennett, M.V.L., 1986. Gap junctional conductance and permeability are linearly related. Science 234, 461–464.

Vorndran, C., Minta, A., Poenie, M., 1995. New fluorescent calcium indicators designed for cytosolic retention or measuring calcium near membranes. Biophys. J. 69, 2112–2124.
Vreugdenhil, E., Jackson, J.F., Bouwmeester, T., Smit, A.B., Van Minnen, J., Van Heerikhuizen, H., Klootwijk, J., Joosse, J., 1988. Isolation, characterization, and evolutionary aspects of a cDNA clone encoding multiple neuropeptides involved in the stereotyped egg-laying behavior of the freshwater snail *Lymnaea stagnalis*. J. Neurosci. 8, 4184–4191.
Wendelaar Bonga, S.E., 1971. Formation, storage, and release of neurosecretory material studied by quantitative electron microscopy in the freshwater snail *Lymnaea stagnalis* (L.). Z. Zellforsch 113, 490–517.
Weyer, F., 1935. Ueber driisenartige nervenzellen im gehirn der honigbiene, *Apis mellifica* L. Zool. Anz 112, 137–141.
White, S.H., Magoski, N.S., 2012. Acetylcholine-evoked afterdischarge in *Aplysia* bag cell neurons. J. Neurophysiol. 107, 2672–2685.
Willenberg, H.S., Schott, M., Saeger, W., Tries, A., Scherbaum, W.A., Bornstein, S.R., 2006. Expression of connexins in chromaffin cells of normal human adrenals and in benign and malignant pheochromocytomas. Ann. N.Y. Acad. Sci. 1073, 578–583.
Yang, Q.Z., Hatton, G.I., 1988. Direct evidence for electrical coupling among rat supraoptic nucleus neurons. Brain Res. 463, 47–56.
Zhang, Y., Kaczmarek, L.K., 2008. Bag cell neurons. Scholarpedia 3, 4095.

CHAPTER 8

Electrical Synapses in Fishes: Their Relevance to Synaptic Transmission

A.E. Pereda, M.V.L. Bennett
Albert Einstein College of Medicine, Bronx, NY, United States

> *Although there is every indication that chemical transmission occurs across those junctions which have been discussed in this review and which are most familiar to the physiologist, it is probable that electrical transmission occurs at certain other junctions.*
>
> ***Paul Fatt (1954)***

The elucidation of the mechanisms of communication between neurons involved the use of identifiable cell contacts from various animal species, which offered the required experimental accessibility for these investigations. Within this context seminal observations in fishes were foundational in what we recognize today as " synaptic electrical transmission." Moreover, fish electrical synapses first revealed the functional contributions of this modality of transmission to vertebrate neural circuits and provided early evidence of their plastic properties. Present and future investigations in genetically tractable species will shed light on the molecular mechanisms and functional diversity of electrical transmission in the vertebrate brain. This chapter discusses the past, present, and potential future contributions of fish electrical synapses to the understanding of synaptic communication.

1. INTRODUCTION: THE DISCOVERY OF ELECTRICAL TRANSMISSION

Early considerations of the mechanism of synaptic transmission contrasted chemically vs electrically mediated forms of communication. Electrical transmission was then conceived to result from the ability of the electric field generated by a presynaptic action potential to increase (or decrease) the excitability of the postsynaptic neuron. Because at most cell contacts initially examined (frog neuromuscular junction, sympathetic ganglia of various species, electric organs in fishes, crustacean neuromuscular junction, cat spinal motoneurons) (Fatt, 1954) transmission was found to be chemically mediated, it was widely accepted to be the sole mechanism for synaptic communication. Challenging

Network Functions and Plasticity
ISBN 978-0-12-803471-2, http://dx.doi.org/10.1016/B978-0-12-803471-2.00008-4

this view, Paul Fatt (1954) warned that electrical transmission could still occur at cell contacts at which conditions appeared more favorable for this form of communication and very likely was the mode of transmission between segments of various septate axons of invertebrates, which conduct impulses in both directions (Kao and Grundfest, 1957; Kao, 1960). He speculated that, in contrast to the examples examined until then at which small axons transmitting electrically would not be as effective in generating a postsynaptic response in a large postsynaptic cell, pre- and postsynaptic neurons are in these cases of comparable size. Fatt also noted, insightfully, that having pre- and postsynaptic membranes close together and of higher conductance than surrounding membrane would favor electrical transmission, thus anticipating the condition by which channels that directly connect the interiors of the joined cells mediate transmission. Following Bernard Katz's suggestion, seminal work by Furshpan and Potter (1959) provided unambiguous electrophysiological evidence for the presence of electrical transmission between giant and motor fibers of the abdominal nerve cord of the crayfish, a synapse embedded in a network mediating escape responses. Consistent with a low resistance connection, "electrotonic" transmission of subthreshold depolarizations was observed between pre- and postsynaptic fibers, a property in striking contrast to the situation at chemical synapses. However, transmission was polarized. Depolarizations were transmitted from the presynaptic fiber to the postsynaptic fiber to much greater extent than hyperpolarizations whereas the converse was true for antidromic transmission, a rectifying property that accounted for unidirectional transmission of action potentials. These studies were followed by investigations in fishes (see later discussion), which provided not only the early support for the presence of electrical transmission in vertebrates but also, more importantly, of the gap junction as its usual structural basis. Fish electrical synapses were instrumental in our understanding of the functional properties and structural determinants of electrical transmission in the vertebrate brain. Now it is generally accepted that both chemical and electrical transmission can occur at synapses. By reviewing key experimental observations, this article aims to highlight the contributions of fish electrical synapses to our current understanding of electrical transmission and synaptic communication in general.

2. SUPRAMEDULLARY NEURONS IN THE PUFFER FISH: THE FIRST EVIDENCE OF ELECTRICAL COUPLING BETWEEN VERTEBRATE NEURONS

Studies at invertebrate junctions (Watanabe, 1958; Furshpan and Potter, 1959) were followed by evidence at supramedullary neurons in puffer fish (Fig. 8.1A), for the large size of these clustered neurons facilitated simultaneous recording from multiple cells (Fig. 8.1B–C). Supramedullary neurons sit dorsally on the spinal cord or somewhat

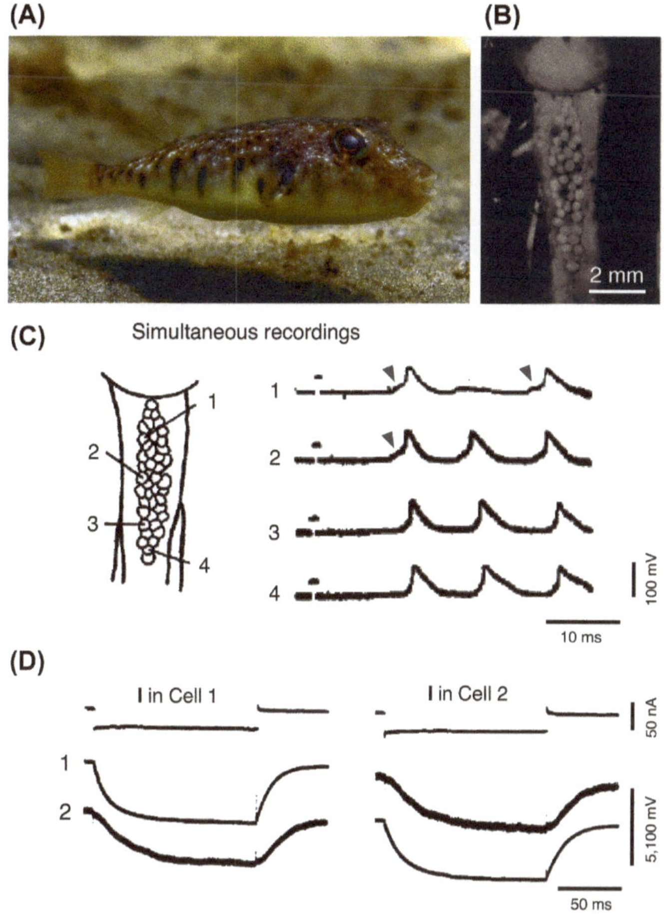

Figure 8.1 ***Early evidence of electrical coupling in the puffer fish.*** Supramedullary neurons in the puffer fish (*Spheroides maculatus*) fire synchronously and are electrically coupled. (A) The fish, 15–20 cm in length. *(Reproduced from SmugMug, with permission.)* (B) Supramedullary neurons lie on the dorsal surface of the spinal cord just behind the cerebellum (at the top of the image). The neuronal somata are spherical and 200–300 μm in diameter. Two dorsal roots are visible on the left side of the cord, which is quite short, and most of the spinal canal is occupied by dorsal and ventral roots (cauda equina). The unfixed cord was stained with toluidine blue; cells that had been injured by microelectrodes are darkly stained. The cytoplasm of intact cells is unstained, and the cells are outlined by labeling of surface associated tissue, such as blood vessels and ependyma. *(Modified from Bennett, M.V., Crain, S.M., Grundfest, H., 1959b. Electrophysiology of supramedullary neurons in Spheroides maculatus. I. Orthodromic and antidromic responses. J. Gen. Physiol. 43, 159–188, with permission.)* (C) Diagram of the cluster showing placement of four electrodes recording from the cell responses to a cranial nerve stimulus. Each cell shows three spikes, although the second spike failed to invade the soma of Cell 1. The inflection on the rising phase (*arrowheads*) of the other spikes is due to delayed invasion of the soma by the spike in the axon. Responses of Cells 3 and 4 are delayed compared with those of the more rostral cells. *(Modified from Bennett, M.V., Crain, S.M., Grundfest, H., 1959a. Electrophysiology of supramedullary neurons in Spheroides maculatus. III. Organization of the supramedullary neurons. J. Gen. Physiol. 43, 221–250, with permission.)* (D) Electrical coupling of adjacent cells in another preparation. The upper trace shows the applied hyperpolarizing current in Cell 1 (left record) and in Cell 2 (right record), the middle trace shows the voltage in Cell 1, and the lower trace shows the voltage in Cell 2; the voltage in the cell in which current was applied is shown at lower gain. The coupling was quite symmetrical in the two directions. *(Modified from Bennett, M.V., 1966. Physiology of electrotonic junctions. Ann. N.Y. Acad. Sci. 137, 509–539, with permission.)*

embedded in the dorsal midline (Fig. 8.1B) and spinal, cranial nerve, and cutaneous stimulation evoke spike responses (Bennett et al., 1957, 1959a,b). The function of these neurons remains unknown. Trying to get two electrodes into one cell (for voltage recording and current application) and instead impaling neighboring cells led to the demonstration that when one cell fires, they all do. Impulses arise distant from the cell bodies, more like many invertebrate neurons where synapses are on cell processes and not on the somata. Synchronized action potentials were preceded by a depolarization (Fig. 8.1C, arrowheads). Although one of us involved in these studies (M.V.L.B.) thought that the potential initiating the soma spike was a synaptic potential that drove synchronous firing, John C. Eccles, who visited Harry Grundfest's laboratory during one of the recording sessions, looked at the oscilloscope and suggested it was instead an axon spike comparable with the initial segment response in cat motoneurons (Brock et al., 1953). Moreover, he suggested that pushing an electrode into the neuropil beneath the large cells would record the axon spike. Sure enough, Eccles was right, but beyond its original purpose, the procedure led to a much more important discovery. In these experiments, on finding a two-peaked spike (like those observed in motoneuron axons by Eccles and collaborators) (Brock et al., 1953), M.V.L.B. polarized overlying cells to find the soma of origin. It quickly became obvious that a soma not giving rise to the axon (and having an independent spike) could be electrically ("electrotonically") coupled to the axon recorded from. Increasing the gain of recordings from neighboring somata showed that there was coupling between them that had been missed, because the coupling potential was much smaller than in the axons closer to the sites of electrical coupling. Later work (see later discussion) revealed the presence of gap junctions between the cells' axons that were far enough away from the somata that the electrotonic potentials generated there were easy to miss during somatic recordings (Bennett et al., 1967a). Hyperpolarization and depolarization of one supramedullary neuron caused corresponding polarization of the same sign in adjacent neurons, this effect was bidirectional (Fig. 8.1D), and the transfer resistances (V_1/i_2 and V_2/i_1) were equal in the two directions (Bennett, 1966; Bennett et al., 1967c). These properties indicated transmission of electrical signals via a low resistance pathway and the existence of electrical connections between supramedullary neurons. This form of transmission (and the strength of coupling) clearly underlies the synchronized firing of the neurons (Fig. 8.1C). Thus these early observations in puffer fish not only provided the first evidence for electrical synaptic transmission in vertebrates but also suggested that a selective value of reciprocal electrical transmission could be synchronization of neuronal activity. A second possible selective advantage of electrical transmission is its shorter latency, which appears important in the escape networks of the septate axons and the Mauthner cell. And as will be discussed later, both reciprocal transmission and shorter delay are important in many of the electric organ control systems.

3. ELECTRIC FISHES: CONTRIBUTION OF ELECTRICAL SYNAPSES TO SYNCHRONIZED NEURONAL ACTIVITY

There are strongly and weakly electric fishes. Strongly electric fishes can generate extraordinarily large voltages (electric eel: several hundred volts in fresh water, kilowatt peak power; torpedo: 50 V in sea water, kilowatt peak power; electric catfish: several hundred volts, power not measured) (Bennett, 1971a). Power output is a result of many cells in series and parallel, and requires synchronization of the many generating cells. The high frequencies observed in bursts, 100–1000 Hz, can only be maintained for short periods. Weakly electric fishes generate much smaller voltages and many can maintain discharges indefinitely, even at frequencies in excess of 1 kHz. Again, synchrony of discharge increases power output. The pulses serve in communication and in an active sensory system that detects spatial differences in conductance in the environment. Weakly electric fishes are generally very sensitive to their own discharges and to external electric fields and may have separate sets of electroreceptors specialized for different frequency signals (Bennett, 1971b).

Electric pulses are generated by electric organs distributed along the fish body (to an extent that differs among species) and the generating cells, electrocytes, are excited synchronously by neural signals originating in a central command nucleus. The mormyrids (Fig. 8.2A), widely distributed in Africa, were of particular interest because of the brevity of the organ discharge (biphasic and about 0.2 ms in duration) in some groups (Bennett, 1971a). In these fishes, the command signal is highly synchronous in the medullary pacemaker nucleus (Elekes and Szabo, 1985), which initiates the command and transmits it to the medullary relay located just dorsally (Bennett, 1971a), which in turn sends axons down the spinal cord to the spinal electromotor neurons, which innervate the electrocytes. Neurons in both medullary and spinal relay nuclei are strongly electrically coupled (Fig. 8.2B), and impulses directly evoked in one cell often propagate to all the other cells (Fig. 8.2C). The electrical coupling of the controlling nuclei ultimately contributes to the synchronization of electric organs with a precision that could not be mediated by chemical synapses between neurons in the medullary relay or spinal electromotor nuclei. Studies of several other groups of electric fishes supported the role of electrical synapses in synchronization (gymnotids, electric catfish, stargazer/*Astroscopus*) (Bennett, 1971a). Further examination in other fast motor systems in fishes demonstrated electrical coupling was also present between neurons controlling fast muscle systems including extraocular (Kriebel et al., 1969) and sonic muscles (Pappas and Bennett, 1966). This work established the role of electrical transmission in synchronizing neuronal activity, a property repeatedly found to be relevant to neuronal networks in general and for a wide range of species, including mammals. Although speed of electrical transmission was required for the degree of synchronization observed in these fast systems and synchrony was an indicator that electrical synapses would be found, electrical synapses

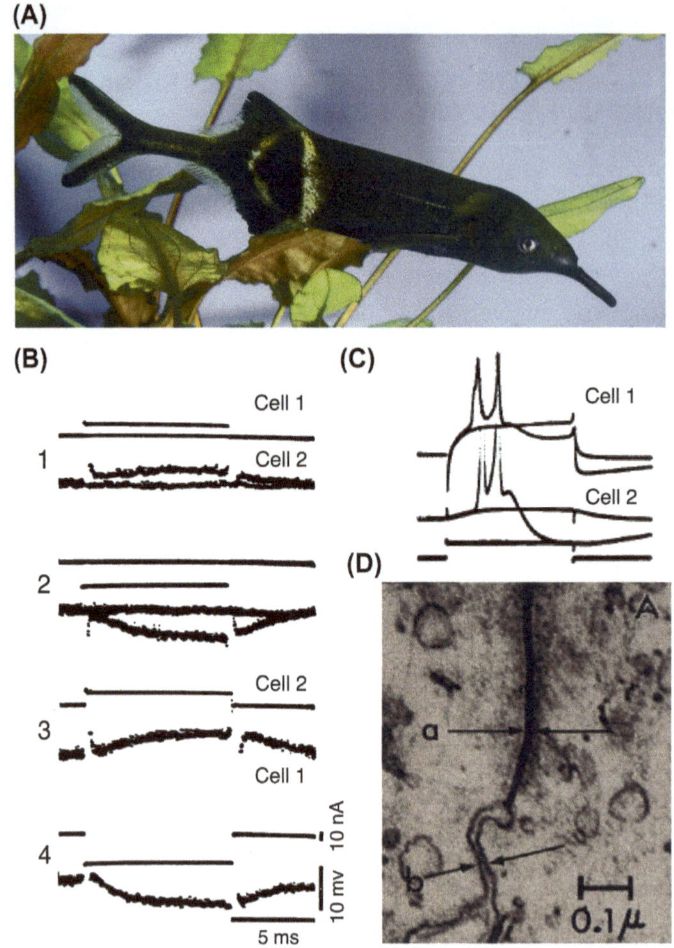

Figure 8.2 ***Coupling of electromotor neurons in the mormyrid electric fish,* Gnathonemus petersii.** (A) The fish, 15–20 cm as used in these studies. *(Reproduced from Alamy, with permission.)* (B) Electrical coupling of two neurons, separated by ~300 μm along the rostral caudal axis. In frames 1 and 2, current was passed in Cell 1 and both de- and hyperpolarization were transmitted to Cell 2 with comparable transfer resistances for the two polarities. Cell 1 was in a bridge circuit, and bridge imbalance prevented recording of the potential. A superimposed trace without current shows the base line. In frames 3 and 4, current was passed in Cell 2 and voltage was recorded in Cell 1. A similar degree of transmission of de- and hyperpolarization was observed. Transfer resistances in the two directions were comparable. *(Modified from Bennett, M.V., Aljure, E., Nakajima, Y., Pappas, G.D., 1963. Electrotonic junctions between teleost spinal neurons: electrophysiology and ultrastructure. Science 141, 262–264, with permission.)* (C) Directly evoked impulses in a different pair of spinal electromotor neurons that were less than 100 μm apart. Depolarization in Cell 1 was suprathreshold for impulses; the spinal neurons tend to fire two or three spikes in response to a just suprathreshold stimulus. The two-spike response occurred in both cells. *(Modified from Bennett, M.V., Pappas, G.D., Aljure, E., Nakajima, Y., 1967c. Physiology and ultrastructure of electrotonic junctions. II. Spinal and medullary electromotor nuclei in mormyrid fish. J. Neurophysiol. 30, 180–208, with permission.)* (D) A putative gap junction between two thick processes of the spinal neurons. The overall thickness of the junction (*arrows at a*) is less than at other regions of apposition (*arrows at b*), and there is dense material in the adjacent cytoplasm. Failure to see a gap may have resulted from staining artifact and/or tilt of the section, which is considerably thicker than the width of the gap in other preparations. *(Modified from Bennett, M.V., Aljure, E., Nakajima, Y., Pappas, G.D., 1963. Electrotonic junctions between teleost spinal neurons: electrophysiology and ultrastructure. Science 141, 262–264, with permission.)*

occur in other systems, where the degree of synchronization is much less and the speed of electrical transmission is not required, eg, the supramedullary neurons and generators of respiratory rhythm in mammals (Zavala-Tecuapetla et al., 2014).

Beyond characterizing electrical synapses and their functional relevance in vertebrates, work on electric fishes provided evidence for the anatomical basis of electrical transmission. Ultrastructural analysis in mormyrids showed that electrical coupling was associated with areas of close membrane apposition between the coupled cells (Fig. 8.2D; Bennett et al., 1963). Similar areas of close membrane apposition were also found in the electromotor control systems of gymnotid fish (Bennett et al., 1967b) and the electric catfish *Malapterurus electricus* (Bennett et al., 1967b) and between supramedullary neurons of the puffer fish. Similar appositions, which we now term gap junctions, were found in the sonic muscle control systems (Pappas and Bennett, 1966) and in the oculomotor nuclei (Kriebel et al., 1969). Together with evidence from mixed (electrical and chemical) synapses in goldfish (see later discussion), these studies demonstrate that gap junctions constitute the structural basis for this form of electrical transmission. It should be mentioned that, a second form of electrical transmission, which is inhibitory, was shown to occur via an electric field effect (Furukawa and Furshpan, 1963), a mechanism that is also known as "ephaptic" interaction (Arvanitaki et al., 1964) (see later discussion).

4. CLUB ENDINGS IN GOLDFISH: ELECTRICAL AND CHEMICAL SYNAPSES CAN INTERACT

Club endings, which are synaptic contacts on the goldfish Mauthner cell formed by auditory afferents (Fig. 8.3A), were investigated more or less contemporaneously with the electric organ control systems and these studies reached similar conclusions with respect to electrical coupling mediated by gap junctions (Robertson, 1963; Robertson et al., 1963; Furshpan, 1964). The focus was on an identifiable synapse that is accessible for ultrastructural and physiological analysis, the large myelinated club endings (Club endings) (Bartelmez and Hoerr, 1933). These contacts are terminations of auditory afferents on the lateral dendrite of the Mauthner cells, a bilateral pair of large reticulospinal neurons, one on each side of the medulla of fishes and some amphibia, that excite motoneurons in escape behaviors (Korn and Faber, 2005). J. David Robertson et al. (1963) argued that given the involvement of the Mauthner cell in escape responses, transmission at these synapses might be electrical as found to be the case in the crayfish escape network. Edwin Furshpan had already reported that activation of these afferents evokes a postsynaptic response that is very short latency (Furshpan and Furukawa, 1962) and had inferred that transmission was electrical as at the motor giant synapses in crayfish. Moreover, the synapses are morphologically "mixed" in having both gap junctions and sites characteristic of chemical transmission. Stimulation of the posterior branch of the eighth nerve (where these afferents run) evokes a mixed response

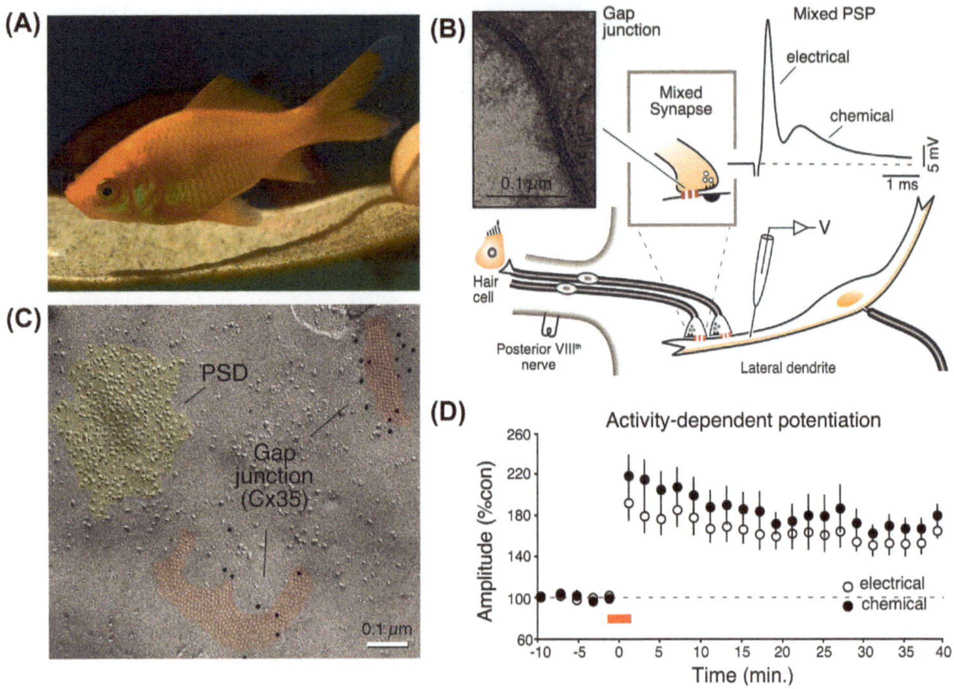

Figure 8.3 ***The large myelinated club endings.*** (A) Common goldfish (*Carassius auratus*), 10–15 cm in length *(Reproduced from Alamy, with permission.)* were generally used in these studies. (B) Large myelinated club endings (Club endings) are terminations of afferents from the sacculus (where hair cells are located) on the lateral dendrite of the Mauthner cell. Electrical and chemical transmission coexist at these terminals and therefore the synapse is mixed [upper middle inset: mixed synapse where two gap junction channels are shown in red, and an active zone and postsynaptic density (PSD) are indicated]. The upper left inset shows a gap junction in a Club ending (presynaptic axon above). Stimulation of the afferents in the posterior VIII nerve evokes a synaptic response composed of an early and brief large depolarization representing the coupling of presynaptic action potentials (electrical component) and small, delayed, chemical response mediated by glutamate (chemical). *(Modified from Pereda and Faber (2011); Pereda, A.E., Faber, D.S., 2011. Physiology of the Mauthner Cell: Discovery and Properties. Elsevier Inc., with permission;* Gap junction inset modified from *Robertson, J.D., 1963. The occurrence of a subunit pattern in the unit membranes of club endings in Mauthner cell synapses in Goldfish brains. J. Cell Biol. 19, 201–221; Robertson, J.D., Bodenheimer, T.S., Stage, D.E., 1963. The ultrastructure of Mauthner cell synapses and nodes in goldfish brains. J. Cell Biol. 19, 159–199 with permission.)* (C) Freeze fracture replica immunolabeling of a goldfish Club ending. The image shows two gap junction plaques (pink) with labeling for Cx35 (using a Cx36 antibody that labels all mouse Cx36-related fish connexins) and a PSD (yellow) of a glutamatergic synapse. Note the proximity between specializations for chemical and electrical transmission (unpublished image kindly provided by Dr. John Rash, Colorado State University). (D) Activity-dependent potentiation of electrical and chemical transmission. High-frequency stimulation of the posterior VIII nerve (100 Hz for 1 s combined with postsynaptic depolarization (see Cachope et al., 2007 for details) led to long-term potentiation of the electrical and chemical components of the mixed EPSP. *(Modified from Cachope, R., Mackie, K., Triller, A., O'Brien, J., Pereda, A.E., 2007. Potentiation of electrical and chemical synaptic transmission mediated by endocannabinoids. Neuron 56, 1034–1047, with permission.)*

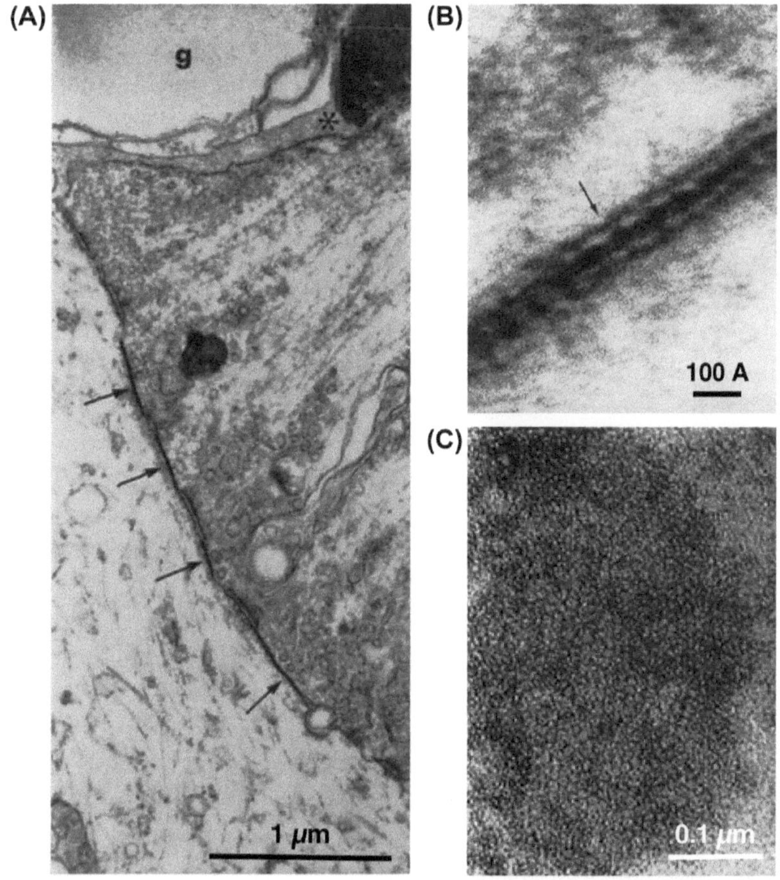

Figure 8.4 ***Evidence for the morphological basis of electrical transmission obtained at large myelinated club endings.*** (A) Image of a Club ending at which "synaptic discs" (gap junctions) are indicated with arrows. *g*, glial process. *(Modified from Robertson, J.D., 1963. The occurrence of a subunit pattern in the unit membranes of club endings in Mauthner cell synapses in Goldfish brains. J. Cell Biol. 19, 201–221; Robertson, J.D., Bodenheimer, T.S., Stage, D.E., 1963. The ultrastructure of Mauthner cell synapses and nodes in goldfish brains. J. Cell Biol. 19, 159–199, with permission.)* (B) High magnification of a synaptic disc. (C) Frontal view of a synaptic disc. *(Panels B and C are modified from Robertson, J.D., Bodenheimer, T.S., Stage, D.E., 1963. The ultrastructure of Mauthner cell synapses and nodes in goldfish brains. J. Cell Biol. 19, 159–199; Robertson, J.D., 1963. The occurrence of a subunit pattern in the unit membranes of club endings in Mauthner cell synapses in Goldfish brains. J. Cell Biol. 19, 201–221, with permission.)*

composed of a fast early electrical component and a smaller, delayed chemical component (Fig. 8.3B). Chemical transmission was unidirectional, whereas electrical transmission was bidirectional and had a much shorter delay (Furshpan, 1964). From the anatomical point of view, together with specializations characteristic of chemical transmission, ultrastructural analysis of these terminals (Robertson, 1963; Robertson et al., 1963) revealed areas of close membrane apposition (Fig. 8.4A) of pentalaminar (Fig. 8.4B) or septalaminar appearance with permanganate and osmium fixation, respectively. Many of these areas were more or less circular in "en face" views, and Robertson called them "synaptic discs" (Fig. 8.4C; Robertson, 1963). Robertson's report provides one of the earliest descriptions of the substructure of what we know today as "gap junction plaques," or simply "gap junctions". The correlation of these structures and electrical transmission provided additional evidence that these gap junctions were the morphological substrate for electrical coupling. The coexistence of both forms of transmission at these terminals made it easy later to differentiate the properties of electrical and chemical transmission and their interactions (see later discussion) (Pereda et al., 2004).

A second form of electrical transmission occurs at the Mauthner cell. Interneurons generate a positive extracellular electric field that reduces the excitability of underlying membrane of the axon hillock, which in the Mauthner cell is the site of impulse initiation (Furukawa and Furshpan, 1963; Faber and Korn, 1973; Korn et al., 1978). Also characterized as an "ephaptic" interaction (Arvanitaki et al., 1964), this form of electrical communication does not involve intercellular channels but afferent terminals that pass only outward current and presumably are devoid of sodium channels. Areas of higher resistance in the surrounding parenchyma, the axon cap, increase the positivity (see Faber and Korn, 1989 for more information). As a consequence of this anatomical arrangement, an action potential in the Mauthner cell causes reciprocal inhibition of these interneurons (Faber and Korn, 1973). A similar form of inhibition of cerebellar Purkinje cells has been reported (Korn and Axelrad, 1980). These interactions demonstrated that electrical transmission, as initially speculated, can occur via electrical field effects given the right membrane specializations.

Club endings later provided additional relevant insights into the properties of electrical transmission in respect to plasticity and interactions with chemical synapses. High-frequency activation of Club endings led to long-term potentiation of both the electrical and chemical components of the mixed excitatory postsynaptic potential (EPSP) (Yang et al., 1990). Thus strikingly, both electrical and chemical forms of transmission were capable of expressing activity-dependent long-term potentiation. Interestingly, potentiation of electrical transmission results from interactions with the chemical synapses localized in the same contact, which were found to be glutamatergic (Yang et al., 1990; Pereda and Faber, 1996; Pereda et al., 1998, 2003). Activation of *N*-methyl-D-aspartate (NMDA) receptors by stimulation of the afferents with bursts of 500 Hz caused an increase in postsynaptic Ca^{2+} and activation of Ca^{2+}/calmodulin-dependent kinase II (CaMKII), which led to the potentiation of both chemical and electrical transmission (Pereda and Faber, 1996; Pereda et al., 1998). This pathway is also capable of triggering short-term potentiation of the mixed EPSP (Pereda and Faber, 1996) and demonstrates an intimate functional relationship between chemical and electrical mechanisms of transmission. These interactions are short range, as they are restricted to the stimulated terminal (Pereda et al., 2003) in which the two forms of transmission are in close proximity (Fig. 8.3C). Because the induction of plasticity is triggered postsynaptically (ie, interactions between chemical and electrical synapses occur at the postsynaptic cells) this property not only pertains to mixed synapses but can also occur heterosynaptically, where NMDA receptors (located at postsynaptic densities, PSDs, or extrasynaptically) are in close proximity to neuronal gap junctions. Accordingly, heterosynaptic NMDA-CaMKII—mediated potentiation of electrical transmission was reported in the mammalian retina (Kothmann et al., 2012) and the inferior olive (Turecek et al., 2014; Welsh and Turecek 2017 (Chapter 12)).

A second form of activity-dependent potentiation of electrical (and chemical) transmission at Club endings also requires activation of glutamate receptors (Cachope et al., 2007). That is, 1 s of high-frequency (100 Hz) stimulation of Club endings leads to potentiation of both forms of transmission that requires coincident activation of mGluR1 receptors and postsynaptic depolarization (Cachope et al., 2007) (Fig. 8.3D). This leads to the production and release of endocannabinoids, which in turn trigger release of dopamine from nearby varicosities belonging to dopaminergic neurons. Dopamine ultimately leads to potentiation via activation of postsynaptic dopamine D1 receptors and protein kinase A (PKA; Pereda et al., 1992, 1994; Cachope et al., 2007). Club endings were also reported to undergo depression of both forms transmission via a still unidentified mechanism (Yang and Faber, 1991). Thus the coexistence of chemical and electrical synapses at Club endings provided the opportunity not only to compare both forms of transmission but also to expose relevant interactions between them.

Finally, antidromic electrical transmission at Club endings (from the Mauthner cell to the presynaptic ending) was found to be amplified by a persistent Na^+ current in a voltage-dependent fashion during presynaptic depolarization (Curti and Pereda, 2004). This interaction between electrical coupling and subthreshold Na^+ currents was reported to occur at electrical synapses in the cerebellum (Mann-Metzer and Yarom, 1999; Dugué et al., 2009) and thalamus (Haas and Landisman, 2011), and likely represents another widespread property of electrical synapses elsewhere. Thus investigations in Club endings added to our knowledge of electrical transmission in the vertebrate brain and its anatomical bases and revealed plastic properties.

5. RETINA: MODULATION OF ELECTRICAL TRANSMISSION AND THE IDENTIFICATION OF NEURONAL CONNEXINS

Because of its cellular connectivity, layered organization, and repeating structure, the retina is an ideal preparation for the study of synaptic communication. Not surprisingly, electrical synapses in fish retina greatly contributed to our understanding of electrical transmission. We will focus in this review on three cardinal examples: the modulation of electrical coupling and its mechanisms, the first observations indicating the functional relevance of gap junction hemichannels, and, finally, the cloning of the first member of a key family of neuronal gap junction proteins.

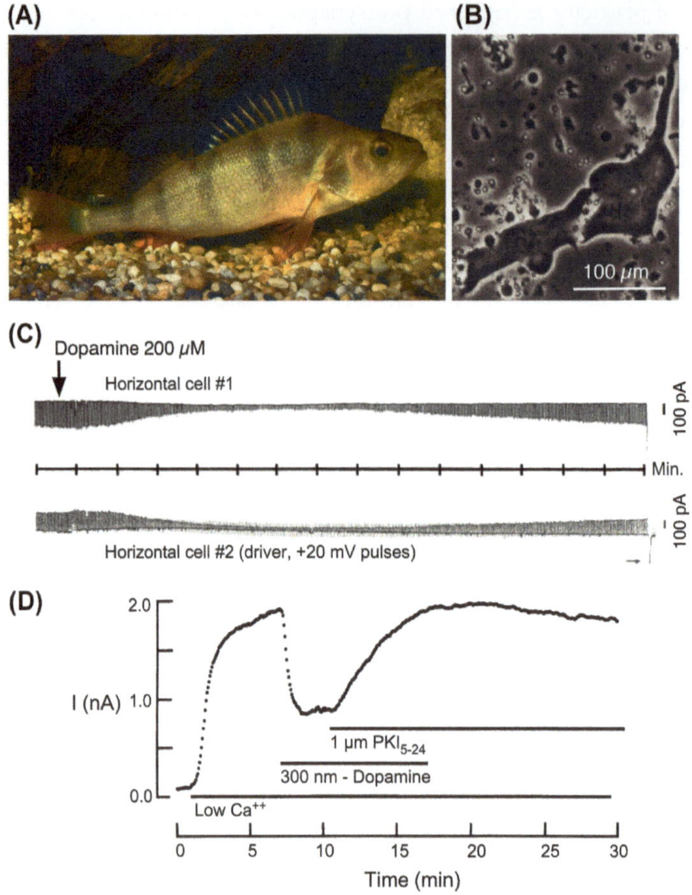

Figure 8.5 ***Modulation of electrical transmission in fish retina.*** (A) White perch (*Roccus americana*), about 15 cm long *(Reproduced from Alamy, with permission.)*, was one of the species used for studies of electrical transmission in fish retina. (B) Pair of dissociated horizontal H2 cells from white perch retina. (C) Application of dopamine (200 μM) uncoupled a pair of H2 cells in culture. Voltage steps were applied to a voltage-clamped cell (driver cell, *lower trace*), whereas coupling was monitored in a second cell under current clamp configuration (follower cell, *upper trace*). *(Panels B and C are from Lasater, E.M., Dowling, J.E, 1985. Dopamine decreases conductance of the electrical junctions between cultured retinal horizontal cells. Proc. Natl. Acad. Sci. U.S.A. 82, 3025–3029, with permission.)* (D) Voltage clamp recording in a single isolated horizontal cell from catfish retina. A large membrane current was observed when the cell was bathed with a medium containing a reduced concentration of Ca^{2+} (Low Ca^{2+}), which causes opening of gap junction hemichannels (membrane voltage was maintained at 20 mV). Dopamine suppressed the low Ca^{2+}-induced current increase (300 nM dopamine), and this effect was reversed by intracellular application of a peptide (PKI 5-24) that inhibits cyclic AMP-dependent protein kinase (PKA). *(Modified from DeVries, S.H., Schwartz, E.A., 1992. Hemi-gap-junction channels in solitary horizontal cells of the catfish retina. J. Physiol. 445, 201–230, with permission.)*

In early studies, transmission through electrical synapses was relatively fixed and unmodifiable, perhaps because in the synchronously firing systems, it was functionally relevant that transmission was constant. Then, work in turtle retina showed that both the receptive field of retinal horizontal cells and the spread of Lucifer yellow between these cells were decreased by dopamine (Piccolino et al., 1984). The effect required the activation of dopamine D1 receptors and involved cyclic AMP (cAMP) (Piccolino et al., 1984), suggesting that dopamine dynamically regulates the conductance of electrical synapses between these cells through PKA. This work was followed by observations in the white perch (Fig. 8.5A) in which pairs of isolated horizontal cells were simultaneously voltage clamped (Fig. 8.5B; Lasater and Dowling, 1985). Electrical coupling was directly measured by passing voltage pulses in one of the cells while recording current changes in the other cell. Addition of dopamine to the bath solution led to decoupling of the cells (Fig. 8.5C), an effect that was mimicked by bath application of 8-bromo-cyclic AMP, a membrane permeable form of cAMP (Lasater and Dowling, 1985). Dopamine was later found to be released by interplexiform cells and to mediate its effects via activation of PKA (Lasater et al., 1987). Dopamine not only decreased electrical coupling of horizontal cells but also enhanced their responses to glutamate (Knapp and Dowling, 1987). Also, experiments in voltage-clamped pairs of white perch horizontal cells revealed that dopamine decreases open probability and duration of openings of the gap junction channels rather than decreasing their open state conductance (McMahon et al., 1989). These data remain the best example of physiological modulation of single channel properties at an electrical synapse. Dopamine was later found to regulate electrical transmission via a cAMP-dependent pathway in other systems (Pereda et al., 2013) including an invertebrate (Johnson et al., 1993, 2005).

Intercellular channels are formed by the apposition of two "hemichannels." Exploring the properties of electrical coupling in pairs of catfish isolated horizontal cells, DeVries and Schwartz (DeVries and Schwartz, 1992) made a striking observation. They observed in single (thus uncoupled) horizontal cells that membrane conductance increased when extracellular Ca^{2+} was decreased (Fig. 8.5D). The properties of this current, including the actions of dopamine and PKA blockers, were consistent with opening of gap junction hemichannels (Fig. 8.5D; Lasater et al., 1987). This landmark study represents one of the first demonstrations that the channels that form gap junctions or "hemichannels," one contributed by each coupled cell to form a gap junction channel, can also function independently under the right conditions, although their high permeability to, eg, Ca^{2+}, can be toxic (Paul et al., 1991) and they are generally closed under physiological conditions. However, we know now that opening of gap junction hemichannels can occur under both normal and pathological conditions (Bennett et al., 2003; Orellana et al., 2012).

Gap junction channels are formed in vertebrates by proteins called connexins, which represent a family of 21 genes in humans (Goodenough and Paul, 2009). However, only a few of them are expressed in neurons (Söhl et al., 2005). Cx36 is the main connexin mediating electrical transmission between neurons in the mammalian brain (Söhl et al., 2005). Despite its relevance, identification of Cx36 came relatively late in comparison with other connexins (ie, Cx32, Cx43), and its ortholog in fishes was actually identified earlier (O'Brien et al., 1996). O'Brien and colleagues reported in 1996 the cloning of a new connexin, Cx35, from a skate retinal cDNA library (O'Brien et al., 1996). This initial observation was followed by the identification of Cx35 from perch retina, together with a second closely related connexin, Cx34.7 (O'Brien et al., 1998). This seminal study paralleled the initial description of Cx36, which was cloned by degenerate reverse transcription-polymerase chain reaction from microdissections of rat inferior olive (Condorelli et al., 1998), another brain structure that heavily relies on electrical coupling of neighboring cells. Channels formed by these three connexins exhibited similar biophysical properties, namely, small unitary conductance and weak voltage dependence (O'Brien et al., 1998; Teubner et al., 2000). Sequence comparison combined with analysis of gene structure led to the identification of a novel connexin subgroup (O'Brien et al., 1998), the gamma subgroup, formed by these neuronal connexins. Moreover, the cloning of Cx35 in fish retina also contributed to identification of consensus sequences for the regulatory actions of PKA and other kinases that are likely to participate in plastic changes (Kothmann et al., 2007; Ouyang et al., 2005; Li et al., 2009).

Finally, the identification of these fish connexins led to the discovery that they are differentially segregated to pre- and postsynaptic hemiplaques at club endings (Rash et al., 2013) and other mixed synapses in goldfish hindbrain (Rash et al., 2015). Freeze-fracture immunogold labeling revealed that while Cx35 was restricted to presynaptic hemiplaques, Cx34.7 was restricted to postsynaptic hemiplaques (Rash et al., 2013, 2015). This observation constituted the first description of heterotypic neuronal gap junctions in the vertebrate brain and it is associated in club endings with rectification of electrical transmission (Rash et al., 2013). Thus, investigations in fish were instrumental in demonstrating that vertebrate electrical synapses can be molecularly and functionally asymmetric.

Although gap junctions mediate electrical coupling in both vertebrates and invertebrates, the junctions are formed by different protein families in the two groups and are distinguishable with electron microscopy. It is a remarkable example of convergent evolution in terms of structure and even pharmacology. The vertebrate (and ascidian) gap junctions are formed of connexins, a gene family distinct from pannexins (Panchin et al., 2000), or innexins (Phelan and Starich, 2001; Phelan et al., 2017 (Chapter 2)), which form gap junctions in protostomes and some deuterostomes. Vertebrates also have pannexins, but they are not known to form gap junctions and probably only function as channels across single membranes.

6. ZEBRAFISH: CONNEXIN DIVERSITY AND COMMON DEVELOPMENTAL STEPS FOR CHEMICAL AND ELECTRICAL SYNAPSES

The zebrafish (*Danio rerio*) (Fig. 8.6A) is a fresh water fish that has been firmly established as a valuable experimental model. Because of its amenity for molecular genetic analysis,

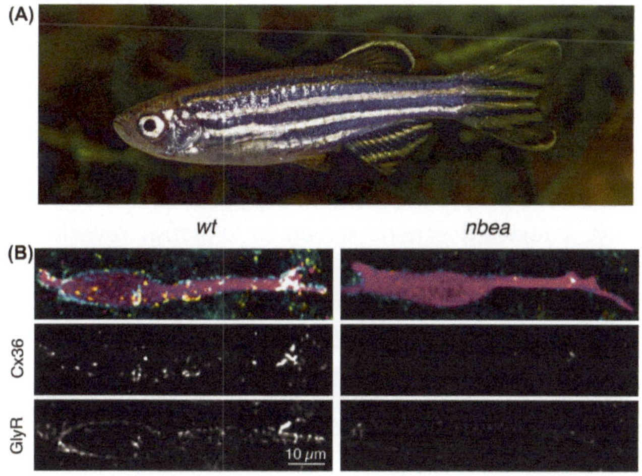

Figure 8.6 ***Developmental studies of electrical transmission in zebrafish.*** (A) Adult zebrafish (*Danio rerio*), about 4 cm long. *(Reproduced from Alamy, with permission.)* (B) Staining of the Mauthner cell with a mouse Cx36 antibody (that labels all fish Cx36-related connexins, yellow), a glycine receptor (GlyR, cyan) antibody, and GFP (magenta) antibody in a 5 dpf wild-type zebrafish (*wt*) and a fish with a loss of function mutation in neurobeachin (*nbea*), which causes defects in the formation of both electrical and chemical synapses. Top panel: merge. Lower panels: individual Cx36 and GlyR channels. *(Modified from Miller, A.C., Voelker, L.H., Shah, A.N., Moens, C.B., 2015. Neurobeachin is required postsynaptically for electrical and chemical synapse formation. Curr. Biol. 25, 16–28, with permission.)*

and transparency that facilitates in vivo imaging, the zebrafish has become an important animal model to study various aspects of nervous system function, both at the cellular and systems level (Portugues and Engert, 2009; Ahrens et al., 2012; Friedrich et al., 2013; Feierstein et al., 2015). We will focus on two main aspects of electrical synapses in zebrafish: the diversity of neuronal connexins and the identification of common developmental regulatory mechanisms for chemical and electrical transmission derived from forward genetic screens. More than 36 connexin genes have now been identified in zebrafish (Eastman et al., 2006). The higher number of connexins (21 in humans) reflects additional genome duplication that teleosts underwent during evolution (Volff, 2005). Cloning of connexins in zebrafish led to the identification of the connexins that form gap junctions between retinal horizontal cells. Although retinal horizontal cells provided the first description and demonstration of dynamic changes in gap junctional conductance between neurons (see earlier discussion), transmission between these cells was found not to be mediated by fish homologs of Cx36, suggesting the presence of additional neuronal connexins. The early identification of Cx55.5 in zebrafish (Dermietzel et al., 2000) led to the cloning of a second connexin in zebrafish horizontal cells, Cx52.6 (Zoidl et al., 2004). The initial cloning of Cx55.5 made possible the identification, based on sequence similarity, of Cx57 in mouse retina (Hombach et al., 2004), which also expresses Cx50. Cx57 was found to be orthologous to zebrafish Cx52.6. In contrast, Cx59, which is orthologous to Cx55.5, was only detected at low levels in human retina (Söhl et al., 2010). Thus the identification of neuronal connexins at zebrafish electrical synapses led to a more complete description of the diversity of neuronal connexins.

The multidomain scaffolding protein Neurobeachin (Nbea) is involved in neuronal post-Golgi membrane trafficking. Its presence is essential for normal chemical synaptic transmission (Wang et al., 2000; Nair et al., 2013), and Nbea null-mutant mice exhibit reduced synaptic responses at central synapses and a reduction in the surface expression of both glutamatergic and GABAergic receptors (Lauks et al., 2012; Nair et al., 2013). Not surprisingly, these animals are paralyzed and die perinatally (Su et al., 2004; Nair et al., 2013). Remarkably, a forward genetic screen in zebrafish revealed that Nbea is also required for the formation of electrical synapses (Miller et al., 2015). Zebrafish lacking Nbea exhibit a marked reduction in the staining of neuronal connexins, detected in this case with an anti-mouse Cx36 antibody that recognizes all Cx36-related connexins, including those at the Club endings and other synapses on the Mauthner cell (Fig. 8.6B). Consistent with Nbea's role in the formation of chemical synapses, these mutants exhibited a marked reduction in staining for glycine receptors (Fig. 8.6B). Moreover, chimera analysis, which allows tracking of mutant and wild-type neurons, indicated that Nbea was required only postsynaptically for the formation of electrical and chemical synapses (Miller et al., 2015). This finding indicates that despite marked differences in the mechanisms of cellular interaction, electrical and chemical transmission share common regulatory steps. Furthermore, it once again indicates that synaptic communication within the brain circuits may be achieved by combining electrical and chemical transmission.

7. CONCLUSIONS

Fish electrical synapses critically contributed to the description of electrical transmission. Although early evidence for the existence of electrical transmission originated in invertebrates, experiments in fishes demonstrated its presence in vertebrates. More importantly, correlation of electrophysiological and ultrastructural analysis in fishes led to the discovery of the anatomical basis for this form of communication. The contributions of fish electrical synapses were not restricted to the initial period of exploration of the mechanisms of synaptic transmission. Cloning of new gap junction proteins in fish led to the identification of the first Cx36-like gene (O'Brien et al., 1996), a neuronal connexin of fundamental importance in the mammalian brain (Söhl et al., 2005). From an evolutionary point of view, fish electrical synapses exposed the role of electrical coupling in promoting neuronal synchronization, which is a common, but not universal, function of this modality of transmission (Bennett and Zukin, 2004). Fish electrical synapses also provided the initial evidence of the rich plastic properties of electrical communication, a feature that had been perceived to be restricted to chemical synapses.

Studies on fish mixed synapses also exposed a close functional relationship between their chemical and electrical components (Pereda, 2014) and led to the identification of similar interactions in the mammalian brain. Also, forward genetic screens revealed that electrical and chemical synapses share common developmental mechanisms (Miller et al., 2015). Together, these two findings contributed to the blurring of the conceptual antagonism between chemical and electrical synapses and indicate that synaptic transmission within neural circuits, as in fish mixed synapses, can occur via (at least) two complementary forms of communication. Paradoxically then, although fish electrical synapses were instrumental for differentiating electrical vs chemical transmission, they also provided with the first evidence that synaptic communication may be achieved by combining electrical and chemical synapses, thus serving a wider range of functional requirements.

There are many reasons to think that electrical synapses in fishes will continue to provide us in the future with valuable information on the properties and mechanisms of synaptic communication. The development of the zebrafish model brought a set of very powerful tools to address the issues. The combination of molecular and forward genetics with cutting edge imaging analysis of increasing resolution is likely to yield detailed information on the molecular organization and mechanisms of electrical transmission. Interestingly, the involvement of electrical synapses has been reported in the hypothalamus of cichlids (Ma et al., 2015). The exceptional rate of speciation of cichlids makes them the largest and more diverse vertebrate family in the world (Henning and Meyer, 2014). Comparative analysis between some of these species could lead to interesting insights into evolutionary aspects of electrical transmission. The use of these two (and other) advantageous fish models is likely to become a source of excitement and discovery, and guarantee that fish, again and again, will keep surprising us with their revelations.

ACKNOWLEDGMENTS

We thank John O'Brien for helpful discussions. Supported by National Institutes of Health (NIH) grants NIH DC03186, DC011099, NS055726, NS085772, and NS0552827 to A.E.P, and NS055363 to M.V.L B.

REFERENCES

Ahrens, M.B., Li, J.M., Orger, M.B., Robson, D.N., Schier, A.F., Engert, F., Portugues, R., 2012. Brain-wide neuronal dynamics during motor adaptation in zebrafish. Nature 485, 471–477.

Arvanitaki, A., Chalozonitis, N., Costa, H., 1964. Excitation of the giant neuron by crossed exponential transmembranal currents (*Aplysia fasciata*). C. R. Seances Soc. Biol. Fil. 158, 2373–2377.

Bartelmez, G.W., Hoerr, N.L., 1933. The vestibular club endings in ameiurus. Further evidence on the morphology of the synapse. J. Comp. Neurol. 57, 401–428.

Bennett, M.V.L., 1971a. Electric organs. In: Hoar, W.S., Randall, D.J. (Eds.), Fish Physiology, vol. 5. Academic Press, New York, pp. 347–491.

Bennett, M.V.L., 1971b. Electroreceptors. In: Hoar, W.S., Randall, D.J. (Eds.), Fish, Physiology, vol. 5. Academic Press, New York, pp. 493–574.

Bennett, M.V.L., Contreras, J.E., Bukauskas, F.F., Sáez, J.C., 2003. New roles for astrocytes: gap junction hemichannels have something to communicate. Trends Neurosci. 26, 610–617.

Bennett, M.V.L., Crain, S.M., Grundfest, H., 1957. Patterns of response and neural organization of supramedullary neurons of puffer (blowfish), *Spheroides maculatus*. Biol. Bull. 325–326.

Bennett, M.V.L., Zukin, R.S., 2004. Electrical coupling and neuronal synchronization in the mammalian brain. Neuron 41, 495–511.

Bennett, M.V., 1966. Physiology of electrotonic junctions. Ann. N.Y. Acad. Sci. 137, 509–539.

Bennett, M.V., Aljure, E., Nakajima, Y., Pappas, G.D., 1963. Electrotonic junctions between teleost spinal neurons: electrophysiology and ultrastructure. Science 141, 262–264.

Bennett, M.V., Crain, S.M., Grundfest, H., 1959a. Electrophysiology of supramedullary neurons in *Spheroides maculatus*. III. Organization of the supramedullary neurons. J. Gen. Physiol. 43, 221–250.

Bennett, M.V., Crain, S.M., Grundfest, H., 1959b. Electrophysiology of supramedullary neurons in *Spheroides maculatus*. I. Orthodromic and antidromic responses. J. Gen. Physiol. 43, 159–188.

Bennett, M.V., Nakajima, Y., Pappas, G.D., 1967a. Physiology and ultrastructure of electrotonic junctions. I. Supramedullary neurons. J. Neurophysiol. 30, 161–179.

Bennett, M.V., Nakajima, Y., Pappas, G.D., 1967b. Physiology and ultrastructure of electrotonic junctions. 3. Giant electromotor neurons of *Malapterurus electricus*. J. Neurophysiol. 30, 209–235.

Bennett, M.V., Pappas, G.D., Aljure, E., Nakajima, Y., 1967c. Physiology and ultrastructure of electrotonic junctions. II. Spinal and medullary electromotor nuclei in mormyrid fish. J. Neurophysiol. 30, 180–208.

Brock, L.G., Coombs, J.S., Eccles, J.C., 1953. Intracellular recording from antidromically activated motoneurones. J. Physiol. 122, 429–461.

Cachope, R., Mackie, K., Triller, A., O'Brien, J., Pereda, A.E., 2007. Potentiation of electrical and chemical synaptic transmission mediated by endocannabinoids. Neuron 56, 1034–1047.

Condorelli, D.F., Parenti, R., Spinella, F., Trovato Salinaro, A., Belluardo, N., Cardile, V., Cicirata, F., 1998. Cloning of a new gap junction gene (Cx36) highly expressed in mammalian brain neurons. Eur. J. Neurosci. 10, 1202–1208.

Curti, S., Pereda, A.E., 2004. Voltage-dependent enhancement of electrical coupling by a subthreshold sodium current. J. Neurosci. 24, 3999–4010.

Dermietzel, R., Kremer, M., Paputsoglu, G., Stang, A., Skerrett, I.M., Gomes, D., Srinivas, M., Janssen-Bienhold, U., Weiler, R., Nicholson, B.J., Bruzzone, R., Spray, D.C., 2000. Molecular and functional diversity of neural connexins in the retina. J. Neurosci. 20, 8331–8343.

DeVries, S.H., Schwartz, E.A., 1992. Hemi-gap-junction channels in solitary horizontal cells of the catfish retina. J. Physiol. 445, 201–230.

Dugué, G.P., Brunel, N., Hakim, V., Schwartz, E., Chat, M., Lévesque, M., Courtemanche, R., Léna, C., Dieudonné, S., 2009. Electrical coupling mediates tunable low-frequency oscillations and resonance in the cerebellar Golgi cell network. Neuron 61, 126–139.

Eastman, S.D., Chen, T.H.-P., Falk, M.M., Mendelson, T.C., Iovine, M.K., 2006. Phylogenetic analysis of three complete gap junction gene families reveals lineage-specific duplications and highly supported gene classes. Genomics 87, 265–274.

Elekes, K., Szabo, T., 1985. The mormyrid brainstem—III. Ultrastructure and synaptic organization of the medullary "pacemaker" nucleus. Neuroscience 15, 431–443.

Faber, D.S., Korn, H., 1973. A neuronal inhibition mediated electrically. Science 179, 577–578.

Faber, D.S., Korn, H., 1989. Electrical field effects: their relevance in central neural networks. Physiol. Rev. 69, 821–863.

Fatt, P., 1954. Biophysics of junctional transmission. Physiol. Rev. 34, 674–710.

Feierstein, C.E., Portugues, R., Orger, M.B., 2015. Seeing the whole picture: a comprehensive imaging approach to functional mapping of circuits in behaving zebrafish. Neuroscience 296, 26–38.

Friedrich, R.W., Genoud, C., Wanner, A.A., 2013. Analyzing the structure and function of neuronal circuits in zebrafish. Front. Neural Circuits 7, 71.

Furshpan, E.J., 1964. "Electrical transmission" at an excitatory synapse in a vertebrate brain. Science 144, 878–880.

Furshpan, E.J., Furukawa, T., 1962. Intracellular and extracellular responses of the several regions of the Mauthner cell of the goldfish. J. Neurophysiol. 25, 732–771.

Furshpan, E.J., Potter, D.D., 1959. Transmission at the giant motor synapses of the crayfish. J. Physiol. 145, 289–325.

Furukawa, T., Furshpan, E.J., 1963. Two inhibitory mechanisms in the Mauthner neurons of goldfish. J. Neurophysiol. 26, 140–176.

Goodenough, D.a, Paul, D.L., 2009. Gap junctions. Cold Spring Harb. Perspect. Biol. 1, a002576.

Haas, J.S., Landisman, C.E., 2011. State-dependent modulation of gap junction signaling by the persistent sodium current. Front. Cell. Neurosci. 5, 31.

Henning, F., Meyer, A., 2014. The evolutionary genomics of cichlid fishes: explosive speciation and adaptation in the postgenomic era. Annu. Rev. Genom. Hum. Genet. 15, 417–441.

Hombach, S., Janssen-Bienhold, U., Söhl, G., Schubert, T., Büssow, H., Ott, T., Weiler, R., Willecke, K., 2004. Functional expression of connexin57 in horizontal cells of the mouse retina. Eur. J. Neurosci. 19, 2633–2640.

Johnson, B.R., Peck, J.H., Harris-Warrick, R.M., 1993. Amine modulation of electrical coupling in the pyloric network of the lobster stomatogastric ganglion. J. Comp. Physiol. A 172, 715–732.

Johnson, B.R., Schneider, L.R., Nadim, F., Harris-Warrick, R.M., 2005. Dopamine modulation of phasing of activity in a rhythmic motor network: contribution of synaptic and intrinsic modulatory actions. J. Neurophysiol. 94, 3101–3111.

Kao, C.Y., 1960. Postsynaptic electrogenesis in septate giant axons. II. Comparison of medial and lateral giant axons of crayfish. J. Neurophysiol. 23, 618–635.

Kao, C.Y., Grundfest, H., 1957. Postsynaptic electrogenesis in septate giant axons. I. Earthworm median giant axon. J. Neurophysiol. 20, 553–573.

Knapp, A.G., Dowling, J.E., 1987. Dopamine enhances excitatory amino acid-gated conductances in cultured retinal horizontal cells. Nature 325, 437–439.

Korn, H., Axelrad, H., 1980. Electrical inhibition of Purkinje cells in the cerebellum of the rat. Proc. Natl. Acad. Sci. U.S.A. 77, 6244–6247.

Korn, H., Faber, D.S., 2005. The Mauthner cell half a century later: a neurobiological model for decision-making? Neuron 47, 13–28.

Korn, H., Triller, A., Faber, D.S., 1978. Structural correlates of recurrent collateral interneurons producing both electrical and chemical inhibitions of the Mauthner cell. Proc. R. Soc. Lond. Ser. B Biol. Sci. 202, 533–538.

Kothmann, W.W., Li, X., Burr, G.S., O'Brien, J., 2007. Connexin 35/36 is phosphorylated at regulatory sites in the retina. Vis. Neurosci. 24, 363–375.

Kothmann, W.W., Trexler, E.B., Whitaker, C.M., Li, W., Massey, S.C., O'Brien, J., 2012. Nonsynaptic NMDA receptors mediate activity-dependent plasticity of gap junctional coupling in the AII amacrine cell network. J. Neurosci. 32, 6747–6759.

Kriebel, M.E., Bennett, M.V., Waxman, S.G., Pappas, G.D., 1969. Oculomotor neurons in fish: electrotonic coupling and multiple sites of impulse initiation. Science 166, 520–524.

Lasater, E.M., Dowling, J.E., 1985. Dopamine decreases conductance of the electrical junctions between cultured retinal horizontal cells. Proc. Natl. Acad. Sci. U.S.A. 82, 3025–3029.

Lasater, E.M., Gonzalez, A., Fischer, C., Krebs, E.H., 1987. Retinal horizontal cell gap junctional conductance is modulated by dopamine through a cyclic AMP-dependent protein kinase. Neurobiology 84, 7319–7323.

Lauks, J., Klemmer, P., Farzana, F., Karupothula, R., Zalm, R., Cooke, N.E., Li, K.W., Smit, A.B., Toonen, R., Verhage, M., 2012. Synapse associated protein 102 (SAP102) binds the C-terminal part of the scaffolding protein neurobeachin. PLoS One 7, e39420.

Li, H., Chuang, A.Z., O'Brien, J., 2009. Photoreceptor coupling is controlled by connexin 35 phosphorylation in zebrafish retina. J. Neurosci. 29, 15178–15186.

Ma, Y., Juntti, S.A., Hu, C.K., Huguenard, J.R., Fernald, R.D., 2015. Electrical synapses connect a network of gonadotropin releasing hormone neurons in a cichlid fish. Proc. Natl. Acad. Sci. U.S.A. 112, 3805–3810.

Mann-Metzer, P., Yarom, Y., 1999. Electrotonic coupling interacts with intrinsic properties to generate synchronized activity in cerebellar networks of inhibitory interneurons. J. Neurosci. 19, 3298–3306.

McMahon, D.G., Knapp, A.G., Dowling, J.E., 1989. Horizontal cell gap junctions: single-channel conductance and modulation by dopamine. Proc. Natl. Acad. Sci. U.S.A. 86, 7639–7643.

Miller, A.C., Voelker, L.H., Shah, A.N., Moens, C.B., 2015. Neurobeachin is required postsynaptically for electrical and chemical synapse formation. Curr. Biol. 25, 16–28.

Nair, R., Lauks, J., Jung, S., Cooke, N.E., de Wit, H., Brose, N., Kilimann, M.W., Verhage, M., Rhee, J., 2013. Neurobeachin regulates neurotransmitter receptor trafficking to synapses. J. Cell Biol. 200, 61–80.

O'Brien, J., al-Ubaidi, M.R., Ripps, H., 1996. Connexin 35: a gap-junctional protein expressed preferentially in the skate retina. Mol. Biol. Cell 7, 233–243.

O'Brien, J., Bruzzone, R., White, T.W., Al-Ubaidi, M.R., Ripps, H., 1998. Cloning and expression of two related connexins from the perch retina define a distinct subgroup of the connexin family. J. Neurosci. 18, 7625–7637.

Orellana, J.A., von Bernhardi, R., Giaume, C., Sáez, J.C., 2012. Glial hemichannels and their involvement in aging and neurodegenerative diseases. Rev. Neurosci. 23, 163–177.

Ouyang, X., Winbow, V.M., Patel, L.S., Burr, G.S., Mitchell, C.K., O'Brien, J., 2005. Protein kinase A mediates regulation of gap junctions containing connexin35 through a complex pathway. Brain Res. Mol. Brain Res. 135, 1–11.

Panchin, Y., Kelmanson, I., Matz, M., Lukyanov, K., Usman, N., Lukyanov, S., 2000. A ubiquitous family of putative gap junction molecules. Curr. Biol. 29, R473–R474.

Pappas, G.D., Bennett, M.V., 1966. Specialized junctions involved in electrical transmission between neurons. Ann. N.Y. Acad. Sci. 137, 495–508.

Paul, D.L., Ebihara, L., Takemoto, L.J., Swenson, K.I., Goodenough, D.A., 1991. Connexin46, a novel lens gap junction protein, induces voltage-gated currents in nonjunctional plasma membrane of *Xenopus* oocytes. J. Cell Biol. 115, 1077–1089.

Pereda, A.E., Bell, T.D., Chang, B.H., Czernik, A.J., Nairn, A.C., Soderling, T.R., Faber, D.S., 1998. Ca2+/calmodulin-dependent kinase II mediates simultaneous enhancement of gap-junctional conductance and glutamatergic transmission. Proc. Natl. Acad. Sci. U.S.A. 95, 13272–13277.

Pereda, A., Brien, J.O., Nagy, J.I., Smith, M., Bukauskas, F., Davidson, K.G.V., Kamasawa, N., Yasumura, T., Rash, J.E., 2003. Short-range functional interaction between Connexin35 and neighboring chemical synapses. Cell Commun. Adhes. 419–423.

Pereda, A., Triller, A., Korn, H., Faber, D.S., 1992. Dopamine enhances both electrotonic coupling and chemical excitatory postsynaptic potentials at mixed synapses. Proc. Natl. Acad. Sci. U.S.A. 89, 12088–12092.

Pereda, A.E., 2014. Electrical synapses and their functional interactions with chemical synapses. Nat. Rev. Neurosci. 15, 250–263.

Pereda, A.E., Curti, S., Hoge, G., Cachope, R., Flores, C.E., Rash, J.E., 2013. Gap junction-mediated electrical transmission: regulatory mechanisms and plasticity. Biochim. Biophys. Acta 1828, 134–146.

Pereda, A.E., Faber, D.S., 1996. Activity-dependent short-term enhancement of intercellular coupling. J. Neurosci. 16, 983–992.

Pereda, A.E., Faber, D.S., 2011. Physiology of the Mauthner Cell: Discovery and Properties. Elsevier Inc.

Pereda, A.E., Nairn, A.C., Wolszon, L.R., Faber, D.S., 1994. Postsynaptic modulation of synaptic efficacy at mixed synapses on the Mauthner cell. J. Neurosci. 14, 3704–3712.

Pereda, A.E., Rash, J.E., Nagy, J.I., Bennett, M.V.L., 2004. Dynamics of electrical transmission at club endings on the Mauthner cells. Brain Res. Brain Res. Rev. 47, 227–244.

Phelan, P., Bacon, J.P., Blagburn, J.M., 2017. Neural circuits underlying escape behavior in Drosophila: focus on electrical signaling. In: Jing, J. (Ed.), Network Functions and Plasticity. Elsevier (Chapter 2).

Phelan, P., Starich, T.A., 2001. Innexins get into the gap. Bioessays 23, 388–396.

Piccolino, M., Neyton, J., Gerschenfeld, H.M., 1984. Decrease of gap junction permeability induced by dopamine and cyclic adenosine 3′:5′-monophosphate in horizontal cells of turtle retina. J. Neurosci. 4, 2477–2488.

Portugues, R., Engert, F., 2009. The neural basis of visual behaviors in the larval zebrafish. Curr. Opin. Neurobiol. 19, 644–647.

Rash, J.E., Curti, S., Vanderpool, K.G., Kamasawa, N., Nannapaneni, S., Palacios-Prado, N., Flores, C.E., Yasumura, T., O'Brien, J., Lynn, B.D., Bukauskas, F.F., Nagy, J.I., Pereda, A.E., 2013. Molecular and functional asymmetry at a vertebrate electrical synapse. Neuron 79, 957–969.

Rash, J.E., Kamasawa, N., Vanderpool, K.G., Yasumura, T., O'Brien, J., Nannapaneni, S., Pereda, A.E., Nagy, J.I., 2015. Heterotypic gap junctions at glutamatergic mixed synapses are abundant in goldfish brain. Neuroscience 285, 166–193.

Robertson, J.D., 1963. The occurrence of a subunit pattern in the unit membranes of club endings in Mauthner cell synapses in Goldfish brains. J. Cell Biol. 19, 201–221.

Robertson, J.D., Bodenheimer, T.S., Stage, D.E., 1963. The ultrastructure of Mauthner cell synapses and nodes in goldfish brains. J. Cell Biol. 19, 159–199.

Söhl, G., Joussen, A., Kociok, N., Willecke, K., 2010. Expression of connexin genes in the human retina. BMC Ophthalmol. 10, 27.

Söhl, G., Maxeiner, S., Willecke, K., 2005. Expression and functions of neuronal gap junctions. Nat. Rev. Neurosci. 6, 191–200.

Su, Y., Balice-Gordon, R.J., Hess, D.M., Landsman, D.S., Minarcik, J., Golden, J., Hurwitz, I., Liebhaber, S.A., Cooke, N.E., 2004. Neurobeachin is essential for neuromuscular synaptic transmission. J. Neurosci. 24, 3627–3636.

Teubner, B., Degen, J., Sohl, G., Guldenagel, M., Bukauskas, F.F., Trexler, E.B., Verselis, V.K., De Zeeuw, C.I., Lee, C.G., Kozak, C.A., Petrasch-Parwez, E., Dermietzel, R., Willecke, K., 2000. Functional expression of the murine connexin 36 gene coding for a neuron-specific gap junctional protein. J. Memb. Biol. 176, 249–262.

Turecek, J., Yuen, G.S., Han, V.Z., Zeng, X.-H., Bayer, K.U., Welsh, J.P., 2014. NMDA receptor activation strengthens weak electrical coupling in mammalian brain. Neuron 81, 1375–1388.

Volff, J.-N., 2005. Genome evolution and biodiversity in teleost fish. Heredity (Edinb) 94, 280–294.

Wang, X., Herberg, F.W., Laue, M.M., Wullner, C., Hu, B., Petrasch-Parwez, E., Kilimann, M.W., 2000. Neurobeachin: a protein kinase A-anchoring, beige/Chediak-higashi protein homolog implicated in neuronal membrane traffic. J. Neurosci. 20, 8551–8565.

Watanabe, A., 1958. The interaction of electrical activity among neurons of lobster cardiac ganglion. Jpn. J. Physiol. 8, 305–318.

Welsh, J.P., Turecek, J., 2017. Implications of electrical synapse plasticity in the inferior olive. In: Jing, J. (Ed.), Network Functions and Plasticity. Elsevier (Chapter 12).

Yang, X.D., Faber, D.S., 1991. Initial synaptic efficacy influences induction and expression of long-term changes in transmission. Proc. Natl. Acad. Sci. U.S.A. 88, 4299–4303.

Yang, X.D., Korn, H., Faber, D.S., 1990. Long-term potentiation of electrotonic coupling at mixed synapses. Nature 348, 542–545.

Zavala-Tecuapetla, C., Tapia, D., Rivera-Angulo, A.J., Galarraga, E., Peña-Ortega, F., 2014. Morphological characterization of respiratory neurons in the pre-Bötzinger complex. Prog. Brain Res. 209, 39–56.

Zoidl, G., Bruzzone, R., Weickert, S., Kremer, M., Zoidl, C., Mitropoulou, G., Srinivas, M., Spray, D.C., Dermietzel, R., 2004. Molecular cloning and functional expression of zfCx52.6: a novel connexin with hemichannel-forming properties expressed in horizontal cells of the zebrafish retina. J. Biol. Chem. 279, 2913–2921.

CHAPTER 9

Dynamic Properties of Electrically Coupled Retinal Networks

S. Trenholm[1], G.B. Awatramani[2]
[1]Friedrich Miescher Institute for Biomedical Research, Basel, Switzerland; [2]University of Victoria, Victoria, BC, Canada

1. INTRODUCTION

The vertebrate retina comprises over 60 different cell types that make specific connections with one another in distinct retinal layers (Masland, 2012). These different cell types can be functionally grouped into approximately 30 channels that differentially process incoming visual signals, and these diverse channels are then relayed to a host of higher visual areas (Dacey, 2004; Azeredo da Silveira and Roska, 2011; Robles et al., 2014; Sanes and Masland, 2015; Baden et al., 2016). As each retinal channel exhibits a unique pattern of gap junction coupling (Völgyi et al., 2009), it is likely that coupling plays diverse functional roles in each of these channels. Furthermore, as each retinal channel exhibits a distinct ensemble of excitatory and inhibitory chemical synapses, each channel likely displays unique interactions between electrical and chemical synaptic transmission, yielding a huge variety of neuronal response properties (Pereda, 2014). Here, we review both the classical and recent literature to provide an overview of the assorted functional circuit properties resulting from gap junction coupling in the retina and we describe how these circuit properties arise from interactions between electrical and chemical synaptic signals.

2. OVERVIEW OF THE SYNAPTIC ARCHITECTURE OF THE RETINA

Classically, the vertebrate retina is described as comprising five principal neuron classes: photoreceptors (rods and cones), horizontal cells, bipolar cells, amacrine cells, and ganglion cells. We summarize the synaptic connectivity patterns mediating signals between these different retinal neurons (Fig. 9.1). For electrical synapses, we indicate the connexin subtypes that are involved in forming gap junctions, as well as some of the biophysical properties of these connections.

Vision starts in rod and cone photoreceptors, where light is converted into electrochemical signals. Rod and cone photoreceptors form glutamatergic synapses with downstream bipolar and horizontal cells: sign-conserving ionotropic glutamate receptors on horizontal cells and OFF bipolar cells, and sign-inverting metabotropic glutamate receptors on ON bipolar cells (Masland, 2012). In addition, rods and

Network Functions and Plasticity
ISBN 978-0-12-803471-2, http://dx.doi.org/10.1016/B978-0-12-803471-2.00009-6

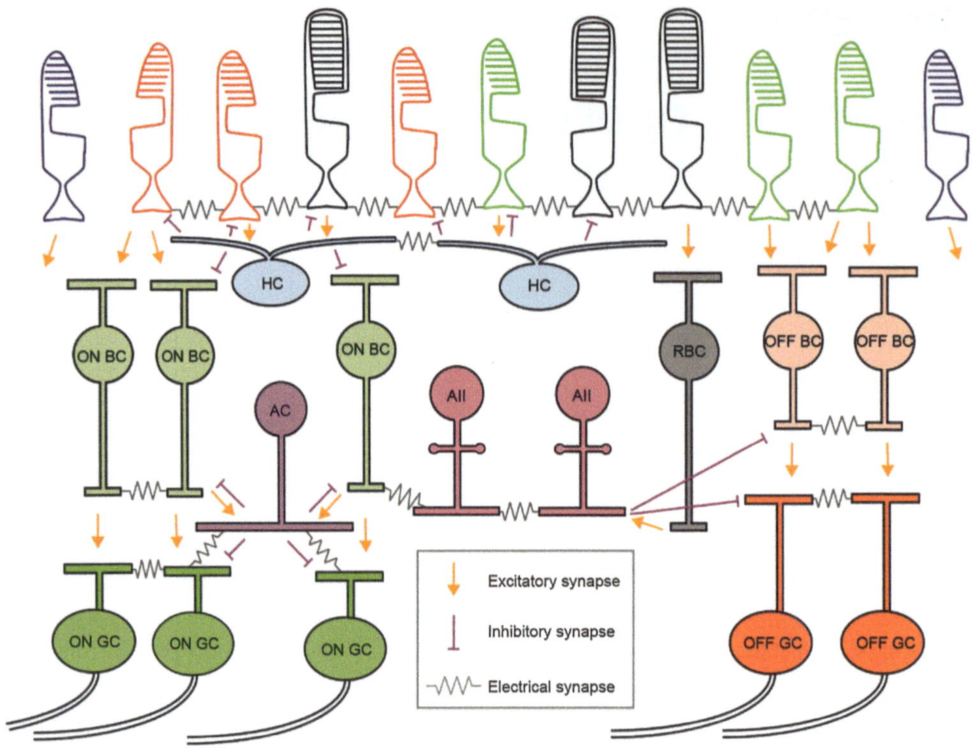

Figure 9.1 ***Synaptic architecture of the retina.*** The diagram represents a simplified view of the mammalian retina. Photoreceptors are on the top of the diagram, with *rods* colored black, and *cones* colored based on their absorption wavelength (note that *blue cones* do not appear to be coupled; Hornstein et al., 2004; Li and DeVries, 2004). Horizontal cells (HC), ON and OFF bipolar cells (BC) and a rod bipolar cell (RBC) are shown in the middle, as are AII amacrine cells, as well as a generic amacrine cell (AC). ON and OFF ganglion cells (GC) are on the bottom of the diagram. Note that negative feedback from horizontal cells to photoreceptors may not involve a classical synapse (Thoreson and Mangel, 2012).

cones form electrical synapses both between themselves (cone–cone, rod–rod) and between each other (rod–cone; Baylor et al., 1971; Custer, 1973; Raviola and Gilula, 1973; Copenhagen and Owen, 1976; Tsukamoto et al., 2001; DeVries et al., 2002; Hornstein et al., 2004, 2005; Li et al., 2012). Cone gap junctions are thought to comprise connexin 36 (Cx36) in mammals (Lee et al., 2003; Feigenspan et al., 2004; O'Brien et al., 2012; Kántor et al., 2015) and Cx35 in zebrafish (Li et al., 2009), but the connexin involved in rod electrical synapses has not been definitively identified (Lee et al., 2003; but see also Zhang and Wu, 2004; Li et al., 2013). For both cones (Raviola and Gilula, 1973; Tsukamoto et al., 1992; Li and DeVries, 2004) and rods (Tsukamoto et al., 2001), gap junctions are localized near the axon terminal. The gap junction conductance between cones has a range of measured values between 200 and 800 pS and is bidirectionally symmetric (DeVries et al., 2002; Hornstein et al., 2004; Li and DeVries, 2004), similar to what has been found for rod–rod coupling (Zhang and Wu, 2005; Li et al., 2012; Jin and Ribelayga, 2016), whereas coupling between rods and cones has a conductance of 40–200 pS and is bidirectionally symmetrical (Gao et al., 2013). The coupling coefficient between cones (ie, the change in the membrane potential of the postjunctional cell divided by the change in the membrane potential in the prejunctional cell when the membrane potential of the prejunctional cell is experimentally altered) is ~0.15 (Hornstein et al., 2004) similar to what has been described between rods (Werblin, 1978), whereas the coupling coefficient between rods and cones appears to be ~0.04 (Attwell et al., 1984). Thus homologous photoreceptor coupling (ie, rod–rod and cone–cone) appears to be stronger than heterologous coupling (ie, rod–cone). From tracer dye experiments, it has been shown that rod–rod coupling appears to be much more common than rod–cone coupling, and some rods may be uncoupled (Li et al., 2012). For cones, being exclusively coupled to rods appears to be more common than being coupled to both rods and cones or being coupled only to other cones, and some cones may be uncoupled (Hornstein et al., 2005). Importantly, as will be discussed later, coupling strength is not fixed, but instead varies depending on time of day and light conditions.

Horizontal cells provide feedback inhibition to photoreceptors and feedforward inhibition to bipolar cells, although the exact nature of the mechanism mediating these interactions remains unclear (Thoreson and Mangel, 2012). Horizontal cells are connected to one another via gap junctions (Raviola and Gilula, 1975; Kolb, 1977). Tracer dye injected into a single horizontal cell can spread into more than 1000 surrounding horizontal cells extending across a distance of over 1 mm (Xin and Bloomfield, 1999), indicating that coupling between horizontal cells is likely very strong. Indeed, physiological recordings have found the junctional conductance to be very large (>1 nS), with a coupling coefficient greater than 0.5 (often greater than 0.9) and with cells being connected bidirectionally and symmetrically (Lasater and Dowling, 1985; DeVries and Schwartz, 1989; McMahon, 1994; McMahon and Mattson, 1996; note that these studies were performed in cultured horizontal cells). Horizontal cells can form both axoaxonic and dendrodendritic gap junctions. In mouse, Cx57 is present in both dendritic and axonic gap junctions (Hombach et al., 2004; Janssen-Bienhold et al., 2009), whereas Cx50 appears to be restricted to axonal gap junctions (Dorgau et al., 2015). In rabbit A-type horizontal cells, Cx50 mediates dendritic gap junctions (O'Brien et al., 2006), whereas Cx57 appears restricted to axons in B-type horizontal cells (Pan et al., 2012). In zebrafish, gap junctions are thought to comprise Cx52.6, 52.9, and 55.5 (Shields et al., 2007; Klaassen et al., 2011).

Bipolar cells, of which there are approximately 10 types (Wässle et al., 2009), pass photoreceptor signals onto ganglion cells via glutamate release, and onto amacrine cells via glutamate release and gap junction-mediated signals. Both ON and OFF bipolar cells can also be electrically coupled to other bipolar cells with a bidirectional and symmetrical coupling conductance of ~1 nS and a coupling coefficient of ~0.3 (Kujiraoka and Saito, 1986; Umino et al., 1994; Mills, 1999; Dacey et al., 2000; Arai et al., 2010). Most types of ON cone bipolar cells are also electrically coupled to AII amacrine cells, possibly via Cx36/45 heterologous gap junctions, with a junctional conductance of approximately 1 nS, which is bidirectional but asymmetric, and with a coupling coefficient between 0.3 and 0.6 (Kolb and Famiglietti, 1974; Deans et al., 2002; Veruki and Hartveit, 2002a; Marc et al., 2014). The measured asymmetry in the junctional conductance can be well accounted for by the impedance mismatch between AII amacrine and bipolar cells (Veruki and Hartveit, 2002a). Additionally, recent work has shown that certain ON cone bipolar cells are coupled to non-AII amacrine cells (Farrow et al., 2013; Lee et al., 2015). Bipolar cells in mouse can express either Cx36 or Cx45 (Feigenspan et al., 2004; Han and Massey, 2005; Maxeiner et al., 2005; Hilgen et al., 2011). Gap junctions between bipolar cells can be localized in the dendrites (Arai et al., 2010) or axon terminals (Marc et al., 1988; Jacoby and Marshak, 2000), whereas gap junctions between ON cone bipolar cells and AII amacrine cells are localized in the axon terminal (Strettoi et al., 1992).

Amacrine cells, of which there appear to be more than 30 types (Masland, 2012), provide inhibition, via gamma aminobutyric acid (GABA) or glycine release, to bipolar cells, other amacrine cells, and ganglion cells. Certain amacrine cells also signal via release of acetylcholine (Masland and Mills, 1979; Lee et al., 2010), dopamine (Zhang et al., 2007) or glutamate (Baden and Euler, 2016). Amacrine cells can be electrically coupled to each other (Xin and Bloomfield, 1997; Li et al., 2002; Veruki and Hartveit, 2002b; Abdel-Majid et al., 2005; Marc et al., 2014), to bipolar cells, as described earlier, or to ganglion cells (Xin and Bloomfield, 1997; Völgyi et al., 2009). Perhaps best well characterized is homologous coupling between AII amacrine cells. Coupling between AII amacrine cells is bidirectional and symmetric, with a junctional conductance of ~700 pS and a coupling coefficient of ~0.3 (Veruki and Hartveit, 2002b). AII amacrine cells form gap junctions with one another in the dendritic arbors (Strettoi et al., 1992), and other amacrine cell types likely also form dendrodendritic gap junctions with other amacrine and ganglion cells, since this is often the only location where their processes overlap. AII amacrine cells are thought to connect to one another and to ON cone bipolar cells with Cx36 containing gap junctions, connections that are crucial for mediating rod vision (discussed in a later section). Other amacrine cell types can be coupled to ganglion cells and other amacrine cells with either Cx36- or Cx45-containing gap junctions (Pan et al., 2010; Pang et al., 2013).

Finally, ganglion cells, of which there are ~30 types (Sanes and Masland, 2015; Baden et al., 2016), receive excitatory glutamatergic input from bipolar cells and inhibitory GABAergic and glycinergic input from amacrine cells, and pass signals along their axons to diverse higher visual areas (Robles et al., 2014). Ganglion cells can also be connected to amacrine and other ganglion cells via gap junctions (Vaney, 1991, 1994; Xin and Bloomfield, 1997; Völgyi et al., 2009; Müller et al., 2010a; Reifler et al., 2015). Gap junctions belonging to ganglion cells predominantly comprise Cx36 (Schubert et al., 2005a; Pan et al., 2010), although some appear to comprise Cx45 (Schubert et al., 2005b) and possibly also Cx30.2 (Müller et al., 2010b). Different subtypes of ganglion cells have unique coupling patterns (Völgyi et al., 2009), suggesting there exists a diversity of possible functional roles for gap junction signaling in these parallel visual channels. Furthermore, ganglion cell–ganglion cell coupling appears to be exclusively homologous, with no coupling between different ganglion cell types (Vaney, 1994; Hu and Bloomfield, 2003; Völgyi et al., 2009; Trenholm et al., 2013a). Gap junctions between ganglion cells appear to be dendrodendritic and pass current bidirectionally and symmetrically, with a junctional conductance of ~1 nS and a coupling coefficient of ~0.15 (Hidaka et al., 2004; Trenholm et al., 2013a).

3. GAP JUNCTION COUPLING DIFFERENTIALLY MODIFIES RECEPTIVE FIELD SIZE IN DIFFERENT RETINAL CELL TYPES

Altering the extent to which retinal neurons are electrically coupled to their surrounding neighbors can significantly affect the size of their receptive fields. For most retinal neurons, changes in time of day or in light adaptation have been shown to regulate the strength and extent of coupling. The molecular mechanisms regulating these changes in coupling have been studied in detail and are reviewed by Ribelayga and O'Brien, 2017 (Chapter 10). In this section, we describe the effects of gap junction signaling, and modulation of gap junction signaling, on receptive fields of different retinal cell types.

In photoreceptors, circadian time and light adaptation appear to be the predominant cues for modifying coupling strength (Ribelayga et al., 2008; Gao et al., 2013; Jin and Ribelayga, 2016). Modulating gap junction strength is important as photoreceptor coupling appears to affect receptive field properties. In states where rod–cone gap junctions are more strongly coupled, cones respond at lower light intensities and their receptive fields are larger (Ribelayga et al., 2008). Additionally, when rods are more strongly coupled, their receptive fields are larger and the trial-to-trial variability in their responses to dim flashes is decreased (Li et al., 2012). Therefore increased coupling allows photoreceptors to pool signals with their neighbors more effectively. Furthermore, as gap junctions allow rod responses to pass into cones (Nelson, 1977; Schneeweis and Schnapf, 1995; Hornstein et al., 2005; Asteriti et al., 2014) and cone responses to pass into rods (Wu and Yang, 1988), strengthening electrical coupling should extend the range of light intensities over which both rod and cone circuits function.

In horizontal cells, unlike in photoreceptors, circadian rhythm does not appear to control gap junction coupling (Ribelayga and Mangel, 2003, 2007). In contrast, horizontal cell coupling seems to exhibit a triphasic relationship with ambient illumination: tracer coupling is minimal in the dark adapted or "dark-suppressed" state; dramatically increased in midlight levels, sometimes referred to as a "light-sensitized" state; again reduced to minimal levels in the light adapted state (Xin and Bloomfield, 1999; Baldridge, 2001). The extent of horizontal cell coupling appears to strongly regulate the receptive field size (Mangel and Dowling, 1985), with receptive fields being roughly three times larger in light-sensitized states (Xin and Bloomfield, 1999). Furthermore, knocking out Cx57 decreases horizontal cell receptive field size (Shelley et al., 2006). As horizontal cell feedback to photoreceptors has been suggested to underlie surround receptive fields in certain ganglion cells (Werblin and Dowling, 1969), it might be expected that modifying horizontal cell coupling strength would alter ganglion cell receptive field size. However, this remains controversial, and as will be described later, at least for some ganglion cell types, horizontal cell signaling does not appear to be the primary regulator of receptive field surround size (Farrow et al., 2013; Hoggarth et al., 2015). Additionally, Cx57 knockout has little effect on spatial tuning of ganglion cells and on mouse psychophysical behavioral tests of spatial acuity (Dedek et al., 2008), suggesting that horizontal cells coupling does not play a major role in shaping spatial tuning in ganglion cells. However, horizontal cells may play an important role in temporal tuning of ganglion cells. It has been suggested that strong electrical coupling may serve as a means to functionally remove horizontal cells from the retinal circuit, as strong coupling could shunt their input current from photoreceptors. As such, changing the strength of coupling could play a role in switching the retina from a mode of short integration time in daylight conditions to a mode of long integration time in night-light conditions (Pandarinath et al., 2010).

For bipolar cell—bipolar cell coupling, the gap junction conductance increases in the light adapted as compared with dark adapted retina (at least for goldfish Mb-1 bipolar cells; Arai et al., 2010). Although coupling between bipolar cells has been suggested to increase their receptive field sizes, which can be much larger than their dendritic fields (Kujiraoka and Saito, 1986; Saito and Kujiraoka, 1988; Dacey et al., 2000; but see also Berntson and Taylor, 2000), to our knowledge light adaptation-driven changes in receptive field size resulting from changes in coupling have never been explicitly tested in bipolar cells. Furthermore, it has been found that bipolar cell antagonistic receptive field surrounds can strengthen in light-adapted retinal preparations (Werblin, 1970), which would likely counteract the effects of a concomitant increase in bipolar cell coupling. Thus the overall effects of light adaptation and gap junction modulation on bipolar cell receptive field size require further investigation.

In amacrine cells, coupling also appears to be controlled by the light adaptation state (Bloomfield et al., 1997). AII amacrine cells exhibit a similar light adaptation/

coupling relationship as horizontal cells, with coupling being minimal during very dim and bright conditions, and enhanced during midlight conditions (Bloomfield et al., 1997). Furthermore, the receptive field size of AII amacrine cells appears to be strongly influenced by coupling, as the receptive field can be two to three times larger than the dendritic field in light-sensitized states (Bloomfield and Xin, 1997). In contrast, unlike AII amacrine cells, other types of amacrine cells, which are also extensively coupled, do not exhibit obvious modification of their receptive field size from gap junction inputs (Bloomfield and Xin, 1997). Why there is a difference in the effect of coupling on receptive fields of different types of amacrine cells remains unclear.

In ganglion cells, there appears to be a biphasic relationship between coupling strength and light adaptation, with coupling being reduced in the dark-adapted retina and increased in brighter conditions (at least for OFF α cells in the rabbit retina; Hu et al., 2010). However, increased coupling strength does not appear to alter the size of the classical (ie, spiking) receptive field (Hu et al., 2010). Interestingly, this does not mean that gap junction inputs do not affect ganglion cell receptive fields. Coupled inputs from neighboring ganglion cells provide coupled ganglion cells with extensive excitatory subthreshold receptive fields, which surround the receptive field center (Trenholm et al., 2013a). Furthermore, in contrast to the expected increase in receptive field size arising from increased coupling, for many ganglion cell types the size of their classical receptive field actually decreases in bright light conditions (Barlow et al., 1957; Peichl and Wässle, 1983; Muller and Dacheux, 1997). This decrease is thought to arise from a strengthening of an inhibitory surround arising from either horizontal and/or amacrine cell signaling. Recent work has shown that for certain ganglion cells types, including mouse ON and OFF α-like ganglion cells (Farrow et al., 2013) and directionally selective (DS) ganglion cells (Hoggarth et al., 2015), surround inhibition is driven by wide-field amacrine cells, which can be switched on at cone threshold light intensities. Knocking out Cx36 eliminates the bright light-induced reduction in receptive field size (Farrow et al., 2013), indicating that coupling, possibly between wide-field amacrine cells and ON cone bipolar cells, plays an integral role in the receptive field switch mechanism. Furthermore, this switch between weak and strong surround inhibition can be toggled back and forth by changing the ambient illumination on the timescale of seconds. Therefore, although light adaptation changes ganglion cell coupling patterns, classical receptive field size appears to be more strongly regulated by surround inhibition.

4. GAP JUNCTIONS ARE REQUIRED FOR NIGHTTIME VISION

Gap junctions play several prominent roles in visual processing during dark (scotopic) light conditions when rods are the primary active photoreceptors. In photoreceptors, intrinsic noise is unique to individual photoreceptors, whereas coupling allows light-driven signals to be shared. Coupling therefore allows for more effective detection of moderately weak signals and a decrease in response variability, although these gains appear to come with a loss of absolute sensitivity and resolution (Fain, 1975; Lamb and Simon, 1976; Attwell et al., 1984; Tessier-Lavigne and Attwell, 1988; DeVries et al., 2002; Hornstein et al., 2005; Li et al., 2012). Additionally, as there appears to be significant signal clipping, or saturation of synaptic transmission, between rods and rod bipolar cells, it has been suggested that coupling could permit small visual stimuli to effectively activate several rods, via gap junction spread, and that this group of activated rods would then more effectively drive downstream rod bipolar cells (Attwell et al., 1987).

Gap junctions also relay rod-mediated signals throughout the retina. First, rods can pass their signals directly onto cones via rod–cone gap junctions, and these signals can then be passed to the rest of the retina via the cone circuitry. This is an important pathway whereby rod signals reach the inner retina (Völgyi et al., 2004; Trexler et al., 2005; Abd-El-Barr et al., 2009). Second, some rods appear to directly form synapses with certain OFF bipolar cells (Soucy et al., 1998; Tsukamoto et al., 2001; Li et al., 2004). Signal pooling between coupled rods is thought to provide this pathway with its heightened sensitivity. Third, gap junctions play a crucial role in the classical rod pathway. In this circuit, rod signals are first passed to downstream rod ON bipolar cells. These rod bipolar cells then provide glutamatergic synaptic input to AII amacrine cells, which in turn pass rod signals to ON and OFF cone bipolar cells and OFF ganglion cells: to ON bipolar cells via gap junctions, and to OFF bipolar and ganglion cells via glycine release (Kolb and Famiglietti, 1974; Völgyi et al., 2004; Ivanova et al., 2006; Münch et al., 2009; Marc et al., 2014). Consistent with these findings, knockout of Cx36 greatly reduces scotopic retinal responses (Deans et al., 2002; Völgyi et al., 2004; Abd-El-Barr et al., 2009).

5. GAP JUNCTIONS PROMOTE SPONTANEOUS ACTIVITY DURING RETINAL DEGENERATION

One outcome of the extensive coupling between AII amacrine and ON cone bipolar cells is that, upon photoreceptor degeneration, this network drives spontaneous neuronal oscillations (Fig. 9.2). Currently there are several rodent models for the disease retinitis pigmentosa, a retinal degeneration (rd) disease in which photoreceptors degenerate, leading to vision impairment (for a review, see Trenholm and Awatramani, 2015). One remarkable feature of the blind rd retina is that instead of being silent, it is highly spontaneously active. In fact, many ganglion cells in mouse models for rd exhibit spontaneous activity with regular oscillations occurring with a frequency of ~10 Hz (Margolis et al., 2008; Stasheff, 2008; Borowska et al., 2011; Menzler and Zeck, 2011; Stasheff et al., 2011; Yee et al., 2012; Biswas et al., 2014). Since many strategies are currently being

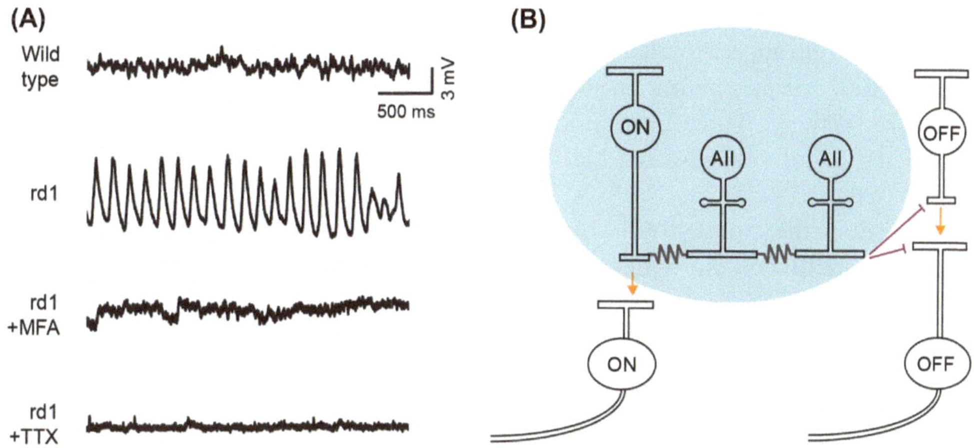

Figure 9.2 ***The electrically coupled network of AII amacrine/ON cone bipolar cells drives spontaneous pacemaker activity in the photoreceptor degenerating retina.*** (A) Recordings of the membrane potential from AII amacrine cells in the whole mount retina in control, rd1 retina, rd1 retina upon application of the gap junction blocker meclofenamic acid (MFA), and rd1 retina upon application of the voltage-gated Na^+ channel blocker tetrodotoxin (TTX). *(Adapted from Trenholm, S., Awatramani, G., 2015. Origins of spontaneous activity in the degenerating retina. Front. Cell. Neurosci. 9, 277.)* (B) The simplified circuit underlying the generation of the spontaneous oscillations in the rd1 retina. Oscillations arise from cross-junctional interactions between electrically coupled AII amacrine/ON cone bipolar cells. This activity then passes to both ON and OFF ganglion cells, resulting in oscillatory spike activity.

designed in attempt to restore vision to photoreceptor degenerated retinae, including techniques using electrical implants (Weiland and Humayun, 2014), stem cells (Singh et al., 2013), gene therapy (Sahel and Roska, 2013), and optogenetic approaches (Busskamp et al., 2012), it is important to understand the biophysical properties of the rd retina since any spontaneous noise will likely reduce the efficacy of therapeutic approaches.

Retinal ganglion cells in rd retinae receive oscillatory excitatory and inhibitory inputs (Margolis et al., 2008; Borowska et al., 2011; Yee et al., 2012, 2014) and become relatively silent when a cocktail of excitatory and inhibitory synaptic blockers is applied to the retina (Borowska et al., 2011; Menzler and Zeck, 2011). This indicates that the spontaneous activity is not intrinsic to rd ganglion cells (but see also Sekirnjak et al., 2011). Subsequent recordings from bipolar cells and AII amacrine cells in intact retinal preparations revealed that they also exhibit ~10-Hz membrane oscillations in degenerating retinae (Borowska et al., 2011). Unlike ganglion cells, AII amacrine cells and ON cone bipolar cells continue to oscillate in the presence of a cocktail of chemical synaptic blockers (Borowska et al., 2011). Thus the spontaneous oscillations appear to be intrinsic to the electrically coupled network of ON cone bipolar cells and AII amacrine cells. Pharmacologically blocking gap junctions inhibits ganglion cell oscillations in the degenerating retina (Menzler and Zeck, 2011) and also blocks oscillations in AII amacrine and

ON cone bipolar cells (Trenholm et al., 2012). Thus neither AII amacrine cells nor ON cone bipolar cells appear to oscillate independently in the degenerating retina, indicating that electrical synapses permit cross-junctional interactions that are required for driving pacemaker activity. These cross-junctional interactions have also been shown to rely on the activity of voltage-gated Na^+ channels in AII amacrine cells (Trenholm et al., 2012).

Why does the coupled AII amacrine/ON cone bipolar cell network not spontaneously oscillate in the wild-type retina? Experiments in wild-type retinae have shown that either pharmacologically blocking photoreceptor output (Trenholm et al., 2012) or photobleaching the retina with bright light (Menzler et al., 2014) can lead to rhythmic activity with similar properties to those found in rd retinae. These findings suggest that no significant circuit rewiring is required upon photoreceptor degeneration to initiate spontaneous activity and instead indicate that a network hyperpolarization is the prime trigger for putting the network into an oscillatory state (in wild-type retinae, blocking photoreceptor output hyperpolarizes the AII amacrine/ON cone bipolar cell network). Indeed, the resting membrane potential of AII amacrine cells in degenerating retinae appears to be more hyperpolarized than in wild-type retinae (Choi et al., 2014). Another work indicates that coupling strength may also be increased in the rd retina (Ivanova et al., 2015), and this could promote oscillations in the AII amacrine/ON cone bipolar cell network.

6. ELECTRICAL SYNAPSES ARE IMPORTANT FOR SIGNALING VISUAL MOTION

Coupling between AII amacrine cells and ON cone bipolar cells, which is important for rod-mediated vision, appears to play a distinct role during cone-mediated vision, demonstrating the remarkable flexibility in circuit computations offered by electrochemical signaling. In bright light conditions, where ON cone bipolar cells are more strongly activated relative to AII amacrine cells, the direction of signal flow through the gap junction is reversed. In mouse retina, this allows ON cone bipolar cells to mediate a rapid "crossover" inhibition to a subset of OFF ganglion cells (the PV-5 cell type), and this inhibition is important for their ability to detect looming stimuli (Münch et al., 2009). In this cell type, stationary flashes of light evoke transient OFF excitation and fast, transient ON inhibition mediated by AII amacrine cells (Münch et al., 2009). A receding dark stimulus (for instance, a large dark spot that decreases in size over time) produces only ON-driven inhibition and thus no spiking response. In contrast, an approaching dark stimulus (for instance, a small dark spot that increases in size over time) produces exclusively OFF-driven excitation, and thus results in a strong spiking response. To make this cell more selective for expanding versus laterally moving stimuli (which would evoke both excitation and inhibition), it receives fast ON-driven inhibitory inputs that arrive via the electrically coupled AII amacrine/ON cone bipolar cell network and weaken the effect of excitatory OFF inputs. Knocking out Cx36 results in PV-5 cells that no longer respond preferentially to approaching stimuli, making them less looming selective. Subsequent behavioral experiments have shown that mice have an innate freezing/fleeing response to dark looming stimuli above their heads (Yilmaz and Meister, 2013), and this reflex appears to rely on activity of alpha-like ganglion cells in the retina (Yilmaz et al., 2014), of which the PV-5 cell type is included. The ability to detect looming motion, such as an approaching predator, is behaviorally important. Approaching object motion drives behavioral reflexes in many different animal species (Schiff et al., 1962; Ishikane et al., 2005; Temizer et al., 2015), including humans (Ball and Tronick, 1971; King et al., 1992). Visual looming sensitive neurons have been described in many species (Hatsopoulos et al., 1995; Sun and Frost, 1998; Münch et al., 2009; Zhao et al., 2014; Temizer et al., 2015) but the commonalities in the circuit mechanisms driving looming responses in different species remain to be tested.

Other than regulating looming sensitive responses, gap junction signaling can modulate the latency of responses of certain ganglion cells to moving stimuli. Due predominantly to the relatively slow speed of phototransduction in rods and cones, the retina has an inherent problem in reliably indicating where a moving image is in space. When an image is flashed on the retina, it takes approximately 50 ms before a ganglion cell will fire an action potential. For a static image, this delay does not cause a problem. However, if an image is moving across the retina, by the time ganglion cells begin to signal the presence of the image via the generation of action potentials, the image will have moved from the location where it initiated those initial action potentials, and the ganglion cells will therefore be signaling for something that happened in the past (ie, they will be indicating that the image is behind where it actually is in space). This lag, the discrepancy between where a ganglion cell indicates an image is vs. where the image actually is in space, increases as the image moves faster and faster. One way that the retina appears to deal with this problem is to shift the center of a ganglion cell's receptive field toward the leading edge of a moving stimulus (Berry et al., 1999; Johnston and Lagnado, 2015). In brief, when a classical receptive field of a ganglion cell is measured by stimulation with static flashes of light, and subsequently the receptive field of the same ganglion cell is measured with a laterally moving stimulus, the center of the receptive field shifts slightly toward the leading edge of the moving stimulus (Berry et al., 1999; Johnston and Lagnado, 2015). If downstream targets of retinal ganglion cells can read out activity based on peak spike rates, this strategy could provide the retina with a trick such that it could compensate for lags in neuronal processing and provide more reliable localization of moving images. Another way to compensate for this lag, when coding for moving stimuli, involves electrically coupling ganglion cells to one another, thus enabling ganglion cells to prime their neighbors to the presence of an upcoming moving image. This phenomenon, termed lag normalization, has been described in a population of

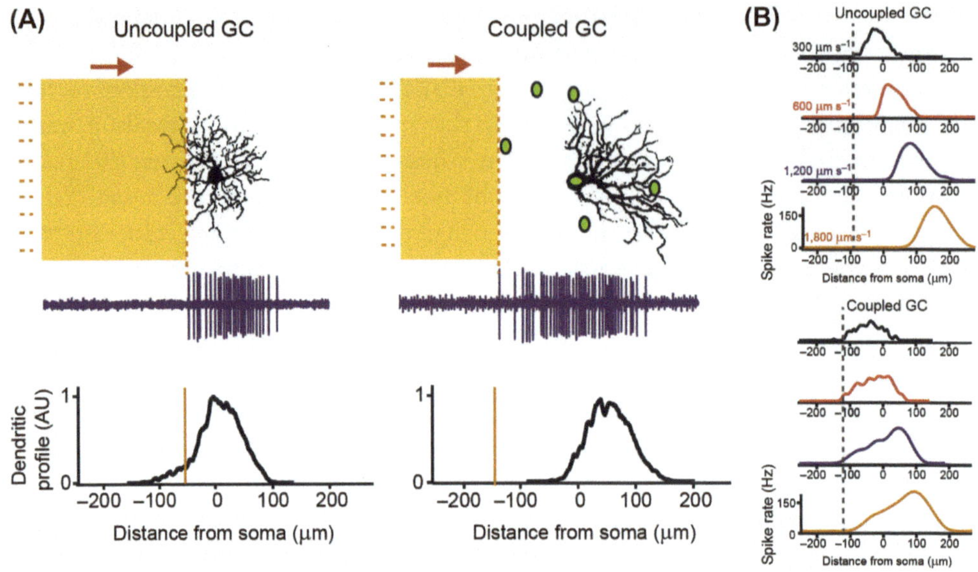

Figure 9.3 ***Lag normalization in coupled retinal ganglion cells.*** (A) A moving bar of light stimulates spikes in an uncoupled (left) and a coupled (right) ganglion cell (GC; the location of the somata of nearby coupled ganglion cells are shown in green). The spike train indicating the light response is shown in blue below each cell. The light stimuli (yellow) are shown at the spatial positions where they initiated responses while moving across the cells. Below are shown the average positions where spiking responses initiated (*orange vertical line*) and the average dendritic spread (*black lines*) of nine uncoupled (left) and nine coupled (right) ganglion cells. (B) Plots of Gaussian fits of the spike rate of uncoupled (top) and coupled (bottom) ganglion cells resulting from bars of light moving at four different speeds (as indicated). The spike rate is plotted with respect to where in space the leading edge of the moving stimulus was with respect to the soma. *(Adapted from Trenholm, S., Schwab, D.J., Balasubramanian, V., Awatramani, G.B., 2013b. Lag normalization in an electrically coupled neural network. Nat. Neurosci. 16, 154–156.)*

homologously coupled DS ganglion cells (Fig. 9.3; Trenholm et al., 2013b). These cells, as described earlier in the receptive field section, have an extensive subthreshold excitatory receptive field that surrounds their receptive field center, due to gap junction inputs from their surrounding neighbors. For static flashes, electrical synaptic inputs are not strong enough, even when combined with weak bipolar cell inputs at the edge of the classical receptive field, to significantly expand the classical receptive field (Hu et al., 2010; Trenholm et al., 2013a). However, for moving images, the gap junction input builds up over several connected cells such that it can then synergistically combine with chemical synaptic inputs at the edge of the classical receptive field to produce spiking responses much farther from the edge of the dendritic tree than is seen for uncoupled cells, thus providing coupled cells with the ability to effectively anticipate upcoming moving images (Trenholm et al., 2013b). Interestingly, this means that diverse stimuli (in this case static vs. moving stimuli) are dynamically coded by the same set of cells. Most notable, however, is the fact that when these coupled cells are stimulated with images moving at a variety of speeds, their responses initiate near the same spatial location (ie, they can "lag normalize"), whereas when the same experiment is performed with uncoupled ganglion cells the responses initiate at positions further and further in space at faster and faster speeds (Trenholm et al., 2013b). The population of coupled DS ganglion cells that have been shown to lag normalize respond preferentially to images moving ventrally on the retina (upward in the visual field) and have been shown to send axons to accessory optic brain areas (Kay et al., 2011) that are involved in driving visual reflexes (Dhande et al., 2013), raising the possibility that lag normalization plays an important role in certain reflexive behaviors, although this remains to be established.

7. FAST GAP JUNCTION SIGNALS DRIVE FINE-SCALE CORRELATED SPIKE OUTPUT

Gap junctions between neurons tend to act as low-pass filters (Connors and Long, 2004; Pereda et al., 2013), and thus slow components of the prejunctional membrane response (eg, excitatory postsynaptic potentials) are transmitted with much more fidelity than fast components (eg, action potentials). Most of the circuit properties we have described so far likely arise in large part due to the transmission of the slow component of the prejunctional signal. In this final section, we focus on gap junction-mediated spike synchrony, which arises primarily from the fast, brief, and highly attenuated component of the coupled input: the coupled spikelet.

Seminal work in the 1980s revealed that neighboring retinal ganglion cells can exhibit significant spike synchrony (Arnett and Spraker, 1981; Mastronarde, 1983a,b,c). Simultaneous recordings from multiple ganglion cells subsequently revealed that correlated activity was a predominant feature of ganglion cell population signaling (Meister et al., 1995). Further anatomical, physiological, and pharmacological work verified that gap junction coupling can lead to fine-scale correlations between retinal ganglion cells across many different species (Brivanlou et al., 1998; DeVries, 1999; Hu and Bloomfield, 2003; Greschner et al., 2011; Völgyi et al., 2013; Xiao et al., 2013; Trenholm et al., 2014).

Correlations between pairs of ganglion cells can be separated, based on their temporal characteristics, into three distinct groups: narrow, medium, and broad (Fig. 9.4; Mastronarde, 1989; Brivanlou et al., 1998; Völgyi et al., 2013). Narrow synchrony occurs within ± 2 ms, with the peaks in the cross-correlogram displaced on either side of 0 ms delay. Medium and broad correlations exhibit a peak in the cross-correlogram at 0 ms delay, although there has been some spread in the reported values of their timescales, with medium correlations being reported as occurring between 2 and 10 ms (Mastronarde, 1989), between 10 and 50 ms (Brivanlou et al., 1998), and at less than 100 ms (Völgyi et al., 2013), whereas broad correlations

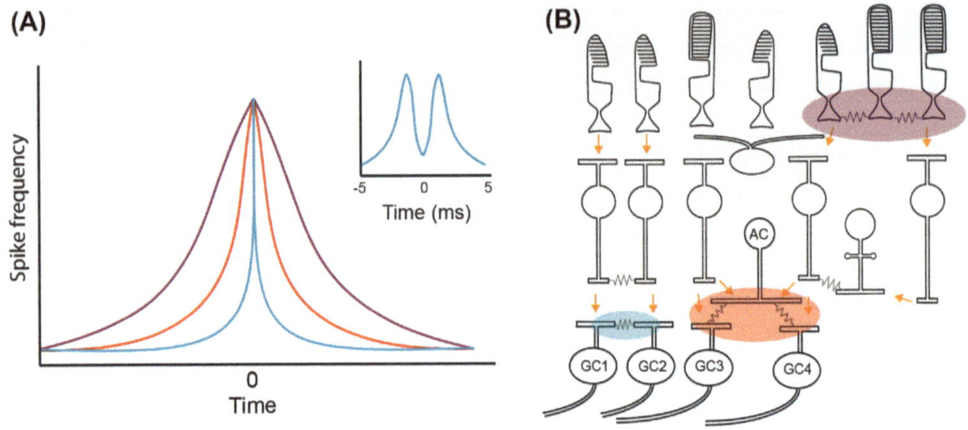

Figure 9.4 ***Synchronous ganglion cell spike activity occurs at three distinct timescales via three distinct mechanisms.*** (A) A simplified cross-correlogram, indicating the relative timing of broad (purple), medium (red), and narrow (blue) spike correlations. The inset shows a higher resolution view of the narrow correlations, revealing bimodal peaks on either side of 0-ms delay. (B) A diagram of a simplified retinal circuit in which the synaptic locations where broad, medium, and narrow scale correlations arise are indicated: purple for broad, red for medium, and blue for narrow. In this diagram, narrow correlations are present between GC1 and GC2, whereas medium correlations are present between GC3 and GC4.

have been reported as occurring at around 50 ms (Mastronarde, 1989), greater than 50 ms (Brivanlou et al., 1998), and greater than 100 ms (Völgyi et al., 2013). Regardless of these small dissimilarities, the existence of three general types of synchrony, and their different mechanistic underpinnings, appears to be conserved across species. Initial pharmacological studies found that narrow and medium correlations persisted in the presence of chemical synaptic blockers, yet were antagonized in the presence of gap junction blockers (Brivanlou et al., 1998). Subsequent experiments using Cx36 knockout mice revealed that broad correlations may also require gap junction signaling between photoreceptors (Völgyi et al., 2013). Therefore broad correlations appear to arise in photoreceptors and are passed onto ganglion cells via input from bipolar cells. Medium correlations appear to arise when pairs of ganglion cells are coupled via an intermediary amacrine cell (Mastronarde, 1989; Brivanlou et al., 1998; DeVries, 1999; Schnitzer and Meister, 2003; Völgyi et al., 2013), whereas narrow correlations arise from reciprocally coupled ganglion cells (Mastronarde, 1983c; Brivanlou et al., 1998; DeVries, 1999; Hu and Bloomfield, 2003; Hidaka et al., 2004; Völgyi et al., 2013; Trenholm et al., 2014). In the following paragraphs we focus on narrow correlations arising from reciprocal ganglion cell gap junction connectivity.

As described earlier, reciprocal ganglion cell coupling results in correlations that occur within ±2 ms. This form of synchrony is thought to arise when one cell spikes, thus biasing its coupled neighbor to fire, and vice versa. However, until recently it remained unclear how exactly gap junction inputs drove fine-scale correlated activity, and it was even questioned whether relatively weak electrical synapses could be responsible for robust correlated spiking (Trong and Rieke, 2008). Although some attempts to test whether an action potential in a prejunctional cell could directly elicit an action potential in a postjunctional cell suggested that it was possible (Hu and Bloomfield, 2003), subsequent experiments revealed that action potentials in prejunctional ganglion cells drive almost exclusively subthreshold signals in postjunctional neighbors (Trenholm et al., 2014). This finding is consistent with reports that ganglion cell classical receptive fields are not strongly altered by the presence of gap junctions (Hu et al., 2010; Trenholm et al., 2013a; but see also DeVries, 1999), as well as with early work that suggested a spike in a prejunctional ganglion cell directly drives a spike in a postjunctional neighbor with a success rate of less than 5% (Mastronarde, 1983c). Indeed, in ganglion cells, coupled spikelets, when measured at the postjunctional soma, are relatively small in amplitude ($\sim$1 mV) and arrive with very little delay ($<$1 ms; Trenholm et al., 2014). As such, ganglion cells are similar to many other coupled neurons throughout the CNS in which spiking in a prejunctional neuron drives predominantly subthreshold activity in postjunctional neighbors (Connors and Long, 2004). The question therefore arises: How do subthreshold coupled spikelets drive synchronous activity when they do not directly drive postjunctional spiking?

It was theorized that active dendritic properties, which exist in many neurons in the CNS, could possibly allow for nonlinear summation of chemical and electrical synaptic inputs, thus permitting subthreshold coupled inputs to effectively modulate the timing of action potential output (Brivanlou et al., 1998). ON-OFF DS ganglion cells provide an ideal model for studying the impact of gap junction inputs on active dendritic processing. Previous work has shown that DS ganglion cells exhibit regenerative, tetrodotoxin-sensitive, Na^+-channel-dependent dendritic spikes, and that these dendritic spikes drive action potential generation at the soma/axon hillock with high reliability (Oesch et al., 2005; Sivyer and Williams, 2013). Of the four subtypes of ON-OFF DS ganglion cells, which code the four cardinal directions (Oyster and Barlow, 1967), only a single subtype is homologously electrically coupled in the adult mammalian retina (Vaney, 1994), and this corresponds to the subtype that prefers ventral motion on the retina (upward in the visual field; Trenholm et al., 2013a; Xu et al., 2013). Experiments utilizing this cell type, fluorescently labeled in the Hb9::eGFP mouse line (Trenholm et al., 2011), have revealed that nonlinear interactions between coincident electrical and chemical synaptic inputs in the dendrites of coupled DS ganglion cells underlie synchronous spike activity (Trenholm et al., 2014).

How do active dendritic properties promote synchronous output spiking? By simulating coupled spikelets via current injection at the soma, it was found that coupled spikelets do not modulate action potential timing at the soma, suggesting instead that they act in the dendrites (Trenholm et al., 2014). However, it was then found that coupled spikelets do not directly drive dendritic spikes in postjunctional neurons, indicating that coupled spikelets likely interact in the dendrites with light-driven chemical synaptic inputs to produce dendritic spikes. Indeed, subsequent experiments in which light-driven dendritic spikes were selectively pharmacologically blocked revealed that dendritic spikes were required for action potential synchrony between pairs of coupled ganglion cells (Trenholm et al., 2014). Therefore coupled spikelets interact nonlinearly with chemical synaptic inputs in the dendrites to generate synchrony between coupled ganglion cells. This was further validated with experiments in which light was flashed either at a shared location of receptive overlap between a pair of neighboring coupled ganglion cells, or at two different spatial locations such that each ganglion cell was driven by bipolar cell inputs independently of the other ganglion cell. Using this experimental paradigm it was found that action potentials were only synchronized when a common region of bipolar cell input was activated, indicating that synchrony arises when coincident chemical and electrical synaptic inputs arrive in the overlapping dendritic region between neighboring coupled DS ganglion cells and summate nonlinearly (Fig. 9.5; Trenholm et al., 2014).

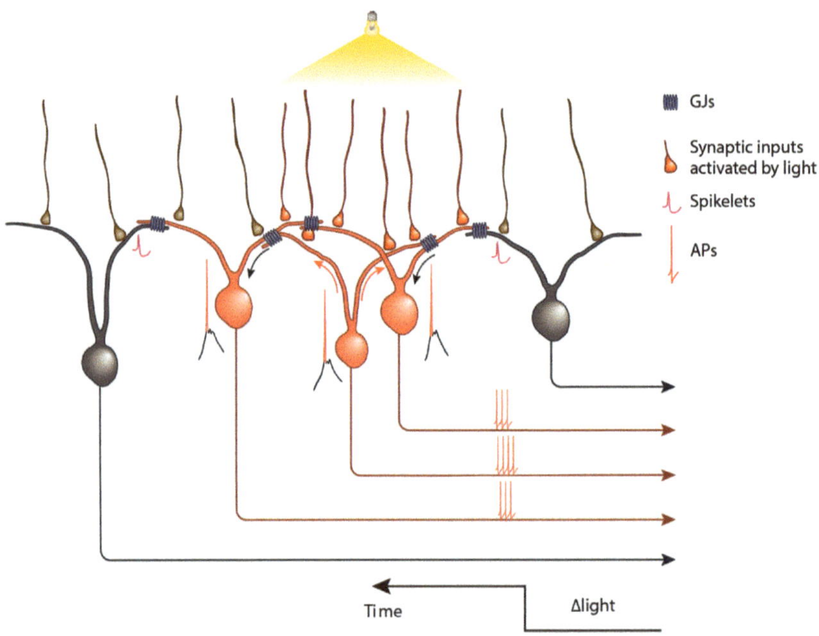

Figure 9.5 ***Coupled spikelets interact nonlinearly with chemical synaptic inputs in coupled ganglion cell dendrites to drive correlated spiking.*** A flash of light drives generation of an action potential (AP) in the first ganglion cell (*center cell*) that back-propagates into its dendritic tree (*red arrows*) and crosses gap junctions (GJs), driving coupled spikelets in postjunctional neighbors (magenta). These coupled spikelets summate with light-driven chemical synaptic input from bipolar cells in the post-junctional ganglion cell dendrites, which then generate dendritic spikes that propagate to the soma (*black arrows*), where they effectively drive synchronous output spiking in the coupled ganglion cells. *(This figure is used, with permission, from Lanore, F., Silver, R.A., 2014. Reading dendritic activity with gap junctions. Nat. Neurosci. 17, 1625–1627.)*

What is the role for correlated ganglion cell activity? Correlated ganglion cell activity could provide computational advantages and could also play an important role in effective transmission of retinal signals to higher visual areas. Computationally, it has been suggested that correlated ganglion cell firing, in which individual ganglion cells can participate in multiple different correlated groups, could greatly enhance the bandwidth of retinal signals leaving the eye via the optic nerve (Meister et al., 1995; Schnitzer and Meister, 2003). For example, as correlated ganglion cell spikes tend to occur when light activates overlapping areas between the receptive fields of neighboring ganglion cells (Meister et al., 1995; Schnitzer and Meister, 2003; Trenholm et al., 2014), correlated activity could relay spatial information that is absent from the spike trains of individual ganglion cells. In addition, as the strength of synchrony between ganglion cells appears to decrease as the spike rate increases (Xiao et al., 2013; Trenholm et al., 2014), correlated activity could relay information independently from the spike rate. Similarly, it has been found that for coupled ON DS cells, which are indirectly coupled via an amacrine cell, moving stimuli drive correlated spiking in every direction except for the null direction (Ackert et al., 2006), and in such a scenario it is possible that noncorrelated spiking could indicate the null direction, whereas the spike rate and correlated spiking would signal the preferred direction. However, such parallel coding strategies would require complex decoders in higher visual areas, which have yet to be described. Another possible role for correlated activity could be to enhance the efficacy of synaptic transmission from retinal ganglion cells to retinorecipient areas. For instance, in the lateral geniculate nucleus (LGN), it has been shown that near-coincident retinal inputs greatly increase synaptic efficacy (Usrey et al., 1998; Carandini et al., 2007). As several retinal ganglion cells are thought to converge onto individual LGN relay neurons (Cleland et al., 1971; Chen and Regehr, 2000), synchrony between neighboring ganglion cells could allow for more effective drive of retinorecipient neurons.

8. SUMMARY AND FUTURE DIRECTIONS

The vertebrate retina is an extensively coupled neuronal tissue, with homologous and heterologous gap junction coupling occurring between cells in all layers. In photoreceptors, gap junctions allow for pooling of signals in nighttime when photons are scarce and they extend the range of rod and cone functioning. In horizontal cells, gap junctions allow for strong modulation of receptive field size, although the computational outcomes of this modulation are still unclear. In bipolar cells and amacrine cells, gap junctions allow for the spread of signals from rod to cone circuits and allow for synchrony between coupled networks. In ganglion cells, gap junctions promote synchronous spike activity and endow cells with dynamic capabilities to respond differentially to static and moving visual stimuli. Gap junctions also play a role in retinal disease models by promoting spontaneous activity upon photoreceptor degeneration.

In recent years, the identification of transgenic mouse lines with labeled retinal cell types (Seung and Sümbül, 2014) and advances in multielectrode array and imaging technologies (Pillow et al., 2008; Farrow and Masland, 2011; Baden et al., 2016) have allowed for significant advances in recording from all the different retinal cell types. Still, we are in the early days of cataloging the unique characteristics of each of the ~30 retinal channels. Future studies will elucidate how chemical and electrical synapses interact in the circuits underlying each retinal channel, and how these interactions drive dynamic responses to both static and moving stimuli, and affect the overall sensitivity of specific circuits (Murphy and Rieke, 2011; Kuo et al., 2016). Examining how specific retinal computations, such as looming detection or lag normalization, are regulated in different ambient light levels, when receptive fields of different cell types in the underlying circuits are differentially modulated both by changes in gap junction coupling and surround inhibition, will likely reveal a remarkable flexibility in how the retina can respond to the same image in different environments. Furthermore, the role of active dendritic properties, which are common throughout the CNS (Grienberger et al., 2015), and how they combine with mixed electric-chemical signaling in these different retinal channels remain to be explored. Additionally, although gap junctions are known to play many important roles during retinal development (Cook and Becker, 2009), how they differentially affect the development of different retinal circuits remains unclear (see Arroyo et al., 2016, for a recent study on cell type specific gap junction connectivity during retinal development). Future work will also elucidate how gap junctions are involved in permitting secondary cell loss of specific retinal cell types in diverse retinal disease models (Striedinger et al., 2005; Paschon et al., 2012; Kranz et al., 2013; Akopian et al., 2014). Finally, as recent work has shown that gap junction coupling can be modified directly by neuronal activity (Haas et al., 2011; Kothmann et al., 2012; Trenholm et al., 2013a; Lefler et al., 2014; Mathy et al., 2014; Turecek et al., 2014), it will be of great interest to see how activity modulates coupling, and in turn modifies circuit function, in different retinal channels.

ACKNOWLEDGMENTS

We thank Dr. William H Baldridge for helpful comments on the manuscript. Fig. 9.5 is used with permission (license #3685341151944). We acknowledge the following funding sources: Human Frontier Science Program Long-Term Fellowship (LT000173/2013) to ST and Foundation Fighting Blindness to GBA.

REFERENCES

Abdel-Majid, R.M., Archibald, M.L., Tremblay, F., Baldridge, W.H., 2005. Tracer coupling of neurons in the rat retina inner nuclear layer labeled by Fluorogold. Brain Res. 1063, 114–120.

Ackert, J.M., Wu, S.H., Lee, J.C., Abrams, J., Hu, E.H., Perlman, I., Bloomfield, S.A., 2006. Light-induced changes in spike synchronization between coupled ON direction selective ganglion cells in the mammalian retina. J. Neurosci. 26, 4206–4215.

Akopian, A., Atlasz, T., Pan, F., Wong, S., Zhang, Y., Völgyi, B., Paul, D.L., Bloomfield, S.A., 2014. Gap junction-mediated death of retinal neurons is connexin and insult specific: a potential target for neuroprotection. J. Neurosci. 34, 10582–10591.

Arai, I., Tanaka, M., Tachibana, M., 2010. Active roles of electrically coupled bipolar cell network in the adult retina. J. Neurosci. 30, 9260–9270.
Arroyo, D.A., Kirkby, L.A., Feller, M.B., 2016. Retinal waves modulate an intraretinal circuit of intrinsically photosensitive retinal ganglion cells. J. Neurosci. 36 (26), 6892–6905.
Arnett, D., Spraker, T.E., 1981. Cross-correlation analysis of the maintained discharge of rabbit retinal ganglion cells. J. Physiol. 317, 29–47.
Asteriti, S., Gargini, C., Cangiano, L., 2014. Mouse rods signal through gap junctions with cones. eLife 3, e01386.
Attwell, D., Wilson, M., Wu, S.M., 1984. A quantitative analysis of interactions between photoreceptors in the salamander (Ambystoma) retina. J. Physiol. 352, 703–737.
Attwell, D., Borges, S., Wu, S.M., Wilson, M., 1987. Signal clipping by the rod output synapse. Nature 328, 522–524.
Azeredo da Silveira, R., Roska, B., 2011. Cell types, circuits, computation. Curr. Opin. Neurobiol. 21, 664–671.
Abd-El-Barr, M.M., Pennesi, M.E., Saszik, S.M., Barrow, A.J., Lem, J., Bramblett, D.E., Paul, D.L., Frishman, L.J., Wu, S.M., 2009. Genetic dissection of rod and cone pathways in the dark-adapted mouse retina. J. Neurophysiol. 102, 1945–1955.
Baden, T., Euler, T., 2016. Retinal physiology: non-bipolar-cell excitatory drive in the inner retina. Current Biology 26 (15), R706–R708.
Baden, T., Berens, P., Franke, K., Román Rosón, M., Bethge, M., Euler, T., 2016. The functional diversity of retinal ganglion cells in the mouse. Nature 529 (7586), 345–350.
Baldridge, W.H., 2001. Triphasic adaptation of teleost horizontal cells. Prog. Brain Res. 131, 437–449.
Ball, W., Tronick, E., 1971. Infant responses to impending collision: optical and real. Science 171, 818–820.
Barlow, H.B., Fitzhugh, R., Kuffler, S.W., 1957. Change of organization in the receptive fields of the cat's retina during dark adaptation. J. Physiol. 137, 338–354.
Baylor, D.A., Fuortes, M.G., O'Bryan, P.M., 1971. Receptive fields of cones in the retina of the turtle. J. Physiol. 214, 265–294.
Berntson, A., Taylor, W.R., 2000. Response characteristics and receptive field widths of on-bipolar cells in the mouse retina. J. Physiol. 524, 879–889.
Berry, M.J., Brivanlou, I.H., Jordan, T.A., Meister, M., 1999. Anticipation of moving stimuli by the retina. Nature 398, 334–338.
Biswas, S., Haselier, C., Mataruga, A., Thumann, G., Walter, P., Müller, F., 2014. Pharmacological analysis of intrinsic neuronal oscillations in rd10 retina. PLoS One 9, e99075.
Bloomfield, S.A., Xin, D., 1997. A comparison of receptive-field and tracer-coupling size of amacrine and ganglion cells in the rabbit retina. Vis. Neurosci. 14, 1153–1165.
Bloomfield, S.A., Xin, D., Osborne, T., 1997. Light-induced modulation of coupling between AII amacrine cells in the rabbit retina. Vis. Neurosci. 14, 565–576.
Borowska, J., Trenholm, S., Awatramani, G.B., 2011. An intrinsic neural oscillator in the degenerating mouse retina. J. Neurosci. 31, 5000–5012.
Brivanlou, I.H., Warland, D.K., Meister, M., 1998. Mechanisms of concerted firing among retinal ganglion cells. Neuron 20, 527–539.
Busskamp, V., Picaud, S., Sahel, J.A., Roska, B., 2012. Optogenetic therapy for retinitis pigmentosa. Gene Ther. 19, 169–175.
Carandini, M., Horton, J.C., Sincich, L.C., 2007. Thalamic filtering of retinal spike trains by postsynaptic summation. J. Vis. 7 (20), 1–11.
Chen, C., Regehr, W.G., 2000. Developmental remodeling of the retinogeniculate synapse. Neuron 28, 955–966.
Choi, H., Zhang, L., Cembrowski, M.S., Sabottke, C.F., Markowitz, A.L., Butts, D.A., Kath, W.L., Singer, J.H., Riecke, H., 2014. Intrinsic bursting of AII amacrine cells underlies oscillations in the rd1 mouse retina. J. Neurophysiol. 112, 1491–1504.
Cleland, B.G., Dubin, M.W., Levick, W.R., 1971. Simultaneous recording of input and output of lateral geniculate neurones. Nat. New Biol. 231, 191–192.
Connors, B.W., Long, M.A., 2004. Electrical synapses in the mammalian brain. Annu. Rev. Neurosci. 27, 393–418.
Cook, J.E., Becker, D.L., 2009. Gap-junction proteins in retinal development: new roles for the "nexus". Physiology 24, 219–230.

Copenhagen, D.R., Owen, W.G., 1976. Coupling between rod photoreceptors in a vertebrate retina. Nature 260, 57–59.

Custer, N.V., 1973. Structurally specialized contacts between the photoreceptors of the retina of the axolotl. J. Comp. Neurol. 151, 35–55.

Dacey, D., 2004. Origins of perception: retinal ganglion cell diversity and the creation of parallel visual pathways. In: Gazzaniga, M.S. (Ed.), The Cognitive Neurosciences. MIT Press.

Dacey, D., Packer, O.S., Diller, L., Brainard, D., Peterson, B., Lee, B., 2000. Center surround receptive field structure of cone bipolar cells in primate retina. Vis. Res. 40, 1801–1811.

Deans, M.R., Volgyi, B., Goodenough, D.A., Bloomfield, S.A., Paul, D.L., 2002. Connexin36 is essential for transmission of rod-mediated visual signals in the mammalian retina. Neuron 36, 703–712.

Dedek, K., Pandarinath, C., Alam, N.M., Wellershaus, K., Schubert, T., Willecke, K., Prusky, G.T., Weiler, R., Nirenberg, S., 2008. Ganglion cell adaptability: does the coupling of horizontal cells play a role? PLoS One 3, e1714.

DeVries, S.H., 1999. Correlated firing in rabbit retinal ganglion cells. J. Neurophysiol. 81, 908–920.

DeVries, S.H., Schwartz, E.A., 1989. Modulation of an electrical synapse between solitary pairs of catfish horizontal cells by dopamine and second messengers. J. Physiol. 414, 351–375.

DeVries, S.H., Qi, X., Smith, R., Makous, W., Sterling, P., 2002. Electrical coupling between mammalian cones. Curr. Biol. 12, 1900–1907.

Dhande, O.S., Estevez, M.E., Quattrochi, L.E., El-Danaf, R.N., Nguyen, P.L., Berson, D.M., Huberman, A.D., 2013. Genetic dissection of retinal inputs to brainstem nuclei controlling image stabilization. J. Neurosci. 33, 17797–17813.

Dorgau, B., Herrling, R., Schultz, K., Greb, H., Segelken, J., Ströh, S., Bolte, P., Weiler, R., Dedek, K., Janssen-Bienhold, U., 2015. Connexin50 couples axon terminals of mouse horizontal cells by homotypic gap junctions. J. Comp. Neurol. 523, 2062–2081.

Fain, G.L., 1975. Quantum sensitivity of rods in the toad retina. Science 187, 838–841.

Farrow, K., Masland, R.H., 2011. Physiological clustering of visual channels in the mouse retina. J. Neurophysiol. 105, 1516–1530.

Farrow, K., Teixeira, M., Szikra, T., Viney, T.J., Balint, K., Yonehara, K., Roska, B., 2013. Ambient illumination toggles a neuronal circuit switch in the retina and visual perception at cone threshold. Neuron 78, 325–338.

Feigenspan, A., Janssen-Bienhold, U., Hormuzdi, S., Monyer, H., Degen, J., Söhl, G., Willecke, K., Ammermüller, J., Weiler, R., 2004. Expression of connexin36 in cone pedicles and OFF-cone bipolar cells of the mouse retina. J. Neurosci. 24, 3325–3334.

Gao, F., Pang, J.-J., Wu, S.M., 2013. Sign-preserving and sign-inverting synaptic interactions between rod and cone photoreceptors in the dark-adapted retina. J. Physiol. 591, 5711–5726.

Greschner, M., Shlens, J., Bakolitsa, C., Field, G.D., Gauthier, J.L., Jepson, L.H., Sher, A., Litke, A.M., Chichilnisky, E.J., 2011. Correlated firing among major ganglion cell types in primate retina. J. Physiol. 589, 75–86.

Grienberger, C., Chen, X., Konnerth, A., 2015. Dendritic function in vivo. Trends Neurosci. 38, 45–54.

Haas, J.S., Zavala, B., Landisman, C.E., 2011. Activity-dependent long-term depression of electrical synapses. Science 334, 389–393.

Han, Y., Massey, S.C., 2005. Electrical synapses in retinal ON cone bipolar cells: subtype-specific expression of connexins. Proc. Natl. Acad. Sci. USA 102, 13313–13318.

Hatsopoulos, N., Gabbiani, F., Laurent, G., 1995. Elementary computation of object approach by wide-field visual neuron. Science 270, 1000–1003.

Hidaka, S., Akahori, Y., Kurosawa, Y., 2004. Dendrodendritic electrical synapses between mammalian retinal ganglion cells. J. Neurosci. 24, 10553–10567.

Hilgen, G., von Maltzahn, J., Willecke, K., Weiler, R., Dedek, K., 2011. Subcellular distribution of connexin45 in OFF bipolar cells of the mouse retina. J. Comp. Neurol. 519, 433–450.

Hoggarth, A., McLaughlin, A.J., Trenholm, S., Awatramani, G., 2015. Specific wiring of distinct amacrine cells in the directionally selective retinal circuit permits independent coding of direction and size. Neuron 86, 276–291.

Hombach, S., Janssen-Bienhold, U., Söhl, G., Schubert, T., Büssow, H., Ott, T., Weiler, R., Willecke, K., 2004. Functional expression of connexin57 in horizontal cells of the mouse retina. Eur. J. Neurosci. 19, 2633–2640.
Hornstein, E.P., Verweij, J., Schnapf, J.L., 2004. Electrical coupling between red and green cones in primate retina. Nat. Neurosci. 7, 745–750.
Hornstein, E.P., Verweij, J., Li, P.H., Schnapf, J.L., 2005. Gap-junctional coupling and absolute sensitivity of photoreceptors in macaque retina. J. Neurosci. 25, 11201–11209.
Hu, E.H., Bloomfield, S.A., 2003. Gap junctional coupling underlies the short-latency spike synchrony of retinal alpha ganglion cells. J. Neurosci. 23, 6768–6777.
Hu, E.H., Pan, F., Völgyi, B., Bloomfield, S.A., 2010. Light increases the gap junctional coupling of retinal ganglion cells. J. Physiol. 588, 4145–4163.
Ishikane, H., Gangi, M., Honda, S., Tachibana, M., 2005. Synchronized retinal oscillations encode essential information for escape behavior in frogs. Nat. Neurosci. 8, 1087–1095.
Ivanova, E., Müller, U., Wässle, H., 2006. Characterization of the glycinergic input to bipolar cells of the mouse retina. Eur. J. Neurosci. 23, 350–364.
Ivanova, E., Yee, C.W., Baldoni, R., Sagdullaev, B.T., 2015. Aberrant activity in retinal degeneration impairs central visual processing and relies on Cx36-containing gap junctions. Exp. Eye Res.
Jacoby, R.A., Marshak, D.W., 2000. Synaptic connections of DB3 diffuse bipolar cell axons in macaque retina. J. Comp. Neurol. 416, 19–29.
Janssen-Bienhold, U., Trümpler, J., Hilgen, G., Schultz, K., De Sevilla Müller, L.P., Sonntag, S., Dedek, K., Dirks, P., Willecke, K., Weiler, R., 2009. Connexin57 is expressed in dendro-dendritic and axo-axonal gap junctions of mouse horizontal cells and its distribution is modulated by light. J. Comp. Neurol. 513, 363–374.
Jin, N.G., Ribelayga, C.P., 2016. Direct evidence for daily plasticity of electrical coupling between rod photoreceptors in the mammalian retina. J. Neurosci. 36 (1), 178–184.
Johnston, J., Lagnado, L., 2015. General features of the retinal connectome determine the computation of motion anticipation. eLife 4.
Kántor, O., Benkő, Z., Énzsöly, A., Dávid, C., Naumann, A., Nitschke, R., Szabó, A., Pálfi, E., Orbán, J., Nyitrai, M., et al., 2015. Characterization of connexin36 gap junctions in the human outer retina. Brain Struct. Funct.
Kay, J.N., De la Huerta, I., Kim, I.-J., Zhang, Y., Yamagata, M., Chu, M.W., Meister, M., Sanes, J.R., 2011. Retinal ganglion cells with distinct directional preferences differ in molecular identity, structure, and central projections. J. Neurosci. 31, 7753–7762.
King, S.M., Dykeman, C., Redgrave, P., Dean, P., 1992. Use of a distracting task to obtain defensive head movements to looming visual stimuli by human adults in a laboratory setting. Perception 21, 245–259.
Klaassen, L.J., Sun, Z., Steijaert, M.N., Bolte, P., Fahrenfort, I., Sjoerdsma, T., Klooster, J., Claassen, Y., Shields, C.R., Ten Eikelder, H.M.M., et al., 2011. Synaptic transmission from horizontal cells to cones is impaired by loss of connexin hemichannels. PLoS Biol. 9, e1001107.
Kolb, H., 1977. The organization of the outer plexiform layer in the retina of the cat: electron microscopic observations. J. Neurocytol. 6, 131–153.
Kolb, H., Famiglietti, E.V., 1974. Rod and cone pathways in the inner plexiform layer of cat retina. Science 186, 47–49.
Kothmann, W.W., Trexler, E.B., Whitaker, C.M., Li, W., Massey, S.C., O'Brien, J., 2012. Nonsynaptic NMDA receptors mediate activity-dependent plasticity of gap junctional coupling in the AII amacrine cell network. J. Neurosci. 32, 6747–6759.
Kranz, K., Paquet-Durand, F., Weiler, R., Janssen-Bienhold, U., Dedek, K., 2013. Testing for a gap junction-mediated bystander effect in retinitis pigmentosa: secondary cone death is not altered by deletion of Connexin36 from cones. PLoS One 8, e57163.
Kujiraoka, T., Saito, T., 1986. Electrical coupling between bipolar cells in carp retina. Proc. Natl. Acad. Sci. USA 83, 4063–4066.
Kuo, S.P., Schwartz, G.W., Rieke, F., 2016. Nonlinear spatiotemporal integration by electrical and chemical synapses in the retina. Neuron 90 (2), 320–332.

Lamb, T.D., Simon, E.J., 1976. The relation between intercellular coupling and electrical noise in turtle photoreceptors. J. Physiol. 263, 257–286.
Lanore, F., Silver, R.A., 2014. Reading dendritic activity with gap junctions. Nat. Neurosci. 17, 1625–1627.
Lasater, E.M., Dowling, J.E., 1985. Dopamine decreases conductance of the electrical junctions between cultured retinal horizontal cells. Proc. Natl. Acad. Sci. USA 82, 3025–3029.
Lee, E.-J., Han, J.-W., Kim, H.-J., Kim, I.-B., Lee, M.-Y., Oh, S.-J., Chung, J.-W., Chun, M.-H., 2003. The immunocytochemical localization of connexin 36 at rod and cone gap junctions in the Guinea pig retina. Eur. J. Neurosci. 18, 2925–2934.
Lee, S., Kim, K., Zhou, Z.J., 2010. Role of ACh-GABA cotransmission in detecting image motion and motion direction. Neuron 68, 1159–1172.
Lee, S.C.S., Meyer, A., Schubert, T., Hüser, L., Dedek, K., Haverkamp, S., 2015. Morphology and connectivity of the small bistratified A8 amacrine cell in the mouse retina. J. Comp. Neurol. 523, 1529–1547.
Lefler, Y., Yarom, Y., Uusisaari, M.Y., 2014. Cerebellar inhibitory input to the inferior olive decreases electrical coupling and blocks subthreshold oscillations. Neuron 81, 1389–1400.
Li, W., DeVries, S.H., 2004. Separate blue and green cone networks in the mammalian retina. Nat. Neurosci. 7, 751–756.
Li, H., Chuang, A.Z., O'Brien, J., 2009. Photoreceptor coupling is controlled by connexin 35 phosphorylation in zebrafish retina. J. Neurosci. 29, 15178–15186.
Li, H., Zhang, Z., Blackburn, M.R., Wang, S.W., Ribelayga, C.P., O'Brien, J., 2013. Adenosine and dopamine receptors coregulate photoreceptor coupling via gap junction phosphorylation in mouse retina. J. Neurosci. 33, 3135–3150.
Li, P.H., Verweij, J., Long, J.H., Schnapf, J.L., 2012. Gap-junctional coupling of mammalian rod photoreceptors and its effect on visual detection. J. Soc. Neurosci. 32, 3552–3562.
Li, W., Zhang, J., Massey, S.C., 2002. Coupling pattern of S1 and S2 amacrine cells in the rabbit retina. Vis. Neurosci. 19, 119–131.
Li, W., Keung, J.W., Massey, S.C., 2004. Direct synaptic connections between rods and OFF cone bipolar cells in the rabbit retina. J. Comp. Neurol. 474, 1–12.
Mangel, S.C., Dowling, J.E., 1985. Responsiveness and receptive field size of carp horizontal cells are reduced by prolonged darkness and dopamine. Science 229, 1107–1109.
Marc, R.E., Liu, W.L., Muller, J.F., 1988. Gap junctions in the inner plexiform layer of the goldfish retina. Vis. Res. 28, 9–24.
Marc, R.E., Anderson, J.R., Jones, B.W., Sigulinsky, C.L., Lauritzen, J.S., 2014. The AII amacrine cell connectome: a dense network hub. Front Neural Circuits 8, 104.
Margolis, D.J., Newkirk, G., Euler, T., Detwiler, P.B., 2008. Functional stability of retinal ganglion cells after degeneration-induced changes in synaptic input. J. Neurosci. 28, 6526–6536.
Masland, R.H., 2012. The neuronal organization of the retina. Neuron 76, 266–280.
Masland, R.H., Mills, J.W., 1979. Autoradiographic identification of acetylcholine in the rabbit retina. J. Cell Biol. 83, 159–178.
Mastronarde, D.N., 1983a. Correlated firing of cat retinal ganglion cells. I. Spontaneously active inputs to X- and Y-cells. J. Neurophysiol. 49, 303–324.
Mastronarde, D.N., 1983b. Correlated firing of cat retinal ganglion cells. II. Responses of X- and Y-cells to single quantal events. J. Neurophysiol. 49, 325–349.
Mastronarde, D.N., 1983c. Interactions between ganglion cells in cat retina. J. Neurophysiol. 49, 350–365.
Mastronarde, D.N., 1989. Correlated firing of retinal ganglion cells. Trends Neurosci. 12, 75–80.
Mathy, A., Clark, B.A., Häusser, M., 2014. Synaptically induced long-term modulation of electrical coupling in the inferior olive. Neuron 81, 1290–1296.
Maxeiner, S., Dedek, K., Janssen-Bienhold, U., Ammermüller, J., Brune, H., Kirsch, T., Pieper, M., Degen, J., Krüger, O., Willecke, K., et al., 2005. Deletion of connexin45 in mouse retinal neurons disrupts the rod/cone signaling pathway between AII amacrine and ON cone bipolar cells and leads to impaired visual transmission. J. Neurosci. 25, 566–576.
McMahon, D.G., 1994. Modulation of electrical synaptic transmission in zebrafish retinal horizontal cells. J. Neurosci. 14, 1722–1734.

McMahon, D.G., Mattson, M.P., 1996. Horizontal cell electrical coupling in the giant danio: synaptic modulation by dopamine and synaptic maintenance by calcium. Brain Res. 718, 89–96.
Meister, M., Lagnado, L., Baylor, D.A., 1995. Concerted signaling by retinal ganglion cells. Science 270, 1207–1210.
Menzler, J., Zeck, G., 2011. Network oscillations in rod-degenerated mouse retinas. J. Neurosci. 31, 2280–2291.
Menzler, J., Channappa, L., Zeck, G., 2014. Rhythmic ganglion cell activity in bleached and blind adult mouse retinas. PLoS One 9, e106047.
Mills, S.L., 1999. Unusual coupling patterns of a cone bipolar cell in the rabbit retina. Vis. Neurosci. 16, 1029–1035.
Muller, J.F., Dacheux, R.F., 1997. Alpha ganglion cells of the rabbit retina lose antagonistic surround responses under dark adaptation. Vis. Neurosci. 14, 395–401.
Müller, L.P.de S., Do, M.T.H., Yau, K.-W., He, S., Baldridge, W.H., 2010a. Tracer coupling of intrinsically photosensitive retinal ganglion cells to amacrine cells in the mouse retina. J. Comp. Neurol. 518, 4813–4824.
Müller, L.P.de S., Dedek, K., Janssen-Bienhold, U., Meyer, A., Kreuzberg, M.M., Lorenz, S., Willecke, K., Weiler, R., 2010b. Expression and modulation of connexin 30.2, a novel gap junction protein in the mouse retina. Vis. Neurosci. 27, 91–101.
Münch, T.A., da Silveira, R.A., Siegert, S., Viney, T.J., Awatramani, G.B., Roska, B., 2009. Approach sensitivity in the retina processed by a multifunctional neural circuit. Nat. Neurosci. 12, 1308–1316.
Murphy, G.J., Rieke, F., 2011. Electrical synaptic input to ganglion cells underlies differences in the output and absolute sensitivity of parallel retinal circuits. J. Neurosci. 31 (34), 12218–12228.
Nelson, R., 1977. Cat cones have rod input: a comparison of the response properties of cones and horizontal cell bodies in the retina of the cat. J. Comp. Neurol. 172, 109–135.
O'Brien, J.J., Li, W., Pan, F., Keung, J., O'Brien, J., Massey, S.C., 2006. Coupling between A-type horizontal cells is mediated by connexin 50 gap junctions in the rabbit retina. J. Neurosci. 26, 11624–11636.
O'Brien, J.J., Chen, X., Macleish, P.R., O'Brien, J., Massey, S.C., 2012. Photoreceptor coupling mediated by connexin36 in the primate retina. J. Neurosci. 32, 4675–4687.
Oesch, N., Euler, T., Taylor, W.R., 2005. Direction-selective dendritic action potentials in rabbit retina. Neuron 47, 739–750.
Oyster, C.W., Barlow, H.B., 1967. Direction-selective units in rabbit retina: distribution of preferred directions. Science 155, 841–842.
Pan, F., Paul, D.L., Bloomfield, S.A., Völgyi, B., 2010. Connexin36 is required for gap junctional coupling of most ganglion cell subtypes in the mouse retina. J. Comp. Neurol. 518, 911–927.
Pan, F., Keung, J., Kim, I.-B., Snuggs, M.B., Mills, S.L., O'Brien, J., Massey, S.C., 2012. Connexin 57 is expressed by the axon terminal network of B-type horizontal cells in the rabbit retina. J. Comp. Neurol. 520, 2256–2274.
Pandarinath, C., Bomash, I., Victor, J.D., Prusky, G.T., Tschetter, W., Nirenberg, S., Pandarinath, C., Bomash, I., Victor, J.D., Prusky, G.T., et al., 2010. A novel mechanism for switching a neural system from one state to another. Front. Comput. Neurosci. 4, 2.
Pang, J.-J., Paul, D.L., Wu, S.M., 2013. Survey on amacrine cells coupling to retrograde-identified ganglion cells in the mouse retina. Invest. Ophthalmol. Vis. Sci. 54, 5151–5162.
Paschon, V., Higa, G.S.V., Resende, R.R., Britto, L.R.G., Kihara, A.H., 2012. Blocking of connexin-mediated communication promotes neuroprotection during acute degeneration induced by mechanical trauma. PLoS One 7, e45449.
Peichl, L., Wässle, H., 1983. The structural correlate of the receptive field centre of alpha ganglion cells in the cat retina. J. Physiol. 341, 309–324.
Pereda, A.E., 2014. Electrical synapses and their functional interactions with chemical synapses. Nat. Rev. Neurosci. 15, 250–263.
Pereda, A.E., Curti, S., Hoge, G., Cachope, R., Flores, C.E., Rash, J.E., 2013. Gap junction-mediated electrical transmission: regulatory mechanisms and plasticity. Biochim. Biophys. Acta 1828, 134–146.
Pillow, J.W., Shlens, J., Paninski, L., Sher, A., Litke, A.M., Chichilnisky, E.J., Simoncelli, E.P., 2008. Spatio-temporal correlations and visual signalling in a complete neuronal population. Nature 454, 995–999.

Raviola, E., Gilula, N.B., 1973. Gap junctions between photoreceptor cells in the vertebrate retina. Proc. Natl. Acad. Sci. USA 70, 1677–1681.

Raviola, E., Gilula, N.B., 1975. Intramembrane organization of specialized contacts in the outer plexiform layer of the retina. A freeze-fracture study in monkeys and rabbits. J. Cell Biol. 65, 192–222.

Reifler, A.N., Chervenak, A.P., Dolikian, M.E., Benenati, B.A., Li, B.Y., Wachter, R.D., Lynch, A.M., Demertzis, Z.D., Meyers, B.S., Abufarha, F.S., et al., 2015. All spiking, sustained on displaced amacrine cells receive gap-junction input from melanopsin ganglion cells. Curr. Biol. 25, 2763–2773.

Ribelayga, C., Mangel, S.C., 2003. Absence of circadian clock regulation of horizontal cell gap junctional coupling reveals two dopamine systems in the goldfish retina. J. Comp. Neurol. 467, 243–253.

Ribelayga, C., Mangel, S.C., 2007. Tracer coupling between fish rod horizontal cells: modulation by light and dopamine but not the retinal circadian clock. Vis. Neurosci. 24, 333–344.

Ribelayga, C.P., O'Brien, J., 2017. Circadian and light-adaptive control of electrical synaptic plasticity in the vertebrate retina. In: Jing, J. (Ed.), Network Functions and Plasticity. Elsevier (Chapter 10).

Ribelayga, C., Cao, Y., Mangel, S.C., 2008. The circadian clock in the retina controls rod-cone coupling. Neuron 59, 790–801.

Robles, E., Laurell, E., Baier, H., 2014. The retinal projectome reveals brain-area-specific visual representations generated by ganglion cell diversity. Curr. Biol. 24, 2085–2096.

Sahel, J.-A., Roska, B., 2013. Gene therapy for blindness. Annu. Rev. Neurosci. 36, 467–488.

Saito, T., Kujiraoka, T., 1988. Characteristics of bipolar-bipolar coupling in the carp retina. J. Gen. Physiol. 91, 275–287.

Sanes, J.R., Masland, R.H., 2015. The types of retinal ganglion cells: current status and implications for neuronal classification. Annu. Rev. Neurosci. 38, 221–246.

Schiff, W., Caviness, J.A., Gibson, J.J., 1962. Persistent fear responses in rhesus monkeys to the optical stimulus of "looming". Science 136, 982–983.

Schneeweis, D.M., Schnapf, J.L., 1995. Photovoltage of rods and cones in the macaque retina. Science 268, 1053–1056.

Schnitzer, M.J., Meister, M., 2003. Multineuronal firing patterns in the signal from eye to brain. Neuron 37, 499–511.

Schubert, T., Degen, J., Willecke, K., Hormuzdi, S.G., Monyer, H., Weiler, R., 2005a. Connexin36 mediates gap junctional coupling of alpha-ganglion cells in mouse retina. J. Comp. Neurol. 485, 191–201.

Schubert, T., Maxeiner, S., Krüger, O., Willecke, K., Weiler, R., 2005b. Connexin45 mediates gap junctional coupling of bistratified ganglion cells in the mouse retina. J. Comp. Neurol. 490, 29–39.

Sekirnjak, C., Jepson, L.H., Hottowy, P., Sher, A., Dabrowski, W., Litke, A.M., Chichilnisky, E.J., 2011. Changes in physiological properties of rat ganglion cells during retinal degeneration. J. Neurophysiol. 105, 2560–2571.

Seung, H.S., Sümbül, U., 2014. Neuronal cell types and connectivity: lessons from the retina. Neuron 83, 1262–1272.

Shelley, J., Dedek, K., Schubert, T., Feigenspan, A., Schultz, K., Hombach, S., Willecke, K., Weiler, R., 2006. Horizontal cell receptive fields are reduced in connexin57-deficient mice. Eur. J. Neurosci. 23, 3176–3186.

Shields, C.R., Klooster, J., Claassen, Y., Ul-Hussain, M., Zoidl, G., Dermietzel, R., Kamermans, M., 2007. Retinal horizontal cell-specific promoter activity and protein expression of zebrafish connexin 52.6 and connexin 55.5. J. Comp. Neurol. 501, 765–779.

Singh, M.S., Charbel Issa, P., Butler, R., Martin, C., Lipinski, D.M., Sekaran, S., Barnard, A.R., MacLaren, R.E., 2013. Reversal of end-stage retinal degeneration and restoration of visual function by photoreceptor transplantation. Proc. Natl. Acad. Sci. USA 110, 1101–1106.

Sivyer, B., Williams, S.R., 2013. Direction selectivity is computed by active dendritic integration in retinal ganglion cells. Nat. Neurosci. 16, 1848–1856.

Soucy, E., Wang, Y., Nirenberg, S., Nathans, J., Meister, M., 1998. A novel signaling pathway from rod photoreceptors to ganglion cells in mammalian retina. Neuron 21, 481–493.

Stasheff, S.F., 2008. Emergence of sustained spontaneous hyperactivity and temporary preservation of OFF responses in ganglion cells of the retinal degeneration (rd1) mouse. J. Neurophysiol. 99, 1408–1421.

Stasheff, S.F., Shankar, M., Andrews, M.P., 2011. Developmental time course distinguishes changes in spontaneous and light-evoked retinal ganglion cell activity in rd1 and rd10 mice. J. Neurophysiol. 105, 3002–3009.

Strettoi, E., Raviola, E., Dacheux, R.F., 1992. Synaptic connections of the narrow-field, bistratified rod amacrine cell (AII) in the rabbit retina. J. Comp. Neurol. 325, 152–168.

Striedinger, K., Petrasch-Parwez, E., Zoidl, G., Napirei, M., Meier, C., Eysel, U.T., Dermietzel, R., 2005. Loss of connexin36 increases retinal cell vulnerability to secondary cell loss. Eur. J. Neurosci. 22, 605–616.

Sun, H., Frost, B.J., 1998. Computation of different optical variables of looming objects in pigeon nucleus rotundus neurons. Nat. Neurosci. 1, 296–303.

Temizer, I., Donovan, J.C., Baier, H., Semmelhack, J.L., 2015. A visual pathway for looming-evoked escape in larval zebrafish. Curr. Biol. 25, 1823–1834.

Tessier-Lavigne, M., Attwell, D., 1988. The effect of photoreceptor coupling and synapse nonlinearity on signal: noise ratio in early visual processing. Proc. R. Soc. Lond. B Biol. Sci. 234, 171–197.

Thoreson, W.B., Mangel, S.C., 2012. Lateral interactions in the outer retina. Prog. Retin. Eye Res. 31, 407–441.

Trenholm, S., Awatramani, G., 2015. Origins of spontaneous activity in the degenerating retina. Front. Cell. Neurosci. 9, 277.

Trenholm, S., Johnson, K., Li, X., Smith, R.G., Awatramani, G.B., 2011. Parallel mechanisms encode direction in the retina. Neuron 71, 683–694.

Trenholm, S., Borowska, J., Zhang, J., Hoggarth, A., Johnson, K., Barnes, S., Lewis, T.J., Awatramani, G.B., 2012. Intrinsic oscillatory activity arising within the electrically coupled AII amacrine-ON cone bipolar cell network is driven by voltage-gated Na^+ channels. J. Physiol. 590, 2501–2517.

Trenholm, S., McLaughlin, A.J., Schwab, D.J., Awatramani, G.B., 2013a. Dynamic tuning of electrical and chemical synaptic transmission in a network of motion coding retinal neurons. J. Neurosci. 33, 14927–14938.

Trenholm, S., Schwab, D.J., Balasubramanian, V., Awatramani, G.B., 2013b. Lag normalization in an electrically coupled neural network. Nat. Neurosci. 16, 154–156.

Trenholm, S., McLaughlin, A.J., Schwab, D.J., Turner, M.H., Smith, R.G., Rieke, F., Awatramani, G.B., 2014. Nonlinear dendritic integration of electrical and chemical synaptic inputs drives fine-scale correlations. Nat. Neurosci. 17, 1759–1766.

Trexler, E.B., Li, W., Massey, S.C., 2005. Simultaneous contribution of two rod pathways to AII amacrine and cone bipolar cell light responses. J. Neurophysiol. 93, 1476–1485.

Trong, P.K., Rieke, F., 2008. Origin of correlated activity between parasol retinal ganglion cells. Nat. Neurosci. 11, 1343–1351.

Tsukamoto, Y., Masarachia, P., Schein, S.J., Sterling, P., 1992. Gap junctions between the pedicles of macaque foveal cones. Vis. Res. 32, 1809–1815.

Tsukamoto, Y., Morigiwa, K., Ueda, M., Sterling, P., 2001. Microcircuits for night vision in mouse retina. J. Neurosci. 21, 8616–8623.

Turecek, J., Yuen, G.S., Han, V.Z., Zeng, X.-H., Bayer, K.U., Welsh, J.P., 2014. NMDA receptor activation strengthens weak electrical coupling in mammalian brain. Neuron 81, 1375–1388.

Umino, O., Maehara, M., Hidaka, S., Kita, S., Hashimoto, Y., 1994. The network properties of bipolar-bipolar cell coupling in the retina of teleost fishes. Vis. Neurosci. 11, 533–548.

Usrey, W.M., Reppas, J.B., Reid, R.C., 1998. Paired-spike interactions and synaptic efficacy of retinal inputs to the thalamus. Nature 395, 384–387.

Vaney, D.I., 1991. Many diverse types of retinal neurons show tracer coupling when injected with biocytin or neurobiotin. Neurosci. Lett. 125, 187–190.

Vaney, D.I., 1994. Territorial organization of direction-selective ganglion cells in rabbit retina. J. Neurosci. 14, 6301–6316.

Veruki, M.L., Hartveit, E., 2002a. Electrical synapses mediate signal transmission in the rod pathway of the mammalian retina. J. Neurosci. 22, 10558–10566.

Veruki, M.L., Hartveit, E., 2002b. AII (Rod) amacrine cells form a network of electrically coupled interneurons in the mammalian retina. Neuron 33, 935–946.

Völgyi, B., Deans, M.R., Paul, D.L., Bloomfield, S.A., 2004. Convergence and segregation of the multiple rod pathways in mammalian retina. J. Neurosci. 24, 11182–11192.
Völgyi, B., Chheda, S., Bloomfield, S.A., 2009. Tracer coupling patterns of the ganglion cell subtypes in the mouse retina. J. Comp. Neurol. 512, 664–687.
Völgyi, B., Pan, F., Paul, D.L., Wang, J.T., Huberman, A.D., Bloomfield, S.A., 2013. Gap junctions are essential for generating the correlated spike activity of neighboring retinal ganglion cells. PLoS One 8, e69426.
Wässle, H., Puller, C., Müller, F., Haverkamp, S., 2009. Cone contacts, mosaics, and territories of bipolar cells in the mouse retina. J. Neurosci. 29, 106–117.
Weiland, J.D., Humayun, M.S., 2014. Retinal prosthesis. IEEE Trans. Biomed. Eng. 61, 1412–1424.
Werblin, F.S., 1970. Response of retinal cells to moving spots: intracellular recording in *Necturus maculosus*. J. Neurophysiol. 33, 342–350.
Werblin, F.S., 1978. Transmission along and between rods in the tiger salamander retina. J. Physiol. 280, 449–470.
Werblin, F.S., Dowling, J.E., 1969. Organization of the retina of the mudpuppy, *Necturus maculosus*. II. Intracellular recording. J. Neurophysiol. 32, 339–355.
Wu, S.M., Yang, X.L., 1988. Electrical coupling between rods and cones in the tiger salamander retina. Proc. Natl. Acad. Sci. USA 85, 275–278.
Xiao, L., Zhang, M., Xing, D., Liang, P.-J., Wu, S., 2013. Shifted encoding strategy in retinal luminance adaptation: from firing rate to neural correlation. J. Neurophysiol. 110, 1793–1803.
Xin, D., Bloomfield, S.A., 1997. Tracer coupling pattern of amacrine and ganglion cells in the rabbit retina. J. Comp. Neurol. 383, 512–528.
Xin, D., Bloomfield, S.A., 1999. Dark- and light-induced changes in coupling between horizontal cells in mammalian retina. J. Comp. Neurol. 405, 75–87.
Xu, Z., Zeng, Q., Shi, X., He, S., 2013. Changing coupling pattern of the ON-OFF direction-selective ganglion cells in early postnatal mouse retina. Neuroscience 250, 798–808.
Yee, C.W., Toychiev, A.H., Sagdullaev, B.T., 2012. Network deficiency exacerbates impairment in a mouse model of retinal degeneration. Front. Syst. Neurosci. 6, 8.
Yee, C.W., Toychiev, A.H., Ivanova, E., Sagdullaev, B.T., 2014. Aberrant synaptic input to retinal ganglion cells varies with morphology in a mouse model of retinal degeneration. J. Comp. Neurol. 522, 4085–4099.
Yilmaz, M., Meister, M., 2013. Rapid innate defensive responses of mice to looming visual stimuli. Curr. Biol. 23, 2011–2015.
Yilmaz, M., Duan, X., Sanes, J.R., Meister, M., 2014. Roles of retinal circuits in innate visual behaviors of mice. SFN Abstr. 59, 11/Z6.
Zhang, J., Wu, S.M., 2004. Connexin35/36 gap junction proteins are expressed in photoreceptors of the tiger salamander retina. J. Comp. Neurol. 470, 1–12.
Zhang, J., Wu, S.M., 2005. Physiological properties of rod photoreceptor electrical coupling in the tiger salamander retina. J. Physiol. 564, 849–862.
Zhang, D.-Q., Zhou, T.-R., McMahon, D.G., 2007. Functional heterogeneity of retinal dopaminergic neurons underlying their multiple roles in vision. J. Neurosci. 27, 692–699.
Zhao, X., Liu, M., Cang, J., 2014. Visual cortex modulates the magnitude but not the selectivity of looming-evoked responses in the superior colliculus of awake mice. Neuron 84, 202–213.

CHAPTER 10

Circadian and Light-Adaptive Control of Electrical Synaptic Plasticity in the Vertebrate Retina

C.P. Ribelayga, J. O'Brien
McGovern Medical School and The University of Texas Health Science Center at Houston, Houston, TX, United States

1. INTRODUCTION

Vision allows an organism to collect information about the environment via detection of reflected or emitted light. This aspect of the environment can be incredibly rich in information content and, not surprisingly, some visual systems have evolved to spectacular degrees to collect this information. In vertebrates, the retina is by far the most complex sensory tissue, collecting mapped visual information and dividing it into many parallel streams that are delivered to higher areas of the brain, largely retaining the retinotopic map. The parallel streams contain very different information content, which is extracted from the visual image that is projected onto the photoreceptors through the sophisticated postreceptoral circuitry. Electrical synapses play important and varied roles in most of this circuitry.

Every day, the visual system faces the Herculean task of having to code information useful to the organism at any time, when background light intensity varies over more than a billion-fold range. To meet this challenge the retina splits this range into two sections handled by rod and cone pathways, and adaptation mechanisms are present at every level of retinal circuitry. This chapter focuses on the electrical synaptic mechanisms that contribute to meeting this challenge. These include circadian and light-adaptive mechanisms that cause switching from rod to cone pathways and that optimize retinal circuits to perform under different lighting conditions. We separately review photoreceptor, horizontal cell, and amacrine cell circuits that show the highest degree of plasticity and discuss molecular mechanisms that control this plasticity. Additional information about other electrically coupled retinal circuits and emergent properties of electrical coupling are discussed by Trenholm and Awatramani, 2017 (Chapter 9).

2. PHOTORECEPTORS

Evidence for intercellular coupling between photoreceptors was first suggested by ultrastructural analysis of the retina and the demonstration that points of contact between

Network Functions and Plasticity
ISBN 978-0-12-803471-2, http://dx.doi.org/10.1016/B978-0-12-803471-2.00010-2

photoreceptor cells frequently contained dense material and closely apposed membrane. These interreceptor contacts were first reported between rods and cones in guinea pig (Sjostrand, 1958) and later found between rods and cones and between cones and cones in pigeon, human, gray squirrel, and macaque (Cohen, 1963, 1964, 1965; Missotten et al., 1963). Some of these contacts were made by protruding processes called telodendria extending from the terminals to distant receptors. The nature of these contacts remained obscure, as they lacked clusters of vesicles and therefore did not fit the bona fide criteria for traditional synapses, until Raviola and Gilula (Raviola and Gilula, 1973, 1975) and Witkovsky and colleagues (Witkovsky et al., 1974) presented clear evidence that they were gap junctions.

The first evidence for electrotonic communication between photoreceptors came from Tomita and colleagues (Tomita et al., 1967) who observed that the spectral response curve of carp red cones contained a shoulder at shorter wavelengths that did not fit the Dartnall nomogram of the red pigment, suggesting the possibility of another receptor type contributing to the response. Clear physiological support for electrical synapses between photoreceptors came from work of Baylor and colleagues. They found that the receptive fields of cones were substantially larger than their diameters and further that current injected into one cone was passed to some other cones up to 40 μm away (Baylor et al., 1971). It was noted that this was about the length of the telodendria observed in turtle retina (Lasansky, 1972), suggesting that these processes serve the purpose of direct electrical communication between the receptors, and implicating gap junctions as the route for electrical communication. Additional studies soon followed and brought additional support to a functional basis for the presence of gap junctions between photoreceptors (Attwell and Wilson, 1980; Copenhagen and Owen, 1976; Werblin, 1978). Altogether, these pioneering studies established that vertebrate photoreceptors are electrically coupled through gap junctions.

2.1 Physiological Functions of Electrical Coupling Between Photoreceptors

Soon after the discovery of direct electrical communication between photoreceptors, a number of studies performed in cold-blooded vertebrates (eg, salamander, turtle) laid down the basic properties and functional consequences of photoreceptor electrical coupling. The function of photoreceptor coupling in mammals was investigated more recently and many of its aspects still remain under investigation. Although some photoreceptor telodendria can be chemically postsynaptic to neighboring photoreceptors (cf. Normann et al., 1984), to date all gap junctional connections between photoreceptors appear to function as purely electrical synapses. We review the functional importance of rod–rod, rod–cone, and cone–cone electrical synapses separately.

2.1.1 Rod—Rod Coupling

Electrical coupling allows the propagation of electrical signals within the network of coupled rods and therefore a direct functional consequence of it is the increase in the receptive field size of single rods. This translates into an enhancement of the rod responses upon modest enlargement of the illuminating field (Copenhagen and Owen, 1976; Schwartz, 1973, 1975b; Werblin, 1978) and the possibility for a rod to sense a change in voltage that originates from phototransduction processes in the outer segment of a neighboring rod, even if this change in voltage results from a single photon event (Fain, 1975, 1976; Jin et al., 2015). In this way, rod—rod coupling serves an integrating function by summing rod responses from many coupled rods. The summation area has been reported to be as large as 0.5 mm^2 or including ~8500 rods in the toad retina (Fain, 1975), but usually this area of summation is much smaller in cold-blooded vertebrates (ie, hundreds of rods (Copenhagen and Owen, 1976; Gold, 1979; Leeper et al., 1978)) and even smaller in mammals (ie, <100 rods (Hornstein et al., 2005; Jin et al., 2015; Li et al., 2012)). Time- and voltage-dependent conductances, such as the inward rectifying current I_h, have been proposed to play a role in the propagation and high-pass filtering of electrical signals within the rod network by speeding the offset of responses as distance from the stimulus increases (Attwell and Wilson, 1980; Detwiler et al., 1978; Hestrin, 1987). In addition, other studies have shown that rod responses are in fact band-pass filtered because the membrane capacitive properties set an upper limit to the frequency at which signals can propagate through the network (Armstrong-Gold and Rieke, 2003; Attwell, 1986; Zhang and Wu, 2005). For bright saturating focused stimuli, rod coupling helps circumvent saturation at the rod synapse by distributing the signal over several synapses, thereby extending the operating range of scotopic vision (Hornstein et al., 2005; Tessier-Lavigne and Attwell, 1988). Thus electrical synapses between rods serve an integrating function and extend the dynamic range of rod-mediated vision.

Electrical coupling between rods is also well known to decrease intrinsic and uncorrelated voltage noise that randomly arises in individual rods (Fain, 1975; Schwartz, 1975b; Tessier-Lavigne and Attwell, 1988). The benefit of a decrease in voltage noise in the rod network on the signal:noise ratio at the rod synapse depends, however, on the size and intensity of the light stimulus. For a uniform stimulus (ie, all the rods in the network receive the same stimulus), the signal:noise ratio is increased because noise is reduced by coupling, whereas the signal amplitude does not change because all rods remain isopotential and no net current flows through the gap junctions during illumination (Hornstein et al., 2005; Jin et al., 2015; Li et al., 2012; Tessier-Lavigne and Attwell, 1988). This is also true for localized multiphoton stimuli because coupling decreases noise to a greater extent than signal. However, when only one rod in the network responds to a single photon, the signal:noise ratio is made worse because coupling attenuates the signal to a greater extent than the noise (Hornstein et al., 2005; Jin et al., 2015;

Li et al., 2012; Tessier-Lavigne and Attwell, 1988). The detrimental effect of rod–rod coupling on the signal:noise ratio of single photon events is particularly problematic in mammals where rod-mediated single photon responses set the absolute threshold and are responsible for about half of the operating range of the visual system—the lower half of that range (Field et al., 2005). However, the functional impact of rod coupling will depend on how second-order neurons, and in particular rod bipolar cells, read the signal (Tessier-Lavigne and Attwell, 1988). Using estimates of the rod junctional conductance obtained in guinea pig, Li et al. (2012) calculated that the effect of coupling on human visual performance near absolute threshold should reduce dark-adapted sensitivity by approximately 10%. Although the effect is modest, the modeling supports a counterintuitive view that the retina may not optimally process rod signals near visual threshold. Additional work in this area is needed to fully understand the impact of rod–rod coupling on the physiology of downstream neurons and substantiate theoretical predictions.

2.1.2 Rod–Cone Coupling

Electrical coupling between rods and cones results in the mixing of rod and cone signals. However, the amount of mixing and the influence of one photoreceptor on the other both depend on the anatomical organization of the photoreceptor layer and on the strength of rod–cone coupling at individual gap junctions. In turtle and amphibian retinas, where each rod contacts many cones (Attwell et al., 1984; Raviola and Gilula, 1973), cone signals will influence the light response properties of rods to a greater extent than rod signals will shape cone responses. In these vertebrates, the influence of coupled cones on rods is demonstrated by the presence of additional components in the spectral sensitivity curve of the rods, the presence of kinetics that are characteristics of the cones in the rod response, and the enhancement of the ability of rods to follow high-frequency stimuli (Gold, 1979; Krizaj et al., 1998; Schwartz, 1975a; Wu and Yang, 1988). The significance of rod–cone coupling in these animals is still debated as it appears that rod–cone coupling strength depends on the type of cone as well as that of rod (Gao et al., 2013; Krizaj et al., 1998; Wu and Yang, 1988), but it is legitimate to believe that rod–cone coupling contributes to the spectral and temporal characteristics of the rod response under mesopic conditions, at least in some types of rod.

In retinas where each cone is electrically connected to many rods, rod–cone coupling results in rod signals reaching cones more efficiently than cone signals reaching rods. This anatomical arrangement prevails in most mammals and in fish (O'Brien et al., 2012; Stell, 1967; Tsukamoto et al., 2001). In macaque and goldfish retinas, rod input to cones leads to the appearance of high-sensitivity, slowly recovering light responses in cones (Asteriti et al., 2014; Ribelayga et al., 2008; Schneeweis and Schnapf, 1995). In addition, coupling between rods and cones also alters the spectral characteristics of the cone voltage response (Asteriti et al., 2014; Ribelayga et al., 2008; Schneeweis and Schnapf, 1995). The rod input originating from rod–cone coupling can be detected in postreceptor neurons. It can be observed in cone-connected horizontal cells (Dacheux and Raviola, 1982; Nelson, 1977; Ribelayga and Mangel, 2010; Wang and Mangel, 1996). In the inner retina, cone-mediated rod signals are present in the AII amacrine cells when the AMPA (α-amino-3-hydroxy-5-methyl-4-isoxazolepropionic acid) receptor input between rod bipolar cells and AII amacrine cells is blocked (Trexler et al., 2005). This signal, entering the AII via its gap junction with cone On bipolar cells, has a threshold 2 log units lower than cone threshold. Finally, genetic studies in mice have reported clear evidence that rod signals originating from rod–cone coupling can reach On (Deans et al., 2002) and Off (Volgyi et al., 2004) ganglion cells and can be detected in the electroretinogram (Abd-El-Barr et al., 2009). Thus rod–cone coupling provides a significant pathway for rod signals to activate inner retinal neurons, engaging both On and Off pathways. In mammals, it has been proposed that the rod-to-cone pathway or secondary rod pathway operates under mesopic conditions (Bloomfield and Dacheux, 2001). However, accumulative evidence indicates that rod–cone coupling is dynamically regulated, suggesting that its functional impact on retinal processing may vary accordingly (see Section 2.2 below).

2.1.3 Cone–Cone Coupling

As for rod–rod coupling, cone–cone coupling results in enlargement of the receptive field of single cones (Burkhardt, 1977; Fuortes, 1972), enhancement of responses upon broadening of the illuminating field (Burkhardt, 1977; Fuortes, 1972), and averaging of uncorrelated voltage noise (DeVries et al., 2002; Lamb and Simon, 1976). The extent of coupling between cones has been consistently shown to spread no more than two cone diameters (DeVries et al., 2002; Lamb and Simon, 1976). As cones operate in daylight (ie, in the presence of a photopic light background), coupling is expected to improve the signal:noise ratio at the synapse and thereby increase the fidelity of the synaptic output of the photoreceptor. Cone–cone coupling has been proposed to improve detection of contrast boundaries with only a modest amount of coupling (Lebedev et al., 1998). In high-density photoreceptor arrays, such as the primate fovea, cone coupling is expected to blur the neural representation of the image projected on the retina both optically (DeVries et al., 2002; Schneeweis and Schnapf, 1999) and chromatically (Hornstein et al., 2004; Hsu et al., 2000). However, the optical blur is calculated to be less than that produced by the optics of the eye (DeVries et al., 2002), resulting in a net benefit in acuity. Thus moderate cone–cone coupling markedly improves the signal:noise ratio with little compromise to visual acuity or color vision. It is important to note that cone–cone coupling in mammals is not homogeneous; short (blue)-wavelength sensitive cones are electrically isolated from middle (green)- and long (red)-wavelength sensitive cones (Hornstein et al., 2004; Li and DeVries, 2004). Thus, chromatic resolution in the blue–yellow axis is not compromised.

2.2 Circadian and Light-Adaptive Control of Coupling

The daily variation in ambient light intensity imposes strong constraints on visual function. To operate at any time of the day, the visual system needs to adapt or adjust its sensitivity to light (Dowling, 2012). Adaption in the retina is mediated by the acute effects of ambient light and also by the action of circadian clocks intrinsic to the retina (Dowling, 2012; McMahon et al., 2014). These circadian clocks are time-keeping mechanisms that allow the retina to anticipate and thus prepare for the predictable daily changes in the ambient lighting conditions (McMahon et al., 2014). The interplay between circadian clock activity and acute effects of light and dark on the retina leads to a total reorganization of retinal physiology and function with time of day (Barlow, 2001; Green and Besharse, 2004; Iuvone et al., 2005; Mangel and Ribelayga, 2010; McMahon et al., 2014). Although photoreceptors are arguably the most important neurons in the retina—vision starts within these cells—and gap junctions between photoreceptors are ideally positioned to control retinal processing because photoreceptor function depends at least in part quantitatively on the extent of electrical coupling, physiological observations of plasticity in photoreceptor coupling have been less forthcoming than for other networks of neurons. However, accumulative evidence now indicates that photoreceptor coupling shows a high degree of plasticity controlled by the ambient lighting conditions and/or circadian clocks. We review further the history of the discovery of the plasticity of photoreceptor coupling and summarize recent advances in the field.

Initial evidence that photoreceptor coupling is plastic came from the amphibian retina. Yang and Wu (1989) found that rod–cone coupling in salamander retina increased with background illumination. By recording from a rod and injecting current in an adjacent cone, Yang and Wu (1989) showed that cone input to rods increased as background became brighter. This change in coupling was, however, instantaneous, and is likely not related to modulation of the gap junctional conductance, but rather to reduction of shunting conductances in the rod membrane as light background increased. Clear evidence that amphibian photoreceptor coupling was modulated directly was given by Krizaj et al. (1998) in studies of the effects of dopamine on photoreceptor physiology and coupling. The ability of rods to follow red (660 nm) stimuli above 5 Hz was taken as an index of cone input. Krizaj et al. found that cone input into rods was substantially enhanced by the dopamine D2 receptor agonist quinpirole and was eliminated by D2 receptor antagonist spiperone. Interestingly, only a minority of rods appeared to receive direct cone input and to be regulated by dopamine ligands. In addition, rod–rod coupling appeared to be insensitive to the pharmacological manipulation of D2-like receptor activity. Tracer coupling measurements supported these conclusions. As dopamine plays a central role in reconfiguring retinal function during light adaptation (Witkovsky, 2004), the results from Krizaj et al. suggested that rod–cone coupling would increase with light adaptation and were consistent with Yang and Wu conclusions.

In stark contrast to the results in amphibian photoreceptors, teleost and mammalian photoreceptor networks show the opposite pattern of regulation with light or dopamine. Wang and Mangel (1996) found that the light response properties of goldfish luminosity-type (H1) cone-connected horizontal cells recorded under prolonged dark adaptation during the night resembled those of rods or rod-connected horizontal cells. That is, the light responses of H1 cells had a lower threshold (~2 log units lower) and slower kinetics at night than during the day. In addition, their spectral sensitivity was shifted from ~625 nm during the day to ~500 nm at night (Ribelayga et al., 2002; Wang and Mangel, 1996). These features were consistent with rod input, were not evident following dark adaptation in the daytime, and were quickly converted to cone-type responses by sensitizing flashes of bright photopic light. Wang and Mangel (1996) further demonstrated that the day/night differences in the response properties of H1 cells persisted when the fish were kept in the dark for several days, thereby demonstrating that the appearance of rod signals in the cone-driven cells was regulated by a circadian clock and restricted to the night. Because H1 cells receive synaptic input exclusively from cones (Stell, 1967), one plausible explanation for this finding was that rod–cone gap junctions were open during the subjective night and shut down during the subjective day, altering the rod input into cones and to cone-connected horizontal cells. Similar observations were later made in a mammalian retina: the rod input to cone-connected A-type horizontal cells in rabbit is regulated by a circadian clock and is increased at night (Ribelayga and Mangel, 2010). Ribelayga et al. (2002) and Ribelayga and Mangel (2010) further demonstrated that the circadian regulation of rod input into cone-driven horizontal cells was driven by dopamine through a receptor with D2-like pharmacology in both goldfish and rabbit. Because D2-like receptors are expressed in photoreceptors, whereas horizontal cells express D1-like receptors (Witkovsky, 2004), this again strongly implicated a change in rod–cone coupling as the origin of the change in rod input. Thus the results from goldfish and rabbit retinas indicated that rod–cone coupling should decrease with light adaptation and that maximal state of coupling would occur at night.

Further evidence that rod–cone coupling was modulated by a circadian clock, light, and dopamine came from direct measurements of photoreceptor coupling in goldfish retina (Ribelayga et al., 2008). Ribelayga et al. showed that the light responses of cones resembled those of rods at night or subjective night; that is, their threshold was decreased by ~3 log units, their spectral sensitivity peak shifted toward 500 nm, and their kinetics became much slower. In addition, the receptive field size of dark-adapted cones was ~ fivefold larger at night or subjective night than during daytime or subjective daytime (Ribelayga et al., 2008). The nighttime increase in the extent of tracer coupling between photoreceptors following iontophoresis of biocytin into individual cones further supported a circadian regulation of rod–cone coupling. Both rod and cone photoreceptors were included in the coupled network (Ribelayga et al., 2008). The circadian variations in the cone light response kinetics and tracer coupling were sensitive to the modulation of D2-like receptor activity. Blocking D2-like receptors with spiperone during subjective day induced a nighttime state, whereas activating D2-like receptors with quinpirole at night induced a daytime state. Bright (photopic) light adaptation, but not dim (low mesopic) light adaptation during the subjective night reversed tracer coupling and the light responses of cones to their daytime state (Ribelayga et al., 2008).

The daily or circadian control of tracer coupling among photoreceptors characterized by extensive coupling at night has been observed in mouse (Jin et al., 2015; Li et al., 2013; Ribelayga et al., 2008), rabbit (Ribelayga and Mangel, 2010), and zebrafish (Li et al., 2009). Finally, Jin et al. (2015) demonstrated that rod—rod electrical coupling is regulated by a circadian clock and dopamine in mouse retina in a way very similar to that reported for rod—cone coupling in goldfish. Overall the results from the studies in fish and mammals have led to a view that photoreceptor electrical coupling is homogeneously modulated by light, dopamine, and a circadian clock such that coupling is weak during the day in light, stronger in the dark, and the strongest at night in the dark. The reasons for the discrepancy with the observations made in amphibians remain unknown.

Despite important advances in our understanding of the regulation of photoreceptor electrical coupling, many aspects of its functional impact on retinal processing remain elusive. Likely this functional impact depends on how the physiology of postreceptor neurons is affected by changes in photoreceptor coupling. It is important to note here that the experimental approach should be done without bias, as we tend to think that the retina should always operate optimally. The existence of rod—rod coupling is an example that suggests that the retina may not process rod signals optimally near absolute threshold (Li et al., 2012). Jin et al. (2015) were able to record the light-induced voltage responses of single rods in the intact mouse retina and observe the functional impact of the circadian regulation of rod coupling on rod photoreceptor function. They showed that extensive coupling at night did not prevent the detection of single photons by rods but as expected (Li et al., 2012), coupling was detrimental to the signal:noise ratio of the single photon response (Jin et al., 2015). However, the signal:noise ratio of multiphoton responses was increased at night because of coupling. Thus extensive rod coupling at night may represent a trade-off between optimal processing of single photon events and that of large dim objects. Although counterintuitive, strong rod coupling at night may serve specific needs for retinal function that we do not know of yet. Similarly, the increase in rod—cone coupling at night suggests that the secondary rod pathway in mammals might operate at scotopic intensities at night, but this possibility has yet to be demonstrated.

The functional importance of cone—cone coupling plasticity remains to be established as well. Furthermore, our current knowledge of the light/dark and/or circadian regulation of photoreceptor coupling is almost entirely based on studies that have used derivative measures of coupling. Therefore, the functional impact of the regulation will remain uncertain until we obtain estimates of the junctional conductance between photoreceptors and quantify the range of its daily variations. Finally, dopamine seems to play a central role in the control of photoreceptor coupling, acting both as a light-adaptive neuromodulator and a daytime effector of a circadian clock (Witkovsky, 2004; McMahon et al., 2014). However, other neuromodulators may control photoreceptor coupling as well. In a series of studies, Li et al. (2013) have demonstrated that the purine adenosine controls photoreceptor coupling in mouse and zebrafish (Li et al., 2014). Adenosine acts through A2a receptors that are expressed on photoreceptors and increase coupling, an action opposite to that of dopamine. Interestingly, the extracellular levels of adenosine in the retina vary in opposite phase to those of dopamine, with high levels at night and low levels during the day (Ribelayga and Mangel, 2005). Preliminary evidence also indicates that melatonin, a neuromodulator that is under the control of a circadian clock and whose levels peak at night in the retina, increases electrical coupling between mouse rods (Jin and Ribelayga, 2014). The exact functional roles of dopamine, adenosine, and melatonin in the regulation of photoreceptor coupling remain to be investigated in detail but the possibility that several neuromodulatory systems interact to control coupling suggests that coupling may be functionally important and needs to be tuned with precision according to the time of day and/or environmental lighting conditions.

2.3 Mechanisms of Control

The cellular mechanisms controlling photoreceptor coupling have been studied extensively in recent years. A central element common to all these regulatory mechanisms is connexin35 (Cx35) in cold-blooded vertebrates and its mammalian orthologue Cx36. Cx35/36 is expressed in cones and in rods, although its presence in mammalian rods has been contested (Bloomfield and Volgyi, 2009; Volgyi et al., 2013). In fish, a second connexin (Cx34.7) has been found in cones but its functional role is still unknown (O'Brien et al., 2004). The control of photoreceptor electrical coupling relies primarily on changes in Cx35/36 junction conductance, the major determinant of which is its phosphorylation state (Kothmann et al., 2007; Ouyang et al., 2005). Using phospho-Cx35–specific antibodies against key regulatory sites, Kothmann et al. (2009) and Li et al. (2009, 2013) demonstrated that a highly phosphorylated state of Cx35/36 is directly correlated to extensive tracer coupling, thereby establishing the phosphorylation state of Cx35/36 as an index of gap junction coupling between photoreceptors. Consistent with tracer coupling experiments, Cx35/36 was found to be poorly phosphorylated during the day in the light and highly phosphorylated at night in the dark, in both zebrafish (Li et al., 2009) and mouse (Li et al., 2013; Zhang et al., 2015). A circadian rhythm in Cx36 phosphorylation was observed in mice of the CBA/CaJ strain (Zhang et al., 2015), in which a circadian component in tracer coupling is present (Ribelayga et al., 2008).

Interestingly, large changes in photoreceptor Cx36 phosphorylation that precede light/dark transitions are also observed in C57Bl/6J mice, which lack a circadian rhythm of dopamine secretion (Doyle et al., 2002; Roseboom et al., 1998), implicating control by a circadian rhythm in elements of the signal transduction cascade (Li et al., 2013). Finally, expression of Cx36 transcript and protein have been reported to display daily rhythms with high values measured at night (Katti et al., 2013), although our own published data do not support the presence of a rhythm in Cx36 protein (Li et al., 2013; Zhang et al., 2015). Thus, the phosphorylation state of Cx35/36 at regulatory sites appears to be the major determinant of photoreceptor coupling.

Compelling evidence indicates that adenylate cyclase (AC)/cyclic AMP (cAMP)/protein kinase A (PKA) plays a central role in the control of Cx35/Cx36 phosphorylation. Li et al. (2009, 2013) found that photoreceptor coupling and Cx35/36 phosphorylation depend directly on PKA activity. The PKA activator Sp-8-cpt-cAMPs is sufficient to increase Cx35/36 phosphorylation during the day to nighttime levels, whereas the PKA inhibitor Rp-8-cpt-cAMPs is sufficient to dephosphorylate Cx35/36 at night down to daytime levels (Li et al., 2009, 2013). This situation is dramatically different from that in AII amacrine cells in which PKA activity activates a phosphatase to suppress Cx36 phosphorylation and coupling (see Section 3). The central role of cAMP/PKA in the regulation of Cx35/36 phosphorylation is consistent with the observations that application of the AC activator forskolin during the subjective day increases rod input to goldfish H1 horizontal cells and application of Rp-cAMPs during the subjective night decreases rod input to these cells (Ribelayga et al., 2002). The effects of PKA activators and inhibitors on Cx35/36 phosphorylation are also consistent with their effects on tracer coupling (Li et al., 2009, 2013). The link between AC/cAMP/PKA/Cx36 and dopamine was demonstrated in mouse. It has long been known that the "D2-like" dopamine receptor present in mouse photoreceptors is the D4 receptor (Cohen et al., 1992; Nir et al., 2002). Li et al. (2013) found that activation of the Gi-coupled D4 receptor by dopamine in the daytime suppressed Cx36 phosphorylation (Fig. 10.1), as expected from the results of Ribelayga et al. (2008) or Ribelayga and Mangel (2010). In addition, the effects of a D4 agonist and those of a D4 antagonist could be blocked by Sp-8-cpt-cAMPs and Rp-8-cpt-cAMPs, respectively, demonstrating that the effects of dopamine on Cx36 phosphorylation are primarily mediated by PKA (Li et al., 2013).

Although the concept of regulatory control of photoreceptor coupling by dopamine acting as a mediator of both light and a circadian clock that controls the AC/cAMP/PKA pathway is attractive and highly supported by experimental data, studies have suggested a higher level of complexity in the regulation of photoreceptor coupling. Li et al. (2013) demonstrated that the Gs-coupled adenosine A2a receptor opposes the D4 receptor effects, promoting Cx36 phosphorylation and coupling (Fig. 10.1). In work on zebrafish photoreceptor coupling, Li et al. (2014) have found that the Gi-coupled adenosine A1 receptor is also involved. Li et al. (2014) found that addition of exogenous adenosine

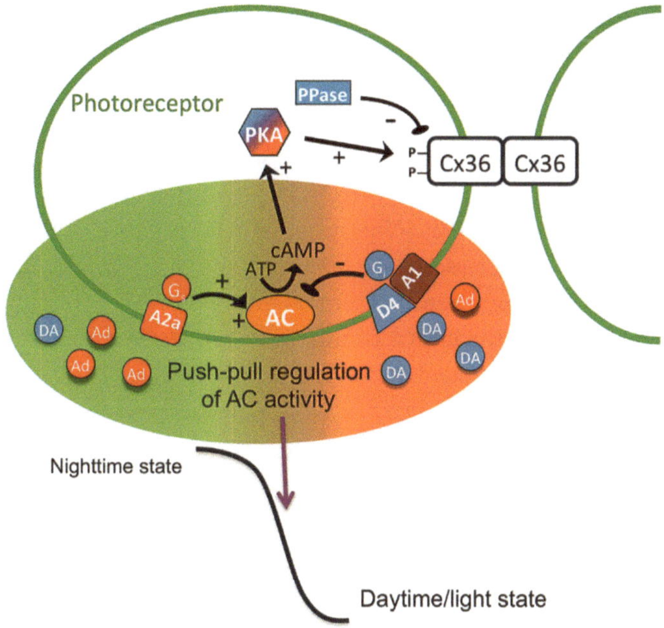

Figure 10.1 ***Signaling pathways that control photoreceptor coupling.*** Gap junctional coupling between photoreceptors is dynamically controlled by phosphorylation of Cx36, which opens the gap junctions. This is accomplished by protein kinase A (PKA) regulated by cyclic AMP (cAMP) concentration, and opposing action of an undescribed phosphatase (PPase). Activation of adenosine A2a receptors (A2a) by adenosine (Ad), favored in darkness during the subjective night, increases Gs-protein signaling to activate adenylate cyclase (AC) and increase cAMP concentration. Subjective daytime and light adaptation increase extracellular dopamine (DA) and decrease extracellular adenosine, favoring activation of dopamine D4 and adenosine A1 receptors. These receptors inhibit AC via Gi-protein signaling. The opposing forces acting upon a single point of regulation (AC) result in a steep biphasic profile for regulation of coupling.

or adenosine receptor agonists in the daytime not only activated A2a receptors, enhancing Cx36 phosphorylation, but also activated A1 receptors having a dampening effect. Furthermore, the A1 receptors were tonically active in the daytime, light-adapted preparation, adding to the suppression of Cx36 phosphorylation driven by the dopamine D4 receptor (Fig. 10.1). Although extracellular adenosine is present at the highest levels at night in dark-adapted retina, it is also present at a lower level during the day and in the light-adapted state (Ribelayga and Mangel, 2005). The affinity of A1 receptors for adenosine is two- to threefold higher than that of A2a receptors (Fredholm et al., 2011). The results of Li et al. (2014) suggest that the low level of extracellular adenosine present in the daytime actually reinforces the inhibitory effect of dopamine D4 receptors to keep photoreceptor coupling to a minimum in the daytime. Their results also suggest that the activity of the A2a receptor, when extracellular adenosine is high enough in dark-adapted conditions at night, is substantially more potent than that of the A1 receptor and drives enhancement of electrical coupling.

Further evidence indicates that melatonin controls photoreceptor coupling. A circadian rhythm of Cx36 phosphorylation or tracer coupling in photoreceptors is clearly observed in the melatonin-proficient CBA/CaJ mouse strain (Jin et al., 2015; Ribelayga and Mangel, 2010; Zhang et al., 2015), whereas it is greatly attenuated in the melatonin-deficient C57Bl/6J (Li et al., 2013; Ribelayga and Mangel, 2010). In addition, exogenously applied melatonin increases rod electrical coupling (Jin and Ribelayga, 2014). Although MT1 and MT2 melatonin receptors are expressed in photoreceptors (Baba et al., 2013), it is likely that the effects of melatonin on Cx36 phosphorylation and rod coupling are indirect and mediated by dopamine (McMahon et al., 2014); however, this remains to be demonstrated. Additional work is needed in this area to clarify the roles of dopamine, adenosine, and melatonin but their effects on photoreceptor coupling may not just reflect their additive actions on the AC/cAMP/PKA pathway. It is indeed possible that A2a and D4 and/or melatonin MT1 and MT2 receptors physically interact with one another to modify their signaling properties (Baba et al., 2013; Kull et al., 1999). In addition, many elements of the signaling pathways that regulate photoreceptor electrical coupling are themselves subject to transcriptional control through the day–night cycle and neuromodulatory signals appear to be required for their rhythmic expression. Both D4 and A2a receptors and adenylate cyclase 1 (AC1)—the predominant AC in photoreceptors—undergo cyclic changes in transcript levels through the day–night and circadian cycle, with typical nocturnal peaks of expression (Jackson et al., 2011; Li et al., 2013). The rhythm of D4 receptor transcript is greatly damped in mice lacking A2a receptors and the rhythm in AC1 is abolished in mice lacking either D4 (Jackson et al., 2011) or A2a (Li et al., 2013) receptors. Thus, it appears that signaling from the extracellular environment through G-protein–coupled receptors controls not only the plasticity of the electrical synapses, but also the expression level and timing of the major components of the signaling cascade.

A number of observations indicate that the regulation of photoreceptor coupling is likely to be even more complex than currently thought and may not be homogenous among photoreceptors. For instance, among mammalian photoreceptors, a circadian clock is likely present in cones and not in rods (Liu et al., 2012) wherein it potentially controls transcriptional activity (Storch et al., 2007). The role of the cone clock in the regulation of the major components of the signaling cascade in the cones and its impact on cone–cone and rod–cone coupling remain to be investigated but suggests that the regulation of photoreceptor coupling among photoreceptors may not be homogenous. The report that *Xenopus* rod–rod coupling was not affected by dopamine receptor agonists or antagonists (Krizaj et al., 1998) strengthens this point. In addition, in situ hybridization of receptor mRNA in mouse retina revealed that dopamine D4 receptor transcripts are widely distributed throughout the outer nuclear layer, whereas adenosine

A2a receptor transcripts are apparently more strongly expressed in cones (Li et al., 2013). It has also been observed that tracer-coupled networks of photoreceptors in nighttime and dark-adapted conditions contain a large proportion of cones (Li et al., 2009, 2013; Ribelayga et al., 2008), determined in mouse to be about twice that expected given their abundance in the retina (Li et al., 2013). These observations suggest that regulatory control of coupling could be stronger in cones and potentially different than that in rods. Finally, many other common molecules, such as endocannabinoids (Cachope et al., 2007), nitric oxide/cGMP (Patel et al., 2006), and protons (Gonzalez-Nieto et al., 2008) have been shown to regulate Cx35/36-mediated coupling. The effects of these molecules on photoreceptor coupling have remained largely unexplored.

3. HORIZONTAL CELLS

Horizontal cells are second-order neurons that spread laterally through the photoreceptor synaptic layer (the outer plexiform layer), generally contacting cones via dendritic processes and rods via axon terminal processes. They provide feedback inhibition to cones and rods and feedforward inhibition to bipolar cells, playing a role in producing the antagonistic surround receptive fields of bipolar and ganglion cells and providing a gain control signal at photoreceptor synapses. Horizontal cells rest at a depolarized potential in the dark, so feedback increases with hyperpolarization and has the counterintuitive effect of increasing photoreceptor synaptic vesicle release. The mechanisms of horizontal cell feedback onto photoreceptors have been controversial for several decades, but consensus suggests that a mechanism regulating proton concentration in the synaptic cleft influencing Ca^{2+} channel activation and a separate ephaptic mechanism altering the apparent voltage across the photoreceptor synaptic plasma membrane both contribute to the feedback (Thoreson and Mangel, 2012). These mechanisms combine to increase photoreceptor Ca^{2+} current and synaptic vesicle release.

Most species of mammals have at least two types of horizontal cell (Masland, 2001). All have one or more types that have a dendritic complex contacting cones and an axon terminal process contacting rods. Fig. 10.2 shows one such horizontal cell from the rabbit retina, known as the B-type horizontal cell. Rabbits also have an axonless horizontal cell, the A-type, whose dendrites contact only cones. Some species of mammals, such as mice, have just a single type of horizontal cell similar to the rabbit B-type. These mammalian horizontal cells have a largely achromatic "luminosity" type response, meaning that they hyperpolarize in response to all visible wavelengths of light (Dowling, 2012).

In contrast to mammals, most other vertebrates have several types of horizontal cell including an axonless rod-specific horizontal cell. Some fish have as many as six physiologically and spectrally identified types of horizontal cell (Connaughton and Nelson, 2010). In these species, several types of horizontal cell show chromatic contrast, responding in a biphasic, triphasic, or even tetraphasic pattern to light across the visible spectrum.

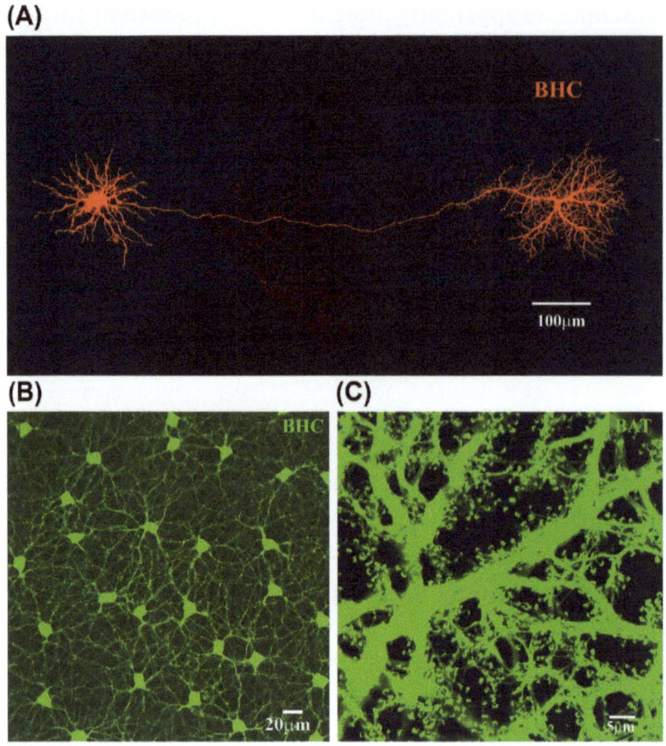

Figure 10.2 ***Morphology and coupling of axon-bearing horizontal cells in the mammalian retina.*** (A) Confocal fluorescent micrograph of a single rabbit B-type horizontal cell (BHC) injected with Neurobiotin; tracer coupling to neighbors was inhibited with meclofenamic acid. The soma and dendritic arbor, which contacts cones, are at the left, whereas the elaborate axon terminal arbor that contacts rods is at the right. (B) Tracer coupling pattern of somata injected with Neurobiotin. Note that little tracer diffuses into the axon terminal arbors as the dendritic gap junctions provide a lower resistance path than do the long, thin axons. (C) Tracer coupling pattern of directly injected axon terminals (BAT). Extensive overlap and strong coupling make it nearly impossible to resolve individual axon terminals. Small hook-shaped extensions each contact a single rod. *(Reproduced with permission from Pan, F., Keung, J., Kim, I.B., Snuggs, M.B., Mills, S.L., O'Brien, J., Massey, S.C., 2012. Connexin 57 is expressed by the axon terminal network of B-type horizontal cells in the rabbit retina. J. Comp. Neurol. 520(10), 2256–2274. http://dx.doi.org/10.1002/cne.23060.)*

In these "chromaticity" types, chromatically specific responses are attributed to cone type-selective wiring (Stell et al., 1975), with hyperpolarizing responses to colors detected by the cones providing the dominant input to the horizontal cell and depolarizing responses driven by feedback from other horizontal cell types to the contacted cones. All species also have "luminosity"-type horizontal cells that hyperpolarize in response to any visible light. The feedback and feedforward inhibition produced by these horizontal cells generate antagonistic and color-opponent surround receptive fields to many neurons, in addition to providing gain control at photoreceptor synapses and sharpening contrast of edges.

3.1 Horizontal Cell Coupling

Horizontal cells are perhaps the most extensively electrically coupled of any type of neuron. Early ultrastructural work on fish horizontal cells found very large cell-to-cell junctions among axon terminals and dendrites (O'Daly, 1967; Yamada and Ishikawa, 1965), which were thought to be similar to electrotonic junctions observed in teleost motor neurons (Pappas and Bennett, 1966). Evidence that horizontal cells were coupled through these junctions came from physiological studies that showed that potentials spread through the network of horizontal cells (Negishi, 1968), and later that there was both direct current flow between pairs and dye diffusion to neighboring horizontal cells (Kaneko, 1971). In species in which there is more than one type of horizontal cell, coupling is always cell type specific. Furthermore, the dendritic complexes and axon terminal complexes of axon-bearing horizontal cells are coupled separately, forming effectively separate networks (Vaney, 1993) (see Fig. 10.2B and C).

Horizontal cell coupling plays a major role in their physiological function. Electrical coupling is the only rapid form of intercellular communication among horizontal cells, and allows light responses to be shared across the network so that the signal used for antagonistic surrounds is integrated over a large area of the visual field. For luminosity-type horizontal cells, this is particularly important for the gain control function so that a regional average light signal is subtracted from the photoreceptor synaptic output, leaving a contrast signal. Changes in horizontal cell coupling alter the area over which this average is calculated, resulting in a more finely resolved contrast signal when coupling is low. A further important effect of coupling is that local light signals affecting a few photoreceptors are dissipated within the coupled network. A modeling study has shown that very extensive horizontal cell coupling essentially removes the feedback to a localized light stimulus from the circuit (Pandarinath et al., 2010).

Studies of horizontal cell gap junctions drove the first efforts to understand the molecular foundations of electrical synapses. Biophysical studies identified gap junction channels with conductances in the range of 40–60 pS in a number of fish species (Lu and McMahon, 1996; McMahon and Brown, 1994; McMahon et al., 1989), but distinct tracer coupling profiles in different mammalian horizontal cells (Mills and Massey, 1998) and channels with distinctly different conductance and voltage gating characteristics in fish horizontal cells (Lu et al., 1999) indicated that there was more than one type of connexin present. Early attempts to clone horizontal cell connexins from retina failed, yielding Cx43 that is present in pigmented epithelium (Wagner et al., 1998) and Cx35 that is present in nearly all neurons except horizontal cells (O'Brien et al., 1996). Work in zebrafish finally identified authentic horizontal cell connexins, including Cx55.5, Cx52.6, and Cx52.9 (Dermietzel et al., 2000; Klaassen et al., 2011; Zoidl et al., 2004) (Fig. 10.3). Mouse Cx57, a homologue of zebrafish Cx52.6, was subsequently identified in mouse horizontal cells (Hombach et al., 2004; Janssen-Bienhold et al., 2009) and axon terminals of rabbit B-type horizontal cells (Pan et al., 2012) (Fig. 10.3). In keeping with the theme of diversity, the lens connexin Cx50 was found to be present in rabbit A-type horizontal cells (O'Brien et al., 2006) (Fig. 10.3) and at a very low level in axon terminals of mouse horizontal cells (Dorgau et al., 2015). Not surprisingly, given the extensive coupling of horizontal cells, these connexins are characterized by relatively large single channel conductance. Mammalian Cx57 displays an average single channel conductance of 57 pS (Palacios-Prado et al., 2009), whereas

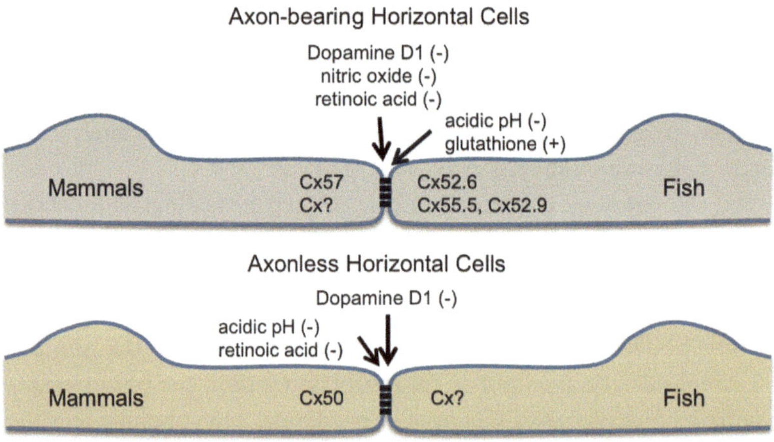

Figure 10.3 ***Signaling pathways that regulate horizontal cell electrical coupling.*** This diagram is not intended to be comprehensive, but provides a summary of signaling mechanisms that have been observed in widely used mammalian (left) and fish (right) model systems. In many cases the same signaling has been observed in other vertebrate taxa. Axon-bearing (top) and axonless (bottom) horizontal cells are treated separately based on known differences in mammals, but these differences may not be so distinct in other taxa. The connexin types expressed by horizontal cells are listed adjacent to the diagrammed gap junction. Cx? represents unknown connexins, eg, from the dendritic gap junctions of rabbit axon-bearing horizontal cells that are neither Cx57 nor Cx50 (Pan et al., 2012), or represents a lack of data. Signaling pathways that are known to regulate both mammalian and fish horizontal cell coupling are listed centered over the gap junctions, whereas those known from one or the other model system are listed over their respective sides. Dopamine-driven reduction of coupling (−) via D1 dopamine receptors has been universally observed, with exceptions noted in the text. It is likely that the other mechanisms observed in only one model system also apply to other vertebrate horizontal cells. These signaling mechanisms may interact, eg, acidic pH and dopamine D1 signaling (Hampson et al., 1994). These signaling pathways encompass a wide range of molecular mechanisms, including protein kinase signaling either linked or not linked to G-protein—coupled cascades, and several potentially direct chemical effects. Notably, the regulation by redox potential is prominent (Zhou et al., 2001), with glutathione enhancing coupling (+), and likely to be applicable to most taxa.

Cx50 produces very large channels with a conductance of 220 pS (Srinivas et al., 1999a). In contrast, Cx35/Cx36, which is used in electrical synapses of most other neurons, displays a very small conductance of 10–15 pS (Srinivas et al., 1999b; Teubner et al., 2000). Thus horizontal cell connexins are specially selected to support a very large degree of electrical coupling compared with other neurons.

3.2 Light-Adaptive Control of Horizontal Cell Coupling

Coupling among many types of horizontal cell is controlled by light adaptation. Research on this topic has been intricately intertwined with studies of the effects of dopamine on coupling. Early studies showed that dopamine enhanced center responses and reduced surround responses of fish horizontal cells, resulting in a reduction of receptive field diameter and dye spread from a recorded cell (Negishi and Drujan, 1979; Teranishi et al., 1983). A variety of studies revealed that the dopamine effects were mediated by D1 dopamine receptors that coupled to AC and increased intracellular cAMP (Lasater and Dowling, 1985; Piccolino et al., 1984; Van Buskirk and Dowling, 1981). The connection between the dopamine and light adaptation effects stemmed from ultrastructural observations that both light adaptation and dopamine application resulted in disorganization and reduction of the packing density of particles in horizontal cell gap junctions (Baldridge et al., 1987, 1989; Kurz-Isler and Wolburg, 1986; Wolburg and Kurz-Isler, 1985). This correlated well with physiological observations that light adaptation increased horizontal cell center response and reduced the receptive field diameter (Baldridge and Ball, 1991; Dong and McReynolds, 1991; Shigematsu and Yamada, 1988).

Modulation of horizontal cell coupling by D1 receptors is a conserved feature of vertebrate retinal organization (Fig. 10.3). In addition to teleost cone-driven horizontal cells discussed ealier, this has also been seen for mudpuppy (Dong and McReynolds, 1991), turtle (Weiler and Akopian, 1992), and mammalian (Hampson et al., 1994) horizontal cells. Curiously, the horizontal cells of the all-rod skate retina showed no such modulation, although they are well coupled (Qian et al., 1993; Qian and Ripps, 1992). In contrast, rod-driven horizontal cells in goldfish are regulated by dopamine D1 receptors in the same manner as the cone-driven horizontal cells (Ribelayga and Mangel, 2007). It should be noted that modulation of horizontal cell coupling is not controlled by a circadian rhythm, as is photoreceptor coupling (Ribelayga and Mangel, 2003, 2007). The D1 receptors on horizontal cells are not activated by endogenous dopamine in the dark-adapted state, even in the subjective daytime. Thus changes in coupling depend primarily on lighting conditions.

A further prominent effect of light adaptation on fish horizontal cells is the elaboration of dendritic spinules that invade cone photoreceptor synapses in the light-adapted state (Wagner, 1980). These processes are normally absent at night and require light adaptation to elaborate, but do so most efficiently in the daytime (Douglas and Wagner, 1983). The presence of spinules correlates well with the expression of feedback onto cones and depends on dopamine D1 receptors in the same way as is reduced coupling (Weiler et al., 1988; Weiler and Wagner, 1984). The similarities in the control of coupling and spinule formation suggest that the two processes have interdependent roles in horizontal cell feedback to photoreceptors. Light-adaptive uncoupling of horizontal cells reduces the size of the surround receptive field that is integrated into the feedback signal to cones, optimizing the system to enhance more local contrast features than in dim light and increasing the locally driven feedback to increase gain control.

The earliest observation of effects of light adaptation on horizontal cell coupling remains the least understood. Prolonged adaptation of the retina to darkness reduced receptive field size and caused uncoupling of fish luminosity-type horizontal cells, just as did dopamine (Mangel and Dowling, 1985; Tornqvist et al., 1988). This result is consistent with observations that goldfish horizontal cell gap junction particle density is low in darkness in the middle of the night, whereas it is significantly higher shortly before light onset, shortly after light offset, or following short-term dark adaptation in the daytime (Kurz-Isler et al., 1992). The mechanism leading to uncoupling during prolonged darkness is not known, although numerous observations indicate that it is not due to release of dopamine (Parkinson and Rando, 1983; Ribelayga and Mangel, 2003; Weiler et al., 1989). Rabbit horizontal cells also show an inverted U-shaped adaptation curve (Xin and Bloomfield, 1999), with low coupling during prolonged dark adaptation, very high coupling with scotopic background illumination, and low coupling again in bright light. As in fish horizontal cells, the bright light—induced uncoupling depends on a D1 dopamine receptor. Understanding of the uncoupling caused by prolonged darkness will require a more detailed understanding of the molecular mechanisms that control coupling.

3.3 Mechanisms of Control

Numerous studies have examined mechanisms that control horizontal cell coupling. As described earlier, a dopamine D1 receptor—driven mechanism has long been known to reduce coupling of horizontal cells. This mechanism requires activation of AC and elevation of cytoplasmic cAMP (Lasater and Dowling, 1985; Piccolino et al., 1984), and depends on activation of PKA (DeVries and Schwartz, 1989; Lasater, 1987), implicating a canonical dopamine D1 receptor pathway. Biophysical studies showed that dopamine reduced the open probability of gap junction channels without altering the single channel conductance (McMahon and Brown, 1994; McMahon et al., 1989).

Although dopamine has potent effects on horizontal cell coupling, some observations indicate that it is not the only agent that mediates light-adaptive changes in coupling. Chief among these is the observation by Baldridge and Ball that light adaptation reduces horizontal cell coupling even in retinas in which dopaminergic amacrine cells have been destroyed via intraocular injections of 6-OH-dopamine (Baldridge and Ball, 1991). Thus other pathways must contribute to light-adaptive changes in coupling. Nitric oxide has been found to reduce coupling of horizontal cells (Miyachi et al., 1990; Pottek et al., 1997; Xin and Bloomfield, 2000) (Fig. 10.3) and some horizontal cells are included among many retinal neurons that contain NADPH (reduced nicotinamide adenine dinucleotide phosphate) diaphorase and nitric oxide synthase, enzymes required for nitric oxide synthesis (Liepe et al., 1994; Osborne et al., 1993; Weiler and Kewitz, 1993). Nitric oxide produces light-adaptive changes in the morphology of photoreceptors and horizontal cell spinules (Greenstreet and Djamgoz, 1994) and physiological feedback to cones (Djamgoz et al., 2000), but curiously light adaptation reduced NADPH diaphorase activity in horizontal cells while increasing it in amacrine cells (Zemel et al., 1996). Thus, nitric oxide may contribute to light-adaptive changes in horizontal cells, but its roles appear to be complex.

Horizontal cells also display distinct regulation by retinoic acid, a light-dependent by-product of phototransduction (Fig. 10.3). This includes both reduction of horizontal cell coupling and elaboration of dendritic spinules in cone terminals that mimic light adaptation (Weiler et al., 1998). The uncoupling effect is stereospecific, requiring all-trans retinoic acid, and has been observed in mouse horizontal cells that use Cx57, in rabbit A-type horizontal cells that use Cx50, and in the dendritic network of carp horizontal cells (Pottek and Weiler, 2000; Weiler et al., 1999; Weiler et al., 1998). Curiously, the axon terminal network of carp horizontal cells showed no regulation of coupling by retinoic acid (Pottek and Weiler, 2000). Although retinoic acid has potent developmental effects through nuclear signaling (Hyatt and Dowling, 1997), its effects on horizontal cell coupling did not depend on transcription (Weiler et al., 1998). Indeed, Zhang and McMahon (2000) found that retinoic acid regulated horizontal cell gap junction channels directly, without the involvement of a retinoic acid receptor. However, this cannot explain the full light-adaptive effect of retinoic acid, as elaboration of horizontal cell spinules requires transduction of a signal.

The molecular basis of the G-protein—coupled receptor-driven changes in horizontal cell coupling is relatively poorly understood. A number of studies have examined the regulation of coupling by Cx50 in the context of lens homeostasis. Cx50 is extensively phosphorylated by protein kinase C gamma in the lens when activity is enhanced by oxidative stress (Lin et al., 2004) or phorbol ester stimulation (Zampighi et al., 2005). This reduces lens cortex dye coupling (Zampighi et al., 2005), but it is not clear whether Cx50 alone or Cx46 as well contributes to this response. In objective studies of posttranslational modifications of bovine Cx50 homologue Cx49 in lens, at least nine phosphorylation sites were identified corresponding to at least five different kinase consensus phosphorylation motifs (Shearer et al., 2008), so there is a high potential for complex regulation. There have been no studies of regulation of Cx57 by phosphorylation, which is expected to be an important mode of control because horizontal cells expressing this connexin show D1 receptor—driven plasticity. However, Cx57 has been found to be very sensitive to uncoupling by low pH with a pKa of 7.4 (Palacios-Prado et al., 2009), suggesting a potentially physiological mechanism to reduce coupling independent of protein phosphorylation. In total, the mechanisms that control horizontal cell coupling appear to be diverse, providing several means to change coupling in response to light adaptation (Fig. 10.3).

4. AMACRINE CELLS

Amacrine cells are the most diverse class of retinal neuron. A minimum estimate of about 30 morphological types has been proposed for the mammalian retina, with substantially more if stratification depth is considered (Masland, 2012). In the fish retina, this minimum estimate is about 40 types, with almost 70 identified when stratification depth is considered (Wagner and Wagner, 1988). Many of these amacrine cell types are electrically coupled, both among themselves and at times to ganglion cells or other amacrine cells, with diverse functional consequences (Bloomfield and Volgyi, 2009) (see also Trenholm and Awatramani, 2017 (Chapter 9)). For this review, we focus on just one type of amacrine cell, the mammalian AII amacrine cell. The AII amacrine cell is a central element of the mammalian rod pathway, pooling excitatory signals from presynaptic rod bipolar cells and passing these to several cone On bipolar cells via electrical synapses (see Fig. 9.1 of Trenholm and Awatramani, 2017 (Chapter 9)). AII amacrine cells are highly coupled to each other via pure electrical synapses, significantly expanding their receptive field in the well-coupled state (Bloomfield et al., 1997).

4.1 Light-Adaptive Control of AII Amacrine Cell Coupling

Plasticity of coupling is a prominent feature of AII amacrine cell physiology. Bloomfield and colleagues observed profound changes in receptive field that correlated with tracer coupling depending on background lighting conditions in rabbit AII amacrine cells (Bloomfield and Volgyi, 2004; Bloomfield et al., 1997). In complete darkness after prolonged dark adaptation, the On center receptive field size and tracer coupled patch of AII amacrine cells were both small, about twice the size of the dendritic arbor. With dim background illumination, the receptive field size and the patch of tracer-coupled cells increased to six to seven times the size of the dark-adapted On center receptive field. Under photopic background illumination, the receptive field size and tracer-coupled patch were again reduced. The resulting inverted U-shaped adaptation curve is strongly reminiscent of the physiological observations of horizontal cell coupling, implying important physiological roles for altered states of coupling during light and dark adaptation. It is generally thought that AII amacrine cell coupling allows rod signals to be pooled, increasing the sensitivity at the level of the AII amacrine cell via convergence. Dunn et al. (2006)

estimated this pooling to be about fourfold in mouse retina compared with that in $Cx36^{-/-}$ retina, based on the reduction in signal:noise ratio in the knockout.

Although the AII amacrine cell is a central element of the rod pathway, in its relatively uncoupled state during photopic light adaptation it is still functional. Manookin et al. (2008) showed that ganglion cells in the Off pathway receive On pathway inhibition that is derived from the transmission of On bipolar cell signals through gap junctions to AII amacrine cells. The inhibitory signals arise from AII glycinergic synapses onto Off ganglion cells. The physiological role of this pathway is most prominent as a disinhibition of the Off pathway upon light decrements and plays a significant role in ganglion cell responses to low contrast stimuli. The light-adaptive reduction in AII amacrine cell coupling may represent a mechanism to switch from a nighttime task integrating rare photon absorption events to a daytime task mediating local crossover inhibition.

4.2 Mechanisms of Control

AII amacrine cell coupling is regulated by dopamine in a manner very similar to that of horizontal cell coupling. Early tracer coupling studies showed that exogenously added dopamine as low as 10 nM reduced the extent of a coupled patch by more than half, with the effect increasing with dopamine concentration up to 10 μM (Hampson et al., 1992). The dopamine effect was blocked by D1 receptor antagonist SCH 23390 and mimicked by D1 receptor agonist SKF 38393, revealing its dependence on D1 dopamine receptors. Mills and Massey (1995) discovered that, although the AII amacrine to AII amacrine cell gap junctions were regulated by dopamine and cAMP signaling as previously described, the AII amacrine to cone On bipolar cell gap junctions were not. Instead, the AII to bipolar cell gap junctions were uncoupled by nitric oxide donors and membrane-permeant cGMP analogs, implicating a nitric oxide signaling pathway involving guanylate cyclase, cGMP, and protein kinase G.

Mills performed a detailed analysis of the regulatory properties of the AII—AII and AII—bipolar cell gap junctions (Xia and Mills, 2004). This work revealed that dopamine did reduce the diffusion coefficient across the AII—bipolar cell gap junctions, but at a reduced efficacy compared with that across AII—AII gap junctions. In contrast, membrane permeant cAMP analogs reduced both equally. The data could be fit with a model in which gates on each side of the gap junctions were sensitive to a cAMP-based signaling pathway. D1 dopamine receptors regulated that pathway on both sides of the AII—AII gap junctions, but on only one side of the AII—bipolar cell gap junctions. Nitric oxide, on the other hand, regulated only the AII—bipolar cell gap junctions. The study also revealed that nitric oxide did not regulate coupling uniformly among the different types of cone On bipolar cells that are coupled to AII amacrine cells. Thus nitric oxide signaling may selectively alter the relative inputs of different bipolar cells to AII amacrine cells, and vice versa.

The molecular mechanisms of AII amacrine cell electrical synaptic plasticity have been studied in detail. AII amacrine cells express Cx36 (Mills et al., 2001), which is regulated by PKA activity and connexin phosphorylation in expression systems (Ouyang et al., 2005). This is consistent with the known regulation of coupling via a dopamine D1 receptor signaling pathway that activates PKA. However, Kothmann and colleagues, using antibodies that recognize the phosphorylated regulatory sites of fish Cx35 (Kothmann et al., 2007), showed that D1 receptor and PKA activation reduced Cx36 phosphorylation, rather than increased it as was predicted (Kothmann et al., 2009). This paradoxical effect was due to activation of protein phosphatase 2A by PKA activity (Fig. 10.4). Kothmann et al. observed an essentially linear relationship between the phosphorylation state of Cx36, as represented by a ratio of phospho-Cx36 to total Cx36 immunostaining on each individual gap junction, and the diffusion coefficient for Neurobiotin through the network of coupled AII amacrine cells. This relationship held over a 20-fold dynamic range of coupling, representing most if not all of the physiological plasticity of the electrical synapse.

The dopamine D1 receptor—mediated dephosphorylation of Cx36 accounted for the photopic light adaptation phase of the AII amacrine cell adaptation curve, but did not account for the low coupling observed in dark-adapted retina. However, the unexpected role of a D1 receptor—activated phosphatase did suggest a potential mechanism: that basal activity of the phosphatase could dephosphorylate the connexin in the absence of a phosphorylating drive. In subsequent experiments, Kothmann et al. showed that such a mechanism was indeed at work. They found that prolonged dark adaptation strongly dephosphorylated Cx36. Brief exposure to light or to metabotropic glutamate receptor antagonists, mimicking light by depolarizing On bipolar cells, triggered extensive Cx36 phosphorylation (Kothmann et al., 2012).

Similar to activity-dependent potentiation of many glutamatergic synapses in the central nervous system, the activity-dependent potentiation of AII amacrine cell electrical synapses depends on NMDA (*N*-methyl-D-aspartate) receptors. AII amacrine cell gap junctions are directly colocalized with nonsynaptic NMDA receptors that sense glutamate spillover from nearby On bipolar cells. Activation of the NMDA receptors by bipolar cell activity triggers CaM kinase II phosphorylation of Cx36, opening the channels (Kothmann et al., 2012) (Fig. 10.4). Thus the absence of nearby On bipolar cell activity during prolonged dark adaptation allows the system to relax to a state in which Cx36 is poorly phosphorylated and coupling is low. At the molecular level, the inverted U-shaped adaptation curve (Fig. 10.4 bottom) arises because of the independence of the pathways that phosphorylate and dephosphorylate Cx36 and their different thresholds of activation.

The physiological significance of AII amacrine cell uncoupling in prolonged darkness is somewhat controversial. The molecular mechanisms that regulate coupling naturally lead to uncoupling in complete darkness due to residual activity of the phosphatase in the absence of activity-dependent drive to phosphorylate Cx36 (Fig. 10.4). In a sense,

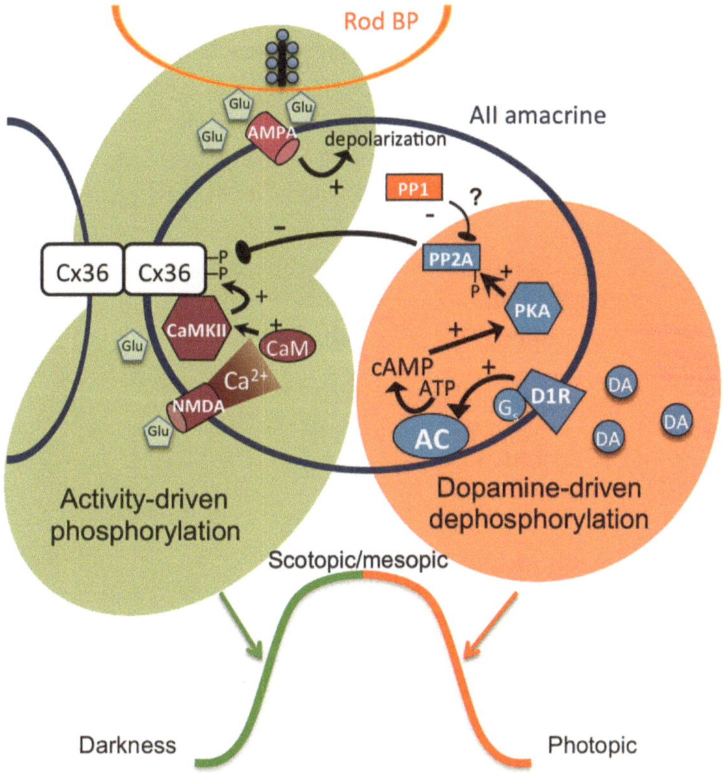

Figure 10.4 ***Signaling pathways that control AII amacrine cell coupling.*** As in photoreceptors, coupling among AII amacrine cells is controlled through a dynamic range of more than an order of magnitude by changes in Cx36 phosphorylation state. Phosphorylation of Cx36 depends on calmodulin (CaM)-dependent protein kinase II (CaMKII) activity driven by Ca^{2+} influx through nonsynaptic NMDA receptors (NMDA) that are activated by spillover glutamate (Glu) derived from light-induced activity of On pathway bipolar cells (rod, cone, or both). Rod pathway driven synaptic activation of AMPA receptors (AMPA) facilitates NMDA receptor activation through depolarizing the cells. At photopic light levels, dopamine (DA) release activates D1 dopamine receptors (D1R), activating a signaling pathway that includes adenylate cyclase (AC), cAMP, and protein kinase A (PKA). PKA in turn activates protein phosphatase 2A (PP2A), which dephosphorylates Cx36 to reduce coupling. Protein phosphatase 1 (PP1) suppresses this pathway. Residual PP2A activity presumably causes the Cx36 dephosphorylation and uncoupling in prolonged darkness in the absence of CaMKII activation.

uncoupling in prolonged darkness could be an artifact of the experimental design that is not physiologically relevant: the retina cannot see anything in the dark. However, some characteristics of signal and noise properties in the rod pathway suggest that there could be a physiological value to dark uncoupling. Scotopic vision presents a particular challenge because single photon signals are extremely small. Synaptic convergence of rod bipolar cells onto AII amacrine cells not pools only scarce photon events, but also the noise within the system. The noise increases by the square root of the number of independent noise sources: by a factor of five for an AII amacrine cell with about 30 presynaptic rod bipolar cells (Smith and Vardi, 1995). Synaptic divergence of each rod bipolar cell introduces correlated synaptic signals to a few neighboring AII amacrine cells. For any single photon event, a small amount of correlation is introduced by the divergence of each rod onto two rod bipolar cells and their subsequent convergence onto one or more neighboring AII amacrine cells. Electrical coupling of the AII amacrine cells increases the signal:noise ratio of correlated signals but decreases their amplitude, whereas active Na^+ and K^+ conductances in AII amacrine cells nonlinearly amplify the correlated signals (Smith and Vardi, 1995). For weak single photon signals, these processes are optimized with weak but nonzero coupling, suggesting that uncoupling in dim light may optimize detection of the weakest signals.

5. CONCLUDING REMARKS

The past two decades have seen tremendous progress in our understanding of the dynamic regulation of electrical coupling between neurons. Much of that progress has resulted from studies conducted in the retina, which has gradually emerged as the model system of choice to study plasticity of electrical synapses in the brain. The regulation of electrical coupling has been demonstrated in all classes of retinal neurons and the importance of light, dopamine, and a circadian clock clearly established. The regulation of Cx35/36 has been by far the most extensively investigated. Persistent changes of coupling associated with light or dark adaptation rely on intracellular pathways that control Cx35/36 phosphorylation state and thereby its junctional conductance. Although dopamine signaling plays a prominent role in this process in many retinal neurons, the regulatory mechanisms involved in the plasticity of electrical coupling have proved to be more complex and diverse than previously thought (eg, Figs. 10.1, 10.3, and 10.4). Many other modulators, such as nitric oxide, adenosine, retinoic acid, and melatonin have become acknowledged key players in the control of coupling. In addition, synaptic activity driving Ca^{2+}/calmodulin-dependent protein kinase II signaling via NMDA receptor activation also plays a major role in light-dependent plasticity in at least one neuron, the AII amacrine cell (Fig. 10.4), and is likely a much more widespread signaling mechanism.

The regulation of other coupling via retinal connexins, such as Cx50 and Cx57 expressed in horizontal cells or Cx45 expressed in some bipolar cells, has been much less studied than that of Cx36 but appears to involve similar pathways. Finally, our knowledge regarding the functional relevance of the regulation of electrical coupling remains limited, even though the complexity and diversity of the regulatory pathways strongly suggest that fine-tuning of coupling may be critical for normal retinal function and physiology. Thus the work performed during the past two decades has helped to establish the dynamic regulatory behavior of electrical synapses and reveal the variety of mechanisms that change electrical coupling. The challenge for the future lies in fully understanding the complexity of the mechanisms that regulate electrical coupling and its functional importance.

OUTSTANDING QUESTIONS

- What is the functional impact of the regulation of electrical coupling in the retina? Although the complex regulation of electrical coupling suggests that plasticity of coupling plays an important role in retinal function, our current understanding of its functional impact remains for a large part hypothetical and the experimental data sometimes appear at odds with the theory. Of note, why rods are coupled at night remains an intriguing unanswered question. Much work remains to be done to clarify the impact of the dynamic modulation of electrical coupling between elements of retinal circuits on the properties of these circuit and retinal processing.
- What are the modulators and associated pathways that control electrical synaptic plasticity in the retina? Many regulatory pathways have been proposed but few have been investigated in detail. Are all of the pathways associated with light or circadian clocks, or do other environmental or physiological influences regulate coupling? An ultimate research goal will be to identify and characterize all the signaling pathways that regulate electrical coupling in the retina.
- What are the effects of these regulatory pathways on the expression, biochemical and biophysical properties of the diverse connexins expressed in retinal cells? Although recent research has made great strides in our understanding of the regulation of Cx36 gap junctions, whether all connexins are similarly regulated by the same pathways will have to be ascertained.

REFERENCES

Abd-El-Barr, M.M., Pennesi, M.E., Saszik, S.M., Barrow, A.J., Lem, J., Bramblett, D.E., Wu, S.M., 2009. Genetic dissection of rod and cone pathways in the dark-adapted mouse retina. J. Neurophysiol. 102 (3), 1945–1955.

Armstrong-Gold, C.E., Rieke, F., 2003. Bandpass filtering at the rod to second-order cell synapse in salamander (*Ambystoma tigrinum*) retina. J. Neurosci. 23 (9), 3796–3806.

Asteriti, S., Gargini, C., Cangiano, L., 2014. Mouse rods signal through gap junctions with cones. eLife 3, e01386.

Attwell, D., Wilson, M., 1980. Behaviour of the rod network in the tiger salamander retina mediated by membrane properties of individual rods. J. Physiol. 309, 287–315.

Attwell, D., Wilson, M., Wu, S.M., 1984. A quantitative analysis of interactions between photoreceptors in the salamander (*Ambystoma*) retina. J. Physiol. 352, 703–737.

Attwell, D., 1986. The Sharpey-Schafer lecture. Ion channels and signal processing in the outer retina. Q. J. Exp. Physiol. 71 (4), 497–536.

Baba, K., Benleulmi-Chaachoua, A., Journe, A.S., Kamal, M., Guillaume, J.L., Dussaud, S., Tosini, G., 2013. Heteromeric MT_1/MT_2 melatonin receptors modulate photoreceptor function. Sci. Signal. 6 (296), ra89.

Baldridge, W.H., Ball, A.K., 1991. Background illumination reduces horizontal cell receptive-field size in both normal and 6-hydroxydopamine-lesioned goldfish retinas. Vis. Neurosci. 7 (5), 441–450.

Baldridge, W.H., Ball, A.K., Miller, R.G., 1987. Dopaminergic regulation of horizontal cell gap junction particle density in goldfish retina. J. Comp. Neurol. 265 (3), 428–436.

Baldridge, W.H., Ball, A.K., Miller, R.G., 1989. Gap junction particle density of horizontal cells in goldfish retinas lesioned with 6-OHDA. J. Comp. Neurol. 287 (2), 238–246 [published erratum appears in J. Comp. Neurol. October 22, 1989. 288(4), 698].

Barlow, R., 2001. Circadian and efferent modulation of visual sensitivity. Prog. Brain Res. 131, 487–503.

Baylor, D.A., Fuortes, M.G., O'Bryan, P.M., 1971. Receptive fields of cones in the retina of the turtle. J. Physiol. 214 (2), 265–294.
Bloomfield, S.A., Dacheux, R.F., 2001. Rod vision: pathways and processing in the mammalian retina. Prog. Retin. Eye Res. 20 (3), 351–384.
Bloomfield, S.A., Volgyi, B., 2004. Function and plasticity of homologous coupling between AII amacrine cells. Vis. Res. 44 (28), 3297–3306.
Bloomfield, S.A., Volgyi, B., 2009. The diverse functional roles and regulation of neuronal gap junctions in the retina. Nat. Rev. Neurosci. 10 (7), 495–506.
Bloomfield, S.A., Xin, D., Osborne, T., 1997. Light-induced modulation of coupling between AII amacrine cells in the rabbit retina. Vis. Neurosci. 14 (3), 565–576.
Burkhardt, D.A., 1977. Responses and receptive-field organization of cones in perch retinas. J. Neurophysiol. 40 (1), 53–62.
Cachope, R., Mackie, K., Triller, A., O'Brien, J., Pereda, A.E., 2007. Potentiation of electrical and chemical synaptic transmission mediated by endocannabinoids. Neuron 56 (6), 1034–1047.
Cohen, A.I., Todd, R.D., Harmon, S., O'Malley, K.L., 1992. Photoreceptors of mouse retinas possess D4 receptors coupled to adenylate cyclase. Proc. Natl. Acad. Sci. U.S.A. 89 (24), 12093–12097.
Cohen, A.I., 1963. The fine structure of the visual receptors of the pigeon. Exp. Eye Res. 2, 88–97.
Cohen, A.I., 1964. Some observations on the fine structure of the retinal receptors of the American gray squirrel. Invest. Ophthalmol. 3, 198–216.
Cohen, A.I., 1965. Some electron microscopic observations on inter-receptor contacts in the human and macaque retinae. J. Anat. 99 (3), 595–610.
Connaughton, V.P., Nelson, R., 2010. Spectral responses in zebrafish horizontal cells include a tetraphasic response and a novel UV-dominated triphasic response. J. Neurophysiol. 104 (5), 2407–2422.
Copenhagen, D.R., Owen, W.G., 1976. Functional characteristics of lateral interactions between rods in the retina of the snapping turtle. J. Physiol. 259 (2), 251–282.
Dacheux, R.F., Raviola, E., 1982. Horizontal cells in the retina of the rabbit. J. Neurosci. 2 (10), 1486–1493.
Deans, M.R., Volgyi, B., Goodenough, D.A., Bloomfield, S.A., Paul, D.L., 2002. Connexin36 is essential for transmission of rod-mediated visual signals in the mammalian retina. Neuron 36 (4), 703–712.
Dermietzel, R., Kremer, M., Paputsoglu, G., Stang, A., Skerrett, I.M., Gomes, D., Spray, D.C., 2000. Molecular and functional diversity of neural connexins in the retina. J. Neurosci. 20 (22), 8331–8343.
Detwiler, P.B., Hodgkin, A.L., McNaughton, P.A., 1978. A surprising property of electrical spread in the network of rods in the turtle's retina. Nature 274 (5671), 562–565.
DeVries, S.H., Schwartz, E.A., 1989. Modulation of an electrical synapse between solitary pairs of catfish horizontal cells by dopamine and second messengers. J. Physiol. 414, 351–375.
DeVries, S.H., Qi, X., Smith, R., Makous, W., Sterling, P., 2002. Electrical coupling between mammalian cones. Curr. Biol. 12 (22), 1900–1907.
Djamgoz, M.B., Sekaran, S., Angotzi, A.R., Haamedi, S., Vallerga, S., Hirano, J., Yamada, M., 2000. Light-adaptive role of nitric oxide in the outer retina of lower vertebrates: a brief review. Philos. Trans. R. Soc. Lond B Biol. Sci. 355 (1401), 1199–1203.
Dong, C.J., McReynolds, J.S., 1991. The relationship between light, dopamine release and horizontal cell coupling in the mudpuppy retina. J. Physiol. 440, 291–309.
Dorgau, B., Herrling, R., Schultz, K., Greb, H., Segelken, J., Stroh, S., Janssen-Bienhold, U., 2015. Connexin50 couples axon terminals of mouse horizontal cells by homotypic gap junctions. J. Comp. Neurol. 523 (14), 2062–2081.
Douglas, R.H., Wagner, H.J., 1983. Endogenous control of spinule formation in horizontal cells of the teleost retina. Cell Tissue Res. 229 (2), 443–449.
Dowling, J.E., 2012. The Retina: An Approachable Part of the Brain (Revised Edition). Belknap of Harvard UP, Cambridge, MA.
Doyle, S.E., Grace, M.S., McIvor, W., Menaker, M., 2002. Circadian rhythms of dopamine in mouse retina: the role of melatonin. Vis. Neurosci. 19 (5), 593–601.

Dunn, F.A., Doan, T., Sampath, A.P., Rieke, F., 2006. Controlling the gain of rod-mediated signals in the mammalian retina. J. Neurosci. 26 (15), 3959–3970.
Fain, G.L., 1975. Quantum sensitivity of rods in the toad retina. Science 187 (4179), 838–841.
Fain, G.L., 1976. Sensitivity of toad rods: dependence on wave-length and background illumination. J. Physiol. 261 (1), 71–101.
Field, G.D., Sampath, A.P., Rieke, F., 2005. Retinal processing near absolute threshold: from behavior to mechanism. Annu. Rev. Physiol. 67, 491–514.
Fredholm, B.B., IJzerman, A.P., Jacobson, K.A., Linden, J., Muller, C.E., 2011. International union of basic and clinical pharmacology. LXXXI. Nomenclature and classification of adenosine receptors—an update. Pharmacol. Rev. 63 (1), 1–34.
Fuortes, M.G.F., 1972. Responses of cones and horizontal cells in the retina of the turtle. Invest. Ophthalmol. 11 (5), 275–284.
Gao, F., Pang, J.J., Wu, S.M., 2013. Sign-preserving and sign-inverting synaptic interactions between rod and cone photoreceptors in the dark-adapted retina. J. Physiol. 591 (Pt 22), 5711–5726.
Gold, G.H., 1979. Photoreceptor coupling in retina of the toad, *Bufo marinus*. II. Physiology. J. Neurophysiol. 42 (1 Pt 1), 311–328.
Gonzalez-Nieto, D., Gomez-Hernandez, J.M., Larrosa, B., Gutierrez, C., Munoz, M.D., Fasciani, I., Barrio, L.C., 2008. Regulation of neuronal connexin-36 channels by pH. Proc. Natl. Acad. Sci. U.S.A. 105 (44), 17169–17174.
Green, C.B., Besharse, J.C., 2004. Retinal circadian clocks and control of retinal physiology. J. Biol. Rhythms 19 (2), 91–102.
Greenstreet, E.H., Djamgoz, M.B., 1994. Nitric oxide induces light-adaptive morphological changes in retinal neurones. Neuroreport 6 (1), 109–112.
Hampson, E.C., Vaney, D.I., Weiler, R., 1992. Dopaminergic modulation of gap junction permeability between amacrine cells in mammalian retina. J. Neurosci. 12 (12), 4911–4922.
Hampson, E.C., Weiler, R., Vaney, D.I., 1994. pH-gated dopaminergic modulation of horizontal cell gap junctions in mammalian retina. Proc. Biol. Sci. 255 (1342), 67–72.
Hestrin, S., 1987. The properties and function of inward rectification in rod photoreceptors of the tiger salamander. J. Physiol. 390, 319–333.
Hombach, S., Janssen-Bienhold, U., Sohl, G., Schubert, T., Bussow, H., Ott, T., Willecke, K., 2004. Functional expression of connexin57 in horizontal cells of the mouse retina. Eur. J. Neurosci. 19 (10), 2633–2640.
Hornstein, E.P., Verweij, J., Schnapf, J.L., 2004. Electrical coupling between red and green cones in primate retina. Nat. Neurosci. 7 (7), 745–750.
Hornstein, E.P., Verweij, J., Li, P.H., Schnapf, J.L., 2005. Gap-junctional coupling and absolute sensitivity of photoreceptors in macaque retina. J. Neurosci. 25 (48), 11201–11209.
Hsu, A., Smith, R.G., Buchsbaum, G., Sterling, P., 2000. Cost of cone coupling to trichromacy in primate fovea. J. Opt. Soc. Am. A Opt. Image Sci. Vis. 17 (3), 635–640.
Hyatt, G.A., Dowling, J.E., 1997. Retinoic acid. A key molecule for eye and photoreceptor development. Invest Ophthalmol. Vis. Sci. 38 (8), 1471–1475.
Iuvone, P.M., Tosini, G., Pozdeyev, N., Haque, R., Klein, D.C., Chaurasia, S.S., 2005. Circadian clocks, clock networks, arylalkylamine *N*-acetyltransferase, and melatonin in the retina. Prog. Retin. Eye Res. 24 (4), 433–456.
Jackson, C.R., Chaurasia, S.S., Hwang, C.K., Iuvone, P.M., 2011. Dopamine D4 receptor activation controls circadian timing of the adenylyl cyclase 1/cyclic AMP signaling system in mouse retina. Eur. J. Neurosci. 34 (1), 57–64.
Janssen-Bienhold, U., Trumpler, J., Hilgen, G., Schultz, K., Muller, L.P., Sonntag, S., Weiler, R., 2009. Connexin57 is expressed in dendro-dendritic and axo-axonal gap junctions of mouse horizontal cells and its distribution is modulated by light. J. Comp. Neurol. 513 (4), 363–374.
Jin, N., Ribelayga, C., 2014. Melatonin modulates rod photoreceptor electrical coupling in the mouse retina. Inves. Ophthalmol. Vis. Sci. 55 (13) eAbstract 5006.

Jin, N.G., Chuang, A.Z., Masson, P.J., Ribelayga, C.P., 2015. Rod electrical coupling is controlled by a circadian clock and dopamine in mouse retina. J. Physiol. 593 (7), 1597–1631.

Kaneko, A., 1971. Electrical connexions between horizontal cells in the dogfish retina. J. Physiol. 213 (1), 95–105.

Katti, C., Butler, R., Sekaran, S., 2013. Diurnal and circadian regulation of connexin 36 transcript and protein in the mammalian retina. Invest. Ophthalmol. Vis. Sci. 54 (1), 821–829.

Klaassen, L.J., Sun, Z., Steijaert, M.N., Bolte, P., Fahrenfort, I., Sjoerdsma, T., Kamermans, M., 2011. Synaptic transmission from horizontal cells to cones is impaired by loss of connexin hemichannels. PLoS Biol. 9 (7), e1001107.

Kothmann, W.W., Li, X., Burr, G.S., O'Brien, J., 2007. Connexin 35/36 is phosphorylated at regulatory sites in the retina. Vis. Neurosci. 24 (3), 363–375.

Kothmann, W.W., Massey, S.C., O'Brien, J., 2009. Dopamine-stimulated dephosphorylation of connexin 36 mediates AII amacrine cell uncoupling. J. Neurosci. 29 (47), 14903–14911.

Kothmann, W.W., Trexler, E.B., Whitaker, C.M., Li, W., Massey, S.C., O'Brien, J., 2012. Nonsynaptic NMDA receptors mediate activity-dependent plasticity of gap junctional coupling in the AII amacrine cell network. J. Neurosci. 32 (20), 6747–6759.

Krizaj, D., Gabriel, R., Owen, W.G., Witkovsky, P., 1998. Dopamine D2 receptor-mediated modulation of rod-cone coupling in the *Xenopus* retina. J. Comp. Neurol. 398 (4), 529–538.

Kull, B., Ferre, S., Arslan, G., Svenningsson, P., Fuxe, K., Owman, C., Fredholm, B.B., 1999. Reciprocal interactions between adenosine A_{2A} and dopamine D_2 receptors in Chinese hamster ovary cells co-transfected with the two receptors. Biochem. Pharmacol. 58 (6), 1035–1045.

Kurz-Isler, G., Wolburg, H., 1986. Gap junctions between horizontal cells in the cyprinid fish alter rapidly their structure during light and dark adaptation. Neurosci. Lett. 67 (1), 7–12.

Kurz-Isler, G., Voigt, T., Wolburg, H., 1992. Modulation of connexon densities in gap junctions of horizontal cell perikarya and axon terminals in fish retina: effects of light/dark cycles, interruption of the optic nerve and application of dopamine. Cell Tissue Res. 268 (2), 267–275.

Lamb, T.D., Simon, E.J., 1976. The relation between intercellular coupling and electrical noise in turtle photoreceptors. J. Physiol. 263 (2), 257–286.

Lasansky, A., 1972. Cell junctions at the outer synaptic layer of the retina. Invest. Ophthalmol. 11 (5), 265–275.

Lasater, E.M., Dowling, J.E., 1985. Dopamine decreases conductance of the electrical junctions between cultured retinal horizontal cells. Proc. Natl. Acad. Sci. U.S.A. 82 (9), 3025–3029.

Lasater, E.M., 1987. Retinal horizontal cell gap junctional conductance is modulated by dopamine through a cyclic AMP-dependent protein kinase. Proc. Natl. Acad. Sci. U.S.A. 84 (20), 7319–7323.

Lebedev, D.S., Byzov, A.L., Govardovskii, V.I., 1998. Photoreceptor coupling and boundary detection. Vis. Res. 38 (20), 3161–3169.

Leeper, H.F., Normann, R.A., Copenhagen, D.R., 1978. Evidence for passive electrotonic interactions in red rods of toad retina. Nature 275 (5677), 234–236.

Li, W., DeVries, S.H., 2004. Separate blue and green cone networks in the mammalian retina. Nat. Neurosci. 7 (7), 751–756.

Li, H., Chuang, A.Z., O'Brien, J., 2009. Photoreceptor coupling is controlled by connexin 35 phosphorylation in zebrafish retina. J. Neurosci. 29 (48), 15178–15186.

Li, P.H., Verweij, J., Long, J.H., Schnapf, J.L., 2012. Gap-junctional coupling of mammalian rod photoreceptors and its effect on visual detection. J. Neurosci. 32 (10), 3552–3562.

Li, H., Zhang, Z., Blackburn, M.R., Wang, S.W., Ribelayga, C.P., O'Brien, J., 2013. Adenosine and dopamine receptors coregulate photoreceptor coupling via gap junction phosphorylation in mouse retina. J. Neurosci. 33 (7), 3135–3150.

Li, H., Chuang, A.Z., O'Brien, J., 2014. Regulation of photoreceptor gap junction phosphorylation by adenosine in zebrafish retina. Vis. Neurosci. 31 (3), 237–243.

Liepe, B.A., Stone, C., Koistinaho, J., Copenhagen, D.R., 1994. Nitric oxide synthase in Muller cells and neurons of salamander and fish retina. J. Neurosci. 14 (12), 7641–7654.

Lin, D., Lobell, S., Jewell, A., Takemoto, D.J., 2004. Differential phosphorylation of connexin46 and connexin50 by H_2O_2 activation of protein kinase Cγ. Mol. Vis. 10, 688–695.

Liu, X., Zhang, Z., Ribelayga, C.P., 2012. Heterogeneous expression of the core circadian clock proteins among neuronal cell types in mouse retina. PLoS One 7 (11), e50602.

Lu, C., McMahon, D.G., 1996. Gap junction channel gating at bass retinal electrical synapses. Vis. Neurosci. 13 (6), 1049–1057.

Lu, C., Zhang, D.Q., McMahon, D.G., 1999. Electrical coupling of retinal horizontal cells mediated by distinct voltage-independent junctions. Vis. Neurosci. 16 (5), 811–818.

Mangel, S.C., Dowling, J.E., 1985. Responsiveness and receptive field size of carp horizontal cells are reduced by prolonged darkness and dopamine. Science 229 (4718), 1107–1109.

Mangel, S.C., Ribelayga, C.P., 2010. The circadian clock in the retina regulates rod and cone pathways. In: Besharse, J., Dartt, D.A., Dana, R. (Eds.), Encyclopedia of the Eye. Elsevier, Oxford, UK, pp. 283–289.

Manookin, M.B., Beaudoin, D.L., Ernst, Z.R., Flagel, L.J., Demb, J.B., 2008. Disinhibition combines with excitation to extend the operating range of the off visual pathway in daylight. J. Neurosci. 28 (16), 4136–4150.

Masland, R.H., 2001. The fundamental plan of the retina. Nat. Neurosci. 4 (9), 877–886.

Masland, R.H., 2012. The neuronal organization of the retina. Neuron 76 (2), 266–280.

McMahon, D.G., Brown, D.R., 1994. Modulation of gap-junction channel gating at zebrafish retinal electrical synapses. J. Neurophysiol. 72 (5), 2257–2268.

McMahon, D.G., Knapp, A.G., Dowling, J.E., 1989. Horizontal cell gap junctions: single-channel conductance and modulation by dopamine. Proc. Natl. Acad. Sci. U.S.A. 86 (19), 7639–7643.

McMahon, D.G., Iuvone, P.M., Tosini, G., 2014. Circadian organization of the mammalian retina: from gene regulation to physiology and diseases. Prog. Retin. Eye Res. 39, 58–76.

Mills, S.L., Massey, S.C., 1995. Differential properties of two gap junctional pathways made by AII amacrine cells. Nature 377 (6551), 734–737.

Mills, S.L., Massey, S.C., 1998. The kinetics of tracer movement through homologous gap junctions in the rabbit retina. Vis. Neurosci. 15 (4), 765–777.

Mills, S.L., O'Brien, J.J., Li, W., O'Brien, J., Massey, S.C., 2001. Rod pathways in the mammalian retina use connexin36. J. Comp. Neurol. 436 (3), 336–350.

Missotten, L., Appelmans, M., Michiels, J., 1963. Ultrastructure of the synapses of the visual cells of the human retina. Bull. Mem. Soc. Fr. Ophtalmol. 76, 59–82.

Miyachi, E., Murakami, M., Nakaki, T., 1990. Arginine blocks gap junctions between retinal horizontal cells. Neuroreport 1 (2), 107–110.

Negishi, K., Drujan, B.D., 1979. Reciprocal changes in center and surrounding S potentials of fish retina in response to dopamine. Neurochem. Res. 4 (3), 313–318.

Negishi, K., 1968. Excitation spread along horizontal and amacrine cell layers in the teleost retina. Nature 218 (5136), 39–40.

Nelson, R., 1977. Cat cones have rod input: a comparison of the response properties of cones and horizontal cell bodies in the retina of the cat. J. Comp. Neurol. 172 (1), 109–135.

Nir, I., Harrison, J.M., Haque, R., Low, M.J., Grandy, D.K., Rubinstein, M., Iuvone, P.M., 2002. Dysfunctional light-evoked regulation of cAMP in photoreceptors and abnormal retinal adaptation in mice lacking dopamine D4 receptors. J. Neurosci. 22 (6), 2063–2073.

Normann, R.A., Perlman, I., Kolb, H., Jones, J., Daly, S.J., 1984. Direct excitatory interactions between cones of different spectral types in the turtle retina. Science 224 (4649), 625–627.

O'Brien, J., Al-Ubaidi, M.R., Ripps, H., 1996. Connexin 35: a gap-junctional protein expressed preferentially in the skate retina. Mol. Biol. Cell 7 (2), 233–243.

O'Brien, J., Nguyen, H.B., Mills, S.L., 2004. Cone photoreceptors in bass retina use two connexins to mediate electrical coupling. J. Neurosci. 24 (24), 5632–5642.

O'Brien, J.J., Li, W., Pan, F., Keung, J., O'Brien, J., Massey, S.C., 2006. Coupling between A-type horizontal cells is mediated by connexin 50 gap junctions in the rabbit retina. J. Neurosci. 26 (45), 11624–11636.

O'Brien, J.J., Chen, X., Macleish, P.R., O'Brien, J., Massey, S.C., 2012. Photoreceptor coupling mediated by connexin36 in the primate retina. J. Neurosci. 32 (13), 4675–4687.

O'Daly, J.A., 1967. ATPase activity at the functional contacts between retinal cells which produced S-potential. Nature 216 (5122), 1329–1331.

Osborne, N.N., Barnett, N.L., Herrera, A.J., 1993. NADPH diaphorase localization and nitric oxide synthetase activity in the retina and anterior uvea of the rabbit eye. Brain Res. 610 (2), 194–198.

Ouyang, X., Winbow, V.M., Patel, L.S., Burr, G.S., Mitchell, C.K., O'Brien, J., 2005. Protein kinase A mediates regulation of gap junctions containing connexin35 through a complex pathway. Brain Res. Mol. Brain Res. 135 (1–2), 1–11.

Palacios-Prado, N., Sonntag, S., Skeberdis, V.A., Willecke, K., Bukauskas, F.F., 2009. Gating, permselectivity and pH-dependent modulation of channels formed by connexin57, a major connexin of horizontal cells in the mouse retina. J. Physiol. 587 (Pt 13), 3251–3269.

Pan, F., Keung, J., Kim, I.B., Snuggs, M.B., Mills, S.L., O'Brien, J., Massey, S.C., 2012. Connexin 57 is expressed by the axon terminal network of B-type horizontal cells in the rabbit retina. J. Comp. Neurol. 520 (10), 2256–2274.

Pandarinath, C., Bomash, I., Victor, J.D., Prusky, G.T., Tschetter, W.W., Nirenberg, S., 2010. A novel mechanism for switching a neural system from one state to another. Front. Comput. Neurosci. 4, 2.

Pappas, G.D., Bennett, M.V., 1966. Specialized junctions involved in electrical transmission between neurons. Ann. N. Y. Acad. Sci. 137 (2), 495–508.

Parkinson, D., Rando, R.R., 1983. Effect of light on dopamine turnover and metabolism in rabbit retina. Invest. Ophthalmol. Vis. Sci. 24 (3), 384–388.

Patel, L.S., Mitchell, C.K., Dubinsky, W.P., O'Brien, J., 2006. Regulation of gap junction coupling through the neuronal connexin Cx35 by nitric oxide and cGMP. Cell Commun. Adhes. 13 (1–2), 41–54.

Piccolino, M., Neyton, J., Gerschenfeld, H.M., 1984. Decrease of gap junction permeability induced by dopamine and cyclic adenosine 3′:5′-monophosphate in horizontal cells of turtle retina. J. Neurosci. 4 (10), 2477–2488.

Pottek, M., Weiler, R., 2000. Light-adaptive effects of retinoic acid on receptive field properties of retinal horizontal cells. Eur. J. Neurosci. 12 (2), 437–445.

Pottek, M., Schultz, K., Weiler, R., 1997. Effects of nitric oxide on the horizontal cell network and dopamine release in the carp retina. Vis. Res. 37 (9), 1091–1102.

Qian, H., Ripps, H., 1992. Receptive field properties of rod-driven horizontal cells in the skate retina. J. Gen. Physiol. 100 (3), 457–478.

Qian, H., Malchow, R.P., Ripps, H., 1993. Gap-junctional properties of electrically coupled skate horizontal cells in culture. Vis. Neurosci. 10 (2), 287–295.

Raviola, E., Gilula, N.B., 1973. Gap junctions between photoreceptor cells in the vertebrate retina. Proc. Natl. Acad. Sci. U.S.A. 70 (6), 1677–1681.

Raviola, E., Gilula, N.B., 1975. Intramembrane organization of specialized contacts in the outer plexiform layer of the retina. A freeze-fracture study in monkeys and rabbits. J. Cell Biol. 65 (1), 192–222.

Ribelayga, C., Mangel, S.C., 2003. Absence of circadian clock regulation of horizontal cell gap junctional coupling reveals two dopamine systems in the goldfish retina. J. Comp. Neurol. 467 (2), 243–253.

Ribelayga, C., Mangel, S.C., 2005. A circadian clock and light/dark adaptation differentially regulate adenosine in the mammalian retina. J. Neurosci. 25 (1), 215–222.

Ribelayga, C., Mangel, S.C., 2007. Tracer coupling between fish rod horizontal cells: modulation by light and dopamine but not the retinal circadian clock. Vis. Neurosci. 24 (3), 333–344.

Ribelayga, C., Mangel, S.C., 2010. Identification of a circadian clock-controlled neural pathway in the rabbit retina. PLoS One 5 (6), e11020.

Ribelayga, C., Wang, Y., Mangel, S.C., 2002. Dopamine mediates circadian clock regulation of rod and cone input to fish retinal horizontal cells. J. Physiol. 544 (Pt 3), 801–816.

Ribelayga, C., Cao, Y., Mangel, S.C., 2008. The circadian clock in the retina controls rod-cone coupling. Neuron 59 (5), 790–801.

Roseboom, P.H., Namboodiri, M.A., Zimonjic, D.B., Popescu, N.C., Rodriguez, I.R., Gastel, J.A., Klein, D.C., 1998. Natural melatonin 'knockdown' in C57bl/6J mice: rare mechanism truncates serotonin *N*-acetyltransferase. Brain Res. Mol. Brain Res. 63 (1), 189–197.

Schneeweis, D.M., Schnapf, J.L., 1995. Photovoltage of rods and cones in the macaque retina. Science 268 (5213), 1053–1056.

Schneeweis, D.M., Schnapf, J.L., 1999. The photovoltage of macaque cone photoreceptors: adaptation, noise, and kinetics. J. Neurosci. 19 (4), 1203–1216.

Schwartz, E.A., 1973. Responses of single rods in the retina of the turtle. J. Physiol. 232 (3), 503–514.

Schwartz, E.A., 1975a. Cones excite rods in the retina of the turtle. J. Physiol. 246 (3), 639–651.

Schwartz, E.A., 1975b. Rod-rod interaction in the retina of the turtle. J. Physiol. 246 (3), 617–638.

Shearer, D., Ens, W., Standing, K., Valdimarsson, G., 2008. Posttranslational modifications in lens fiber connexins identified by off-line-HPLC MALDI-quadrupole time-of-flight mass spectrometry. Invest. Ophthalmol. Vis. Sci. 49 (4), 1553–1562.

Shigematsu, Y., Yamada, M., 1988. Effects of dopamine on spatial properties of horizontal cell responses in the carp retina. Neurosci. Res. Suppl. 8, S69–S80.

Sjostrand, F.S., 1958. Ultrastructure of retinal rod synapses of the guinea pig eye as revealed by three-dimensional reconstructions from serial sections. J. Ultrastruct. Res. 2 (1), 122–170.

Smith, R.G., Vardi, N., 1995. Simulation of the AII amacrine cell of mammalian retina: functional consequences of electrical coupling and regenerative membrane properties. Vis. Neurosci. 12 (5), 851–860.

Srinivas, M., Costa, M., Gao, Y., Fort, A., Fishman, G.I., Spray, D.C., 1999a. Voltage dependence of macroscopic and unitary currents of gap junction channels formed by mouse connexin50 expressed in rat neuroblastoma cells. J. Physiol. 517 (Pt 3), 673–689.

Srinivas, M., Rozental, R., Kojima, T., Dermietzel, R., Mehler, M., Condorelli, D.F., Spray, D.C., 1999b. Functional properties of channels formed by the neuronal gap junction protein connexin36. J. Neurosci. 19 (22), 9848–9855.

Stell, W.K., Lightfood, D.O., Wheeler, T.G., Leeper, H.F., 1975. Goldfish retina: functional polarization of cone horizontal cell dendrites and synapses. Science 190 (4218), 989–990.

Stell, W.K., 1967. The structure and relationships of horizontal cells and photoreceptor-bipolar synaptic complexes in goldfish retina. Am. J. Anat. 121 (2), 401–423.

Storch, K.F., Paz, C., Signorovitch, J., Raviola, E., Pawlyk, B., Li, T., Weitz, C.J., 2007. Intrinsic circadian clock of the mammalian retina: importance for retinal processing of visual information. Cell 130 (4), 730–741.

Teranishi, T., Negishi, K., Kato, S., 1983. Dopamine modulates S-potential amplitude and dye-coupling between external horizontal cells in carp retina. Nature 301 (5897), 243–246.

Tessier-Lavigne, M., Attwell, D., 1988. The effect of photoreceptor coupling and synapse nonlinearity on signal:noise ratio in early visual processing. Proc. R. Soc. Lond. B Biol. Sci. 234 (1275), 171–197.

Teubner, B., Degen, J., Sohl, G., Guldenagel, M., Bukauskas, F.F., Trexler, E.B., Willecke, K., 2000. Functional expression of the murine connexin 36 gene coding for a neuron-specific gap junctional protein. J. Membr. Biol. 176 (3), 249–262.

Thoreson, W.B., Mangel, S.C., 2012. Lateral interactions in the outer retina. Prog. Retin. Eye Res. 31 (5), 407–441.

Tomita, T., Kaneko, A., Murakami, M., Pautler, E.L., 1967. Spectral response curves of single cones in the carp. Vis. Res. 7 (7), 519–531.

Tornqvist, K., Yang, X.L., Dowling, J.E., 1988. Modulation of cone horizontal cell activity in the teleost fish retina. III. Effects of prolonged darkness and dopamine on electrical coupling between horizontal cells. J. Neurosci. 8 (7), 2279–2288.

Trenholm, S., Awatramani, G.B., 2017. Dynamic properties of electrically coupled retinal networks. In: Jing, J. (Ed.), Network Functions and Plasticity. Elsevier (Chapter 9).

Trexler, E.B., Li, W., Massey, S.C., 2005. Simultaneous contribution of two rod pathways to AII amacrine and cone bipolar cell light responses. J. Neurophysiol. 93 (3), 1476–1485.

Tsukamoto, Y., Morigiwa, K., Ueda, M., Sterling, P., 2001. Microcircuits for night vision in mouse retina. J. Neurosci. 21 (21), 8616–8623.

Van Buskirk, R., Dowling, J.E., 1981. Isolated horizontal cells from carp retina demonstrate dopamine-dependent accumulation of cyclic amp. Proc. Natl. Acad. Sci. U.S.A. 78 (12), 7825–7829.

Vaney, D.I., 1993. The coupling pattern of axon-bearing horizontal cells in the mammalian retina. Proc. R. Soc. Lond. B Biol. Sci. 252 (1334), 93–101.

Volgyi, B., Deans, M.R., Paul, D.L., Bloomfield, S.A., 2004. Convergence and segregation of the multiple rod pathways in mammalian retina. J. Neurosci. 24 (49), 11182–11192.

Volgyi, B., Kovacs-Oller, T., Atlasz, T., Wilhelm, M., Gabriel, R., 2013. Gap junctional coupling in the vertebrate retina: variations on one theme? Prog. Retin. Eye Res. 34, 1–18.

Wagner, H.J., Wagner, E., 1988. Amacrine cells in the retina of a teleost fish, the roach (*Rutilus rutilus*): a golgi study on differentiation and layering. Philos. Trans. R. Soc. Lond B Biol. Sci. 321 (1206), 263–324.

Wagner, T.L., Beyer, E.C., McMahon, D.G., 1998. Cloning and functional expression of a novel gap junction channel from the retina of *Danio aquipinnatus*. Vis. Neurosci. 15 (6), 1137–1144.

Wagner, H.J., 1980. Light-dependent plasticity of the morphology of horizontal cell terminals in cone pedicles of fish retinas. J. Neurocytol. 9 (5), 573–590.

Wang, Y., Mangel, S.C., 1996. A circadian clock regulates rod and cone input to fish retinal cone horizontal cells. Proc. Natl. Acad. Sci. U.S.A. 93 (10), 4655–4660.

Weiler, R., Akopian, A., 1992. Effects of background illuminations on the receptive field size of horizontal cells in the turtle retina are mediated by dopamine. Neurosci. Lett. 140 (1), 121–124.

Weiler, R., Kewitz, B., 1993. The marker for nitric oxide synthase, NADPH-diaphorase, co-localizes with gaba in horizontal cells and cells of the inner retina in the carp retina. Neurosci. Lett. 158 (2), 151–154.

Weiler, R., Wagner, H.J., 1984. Light-dependent change of cone-horizontal cell interactions in carp retina. Brain Res. 298 (1), 1–9.

Weiler, R., Kohler, K., Kolbinger, W., Wolburg, H., Kurz-Isler, G., Wagner, H.J., 1988. Dopaminergic neuromodulation in the retinas of lower vertebrates. Neurosci. Res. Suppl. 8 (96), S183–S196.

Weiler, R., Kolbinger, W., Kohler, K., 1989. Reduced light responsiveness of the cone pathway during prolonged darkness does not result from an increase of dopaminergic activity in the fish retina. Neurosci. Lett. 99 (1–2), 214–218.

Weiler, R., Schultz, K., Pottek, M., Tieding, S., Janssen-Bienhold, U., 1998. Retinoic acid has light-adaptive effects on horizontal cells in the retina. Proc. Natl. Acad. Sci. U.S.A. 95 (12), 7139–7144.

Weiler, R., He, S., Vaney, D.I., 1999. Retinoic acid modulates gap junctional permeability between horizontal cells of the mammalian retina. Eur. J. Neurosci. 11 (9), 3346–3350.

Werblin, F.S., 1978. Transmission along and between rods in the tiger salamander retina. J. Physiol. 280, 449–470.

Witkovsky, P., Shakib, M., Ripps, H., 1974. Interreceptoral junctions in the teleost retina. Invest. Ophthalmol. 13 (12), 996–1009.

Witkovsky, P., 2004. Dopamine and retinal function. Doc. Ophthalmol. 108 (1), 17–40.

Wolburg, H., Kurz-Isler, G., 1985. Dynamics of gap junctions between horizontal cells in the goldfish retina. Exp. Brain Res. 60 (2), 397–401.

Wu, S.M., Yang, X.L., 1988. Electrical coupling between rods and cones in the tiger salamander retina. Proc. Natl. Acad. Sci. U.S.A. 85 (1), 275–278.

Xia, X., Mills, S., 2004. Gap junctional regulatory mechanisms in the AII amacrine cell of the rabbit retina. Vis. Neurosci. 21 (5), 791–805.

Xin, D., Bloomfield, S.A., 1999. Dark- and light-induced changes in coupling between horizontal cells in mammalian retina. J. Comp. Neurol. 405 (1), 75–87.

Xin, D., Bloomfield, S.A., 2000. Effects of nitric oxide on horizontal cells in the rabbit retina. Vis. Neurosci. 17 (5), 799–811.

Yamada, E., Ishikawa, T., 1965. The fine structure of the horizontal cells in some vertebrate retinae. Cold Spring Harb. Symp. Quant. Biol. 30, 383–392.

Yang, X.L., Wu, S.M., 1989. Modulation of rod-cone coupling by light. Science 244 (4902), 352–354.

Zampighi, G.A., Planells, A.M., Lin, D., Takemoto, D., 2005. Regulation of lens cell-to-cell communication by activation of PKCγ and disassembly of Cx50 channels. Invest. Ophthalmol. Vis. Sci. 46 (9), 3247–3255.

Zemel, E., Eyal, O., Lei, B., Perlman, I., 1996. NADPH diaphorase activity in mammalian retinas is modulated by the state of visual adaptation. Vis. Neurosci. 13 (5), 863–871.

Zhang, D.Q., McMahon, D.G., 2000. Direct gating by retinoic acid of retinal electrical synapses. Proc. Natl. Acad. Sci. U.S.A. 97 (26), 14754–14759.

Zhang, J., Wu, S.M., 2005. Physiological properties of rod photoreceptor electrical coupling in the tiger salamander retina. J. Physiol. 564 (Pt 3), 849–862.

Zhang, Z., Li, H., Liu, X., O'Brien, J., Ribelayga, C.P., 2015. Circadian clock control of connexin36 phosphorylation in retinal photoreceptors of the cba/caj mouse strain. Vis. Neurosci. 32, E009.

Zhou, Z.Y., Sugawara, K., Hashi, R., Muramoto, K., Mawatari, K., Matsukawa, T., Kato, S., 2001. Reactive oxygen species uncouple external horizontal cells in the carp retina and glutathione couples them again. Neuroscience 102 (4), 959–967.

Zoidl, G., Bruzzone, R., Weickert, S., Kremer, M., Zoidl, C., Mitropoulou, G., Dermietzel, R., 2004. Molecular cloning and functional expression of ZfCx52.6: a novel connexin with hemichannel-forming properties expressed in horizontal cells of the zebrafish retina. J. Biol. Chem. 279 (4), 2913–2921.

CHAPTER 11

Electrical Coupling in the Generation of Vertebrate Motor Rhythms

W.-C. Li[1], J.C. Rekling[2]
[1]The University of St Andrews, St Andrews, United Kingdom; [2]University of Copenhagen, Copenhagen, Denmark

1. INTRODUCTION

Many forms of vertebrate motor activity, such as chewing, breathing, and locomotion are rhythmic. This requires synchronized discharges of motoneurons controlling different muscle groups in an orchestrated manner. We provide a brief review of the presence and role of electrical coupling in a few well-studied systems: the pacemaker nucleus in weakly electric fish; mesencephalic trigeminal nucleus (MesV) involved in chewing rhythms; mammalian spinal motoneurons and excitatory interneurons in the *Xenopus* tadpole swimming circuit, brainstem circuits underlying breathing rhythm, and central respiratory chemosensitivity. Gap junctions in these systems can improve activity synchronization among coupled neurons. However, they do not appear to be essential in the intrinsic pacemaker properties. At the network level, coupling can influence rhythmogenesis by redistributing chemical synaptic potentials. Generally, the role of electrical coupling in vertebrate motor rhythms appears to be critically dependent on developmental age, with more crucial functions in the early postnatal period than in the adult (but see Welsh and Turecek, 2017 (Chapter 12)).

2. ELECTRIC FISH

Weakly electric fish rely on precisely timed electric organ discharges (60–2000 Hz) for electrolocation and electrocommunication. The discharge is generated by specialized motoneuron axon terminals, the electric organ. The command signals are generated by a nucleus in the ventral medulla (pacemaker nucleus), which contains three types of neurons—pacemaker cells, relay cells, and parvocells (Smith et al., 2000), totaling 100–160 neurons. Pacemaker cells and parvocells are interneurons synapsing onto relay cells and themselves. The axons of relay cells project to the spinal cord and make electrical synapses on electromotoneurons. The predominant synaptic connections within the pacemaker nucleus are electrical. One important electrical synaptic connection is among pacemaker cells and another is between pacemaker cells and relay cells. Both pacemaker and relay cells fire 1:1 with electric organ discharge frequency. The mostly axoaxonic coupling between pacemaker cells and relay cells mediates reliable transmission of

Network Functions and Plasticity
ISBN 978-0-12-803471-2, http://dx.doi.org/10.1016/B978-0-12-803471-2.00011-4

pacemaker spiking to relay cells. The relay cells in turn transmit synchronized activity reliably to electromotoneurons, which directly control electric organ discharges. Although electrical and chemical synapses between parvocells and pacemaker/relay cells have been identified, their function is not known (Smith and Zakon, 2000) (Fig. 11.1A). The estimated coupling coefficient for the electrical synapses in the pacemaker nucleus is fairly low (~0.01) and sparse, with each cell coupled to a few others of any type (Moortgat et al., 2000b).

A striking feature for the electric organ discharge rhythms is its high precision in spike timing (Fig. 11.1B). The precision in spike timing is equivalent to a standard deviation in

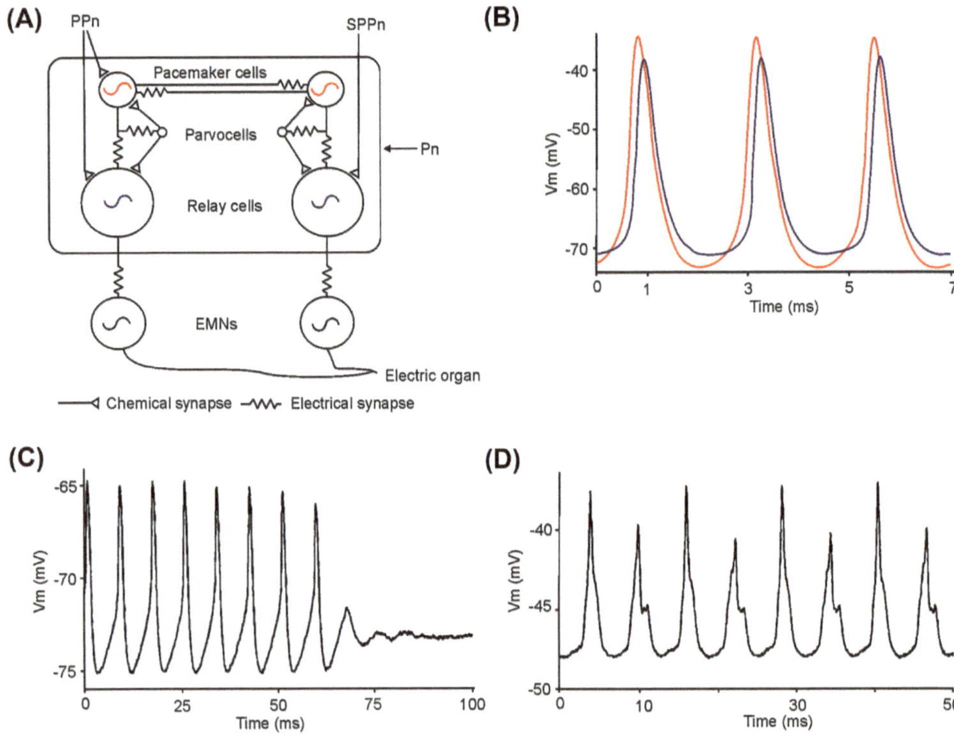

Figure 11.1 ***Electrical synaptic transmission in the pacemaker nucleus (Pn) of a typical weakly electric fish.*** (A) Diagram showing types of cells and synaptic connections within the Pn and electrical synapses between the relay cells and electromotoneurons (EMNs). (*Adapted from Smith, G.T., Lu, Y., Zakon, H.H., 2000. Parvocells: a novel interneuron type in the pacemaker nucleus of a weakly electric fish. J. Comp. Neurol. 423, 427–439*). Circles of different size indicate the size difference in somata between relay cells and the interneurons (pacemaker cells and parvocells). (B) Example recording showing the frequency and phase lock between a pacemaker cell (*red trace*) and a relay cell (*blue trace*). (C) Application of halothane shuts down oscillations in all Pn cells simultaneously. (D) A small subset of Pn cells show continued rhythmic activity in an otherwise silent Pn in the presence of octanol. PPn, prepacemaker nucleus; SPPn, sublemniscal prepacemaker nucleus. *(B–D) is adapted from Moortgat, K.T., Bullock, T.H., Sejnowski, T.J., 2000b. Precision of the pacemaker nucleus in a weakly electric fish: network versus cellular influences. J. Neurophysiol. 83, 971–983.*

cycle periods of as low as 0.14 μs (Bullock, 1970). Such high precision may arise from the electrical coupling among neurons in the circuit, whose main function in many systems has been the synchronization of neural activities. Using gap junction blockers and modeling, however, Moortgat et al. concluded that pacemaker cell firing intrinsically carried high precision with very low coefficient of temporal variation (CV). Removing electrical coupling pharmacologically or in modeling did not have clear effects on the CV in the firing of individual pacemaker cells. Rather, it significantly reduced oscillation frequencies and caused phase shifts between coupled cells (Moortgat et al., 2000a,b). It is not known if electrical coupling is essential in the rhythm generation at the network level. Although the isolated pacemaker nucleus can sustain rhythms without an external drive (Meyer, 1984), physically cutting the pacemaker nucleus in half irreversibly silences the pacemaker nucleus. Blocking electrical coupling with halothane stopped oscillations in pacemaker nucleus neurons simultaneously, whereas octanol and carbenoxolone stopped oscillations in surface neurons leaving rhythms carrying on for minutes in deeper layers of the nucleus (Fig. 11.1C and D). This suggests that at least some pacemaker nucleus cells are spontaneously active (Moortgat et al., 2000b). The role of electrical coupling in the pacemaker nucleus therefore is mostly involved in frequency and phase locking (synchronization).

The nucleus receives descending inputs from the diencephalic prepacemaker nucleus (PPn) and midbrain sublemniscal prepacemaker nucleus (SPPn). Both inputs are glutamatergic. PPn axons synapse onto both pacemaker and relay cells, whereas SPPn mainly projects onto relay cells (Fig. 11.1A). These chemical synapses can transiently modulate electric organ discharge frequency by increasing the firing rates of pacemaker and relay cells that are important in social interactions like mediating the jamming avoidance response (Metzner, 1993; Heiligenberg et al., 1996; Zupanc, 2002).

3. CHEWING

In the brainstem circuits controlling mammalian mastication, electrical coupling has been found between pairs or small clusters of primary afferent neurons in the MesV (Baker and Llinas, 1971; Curti et al., 2012) (Fig. 11.2A). The peripheral neurites of these neurons innervate spindles of jaw-closing muscles and mechanoreceptors of periodontal ligaments. They form synapses onto neurons in the trigeminal motor nucleus, rostral parvocellular reticular formation, and nucleus supratrigeminalis, which are involved in the generation of mastication rhythms. Their cell bodies, interestingly, also receive synaptic inputs from hypothalamus and some brainstem regions (Lazarov, 2002; Verdier et al., 2004). The electrical coupling appears only from postnatal day 8 when chewing behavior begins to be established in the rat. Unlike the usual low-pass filter properties of most electrical synapses, the electrical coupling strength is maximal at 50 Hz, even higher than direct current (DC) coupling. This increase in coupling efficacy can be blocked by the

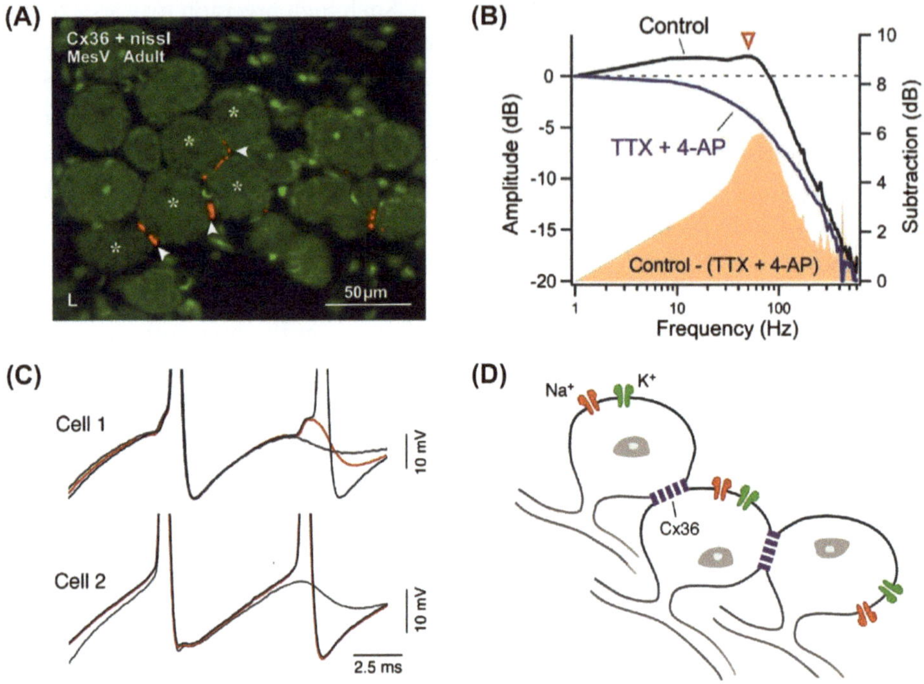

Figure 11.2 ***Gap junction–mediated electrical coupling between mesencephalic trigeminal nucleus (MesV) sensory neurons.*** (A) Immunofluorescence labeling of Cx36 (*red*) counterstained for Nissl (*green*) in adult rat MesV. Asterisks mark a cluster of five cells linked with Cx36 puncta (*arrowheads*). (B) Normalized amplitude of coupled potentials to sine wave current injections at different frequencies into the presynaptic cell. Control is black and blue curve shows the amplitude when tetrodotoxin (TTX) and 4-aminopyridine (4-AP) are applied. *Orange area* shows the difference between control and TTX/4-AP application. *Empty red arrowhead* marks the point with the largest difference (~50 Hz). (C) The electrical coupling between a pair of coupled MesV cells. *Red trace* shows a spikelet in cell 1 resulting from the spiking in cell 2, whereas the *black trace* shows synchronized firing between the two cells resulting from electrical coupling (*gray is subthreshold*). (D) Diagram illustrates the clustering of MesV cells via gap junction connexons in the cell bodies (*blue*) and the promotion of synchronous firing between the coupled cells by Na^+ (*red*) and K^+ (*green*) conductances. *(Modified from Curti, S., Hoge, G., Nagy, J.I., Pereda, A.E., 2012. Synergy between electrical coupling and membrane properties promotes strong synchronization of neurons of the mesencephalic trigeminal nucleus. J. Neurosci. 32, 4341–4359.)*

combined application of tetrodotoxin and 4-aminopyridine. The explanation is that persistent Na^+ currents and some subthreshold K^+ currents like I_A can promote resonant behavior in neurons (Hutcheon and Yarom, 2000), amplifying electrical coupling in a voltage-dependent manner (Fig. 11.2B–D). This type of electrical coupling may increase the excitability of coupled sensory afferents, synchronize their activity, and amplify sensory inputs (Curti et al., 2012). The synchronized firing in subpopulations of MesV neurons and their innervated motoneurons may also be important when strong mastication force is required (Shepherd and Grillner, 2010).

4. GAP JUNCTIONS IN THE SPINAL CORD AND LOCOMOTOR RHYTHMOGENESIS

Motoneurons within the same motor pool in early postnatal animals show electrical coupling (Fulton et al., 1980; Walton and Navarrete, 1991; Perrins and Roberts, 1995b,a). At the interneuron level, mainly excitatory interneurons possess electrical synapses (Saint-Amant and Drapeau, 2001; Parker, 2003; Hinckley and Ziskind-Conhaim, 2006; Li et al., 2009). Interestingly, rhythmogenetic/pacemaker properties have been identified in some excitatory interneurons (Hinckley and Ziskind-Conhaim, 2006; Li et al., 2009, 2010). Although electrical coupling often exists between neurons of different types (Saint-Amant and Drapeau, 2001; Parker, 2003; Serrano-Velez et al., 2014; Song et al., 2016), the coupling appears to be exclusive between Hb9-positive neurons in mice (Hinckley and Ziskind-Conhaim, 2006), type-1 neurons in the pre-Bötzinger complex (preBötC) (Rekling et al., 2000), and between the descending interneurons (dINs) in *Xenopus* tadpoles (Li et al., 2009). The coupling between the excitatory premotor interneurons and motoneurons in adult zebra fish spinal cord has led to the suggestion that motoneurons are an integral part of the locomotor rhythm-generation circuit (Song et al., 2016). Similar interneuron–motoneuron coupling is thought to support the modular organization of the *Aplysia* motor circuit controlling feeding (Jing et al., 2017 (Chapter 5)). It is generally difficult to estimate the extent of coupling, ie, how many other neurons are coupled directly and indirectly to a certain neuron. If the coupling is sparse (Moortgat et al., 2000b), coupling would be difficult to find in random paired recordings. High incidence of electrical coupling, eg, 90% in *Xenopus* tadpole dINs (Li et al., 2009) would suggest the coupling is extensive. The significance of differences in coupling incidence is not known. One possible consequence of high levels of coupling is the collective action of the coupled neurons, meaning the coupled neurons act as one single powerful cell that dictates motor behavior (see text on dINs later in the discussion).

The location of coupling can significantly affect the measurement of coupling strengths and the function of electrical coupling. Most coupling strengths are about 10% measured in paired recordings (Rekling and Feldman, 1997; Nolan et al., 1999; Rekling et al., 2000; Saint-Amant and Drapeau, 2001; Hinckley and Ziskind-Conhaim, 2006; Li et al., 2009). Weak coupling has been found to be very effective in synchronizing moderate firing among neocortex inhibitory interneurons (Galarreta and Hestrin, 1999; Gibson et al., 1999; Bennett and Zukin, 2004). The real coupling strength may be higher when the coupling sites are far away from the recording sites. The coupling via dendrites, which normally have multiple coupling sites, may affect the integration of synaptic potentials. Coupling via axons can be powerful enough to directly evoke spiking to ensure reliable and synchronized activity among coupled cells (Schmitz et al., 2001; Li et al., 2009).

4.1 Gap Junctions Between Spinal Motoneurons

In developing rat spinal cords, gap junctions have been found between motoneurons that are in close proximity (Fulton et al., 1980; Walton and Navarrete, 1991; Chang et al., 1999). The coupling serves as a strong means to synchronize motoneuron activity because the electrical coupling is the strongest within the same motor pool, potentially weak among motoneurons innervating synergistic muscles, and nonexistent between antagonistic pools (Walton and Navarrete, 1991; Perrins and Roberts, 1995b; Chang et al., 1999). However, evidence of gap junctional coupling among adult rat motoneurons is less compelling (Chang et al., 1999), and may consequently have a smaller role in adult motor control at the motoneuronal level.

The coordination between coupled motoneurons can be any low-frequency electrical signals that are not filtered significantly by gap junctions. By blocking action potentials and thus chemical synaptic transmission, Tresch and Kiehn demonstrated some interesting membrane potential oscillations in neonatal rat lumbar spinal motoneurons on the application of N-methyl-D-aspartate (NMDA)/5-hydroxytryptamine (Tresch and Kiehn, 2000). The oscillations in individual motoneurons are phase locked to oscillations in the motor nerves. The gap junction blocker, carbenoxolone, could abolish the oscillations in the motor nerve (ventral root) recordings but intrinsic oscillations in motoneurons remained (Fig. 11.3, also see Nargeot and Bedecarrats, 2017 (Chapter 6)). This is strong evidence that gap junctions can synchronize motoneuron activity. In addition, in *Xenopus* tadpoles, such a synchronizing role by electrical coupling among motoneurons was reported (Perrins and Roberts, 1995a,b; Zhang et al., 2009).

4.2 Is Electrical Coupling Needed for Generating Pacemaker-Like Oscillations?

The presence of gap junctions and the induction of TTX-resistant NMDA-induced oscillations seem to be well correlated in many types of neurons. In TTX, NMDA application could induce both fast and slow oscillations in neonatal rat motoneurons, with the fast ones coupled to DC oscillations in the ventral roots. The fast oscillations are believed to be network based and the slow ones indicate pacemaker properties intrinsic to the recorded motoneuron. Fast oscillations (~2 Hz) were abolished by 100 μM carbenoxolone in neonatal rat motoneurons (Tresch and Kiehn, 2000) but the slower ~0.2-Hz pacemaker-like oscillations remained (Fig. 11.3). In tadpoles, the blockade of electrical coupling by 80% using 18-β glycyrrhetinic acid (18-β-GA) also failed to block 10-Hz oscillations with a slight decrease in oscillation frequency (Li et al., 2009, 2010) (Fig. 11.4). Kepler et al. (1990) have shown that electrical coupling can either increase or decrease the frequency of bursting with varying coupling strengths in a two-neuron model of motor neuron network controlling the foregut movement in crustaceans, depending on the intrinsic state of individual neurons. Therefore, electrical coupling is not an essential part of pacemaker properties.

4.3 The Role of Gap Junctions in Locomotor Rhythmogenesis

Although it is relatively straight forward to reveal the presence of gap junction connexins or electrical coupling, the significance of electrical synaptic transmission in circuit functions is often difficult to ascertain. The function of electrical coupling in synchronizing the activity of coupled neurons has been well established. Indeed, electrical coupling can directly contribute to the phasic excitation during locomotion in *Xenopus* tadpoles

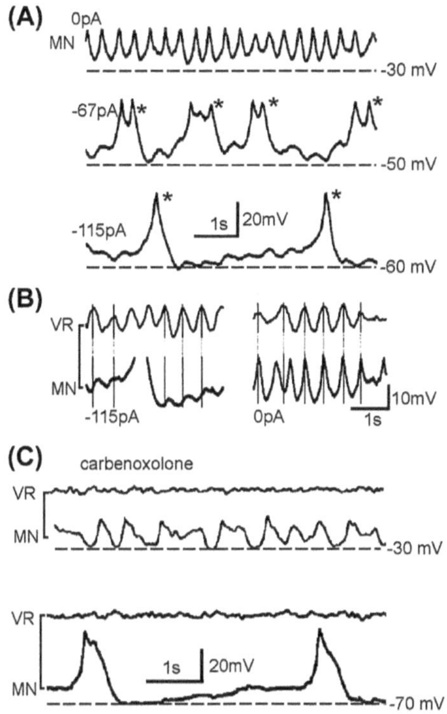

Figure 11.3 ***Oscillations in motoneurons (MN) and ventral roots (VR) during application of 12 μM N-methyl-D-aspartate and 6 μM 5-hydroxytryptamine in the presence of TTX.*** (A) Large intrinsic membrane potential oscillations (*asterisk*) are revealed on top of small oscillations by injecting hyperpolarizing currents (amplitude indicated at the left) into the MN. (B) Coupling of small oscillations in the MN to VR oscillations. MN in the left recording was hyperpolarized to reveal the large oscillation. (C) The gap junction blocker, carbenoxolone, blocks network oscillations in VR but does not affect intrinsic oscillations in the MN. *(Adapted from Tresch, M.C., Kiehn, O., 2000. Motor coordination without action potentials in the mammalian spinal cord. Nat. Neurosci. 3, 593–599.)*

(Perrins and Roberts, 1995a,b; Zhao et al., 1998), lamprey (Parker, 2003), and zebra fish (Kimura et al., 2006). Because pharmacologically blocking gap junctions does not stop intrinsic pacemaker oscillations (Tresch and Kiehn, 2000; Zhang et al., 2009; Li et al., 2010), one would assume that gap junctions may not play a role in the generation of locomotor rhythms. One neglected fact is that most, if not all, pacemaker properties are conditional, eg, upon membrane depolarization provided by chemical synaptic transmission. Electrical coupling allows the redistribution of both excitation and inhibition among coupled neurons (Fig. 11.5), which either directly affects neuronal spiking or alters neuronal pacemaker capabilities. The experimental support of this simple idea, however, has been difficult to gather because of the lack of specific gap junction blockers. Every known gap junction blocker carries some side effects when tested (Rozental et al., 2001; Rouach et al., 2003; Leznik and Llinas, 2005; Wang et al., 2006; Li et al., 2009), although in many studies such side effects are not examined.

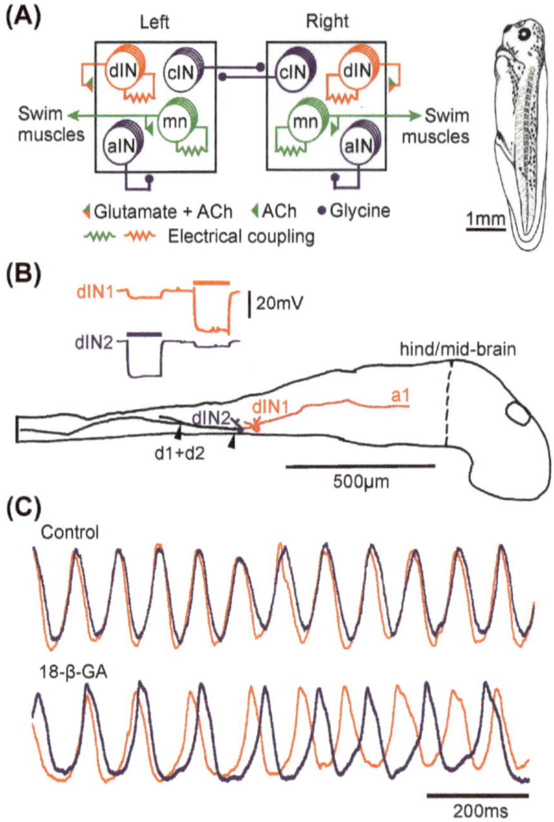

Figure 11.4 ***Electrical coupling between* Xenopus *tadpole descending interneurons (dINs) and the effect of 18-β glycyrrhetinic acid (18-β-GA), a gap junction blocker, on N-methyl-D-aspartate (NMDA)-induced oscillations.*** (A) A tadpole at stage 37/38 (right) and its circuit responsible for generating the swimming rhythms. Synapses on boxes mean all inside neurons receive the inputs. (B) Electrical coupling revealed with current injections (*horizontal bars*, 500 ms) into either dIN1 or dIN2. Both dINs are located in the caudal hindbrain/rostral spinal cord region. The coupling site is likely axonal (*between arrow heads*). a1 is ascending axon. d1 and d2 are descending axons. (C) Membrane oscillations induced by 50 μM NMDA in the presence of tetrodotoxin in two simultaneously recorded dINs (*blue and red traces*). Oscillation amplitudes (25–40 mV) are normalized to help comparison. The gap junction blocker, 18-β-GA at 40 μM, does not block oscillations but desynchronize them. *([(B) is adapted from Li, W.C., Roberts, A., Soffe, S.R., 2009. Locomotor rhythm maintenance: electrical coupling among premotor excitatory interneurons in the brainstem and spinal cord of young* Xenopus *tadpoles. J. Physiol. 587, 1677–1693 and (C) is from Li, W.C., Roberts, A., Soffe, S.R., 2010. Specific brainstem neurons switch each other into pacemaker mode to drive movement by activating NMDA receptors. J. Neurosci. 30, 16609–16620].)*

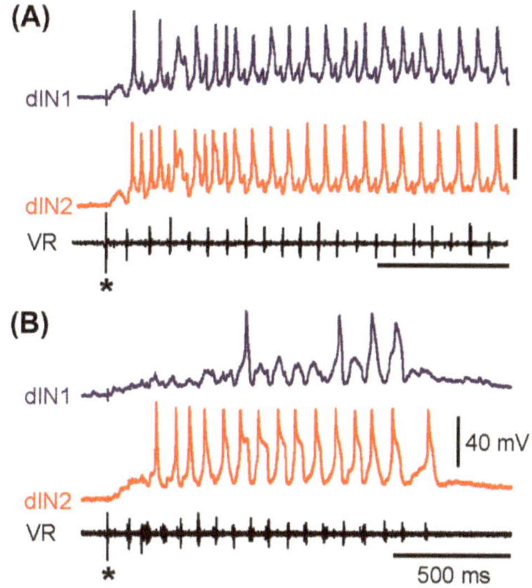

Figure 11.5 ***Effect of 18-β glycyrrhetinic acid (18-β-GA) on descending interneuron (dIN) activity during*** **Xenopus** ***tadpole fictive swimming.*** (A) dIN activity during swimming in control. The two right side neurons, dIN1 and dIN2, are recorded simultaneously with ventral root activity (VR) from the left side. (B) dIN activity after circulating 40 μM 18-β-GA for ~45 min. The asterisk marks the time when a single electrical pulse (1 ms) was applied to skin to start swimming. Note the dIN1 firing has become largely unreliable, the shortening of swimming episode and broadening of the VR bursts on each swimming cycle. *(Adapted from Li, W.C., Roberts, A., Soffe, S.R., 2009. Locomotor rhythm maintenance: electrical coupling among premotor excitatory interneurons in the brainstem and spinal cord of young* Xenopus *tadpoles. J. Physiol. 587, 1677–1693.)*

Some studies make use of several different gap junction blockers, hoping to see a common effect that can be attributed to the blockade of gap junctions. This strategy may work well if the motor circuit is very well understood. Another potential way to consolidate the use of gap junction blockers is to carry out some tedious cross-correlation analyses if the coupling strength can be monitored throughout the blocker application (Li et al., 2009). The fewer side effects the blocker has, the more effective the cross-correlation is. Therefore, it is important to screen blockers first to make sure that they do block gap junctions and the side effects are documented. The onset time of most blocker side effects is quicker than the blockade of the gap junctions themselves. By monitoring electrical coupling strength, neuronal activity, and motor outputs simultaneously, it is possible to calculate the cross-correlation coefficients between pairs of parameters. The effects arising from the blockade of gap junctions should follow a similar time course to the blockade, resulting in high cross-correlation coefficients (Li et al., 2009; Zhang et al., 2009) (Fig. 11.6).

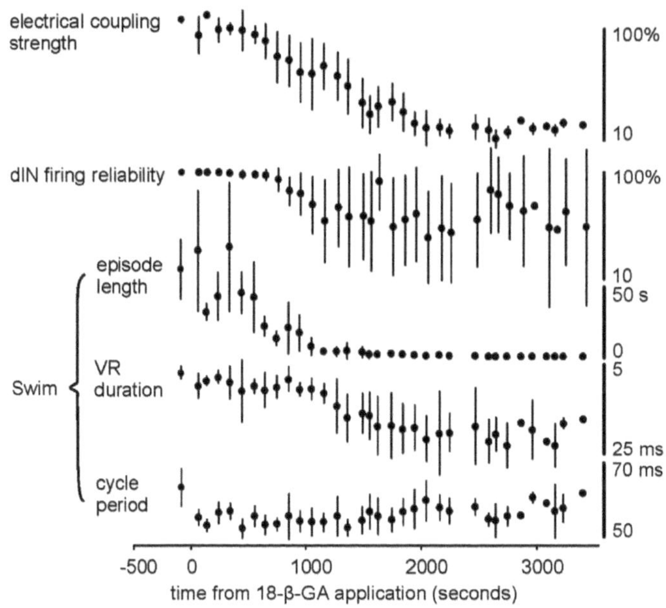

Figure 11.6 ***Time series measurements of electrical coupling strength together with descending interneuron (dIN) firing reliability and swimming parameters throughout 18-β glycyrrhetinic acid (18-β-GA) application.*** Cross-correlation analyses revealed high correlation peaks between electrical coupling strength, dIN firing reliability, episode length, and ventral root (VR) burst duration. *(Figure adapted from Li, W.C., Roberts, A., Soffe, S.R., 2009. Locomotor rhythm maintenance: electrical coupling among premotor excitatory interneurons in the brainstem and spinal cord of young* Xenopus *tadpoles. J. Physiol. 587, 1677–1693 and Zhang, H.Y., Li, W.C., Heitler, W.J., Sillar, K.T., 2009. Electrical coupling synchronises spinal motoneuron activity during swimming in hatchling* Xenopus *tadpoles. J. Physiol. 587, 4455–4466.)*

Using cross-correlation analysis, Li et al. (2009) managed to link the blockade of gap junctions to the loss of firing reliability in dINs, which are extensively electrically coupled. On a similar timescale, ventral root burst duration is prolonged and swimming episode length shortened. The prolongation of ventral root bursts can be explained by increased jittering of dIN and motoneuron spikes after the blockade of electrical coupling. The interpretation of shortened swimming episodes, however, requires some understanding of how tadpole swimming rhythms are maintained. In the tadpole swimming circuit, dINs provide the critical driving force for the swimming circuit (Soffe et al., 2009). dIN activity during swimming is sustained in a regenerative manner by the feedback excitation among dINs themselves (Li et al., 2006; Li, 2011). dINs do not receive evenly distributed chemical synaptic inputs. When electrical coupling is weakened by the blocker18-β-GA, differences in chemical synaptic inputs are revealed (Fig. 11.5B). In dINs where chemical inputs are weaker, spiking fails. The decreased overall dIN activity would weaken the feedback excitation and lead to the collapse of swimming rhythmogenesis mechanisms (Fig. 11.6).

5. INVOLVEMENT OF ELECTRICAL COUPLING IN THE NEURAL CONTROL OF BREATHING

The brainstem contains several distinct groups of neurons that are involved in producing breathing rhythms, and coordinated patterns of motor nerve activity driving pump and airflow resistance muscles (Feldman et al., 2003). Studies of the control of respiration gained considerable momentum when it was reported that highly reduced en bloc or slice preparations from embryonic, newborn, or arterially perfused adult rodents continue to generate respiratory-related rhythmic motor output in vitro (Suzue, 1984; Smith et al., 1991; Paton, 1996). The basic neural structures generating the respiratory rhythm remain intact in these preparations. Thus, spontaneous respiratory-related motor rhythms can be recorded from the cranial and spinal ventral nerve roots that innervate respiratory muscles. Using these preparations, brainstem circuits that are responsible for generating this basic motor rhythm have been studied, with the aim of understanding how rhythmogenic, pattern-forming, and chemosensitive circuits are constructed. These studies have demonstrated that fetal breathing in mice results from the activity of two coupled oscillators, with one, the preBötC, being active shortly before the onset of fetal breathing, and the other, the retrotrapezoid nucleus/parafacial respiratory group (RTN/pFRG), starting activity 1 day earlier (Champagnat et al., 2011). The current understanding of the role in breathing of these two oscillators in the adult is that the preBötC generates inspiratory rhythm, whereas the RTN/pFRG is involved in chemosensation, generating active expiration when metabolic demand increases (Feldman and Kam, 2015; Huckstepp et al., 2015). The exact role of electrical coupling in rhythmogenesis, motor pattern generation, and chemosensation at the level of the brainstem is unclear at present, but data from in vitro en bloc brainstem spinal cord or slice experiments supports its involvement at all three levels (Dean et al., 2002; Solomon and Dean, 2002).

5.1 Rhythmogenesis

Direct evidence for the presence of electrical coupling between respiratory neurons in the preBötC in the newborn brainstem was provided by dual patch clamp recordings (Rekling et al., 2000). The preBötC contains inspiratory neurons that show at least three types of neural discharge patterns, one of which (type-1) is a strong candidate for being an integral part of the respiratory rhythm generator (Rekling et al., 1996b,a). Intracellularly recorded pairs of type-1 neurons showed bidirectional electrical coupling with a low coupling strength (<0.10), and the current passing between two neurons was heavily filtered (cutoff frequency, <10 Hz, Fig. 11.7A and B), which resulted in distorted action potential waveform in postjunctional neurons resembling small-amplitude excitatory synaptic potentials (EPSPs) (Fig. 11.7C). These filter characteristics led to the hypothesis that electrical coupling synchronizes the membrane potential of type-1 neurons at frequencies corresponding to the envelope of the inspiratory EPSP burst (Rekling et al., 2000). In

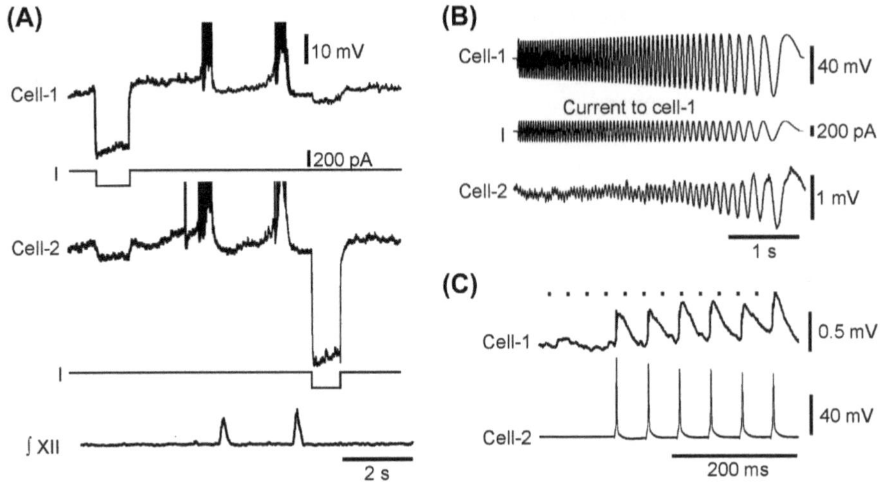

Figure 11.7 ***Bidirectional electrical coupling between type-1 inspiratory neurons in the pre-Bötzinger complex revealed by dual patch recordings.*** (A) Simultaneous recordings from two type-1 neurons (spikes are truncated). A hyperpolarizing pulse in either neuron elicited an attenuated hyperpolarizing response in the other electrically coupled neuron. (B) ZAP current injection into cell 1 elicited an attenuated and filtered response in the coupled cell 2. (C) Train of prejunctional action potentials in cell 2 and corresponding, temporally summated (*dotted line*) postjunctional coupling potential waveforms in a coupled cell 1. *(Modified from Rekling, J.C., Shao, X.M., Feldman, J.L., 2000. Electrical coupling and excitatory synaptic transmission between rhythmogenic respiratory neurons in the preBötzinger complex. J. Neurosci. 20, RC113.)*

line with this idea, substance P, which excites inspiratory neurons in the preBötC, may exert widespread excitatory effects caused by gap junctions, which carry substance P–induced current from neurokinin receptor–expressing preBötC neurons to neighboring inspiratory neurons not expressing neurokinin receptors (Hayes and Del Negro, 2007).

The involvement of electrical coupling in cellular processes generating respiratory rhythms appears to depend critically on the age of the experimental animal. Thus, blockade of gap junctions using carbenoxolone (100 μM), 18-α-glycyrrhetinic acid (25–100 μM), 18-β-GA (25–100 μM), octanol (200–300 μM), or heptanol (200 μM) did not block respiratory rhythmogenesis in an in vitro, arterially perfused adult (ages 35–42 days) rat preparation (Solomon et al., 2003). Instead an increase in phrenic burst frequency was observed, together with a decrease in the peak amplitude of integrated phrenic nerve discharge, a shift from an augmenting phrenic burst pattern to a bell-shaped or square-wave phrenic burst pattern (Fig. 11.8A). Likewise carbenoxolone (100 μM) or octanol (300 μM) did not affect respiratory rhythmogenesis in a more intact in situ arterially perfused juvenile (ages 21–35 days) rat preparation (Rodman et al., 2006). In stark contrast to adult in vitro preparations, experiments on neonatal (P1-P7) en block brainstem-spinal cord preparations or medullary slices

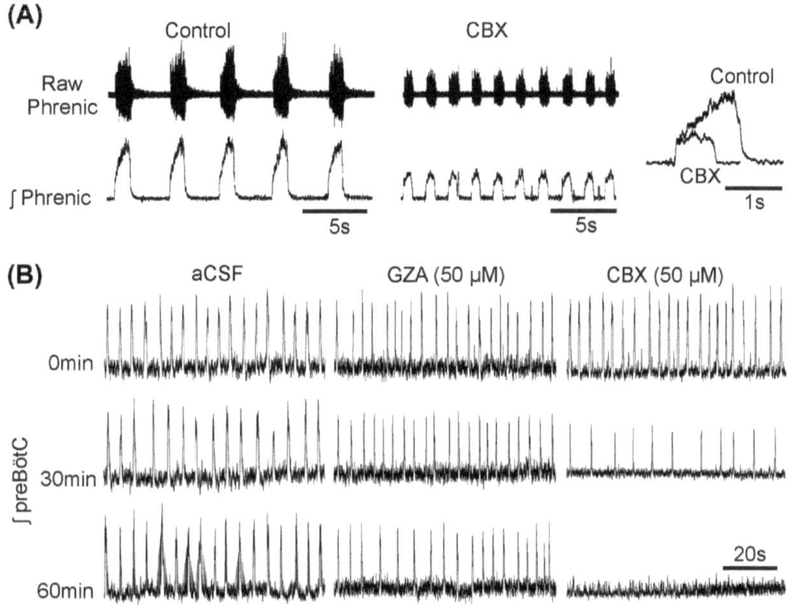

Figure 11.8 ***Gap junction blockade in adult animals does not block respiratory rhythmogenesis, but does so in early postnatal animals.*** (A) Effects of pharmacological blockade of brainstem gap junction coupling with carbenoxolone (CBX, 100 μM) on phrenic nerve discharge in arterially perfused adult rat preparation *(modified from Solomon, I.C., Chon, K.H., Rodriguez, M.N., 2003. Blockade of brain stem gap junctions increases phrenic burst frequency and reduces phrenic burst synchronization in adult rat. J. Neurophysiol. 89, 135–149.).* (B) Time-dependent effects of glycyrrhizic acid (GZA) and CBX on pre-Bötzinger complex (preBötC) burst frequency in an in vitro slice preparation from a neonatal mouse. *(Modified from Elsen, F.P., Shields, E.J., Roe, M.T., Vandam, R.J., Kelty, J.D., 2008. Carbenoxolone induced depression of rhythmogenesis in the pre-Bötzinger Complex. BMC Neurosci. 9, 46.)*

containing the preBötC show a marked effect of gap junction blockers on rhythmogenesis (Rekling et al., 2000; Bou-Flores and Berger, 2001; Elsen et al., 2008). Bath applied or local perfusion over the preBötC of carbenoxolone (50–100 μM), 18-α-glycyrrhetinic acid (50 μM), 18-β-GA (50 μM), heptanol (1 mM), octanol (1 mM), and halothane (10 mM) all reversibly suppressed the frequency of rhythmic respiratory output, recorded on the phrenic nerve, the XII cranial nerve, or field recording in the preBötC (Fig. 11.8B). Full blockade of rhythmogenesis by carbenoxolone requires up to 1 h perfusion of the blocker, and as in many other experiments using these gap junction blockers unspecific effect on neuronal excitability is a confounding issue. However, Elsen et al. (2008) also applied glycyrrhizic acid (GZA, 50 μM), a compound structurally similar to carbenoxolone but without gap junction blocking properties, and found no effect on rhythmogenesis. Notably, GZA, like carbenoxolone, decreases neuronal input resistance and induces an outward holding current, side effects that apparently are not responsible for the effect on rhythmogenesis.

In support for a developmental regulation of electrical coupling in the preBötC a differential expression of connexin 26 (Cx26) and connexin 32 (Cx32) is observed in the preBötC of neonatal and adult rats, with Cx26 labeling seen in a high proportion of preBötC neurons in neonatal rats ≤7 days (postnatal P7) but declining with increasing age. In contrast, Cx32 labeling was sparse in preBötC neurons of neonatal rats ≤P7, but increased with age (Solomon et al., 2001). Gap junctional blockade with carbenoxolone also differentially affects fictive breathing in larval and adult bullfrogs underlining a possible essential difference in the role of electrical coupling between young and adult animals (Winmill and Hedrick, 2003).

5.2 Synchronization of Motoneurons

Dual patch clamp recordings from respiratory motoneurons in the newborn mouse nucleus ambiguus show bidirectional electrical coupling with coupling strengths of 3–14% (Rekling and Feldman, 1997). These brainstem motoneurons were located in the intermediate region of the nucleus ambiguus, which contains motoneurons innervating pharyngeal and intrinsic muscles of the larynx (Hernandez-Morato et al., 2013) involved in maintaining airway patency during breathing. Genioglossal motoneurons in postnatal rats also show electrical coupling as revealed by dye coupling using Lucifer yellow and neurobiotin, and short latency depolarization in response to stimulation of hypoglossal axons (Mazza et al., 1992). The genioglossal muscle is activated just before the diaphragm, and thus also helps maintain upper airway patency (Brouillette and Thach, 1980). No evidence for electrical coupling between genioglossal motoneurons was found in animals older than 8 days. Pharmacological experiments support these observations, because bath application of carbenoxolone (100 μM) affects the degree of synchronization within inspiratory bursts recorded on hypoglossal and phrenic nerves in neonatal mouse medullary slice and en block brainstem-spinal cord preparations (Bou-Flores and Berger, 2001). Counterintuitively, gap junction blockade resulted in an increase in the short-time-scale synchrony observed during phrenic and hypoglossal inspiratory bursts, with up to a fivefold enhancement in the 30–40-Hz peak observed in the power spectrum of the phrenic nerve activity, and a smaller enhancement (~120%) of a 10–20 Hz peak in the power spectrum of the hypoglossal nerve. Carbenoxolone (100 μM) also disrupts NMDA-elicited or Dihydroxyphenylglycine-elicited (DHPG, selective agonist for group I metabotropic glutamate receptors) hypoglossal motoneuron bursting in brainstem slices from newborn (P1–P6) rats (Sharifullina et al., 2008; Cifra et al., 2009). These surprising findings may be explained by the fact that synchrony in neuronal networks can be either increased or decreased by electrical coupling depending on the strength of the electrical coupling, and how the electrically coupled neurons act as an electrical load on the network that is producing the synchronizing activity (Marder, 1998).

Electrical coupling between brainstem motoneurons involved in maintaining an open upper airway may be particularly important in the early postnatal period. Thus, dyscoordination of the genioglossal muscle and the diaphragm may result in upper airway obstruction and cessation of breathing (Remmers et al., 1978). We hypothesize that the weak electrical coupling that exists in the early postnatal period may serve not to synchronize firing on a short timescale, but to synchronize slower depolarization that corresponds to the envelope of the synaptic drive that underlies inspiratory bursts. In this way, electrical coupling may aid group motoneuronal activity in a period when precise patterning of the drive to muscles has not yet developed.

5.3 Chemosensitivity

Central respiratory chemosensitivity, where the brain senses CO_2 and/or H^+ and alters breathing appropriately, involves neurons in the RTN, which is located between the facial motor nucleus and the ventral medullary surface (Guyenet, 2014). At the molecular level, CO_2-stimulated breathing requires the proton-activated receptor GPR4, a pH-sensitive K^+ channel (TASK-2) in RTN neurons, and ATP released from nearby astrocytes (Gourine et al., 2010; Kumar et al., 2015). The embryonic forerunner of the RTN, called the embryonic parafacial (e-pF) population, expresses the transcription factor Phox2b, and collectively functions as a spontaneous oscillator at E14.5, likely mediated by gap junctions between the cells, because carbenoxolone (50 μM) completely abolished spontaneous intercellular burst synchrony (Thoby-Brisson et al., 2009). It is unclear to what extent electrical coupling between RTN neurons persists in adult animals, but again, a developmental effect of gap junction blockade is observed. Thus, in neonatal rats (P7-P12) application of carbenoxolone (100 μM) under in vitro conditions decreased neuronal CO_2/H^+ sensitivity in the RTN (Wenker et al., 2012), and in awake rats less than 10 weeks of age infusion of carbenoxolone (300 μM) into the RTN decreased respiratory responses to CO_2 (Hewitt et al., 2004). However, as the animals matured, the response to carbenoxolone microperfusion reversed, and resulted in stimulation of ventilation after 12 weeks of age. Part of the chemosensitivity in the RTN may depend on Cx26-mediated release of ATP (Huckstepp et al., 2010b; Reyes et al., 2014). Cx26 form hemichannels that are directly sensitive to CO_2 (Huckstepp et al., 2010a), preferentially expressed near the ventral brainstem surface, and interestingly mutations (A88V) in Cx26, which are linked to keratitis-ichthyosis-deafness syndrome in humans, results in CO_2 insensitivity in the hemichannels, and disordered breathing with episodes of central sleep apnea (Meigh et al., 2014).

The locus coeruleus (LC) and the nucleus tractus solitaries (NTS) are also involved in central respiratory chemosensitivity (Guyenet, 2014). More than 80% of LC neurons are chemosensitive, responding to hypercapnia with increased firing rate, and local acidification of noradrenergic neurons in LC increases respiratory frequency (Pineda and

Aghajanian, 1997; Oyamada et al., 1998; Filosa et al., 2002). Bilateral microinjection of carbenoxolone (1 or 3 mM) into the LC in adult rats did not affect resting respiration, but reduced the increase in respiration induced by hypercapnia by ~25% (Fig. 11.9A) (Patrone et al., 2014). Indeed, LC neurons in in vitro slices taken from adult rats show clear evidence for electrical coupling likely located in the dendritic compartment (Ishimatsu and Williams, 1996). Simultaneous whole cell recordings from LC neurons in the en block brainstem-spinal cord or 300-μm slice preparation of the neonatal rat show carbenoxolone-sensitive (Fig. 11.9B and C, 100 μM) electrical coupling between respiratory modulated neurons (Dean et al., 2001; Ballantyne et al., 2004). The coupling is so prominent that the LC neurons can act as an oscillating synchronized ensemble under chemical synaptic blockade (Andrzejewski et al., 2001). Focal perfusion of carbenoxolone (300 μM) into the NTS in awake rats younger than 10 weeks decreased respiratory responses to CO_2, but the response to carbenoxolone was absent in animals 10 weeks of age or older (Parisian et al., 2004).

6. CONCLUDING REMARKS

Electrical coupling has been identified in several important brainstem regions involved in rhythmic motor behavior like electrolocation/electrocommunication, chewing, breathing, and in the spinal cord. There is good evidence that gap junctions can synchronize the activity of coupled neurons. However, they are not essential in the neuronal intrinsic pacemaker properties. At network level, the study from tadpole swimming circuit suggests that electrical coupling can indirectly affect swimming rhythmogenesis by redistributing synaptic potentials among coupled neurons. The roles of electrical coupling in the control of breathing appear to be critically dependent on the age of animals, with regulated, differential expression of connexins at different developmental stages. Their function in the immature neural networks may support vital ensemble behavior before chemical synaptic transmission has fully developed.

OUTSTANDING ISSUES AND FUTURE RESEARCH

- Although a lot of progresses have been made in understanding the role of electrical coupling in motor rhythms, most studies so far heavily rely on the use of gap junction blockers. Unfortunately, all blockers have some side effects and most blockers take long time to work, making the effects of blockers difficult to interpret in most cases. This calls for the design of new blockers that are more specific and can act on a faster timescale. An alternative approach is to generate transgenic animal models where gap junction connexins are knocked out or down. It will be advantageous if such mutation can be targeted to specific sites where gap junctions are found in normal vertebrates.

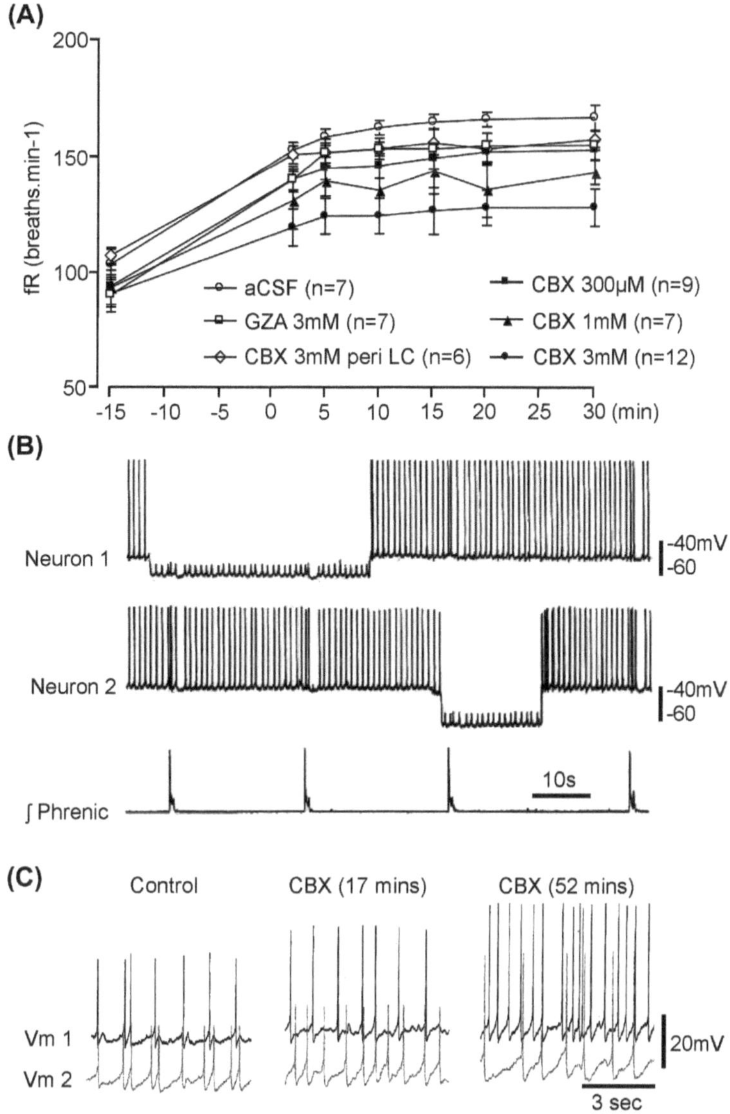

Figure 11.9 ***Electrical coupling between locus coeruleus (LC) neurons is involved in cell–cell synchronization and chemosensation.*** (A) Effect of bilateral intra-LC microinjections of vehicle (artificial Cerebrospinal fluid, aCSF), glycyrrhizic acid (GZA), and carbenoxolone (CBX) on respiratory frequency (f_R) of adult rats in hypercapnia (7% CO_2 at 0 min). Note that CBX (3 mM) significantly decreased hypercapnic ventilator response. (B) Synchronous activity in two simultaneously recorded LC neurons that show respiratory modulation (∫Phrenic—integrated phrenic neurogram). A steady hyperpolarizing current in either neuron uncover subthreshold synchronized rhythmic oscillation of the membrane potential. (C) Paired recording from two LC neurons before and after application of CBX (100 μM), which disrupted synchronized firing, but did not abolish spontaneous activity. *([(A) is adapted from Patrone, L.G., Bicego, K.C., Hartzler, L.K., Putnam, R.W., Gargaglioni, L.H., 2014. Cardiorespiratory effects of gap junction blockade in the locus coeruleus in unanesthetized adult rats. Respir. Physiol. Neurobiol. 190, 86–95, (B) is modified from Ballantyne, D., Andrzejewski, M., Muckenhoff, K., Scheid, P., 2004. Rhythms, synchrony and electrical coupling in the Locus coeruleus. Respir. Physiol. Neurobiol. 143, 199–214, and (C) from Dean, J.B., Kinkade, E.A., Putnam, R.W., 2001. Cell-cell coupling in CO_2/H^+-excited neurons in brainstem slices. Respir. Physiol. 129, 83–100].)*

- Another area that requires more research is the role of electrical coupling in adult brain stem and spinal cord. Owing to the access difficulty in adult vertebrates, most studies have used embryonic or neonatal preparations. There are clear age-dependent changes in the distribution of gap junctions; thus it will remain a big challenge to investigate the roles of electrical coupling in adult vertebrates.
- Gap junctions pass current, but both small and sometimes larger bioactive molecules can pass from one neuron to the other. Much remains to be learned about when neurons use gap junction for exchange of electrical signals, and when exchange of biochemical information is the main purpose of gap junction coupling.
- Neuromodulatory actions on the strength of electrical coupling also need to be studied in more detail, and may involve long-term as well as short-term effects with important consequences for generation and patterned execution of motor rhythms.

REFERENCES

Andrzejewski, M., Muckenhoff, K., Scheid, P., Ballantyne, D., 2001. Synchronized rhythms in chemosensitive neurons of the locus coeruleus in the absence of chemical synaptic transmission. Respir. Physiol. 129, 123–140.

Baker, R., Llinas, R., 1971. Electrotonic coupling between neurones in the rat mesencephalic nucleus. J. Physiol. 212, 45–63.

Ballantyne, D., Andrzejewski, M., Muckenhoff, K., Scheid, P., 2004. Rhythms, synchrony and electrical coupling in the Locus coeruleus. Respir. Physiol. Neurobiol. 143, 199–214.

Bennett, M.V.L., Zukin, R.S., 2004. Electrical coupling and neuronal synchronization in the mammalian brain. Neuron 41, 495–511.

Bou-Flores, C., Berger, A.J., 2001. Gap junctions and inhibitory synapses modulate inspiratory motoneuron synchronization. J. Neurophysiol. 85, 1543–1551.

Brouillette, R.T., Thach, B.T., 1980. Control of genioglossus muscle inspiratory activity. J. Appl. Physiol. Respir. Environ. Exerc. Physiol. 49, 801–808.

Bullock, T.H., 1970. The reliability of neurons. J. Gen. Physiol. 55, 565–584.

Champagnat, J., Morin-Surun, M.P., Bouvier, J., Thoby-Brisson, M., Fortin, G., 2011. Prenatal development of central rhythm generation. Respir. Physiol. Neurobiol. 178, 146–155.

Chang, Q., Gonzalez, M., Pinter, M.J., Balice-Gordon, R.J., 1999. Gap junctional coupling and patterns of connexin expression among neonatal rat lumbar spinal motor neurons. J. Neurosci. 19, 10813–10828.

Cifra, A., Nani, F., Sharifullina, E., Nistri, A., 2009. A repertoire of rhythmic bursting produced by hypoglossal motoneurons in physiological and pathological conditions. Philos. Trans. R. Soc. Lond B Biol. Sci. 364, 2493–2500.

Curti, S., Hoge, G., Nagy, J.I., Pereda, A.E., 2012. Synergy between electrical coupling and membrane properties promotes strong synchronization of neurons of the mesencephalic trigeminal nucleus. J. Neurosci. 32, 4341–4359.

Dean, J.B., Kinkade, E.A., Putnam, R.W., 2001. Cell-cell coupling in CO_2/H^+-excited neurons in brainstem slices. Respir. Physiol. 129, 83–100.

Dean, J.B., Ballantyne, D., Cardone, D.L., Erlichman, J.S., Solomon, I.C., 2002. Role of gap junctions in CO_2 chemoreception and respiratory control. Am. J. Physiol. Lung Cell. Mol. Physiol. 283, L665–L670.

Elsen, F.P., Shields, E.J., Roe, M.T., Vandam, R.J., Kelty, J.D., 2008. Carbenoxolone induced depression of rhythmogenesis in the pre-Bötzinger Complex. BMC Neurosci. 9, 46.

Feldman, J.L., Kam, K., 2015. Facing the challenge of mammalian neural microcircuits: taking a few breaths may help. J. Physiol. 593, 3–23.

Feldman, J.L., Mitchell, G.S., Nattie, E.E., 2003. Breathing: rhythmicity, plasticity, chemosensitivity. Annu. Rev. Neurosci. 26, 239–266.

Filosa, J.A., Dean, J.B., Putnam, R.W., 2002. Role of intracellular and extracellular pH in the chemosensitive response of rat locus coeruleus neurones. J. Physiol. 541, 493–509.

Fulton, B.P., Miledi, R., Takahashi, T., 1980. Electrical synapses between motoneurons in the spinal cord of the newborn rat. Proc. R. Soc. Lond B Biol. Sci. 208, 115–120.

Galarreta, M., Hestrin, S., 1999. A network of fast-spiking cells in the neocortex connected by electrical synapses. Nature 402, 72–75.

Gibson, J.R., Beierlein, M., Connors, B.W., 1999. Two networks of electrically coupled inhibitory neurons in neocortex. Nature 402, 75–79.

Gourine, A.V., Kasymov, V., Marina, N., Tang, F., Figueiredo, M.F., Lane, S., Teschemacher, A.G., Spyer, K.M., Deisseroth, K., Kasparov, S., 2010. Astrocytes control breathing through pH-dependent release of ATP. Science 329, 571–575.

Guyenet, P.G., 2014. Regulation of breathing and autonomic outflows by chemoreceptors. Compr. Physiol. 4, 1511–1562.

Hayes, J.A., Del Negro, C.A., 2007. Neurokinin receptor-expressing pre-botzinger complex neurons in neonatal mice studied in vitro. J. Neurophysiol. 97, 4215–4224.

Heiligenberg, W., Metzner, W., Wong, C.J., Keller, C.H., 1996. Motor control of the jamming avoidance response of *Apteronotus leptorhynchus*: evolutionary changes of a behavior and its neuronal substrates. J. Comp. Physiol. A 179, 653–674.

Hernandez-Morato, I., Pascual-Font, A., Ramirez, C., Matarranz-Echeverria, J., McHanwell, S., Vazquez, T., Sanudo, J.R., Valderrama-Canales, F.J., 2013. Somatotopy of the neurons innervating the cricothyroid, posterior cricoarytenoid, and thyroarytenoid muscles of the rat's larynx. Anat. Rec. 296, 470–479.

Hewitt, A., Barrie, R., Graham, M., Bogus, K., Leiter, J.C., Erlichman, J.S., 2004. Ventilatory effects of gap junction blockade in the RTN in awake rats. Am. J. Physiol. Regul. Integr. Comp. Physiol. 287, R1407–R1418.

Hinckley, C.A., Ziskind-Conhaim, L., 2006. Electrical coupling between locomotor-related excitatory interneurons in the mammalian spinal cord. J. Neurosci. 26, 8477–8483.

Huckstepp, R.T., Eason, R., Sachdev, A., Dale, N., 2010a. CO_2-dependent opening of connexin 26 and related beta connexins. J. Physiol. 588, 3921–3931.

Huckstepp, R.T., id, B.R., Eason, R., Spyer, K.M., Dicke, N., Willecke, K., Marina, N., Gourine, A.V., Dale, N., 2010b. Connexin hemichannel-mediated CO_2-dependent release of ATP in the medulla oblongata contributes to central respiratory chemosensitivity. J. Physiol. 588, 3901–3920.

Huckstepp, R.T., Cardoza, K.P., Henderson, L.E., Feldman, J.L., 2015. Role of parafacial nuclei in control of breathing in adult rats. J. Neurosci. 35, 1052–1067.

Hutcheon, B., Yarom, Y., 2000. Resonance, oscillation and the intrinsic frequency preferences of neurons. Trends Neurosci. 23, 216–222.

Ishimatsu, M., Williams, J.T., 1996. Synchronous activity in locus coeruleus results from dendritic interactions in pericoerulear regions. J. Neurosci. 16, 5196–5204.

Jing, J., Cropper, E.C., Weiss, K.R., 2017. Network functions of electrical coupling present in multiple and specific sites in behavior-generating circuits. In: Jing, J. (Ed.), Network Functions and Plasticity. Elsevier (Chapter 5).

Kepler, T.B., Marder, E., Abbott, L.F., 1990. The effect of electrical coupling on the frequency of model neuronal oscillators. Science 248, 83–85.

Kimura, Y., Okamura, Y., Higashijima, S., 2006. alx, a zebrafish homolog of Chx10, marks ipsilateral descending excitatory interneurons that participate in the regulation of spinal locomotor circuits. J. Neurosci. 26, 5684–5697.

Kumar, N.N., Velic, A., Soliz, J., Shi, Y., Li, K., Wang, S., Weaver, J.L., Sen, J., Abbott, S.B., Lazarenko, R.M., Ludwig, M.G., Perez-Reyes, E., Mohebbi, N., Bettoni, C., Gassmann, M., Suply, T., Seuwen, K., Guyenet, P.G., Wagner, C.A., Bayliss, D.A., 2015. PHYSIOLOGY. Regulation of breathing by CO_2 requires the proton-activated receptor GPR4 in retrotrapezoid nucleus neurons. Science 348, 1255–1260.

Lazarov, N.E., 2002. Comparative analysis of the chemical neuroanatomy of the mammalian trigeminal ganglion and mesencephalic trigeminal nucleus. Prog. Neurobiol. 66, 19–59.

Leznik, E., Llinas, R., 2005. Role of gap junctions in synchronized neuronal oscillations in the inferior olive. J. Neurophysiol. 94, 2447–2456.

Li, W.C., Soffe, S.R., Wolf, E., Roberts, A., 2006. Persistent responses to brief stimuli: feedback excitation among brainstem neurons. J. Neurosci. 26, 4026–4035.

Li, W.C., Roberts, A., Soffe, S.R., 2009. Locomotor rhythm maintenance: electrical coupling among premotor excitatory interneurons in the brainstem and spinal cord of young *Xenopus* tadpoles. J. Physiol. 587, 1677–1693.

Li, W.C., Roberts, A., Soffe, S.R., 2010. Specific brainstem neurons switch each other into pacemaker mode to drive movement by activating NMDA receptors. J. Neurosci. 30, 16609–16620.

Li, W.C., 2011. Generation of locomotion rhythms without inhibition in vertebrates: the search for pacemaker neurons. Integr. Comp. Biol. 51, 879–889.

Marder, E., 1998. Electrical synapses: beyond speed and synchrony to computation. Curr. Biol. 8, R795–R797.

Mazza, E., Nunez-Abades, P.A., Spielmann, J.M., Cameron, W.E., 1992. Anatomical and electrotonic coupling in developing genioglossal motoneurons of the rat. Brain Res. 598, 127–137.

Meigh, L., Hussain, N., Mulkey, D.K., Dale, N., 2014. Connexin26 hemichannels with a mutation that causes KID syndrome in humans lack sensitivity to CO_2. Elife 3, e04249.

Metzner, W., 1993. The jamming avoidance response in Eigenmannia is controlled by two separate motor pathways. J. Neurosci. 13, 1862–1878.

Meyer, J., 1984. Steroid influences upon discharge frequencies of intact and isolated pacemakers of weakly electric fish. J. Comp. Physiol. A 154, 659–668.

Moortgat, K.T., Bullock, T.H., Sejnowski, T.J., 2000a. Gap junction effects on precision and frequency of a model pacemaker network. J. Neurophysiol. 83, 984–997.

Moortgat, K.T., Bullock, T.H., Sejnowski, T.J., 2000b. Precision of the pacemaker nucleus in a weakly electric fish: network versus cellular influences. J. Neurophysiol. 83, 971–983.

Nargeot, R., Bédécarrats, A., 2017. Electrical synapses and learning–induced plasticity in motor rhythmogenesis. In: Jing, J. (Ed.), Network Functions and Plasticity. Elsevier (Chapter 6).

Nolan, M.F., Logan, S.D., Spanswick, D., 1999. Electrophysiological properties of electrical synapses between rat sympathetic preganglionic neurones in vitro. J. Physiol. 519 (Pt. 3), 753–764.

Oyamada, Y., Ballantyne, D., Muckenhoff, K., Scheid, P., 1998. Respiration-modulated membrane potential and chemosensitivity of locus coeruleus neurones in the in vitro brainstem-spinal cord of the neonatal rat. J. Physiol. 513 (Pt. 2), 381–398.

Parisian, K., Wages, P., Smith, A., Jarosz, J., Hewitt, A., Leiter, J.C., Erlichman, J.S., 2004. Ventilatory effects of gap junction blockade in the NTS in awake rats. Respir. Physiol. Neurobiol. 142, 127–143.

Parker, D., 2003. Variable properties in a single class of excitatory spinal synapse. J. Neurosci. 23, 3154–3163.

Paton, J.F., 1996. A working heart-brainstem preparation of the mouse. J. Neurosci. Methods 65, 63–68.

Patrone, L.G., Bicego, K.C., Hartzler, L.K., Putnam, R.W., Gargaglioni, L.H., 2014. Cardiorespiratory effects of gap junction blockade in the locus coeruleus in unanesthetized adult rats. Respir. Physiol. Neurobiol. 190, 86–95.

Perrins, R., Roberts, A., 1995a. Cholinergic and electrical motoneuron-to-motoneuron synapses contribute to on-cycle excitation during swimming in *Xenopus* embryos. J. Neurophysiol. 73, 1005–1012.

Perrins, R., Roberts, A., 1995b. Cholinergic and electrical synapses between synergistic spinal motoneurones in the *Xenopus laevis* embryo. J. Physiol. 485, 135–144.

Pineda, J., Aghajanian, G.K., 1997. Carbon dioxide regulates the tonic activity of locus coeruleus neurons by modulating a proton- and polyamine-sensitive inward rectifier potassium current. Neuroscience 77, 723–743.

Rekling, J.C., Feldman, J.L., 1997. Bidirectional electrical coupling between inspiratory motoneurons in the newborn mouse nucleus ambiguus. J. Neurophysiol. 78, 3508–3510.

Rekling, J.C., Champagnat, J., Denavit-Saubie, M., 1996a. Electroresponsive properties and membrane potential trajectories of three types of inspiratory neurons in the newborn mouse brain stem in vitro. J. Neurophysiol. 75, 795–810.

Rekling, J.C., Champagnat, J., Denavit-Saubie, M., 1996b. Thyrotropin-releasing hormone (TRH) depolarizes a subset of inspiratory neurons in the newborn mouse brain stem in vitro. J. Neurophysiol. 75, 811–819.

Rekling, J.C., Shao, X.M., Feldman, J.L., 2000. Electrical coupling and excitatory synaptic transmission between rhythmogenic respiratory neurons in the preBötzinger complex. J. Neurosci. 20, RC113.

Remmers, J.E., deGroot, W.J., Sauerland, E.K., Anch, A.M., 1978. Pathogenesis of upper airway occlusion during sleep. J. Appl. Physiol. Respir. Environ. Exerc. Physiol. 44, 931–938.

Reyes, E.P., Cerpa, V., Corvalan, L., Retamal, M.A., 2014. Cxs and Panx- hemichannels in peripheral and central chemosensing in mammals. Front. Cell Neurosci. 8, 123.

Rodman, J.R., Harris, M.B., Rudkin, A.H., St-John, W.M., Leiter, J.C., 2006. Gap junction blockade does not alter eupnea or gasping in the juvenile rat. Respir. Physiol. Neurobiol. 152, 51–60.

Rouach, N., Segal, M., Koulakoff, A., Giaume, C., Avignone, E., 2003. Carbenoxolone blockade of neuronal network activity in culture is not mediated by an action on gap junctions. J. Physiol. 553, 729–745.

Rozental, R., Srinivas, M., Spray, D.C., 2001. How to close a gap junction channel. Efficacies and potencies of uncoupling agents. Methods Mol. Biol. 154, 447–476.

Saint-Amant, L., Drapeau, P., 2001. Synchronization of an embryonic network of identified spinal interneurons solely by electrical coupling. Neuron 31, 1035–1046.

Schmitz, D., Schuchmann, S., Fisahn, A., Draguhn, A., Buhl, E.H., Petrasch-Parwez, E., Dermietzel, R., Heinemann, U., Traub, R.D., 2001. Axo-axonal coupling. A novel mechanism for ultrafast neuronal communication. Neuron 31, 831–840.

Serrano-Velez, J.L., Rodriguez-Alvarado, M., Torres-Vazquez, I.I., Fraser, S.E., Yasumura, T., Vanderpool, K.G., Rash, J.E., Rosa-Molinar, E., 2014. Abundance of gap junctions at glutamatergic mixed synapses in adult Mosquitofish spinal cord neurons. Front. Neural Circuits 8.

Sharifullina, E., Ostroumov, K., Grandolfo, M., Nistri, A., 2008. N-methyl-D-aspartate triggers neonatal rat hypoglossal motoneurons in vitro to express rhythmic bursting with unusual Mg^{2+} sensitivity. Neuroscience 154, 804–820.

Shepherd, G.M., Grillner, S. (Eds.), 2010. Handbook of Brain Microcircuits. Oxford University Press, New York.

Smith, G.T., Zakon, H.H., 2000. Pharmacological characterization of ionic currents that regulate the pacemaker rhythm in a weakly electric fish. J. Neurobiol. 42, 270–286.

Smith, J.C., Ellenberger, H.H., Ballanyi, K., Richter, D.W., Feldman, J.L., 1991. Pre-Bötzinger complex: a brainstem region that may generate respiratory rhythm in mammals. Science 254, 726–729.

Smith, G.T., Lu, Y., Zakon, H.H., 2000. Parvocells: a novel interneuron type in the pacemaker nucleus of a weakly electric fish. J. Comp. Neurol. 423, 427–439.

Soffe, S.R., Roberts, A., Li, W.C., 2009. Defining the excitatory neurons that drive the locomotor rhythm in a simple vertebrate: insights into the origin of reticulospinal control. J. Physiol. 587, 4829–4844.

Solomon, I.C., Dean, J.B., 2002. Gap junctions in CO_2-chemoreception and respiratory control. Respir. Physiol. Neurobiol. 131, 155–173.

Solomon, I.C., Halat, T.J., El-Maghrabi, R., O'Neal III, M.H., 2001. Differential expression of connexin26 and connexin32 in the pre-Bötzinger complex of neonatal and adult rat. J. Comp. Neurol. 440, 12–19.

Solomon, I.C., Chon, K.H., Rodriguez, M.N., 2003. Blockade of brain stem gap junctions increases phrenic burst frequency and reduces phrenic burst synchronization in adult rat. J. Neurophysiol. 89, 135–149.

Song, J., Ampatzis, K., Bjornfors, E.R., El Manira, A., 2016. Motor neurons control locomotor circuit function retrogradely via gap junctions. Nature 529, 399–402.

Suzue, T., 1984. Respiratory rhythm generation in the in vitro brain stem-spinal cord preparation of the neonatal rat. J. Physiol. 354, 173–183.

Thoby-Brisson, M., Karlén, M., Wu, N., Charnay, P., Champagnat, J., Fortin, G., 2009. Genetic identification of an embryonic parafacial oscillator coupling to the preBötzinger complex. Nat. Neurosci. 12, 1028–1035.

Tresch, M.C., Kiehn, O., 2000. Motor coordination without action potentials in the mammalian spinal cord. Nat. Neurosci. 3, 593–599.

Verdier, D., Lund, J.P., Kolta, A., 2004. Synaptic inputs to trigeminal primary afferent neurons cause firing and modulate intrinsic oscillatory activity. J. Neurophysiol. 92, 2444–2455.
Walton, K.D., Navarrete, R., 1991. Postnatal changes in motoneurone electrotonic coupling studied in the in vitro rat lumbar spinal cord. J. Physiol. 433, 283–305.
Wang, D., Grillner, S., Wallen, P., 2006. Effects of flufenamic acid on fictive locomotion, plateau potentials, calcium channels and NMDA receptors in the lamprey spinal cord. Neuropharmacology 51, 1038–1046.
Welsh, J.P., Turecek, J., 2017. Implications of electrical synapse plasticity in the inferior olive. In: Jing, J. (Ed.), Network Functions and Plasticity. Elsevier (Chapter 12).
Wenker, I.C., Sobrinho, C.R., Takakura, A.C., Moreira, T.S., Mulkey, D.K., 2012. Regulation of ventral surface CO_2/H^+-sensitive neurons by purinergic signalling. J. Physiol. 590, 2137–2150.
Winmill, R.E., Hedrick, M.S., 2003. Gap junction blockade with carbenoxolone differentially affects fictive breathing in larval and adult bullfrogs. Respir. Physiol. Neurobiol. 138, 239–251.
Zhang, H.Y., Li, W.C., Heitler, W.J., Sillar, K.T., 2009. Electrical coupling synchronises spinal motoneuron activity during swimming in hatchling *Xenopus* tadpoles. J. Physiol. 587, 4455–4466.
Zhao, F.Y., Wolf, E., Roberts, A., 1998. Longitudinal distribution of components of excitatory synaptic input to motoneurones during swimming in young *Xenopus* tadpoles: experiments with antagonists. J. Physiol. 511 (Pt. 3), 887–901.
Zupanc, G.K.H., 2002. From oscillators to modulators: behavioral and neural control of modulations of the electric organ discharge in the gymnotiform fish, *Apteronotus leptorhynchus*. J. Physiol. Paris 96, 459–472.

CHAPTER 12

Implications of Electrical Synapse Plasticity in the Inferior Olive

J.P. Welsh[1,2], J. Turecek[3]
[1]Seattle Children's Research Institute, Seattle, WA, United States; [2]University of Washington, Seattle, WA, United States; [3]Harvard Medical School, Boston, MA, United States

1. INTRODUCTION

The inferior olive (IO) has among the highest density of neuronal gap junctions in the brain of the adult mammal, making it a favorable structure in which to study the physiology and function of electrical synapses. In rats, each neuron within the IO is electrically coupled to 50 other IO neurons (Devor and Yarom, 2002). The electrical synapses occur at convergences of 4–10 dendritic spines that receive inhibitory and excitatory synaptic input wrapped in an astrocytic sheath into a functional unit that is called the "IO glomerulus" (Sotelo et al., 1974; King, 1976; De Zeeuw et al., 1990). Gap junctions among dendritic spines of IO neurons (Rutherford and Gwyn, 1980) and electrical coupling within the IO (Turecek et al., 2016) are present also in nonhuman primates, thereby demonstrating conservation of function through phylogeny and translational relevance to humans. There is no chemical synaptic transmission between IO neurons and there are no interneurons in the IO. Thus all direct communication between IO neurons is via electrical synapses. Unlike many other brain regions that possess electrically coupled neurons at birth, but lose electrical coupling in the early postnatal period (see Li and Rekling, 2017 (Chapter 11)), the opposite is true for the IO. In the rodent IO, electrical synapses are absent at birth but develop shortly thereafter and are maintained into adulthood (Bourrat and Sotelo, 1983).

There is an astounding number of gap junctions within the IO. It is estimated that nearly every dendritic spine within the IO is invested within a glomerulus (De Zeeuw et al., 1998). Golgi staining demonstrated that spines are most heavily present on the distal third of the dendritic arbor where they have an approximate density of two spines per 10 μm (Gwyn et al., 1977), but they can also occur in clusters (Sotelo et al., 1974). Each unilateral IO in the rat contains 24,400 neurons (Schild, 1970) and the mean total length of a rat IO neuron's dendritic arbor is 4.33 mm (Turecek et al., 2016). Thus each IO neuron contains approximately 300 dendritic spines producing a total of 7 million spines in the unilateral IO of the rat. Using an estimate of seven spines per glomerulus (Sotelo et al., 1974), we calculate that there are approximately 1 million glomeruli in the rat IO, each of which provides multiple opportunities for electrical coupling.

Network Functions and Plasticity
ISBN 978-0-12-803471-2, http://dx.doi.org/10.1016/B978-0-12-803471-2.00012-6

An example of an IO glomerulus published by Sotelo et al. (1974) is shown in Fig. 12.1A. The example glomerulus is an ovoid with a major axis of 4.8 μm and contains a core of seven dendritic spines contacted by five synaptic terminals. The immense number of IO glomeruli that support the electrically coupled syncytium, as well as the persistence of gap junctions after the completion of brain development, is testament to the importance of electrical coupling to IO function over the lifetime of the animal.

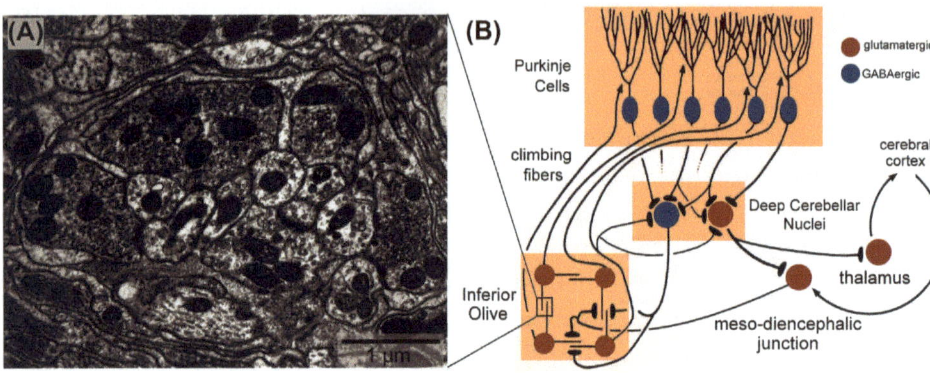

Figure 12.1 The inferior olive (IO) glomerulus and the brain circuitries influenced by the IO. (A) An IO glomerulus showing a core of seven dendritic spines receiving five synaptic terminals encased by an astrocytic sheath. (B) Concentric brain-circuitry loops influenced by IO activity. The innermost loop (*highlighted*) is the olivocerebellar system, which consists of IO neurons, Purkinje cells (PCs), and deep cerebellar nuclei (DCN) neurons. IO neurons are electrically coupled via dendrodendritic gap junctions in IO glomeruli (*box*). IO neurons are glutamatergic and their axons project to PCs as climbing fibers and to the DCNs as axon collaterals. PCs are GABAergic and project to DCNs. Due to electrical coupling within the IO, climbing fibers can trigger synchronous complex spikes in PCs that produce deep hyperpolarization and rebound excitation in DCN neurons. There are two subpopulations of DCN neurons. GABAergic DCN neurons project back to the IO and their terminals innervate dendritic spines within IO glomeruli to regulate the strength of electrical coupling. Glutamatergic DCN neurons participate in a middle and an outer loop by projecting to nuclei in the mesodiencephalic junction and thalamus, respectively. The former projection defines the olivocerebellar-mesodiencephalic loop in which glutamatergic projections from the mesodiencephalic junction return to the IO to innervate IO neuron somata and IO glomeruli. The thalamic projection defines the olivocerebellar-thalamocortical-mesodiencephalic loop in which excitatory projections to motor and nonmotor thalamus polysynaptically influence widespread areas of the cerebral cortex. Cerebral cortex output can influence the IO via descending projections to nuclei within the mesodiencephalic junction that complete the outer loop by projecting back to the IO. GABAergic and glutamatergic inputs to the IO modulate electrical coupling in opposite directions (see Fig. 12.9). *(Modified from (A) Sotelo, C., Llinás, R., Baker, R., 1974. Structural study of inferior olivary nucleus of the cat: morphological correlates of electrotonic coupling. J. Neurophysiol. 37, 541–559, (B) Llinás, R., Welsh, J.P., 1993. On the cerebellum and motor learning. Curr. Opin. Neurobiol. 3, 958–965; McCormick, D.A., 1995. Motor control. The cerebellar symphony. Nature 374, 412–413; De Zeeuw, C.I., Simpson, J.I., Hoogenraad, C.C., Galjart, N., Koekkoek, S.K., Ruigrok, T.J., 1998. Microcircuitry and function of the inferior olive. Trends Neurosci. 21, 391–400.)*

Understanding the functional consequences of electrical synapses in the IO and the plasticity mechanisms that modulate their strength will be essential for understanding cerebellar operations and how the cerebellum regulates motor and nonmotor functions. Fig. 12.1B shows the brain circuitry that is influenced by electrical coupling in the IO. The axonal projections of IO neurons are solely to the cerebellum and give off collaterals that innervate the deep cerebellar nuclei (DCN) before they ascend into the cerebellar cortex as climbing fibers to innervate Purkinje cells. Each Purkinje cell is innervated by one climbing fiber (Szentágothai and Rajkovits, 1959; Hámori and Szentágothai, 1966). Purkinje cells are GABAergic and their output is primarily to the DCN (Ito and Yoshida, 1966) but they also give collaterals that innervate other Purkinje cells and interneurons (Witter et al., 2016). There are two subpopulations of DCN neurons that differ in their neurotransmitter identification and axonal projection. GABAergic neurons within the DCN project exclusively to the IO and innervate dendritic spines within IO glomeruli (Fredette and Mugnaini, 1991). Glutamatergic neurons within the DCN project to the thalamus and to a collection of nuclei at the junction of the midbrain and the diencephalon that includes the parvocellular and magnocellular red nuclei, the tegmental field of Forel, zona incerta, subparafasicularis nucleus, prerubral reticular formation, and the nuclei of Darkschewitz, Bechterew, and interstitialis of Cajal. Many of the nuclei at the mesodiencephalic junction, with the major exception of the magnocellular red nucleus, which projects to the spinal cord, project back to the IO to provide glutamatergic innervation of IO neuron somata and glomeruli (Saint-Cyr and Courville, 1981; Onodera, 1984; Ruigrok and Voogd, 1995). Because a major target of the excitatory neurons within the DCN is the motor and nonmotor thalamus (Middleton and Strick, 1997), electrical coupling within the IO and its influence on cerebellar output also modulates throughput via the thalamocortical system. Many nuclei within the mesodiencephalic junction receive input from the cerebral cortex, allowing the cerebrum to influence IO activity via the subcortical excitatory pathway. The concentric organization of the looped olivocerebellar, olivocerebellar-mesodiencephalic, and olivocerebellar-thalamocortical-mesodiencephalic systems are all influenced by electrical coupling in the IO that ultimately modulates a wide variety of motor and nonmotor functions.

2. ELECTRICAL COUPLING IN THE IO AND MOVEMENT

2.1 The Discovery of Electrical Synapses in the IO

Electrical synapses between neurons in the IO were first reported in a landmark pair of papers by Rodolfo Llinás, Robert Baker, and Constatino Sotelo (Llinás et al., 1974; Sotelo et al., 1974). The papers were noteworthy because they were the first to describe the anatomy of dendrodendritic gap junctions and a consequence of electrical synaptic transmission on neuronal activity in the mammalian brain. The search for electrical synaptic transmission in the mammalian brain was motivated by a fertile period in which the physiology of electrical synapses was studied in the nervous systems of invertebrates (Watanabe, 1958; Furshpan and Potter, 1959; Asada and Bennett, 1971; Spira and Bennett, 1972) and teleosts (Bennett et al., 1963, 1967a,b; Furshpan, 1964; Auerbach and Bennett, 1969; Kriebel et al., 1969; see Pereda and Bennett, 2017 (Chapter 8)). Contemporaneously, the report of gap junctions between bipolar and ganglion neurons in the human retina (Dowling and Boycott, 1966) increased the likelihood that the integrative properties of electrical synapses could be studied in the mammalian brain. In vivo intracellular electrophysiology subsequently identified electrical synapses between neurons in the mesencephalic trigeminal nucleus of the rat, a primary sensory ganglion located within the brain (Baker and Llinás, 1971). A later study demonstrated that electrical transmission existed between axon terminals and their postsynaptic targets on giant neurons in the lateral vestibular nucleus of the rat (Korn et al., 1973). Those terminals were remarkable for containing both a chemical synapse and a gap junction (Sotelo and Palay, 1970) in so-called mixed synapses, which had been first identified in the goldfish medulla (Furshpan, 1964) where they are actively studied today (Pereda et al., 1994, 1998; Pereda and Faber, 1996; Cachope and Pereda, 2015).

The aforementioned studies set the stage for experiments upon the cat IO that provided the first demonstration of electrical synaptic transmission between mammalian brain neurons. The recording of antidromic responses by single intracellular microelectrodes triggered by stimulating a peripheral nerve was the method used to infer electrical synapses in the rat mesencephalic trigeminal nucleus (Baker and Llinás, 1971). Yet a location within the mammalian brain where a discrete electrical stimulus could trigger antidromic responses in a large set of neurons relatively free from orthodromic synaptic input was difficult to locate. A favorable case was the basal white matter of the cerebellum where axons of IO neurons collect before they fan out to innervate the cerebellar cortex. Discrete voltage shocks at that location had been used to activate the axons of IO neurons in cats to study the climbing fiber-Purkinje cell synapse (Eccles et al., 1964, 1966a,b). In the course of those studies, it was observed that the same stimuli also produced a second, longer-latency wave of excitation that was originally thought to be conducted by axon collaterals of the climbing fibers. However, upon further consideration (Llinás et al., 1974), it was deduced that the second wave of excitation was due to the coincident antidromic volley to the IO and the subsequent "reflex" orthodromic activation of Purkinje cells through the full length of the olivocerebellar projection. This second wave of excitation, due to what became termed the "climbing fiber reflex," was more widespread over the cerebellar cortex than the initial orthodromic wave, foreshadowing a stream of experiments on electrical synaptic transmission. The ability to reliably and specifically activate a large set of IO neurons by stimulating their axons in the cerebellar white matter provided the essential groundwork to discover electrical synaptic transmission between mammalian neurons.

Using intracellular recording from IO neurons in anesthetized cats (Llinás et al., 1974), the evidence for electrical synaptic transmission was the occurrence of a "short-latency depolarization" (SLD) occurring 0–300 μs after the initiation of the antidromic spike triggered by cerebellar white matter stimulation (Fig. 12.2A). The SLD was revealed when the stimulus intensity was lowered just below the threshold for triggering the spike and indicated that there was a second driving force for membrane depolarization that had nearly the same latency as the spike but that did not show the latency variation typical for chemical synaptic transmission. The SLD could be recruited to increasingly large amplitudes up to 5 mV from resting membrane potential with graded increases in shock intensity without a change in latency (Fig. 12.2B) and was insensitive to changes in membrane potential induced by intracellular current injection. Thus the SLD was distinct from the all-or-none response of the initial segment and was synchronized to the times in which it was presumed that neighboring neurons fired an

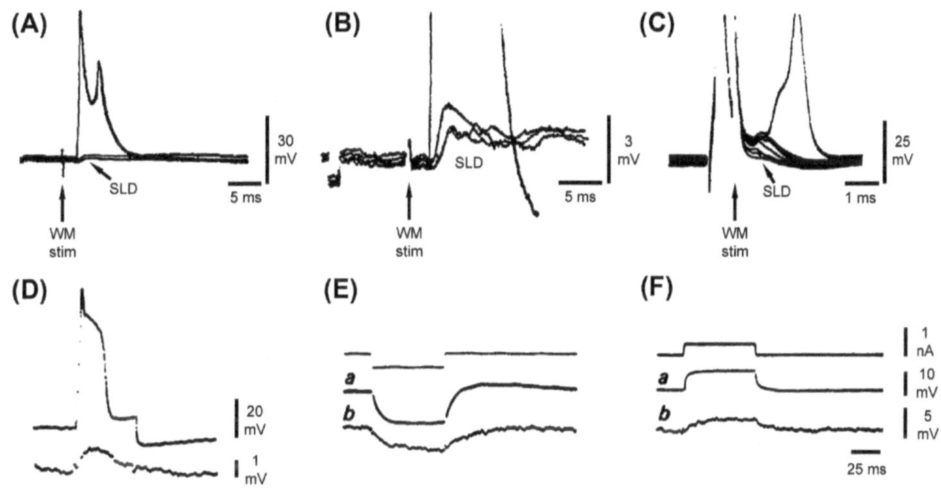

Figure 12.2 Original evidence for electrical coupling in the inferior olive (IO). (A) An intracellularly recorded IO neuron in a cat showing antidromic responses triggered by stimulation of the cerebellar white matter (WM stim). An action potential is triggered above the WM stim threshold. When WM stim is below the spike threshold, short latency depolarizations (SLDs) are evoked. (B) Higher gain shows the graded nature of the SLD and its timing relative to the action potential triggered by suprathreshold WM stim. (C) A large-amplitude SLD triggers an action potential when it occurs as the membrane is depolarized during postspike repolarization. (D) Dual recordings from two IO neurons in a brainstem slice from a guinea pig showing that an action potential in one IO neuron induces an SLD in the other. (E, F) The two-microelectrode test of electrical coupling in neighboring IO neurons reveals coupled potentials in cell b when hyperpolarizing (E) or depolarizing (F) current is injected into cell a. *(Modified from (A–C) Llinás, R., Baker, R., Sotelo, C., 1974. Electrotonic coupling between neurons in cat inferior olive. J. Neurophysiol. 37, 560–571 and (D–F) Llinás, R., Yarom, Y., 1981. Electrophysiology of mammalian inferior olivary neurones in vitro. Different types of voltage-dependent ionic conductances. J. Physiol. 215, 549–567.)*

antidromic spike. By the process of elimination, it was concluded that currents underlying the SLD were conducted via electrical synapses from neighboring neurons having different activation thresholds due to the placement of the stimulating electrode. It was noted that SLDs were observed in nearly all of the recordings, implying that electrical synapses were ubiquitous in the IO. Concurrent ultrastructural studies demonstrated the presence of dendrodendritic gap junctions and identified the IO glomerulus as the integrative unit (Sotelo et al., 1974). Of particular importance for this chapter, the peak depolarization of the SLD measured in vivo was rarely large enough to trigger a spike on its own, but could cross the spike threshold if it occurred when the membrane was already depolarized, such as during postspike repolarization (Fig. 12.2C). In sum, the evidence indicated for the first time that electrical synapses between brain neurons could underlie synchronous spiking.

The deduction that electrical synaptic transmission was ubiquitous in the IO was found to be correct when studies using dual microelectrodes were performed in brainstem slices of the guinea pig (Llinás and Yarom, 1981). The first dual intracellular recordings from the IO demonstrated the SLD in IO neurons when a neighboring neuron fired an action potential that had the same magnitude and time course seen in vivo (Fig. 12.2D). When hyperpolarizing (Fig. 12.2E) or depolarizing (Fig. 12.2F) currents were passed through one of two IO neurons, using the dual stimulation test, coupled potentials were recorded in the other providing direct evidence of electrical coupling. The demonstration of coupled potentials in IO neurons using dual intracellular current injection has been replicated in every species tested, including rats (Devor and Yarom, 2002; Hoge et al., 2011; Turecek et al., 2014), mice (Long et al., 2002; De Zeeuw et al., 2003), and rhesus macaques (Turecek et al., 2016). When connexin36 was discovered to be the primary connexin comprising neuronal gap junctions in mammals (Condorelli et al., 1998; Söhl et al., 1998; Belluardo et al., 1999; Parenti et al., 2000; Rash et al., 2000), it was not surprising that all neurons in all subnuclei of the adult IO showed very high expression, as seen both in the protein and mRNA maps (Belluardo et al., 2000; Condorelli et al., 2000).

2.2 In Vivo Estimates of Electrical Coupling Strength and Patterning in the IO

The finding of prevalent electrical coupling between IO neurons yielded a straightforward hypothesis: spike synchrony was the raison d'être of the olivocerebellar system. Because lesions of the olivocerebellar system had long been known to impair motricity (Murphy and O'Leary, 1971), it was reasonable to hypothesize that spike synchrony within the IO was related to the control of movement. For decades it had been understood that each Purkinje cell received only one climbing fiber (Szentágothai and Rajkovits, 1959) and that the activation of a single climbing fiber triggered a burst of dendritic spikes in a Purkinje cell, termed the "climbing fiber response," which was mechanistically different from the initial segment spike that was triggered by the coincident synaptic input of large numbers of parallel fibers as recorded using sharp intracellular microelectrode (Eccles et al., 1964, 1966a–d; Llinás and Sugimori, 1980). The different responses to climbing fiber and parallel fiber inputs were seen also in extracellular recordings in vivo and thoroughly described (Eccles et al., 1964, 1966a–d), before they were later termed the "complex spike" and "simple spike," respectively (Thach, 1967). The arrangement in which each Purkinje cell received only one climbing fiber (Fig. 12.1B) allowed the occurrence of a complex spike to refer back to an action potential in a single neuron in the IO. Purkinje cell complex spikes occurred infrequently, typically 1–2 Hz on average, which made it difficult to understand how IO neurons could contribute to motor control. Although many studies using single-microelectrode recordings had related the occurrence of complex spikes to short- and long-term rate modulations of simple spiking, especially in vitro (see Ito et al., 2014), the prevalence of electrical synapses in the IO indicated that previous hypotheses of cerebellar operations could be reevaluated within a framework of ensemble coding. This insight brought forth a clear rationale to focus efforts on in vivo electrophysiology using an approach that allowed multiple Purkinje cells to be recorded simultaneously. Thus a multimicroelectrode recording method was developed to allow the isolated recording of complex spikes of 96 individual Purkinje cells within a few square millimeters of a single cerebellar lobule (Sasaki et al., 1989; Llinás and Sasaki, 1989).

The major questions were: (1) was there synchronization of complex spikes in ensembles of Purkinje cells that could be attributed to the presence of electrical synapses in the IO? and (2) was any such synchronization related to an aspect of motor control? In anesthetized rats, the first set of multielectrode studies of the climbing fiber system showed a magnitude of complex spike synchrony among groups of Purkinje cells that was significantly greater than chance (Sasaki et al., 1989; Llinás and Sasaki, 1989). In those experiments, the spatial pattern of complex spike synchrony was observed as bands of Purkinje cells having a distinct rostrocaudal orientation (Fig. 12.3A), indicating that patterns of electrical coupling in the IO were read out onto distinct geometries of synchronized activity in the cerebellum. Of particular interest was that the quantitative analyses showed that complex spike synchrony was relatively infrequent and had cross-correlation values between 10% and 20% when complex spikes within 10 ms of one another were defined to be synchronous (Sasaki et al., 1989) and between 1% and 5% when synchrony was defined as complex spikes occurring within 1 ms of one another (Sugihara et al., 1993; Lang et al., 1996; Yamamoto et al., 2001). From the point of view of the function of electrical synapses in the IO, only values of complex spike synchrony calculated using co-occurrences within 1–2 ms can assess the direct contribution of electrical synapses in the IO, given the maximum 300 μs latency and 1–2 ms rise time of the SLDs in IO neurons (Fig. 12.2) and uniform conduction time of IO axons to the Purkinje cell layer (Sugihara et al., 1993). From this point forward, complex spike "synchrony," as

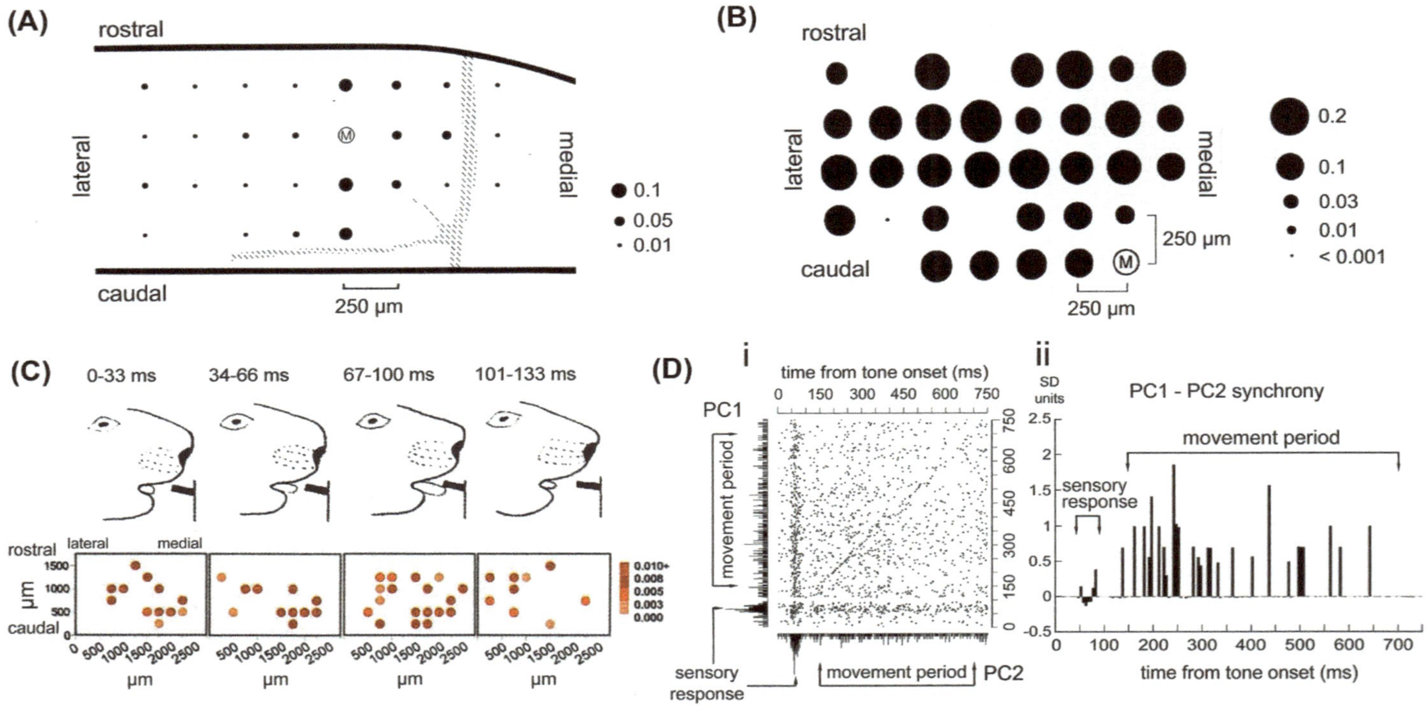

Figure 12.3 Olivocerebellar synchrony in vivo. (A) Plot of a cerebellar folial surface (crus 2a) showing the location of 28 Purkinje cell (PC) recordings in a ketamine-anesthetized rat. Circle diameters indicate degree of complex spike cross-correlation with a reference PC (M, 10 ms bins). (B) Same analysis for 33 PCs in the absence of GABAergic input from the deep cerebellar nuclei (1 ms bins). (C) Similar analysis showing the dynamics of complex spike synchrony in 33-ms epochs during a learned tongue movement in a rat (1 ms bins). (D) Joint-peristimulus time histogram (JPSTH) analysis of the complex spikes of 2 PCs recorded simultaneously during tongue movements triggered by a tone (as in C) at time 0. (Di): Peristimulus time histograms of the complex spikes of 2 PCs along the axes show complex spikes at short latency after tone onset, indicating a sensory response, but low complex spike firing subsequently during tongue movement (movement period). The inner matrix is the JPSTH and plots the co-occurrences of all complex spikes among the 2 PCs at varying time lags relative to one another as a function of time after tone onset. Co-occurrences along the diagonal demonstrate the temporal dynamics of complex spike synchrony after tone onset. (Dii): Values of complex spike synchrony taken from the diagonal of the JPSTH normalized for firing rate show virtual lack of synchrony during the sensory response but significant synchrony during movement. *(Modified from (A) Sasaki, K., Bower, J.M., Llinás, R., 1989. Multiple Purkinje cell recording in rodent cerebellar cortex. Eur. J. Neurosci. 1, 572–586, (B) Lang, E.J., Sugihara, I., Llinás, R., 1996. GABAergic modulation of complex spike activity by the cerebellar nucleoolivary pathway in rat. J. Neurophysiol. 76, 255–275, (C) Welsh, J.P., Lang, E.J., Sugihara, I., Llinás, R., 1995. Dynamic organization of motor control within the olivocerebellar system. Nature 374, 453–457, and (D) Welsh, J.P., 2002. Functional significance of climbing-fiber synchrony: a population coding and behavioral analysis. Ann. NY Acad. Sci. 978, 188–204.)*

an index of electrical synaptic transmission within the IO, will refer exclusively to the co-occurrence of complex spikes within 1 ms of one another.

The findings using multiple microelectrode recordings of complex spikes in cerebellar cortex indicated that synchrony within the IO is highly constrained in both structure and time. The GABAergic projection from the DCN to the IO provides an important structural constraint, as the geometry of Purkinje cells showing complex spike synchrony expanded across the cerebellar cortex after pharmacological blockade of GABAa receptors in the IO or destruction of the GABAergic input from the DCN (Lang et al., 1996, Fig. 12.3B). Moreover, GABAergic synapses also provide an important time constraint for synchronous firing as the prevalence of synchrony in complex spikes increased from 1% to 3% under control conditions to 10–40% after removing GABAergic input to the IO by lesioning the DCN or intraolivary injection of picrotoxin (Lang et al., 1996). Those findings confirmed the long-standing hypothesis that the activation of GABA receptors near the site of gap junctions increases the nonjunctional conductance and uncouples IO neurons by shunting current away from the electrical synapse (Llinás, 1974). It was demonstrated that complex spike synchrony was nearly abolished by pharmacological blockade of gap junctions by local injection of carbenoxolone into the IO of rats (Blenkinsop and Lang, 2006). Moreover, it was subsequently demonstrated that the climbing fiber reflex was abolished in connexin36 knockout mice (Marshall et al., 2007). This last finding completed the loop that was begun by Llinás et al. (1974) by demonstrating that the second activation of Purkinje cells following axonal stimulation was indeed mediated by electrical coupling among IO neurons, as first recognized in the SLD.

In sum, the findings demonstrated that global views of cerebellar operations needed to consider three important facts: (1) electrical coupling with the IO recruits populations of Purkinje cells into well-defined groups of synchronously firing elements, (2) function is distributed over the network in an ensemble code, and (3) the prevalence of complex spike synchrony is low (<10%) as determined by cross-correlation.

2.3 IO Electrical Synapses, Skilled Movement, and Muscle Synergies

The finding of complex spike synchrony during anesthesia demonstrated a functional consequence of electrical coupling in the IO, but left unanswered whether its low probability of occurrence and stereotypic spatial patterning was typical of IO function during sensory processing and movement. Thus, patterns of complex spike synchrony were measured during a skilled tongue movement as thirsty rats licked a water spout at a far distance in response to a tone that signaled water availability (Welsh et al., 1995). The paradigm allowed dozens of glass microelectrodes to be lowered individually into a single cerebellar folium without impeding a natural behavior. The recordings during wakeful movement demonstrated complex spike synchrony in the range of 1–10%, which was not significantly different than the magnitude of synchrony under anesthesia. However, during movement the patterns of complex spike synchrony showed a different geometry than the parasagittal bands seen under anesthesia and took the form of interdigitated clusters that flickered on and off as the tongue was extended and retracted in and out of the mouth (Fig. 12.3C). Each cluster of Purkinje cells showed synchronous complex spikes relating to specific epochs of the movement in which distinct patterns of muscle activation occurred. Moreover, the spatial patterns of complex spike synchrony during movement changed rapidly, on the order of tens of milliseconds (centiseconds). The data indicated that electrically coupled clusters of neurons in the IO reserved their synchronous firing for specific times of movement and that the superposition of the clusters at a given moment represented the specific combination of muscles that was being activated (ie, the "muscle synergy"; see Bernstein, 1967; Welsh and Llinás, 1997). Moreover, the study demonstrated that the spatial patterning of electrical coupling in the IO showed much more flexibility during movement than during rest or under anesthesia.

An important characteristic of complex spike synchrony during wakeful movement was that it occurred against a low population rate of complex spike firing. The presence of synchrony during low population firing differed significantly from the immediate poststimulus period in which population firing was high and complex spike synchrony was low (Welsh, 2002, Fig. 12.3D). This is demonstrated in a representative joint peristimulus time histogram (Fig. 12.3Di) in which the individual complex spike histograms of two Purkinje cells on movement trials showed robust responses evoked by the signaling tone at 50 ms latency (Fig. 12.3Di, sensory response) and a quick return to a low rate as the movement was performed (Fig. 12.3Di, movement period). Within the joint peristimulus time histogram, the diagonal represents the occurrence of complex spike synchrony over time, which is notable during the movement period despite very low firing. When normalized to represent synchrony values exceeding that expected by chance coincidences due to the ongoing firing rate, it was found that complex spike synchrony during the sensory response was not greater than that expected due to the high firing, whereas complex spike synchrony during the movement period far exceeded that expected due to low firing (Fig. 12.3Dii). The analysis provided an early indication that the strength of electrical coupling within the IO was strengthened in real time as the brain transitioned from processing a sensory stimulus to executing a skilled motor response.

The study of complex spike synchrony in awake rats demonstrated that the patterning of electrical coupling within the IO was dynamic and controlled in a way during a skilled movement that specifically related to motricity. It was proposed that synaptic input established the spatial pattern of electrical coupling within the IO prior to the initiation of the movement, thereby assembling the functional clusters that fired synchronously as the movement was executed (Welsh et al., 1995; Welsh and Llinás, 1997). Once the spatial patterning was set, spikes among neurons within the defined clusters would be triggered at the appropriate time relating to muscle firings on the beat of an internal clock that also defined the speed of the movement, with spike intervals between the clusters determined by phase-offsets in the clock.

As we will see later, a reasonable candidate for the hypothesized internal clock is IO neurons' oscillation in membrane potential that occurs below the voltage threshold for triggering spikes. Precisely how such spatial patterning is established is an area of ongoing investigation. At least two processes are involved. The first process is that the pattern is established by real-time modulation of coupling by the GABAergic projection from the DCN, which uses synaptic inhibition to establish boundaries between neuronal clusters within which neurons can fire synchronously (Llinás, 1974; Lang et al., 1996; Lefler et al., 2014). This process is analogous to sculpting. A second process, which we recently demonstrated, is that the pattern of electrotonic coupling is established in the IO via a plasticity process in which electrical synapses are strengthened by glutamatergic synapses (Turecek et al., 2014). This process is analogous to constructing.

2.4 Electrical Synapses in the IO Contribute Centisecond Precision to Muscle Synergies

Although the electrophysiology indicated a role for electrical synapses in the IO for the temporal coordination of motor synergies using correlational analyses, three studies provided confirmatory evidence using genetic knock down of connexin36 in the IO.

The first study used stereotaxic injection of a recombinant lentiviral vector in adult rats to uncouple IO neurons by blocking the trafficking of connexin36 protein to dendritic spines and disrupting the formation of electrical synapses (Placantonakis et al., 2002, 2004, 2006). The initial demonstration expressed an engineered version of connexin36 containing a single amino acid mutation that acted as a dominant negative of wild-type connexin36 in cell cultures (Placantonakis et al., 2002). In such cultures, expression of the mutant connexin36 prevented the formation of gap junctions by trapping wild-type connexin36 in the endoplasmic reticulum and blocking tracer coupling. When the mutant connexin36 was expressed in the IO of adult rats (Placantonakis et al., 2004, 2006), connexin36 was sequestered in dendritic shafts and there was a significant reduction in the presence of connexin36 at dendritic spines (Fig. 12.4A, B). Expression of the mutant connexin36 in the IO of adult rats blocked tracer coupling between IO neurons to an extent equivalent to the block of tracer coupling by carbenoxolone (Fig. 12.4C, D).

The effect of uncoupling IO neurons on behavior was studied using whole-body tremor induced by systemic administration of harmaline (Bruinvels, 1969; Llinás and Volkind, 1973). It is established that harmaline tremor originates in the IO due to the drug's action to induce sustained rebound spiking in IO neurons at the 10-Hz tremor frequency to drive repetitive trains of complex spikes in Purkinje cells and rhythmic cerebellar output (Lamarre et al., 1971; Llinás and Volkind, 1973; De Montigny and Lamarre, 1973, 1975; Llinás and Yarom, 1986). Although harmaline tremor is not a natural

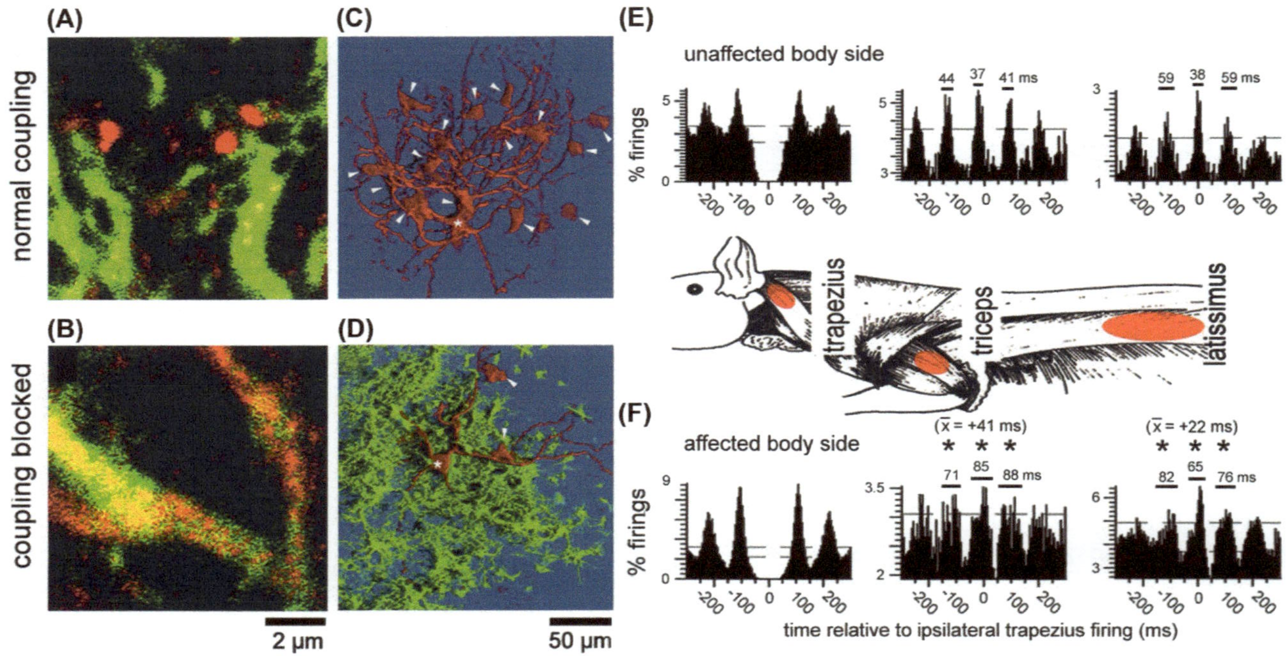

Figure 12.4 Connexin36-mediated coupling in the inferior olive (IO) contributes temporal precision to a muscle synergy. (A) Connexin36 expression (*red*) in the dendrites of green fluorescent protein (GFP)-labeled IO neurons is punctate along the margins of IO dendrites, consistent with localization to dendritic spines. (B) Connexin36 is sequestered within dendritic shafts after expression of a mutant connexin36 having dominant-negative effect on the transport of wild-type connexin36. (C) The typical pattern of tracer coupling in the normal IO. (D) Near block of tracer coupling in IO neuron expressing the mutant connexin36. In C and D arrowheads indicate tracer coupled neurons after injection of Neurobiotin into the neuron with the asterisk. Green shows coexpressed GFP. (E, F) Effect of unilateral uncoupling of IO neurons in adult rats on correlated muscle activity during harmaline tremor for the body sides controlled by a normally coupled IO (E, unaffected body side) and an uncoupled IO (F, affected body side) in the same rat. For each body side, from left to right the histograms show an autocorrelation histogram of trapezius muscle firings and cross-correlation histograms of the triceps and latissimus muscles referenced to the ipsilateral trapezius. Values above the cross-correlograms indicate width at half peak and show that the timing of triceps and latissimus firings relative to trapezius is degraded by 41 and 22 ms on the affected side. *(Modified from (A, B) Placantonakis, D.G., Bukovsky, A.A., Aicher, S.A., Kiem, H.P., Welsh, J.P., 2006. Continuous electrical oscillations emerge from a coupled network: a study of the inferior olive using lentiviral knockdown of connexin36. J. Neurosci. 26, 5008–5016, and (C–F) Placantonakis, D.G., Bukovsky, A.A., Zeng, X.H., Kiem, H.P., Welsh, J.P., 2004. Fundamental role of inferior olive connexin 36 in muscle coherence during tremor. Proc. Natl. Acad. Sci. USA 101, 7164–7169.)*

behavior, it is expressed as a well-organized muscle synergy consisting of muscle firings that are synchronized across spinal levels. Thus harmaline tremor had the advantage of being a well-defined muscle synergy that was induced specifically by synchronous activity in the IO. As such, harmaline tremor provided a strong test paradigm to determine whether electrical coupling of IO neurons contributed to the temporal organization of muscle activation during a whole-body movement.

Uncoupling IO neurons with genetic and anatomical specificity reduced the coherence of muscle firing across spinal levels during tremor (Fig. 12.4E, F). The paradigm unilaterally transduced the IO of adult rats with the mutant connexin36 to block electrical coupling and recorded tremor in the trapezius, triceps brachii, and latissimus dorsi muscles bilaterally. The tremor itself was not prevented by local uncoupling, indicating that repetitive spiking of IO neurons did not require connexin36-mediated coupling, but can be expressed so long as T-type calcium channels are present (Park et al., 2010). The principal finding was that tremor on the body side contralateral to side of the uncoupled IO showed significant temporal dispersion in the onsets of tremor muscle bursts as compared with the unaffected body side, consistent with the exclusively crossed projection of the olivocerebellar pathway and ipsilateral influence of the cerebellum on the body musculature. The tremor bursts on the affected body side were 30–50 ms more variable in time relative to the unaffected body side, which was tightly coherent (compare Fig. 12.4E, F). The experiment provided the first quantitative estimate of the contribution of electrical synapses in the IO for the organization of a movement by demonstrating that they contribute to temporal precision at the centisecond timescale.

Two follow-up studies employed connexin36 knockout mice to examine influences of electrical synapses in the IO on reflex behavior. Mice having global connexin36 knockout showed altered timing of conditioned eyeblinks during Pavlovian conditioning (Van Der Giessen et al., 2008), a paradigm of associative learning that requires an intact IO (Welsh and Harvey, 1998). An advantage of Pavlovian eyeblink conditioning is that the target behavior, closure of the external eyelid, is simple and is expressed through the coordinated activity of only two antagonistic muscles that close and open the eyelid (Manning and Evinger, 1986), which determines the timing of the blink. Mice with global connexin36 knockout closed their eyelids in response to the tone conditioned stimulus approximately 100 ms earlier than wild-type mice (Van Der Giessen et al., 2008). As a control, conditioned eyeblinks were unaffected by brain-wide knockout of connexin36 in GABAergic neurons that did not affect the IO. The absence of a change in eyeblink timing in mice in which connexin36 in the IO was unaffected motivated a conclusion that the change in timing by global knockout was due to a loss of electrical coupling in the IO. In a second study of connexin36-knockout mice (De Gruijl et al., 2014), stereotypic limb and tail reflexes triggered by perturbations while walking on a rotating platform were also altered in timing. Here, the onset latencies of the reflexes were unchanged, but the response topography was more variable beginning 10 ms after movement onset. For both the conditioned eyeblinks and postural reflexes it remains to be determined how the relative timing of responsible muscle activations is affected by uncoupling IO neurons, which will be essential to understand the mechanistic contribution of electrical synapses in the IO to those behaviors.

In sum, behavioral studies indicated that movement timing is degraded on the timescale of centiseconds after the removal of connexin36 and electrical coupling in the IO. Different outcomes were observed depending on the behavior examined and included effects on response onset, response stereotypy, and the relative timing of muscle activation. The latter finding indicates that the temporal precision of muscle synergies during movement requires electrical synapses in the IO, which may explain why conditioned and unconditioned reflexes are altered in timing following connexin36 knockout. The behavioral evidence is consistent with the multielectrode physiology indicating that electrical synapses within the IO provide temporal precision to movement.

3. ELECTRICAL COUPLING AND SUBTHRESHOLD OSCILLATIONS IN THE IO

3.1 Electrical Synapses Are Necessary for Subthreshold Oscillations in the IO

Electrophysiological behaviors that are often correlated with electrical synaptic transmission are subthreshold resonance and oscillation in membrane potential (Traub et al., 2003; Soto-Trevino et al., 2005; Fuentealba and Steriade, 2005; Tchumatchenko and Clopath, 2014). The IO is not an exception, and, indeed, was among the first systems identified in the mammalian brain that contained neurons whose membrane potential oscillated below spike threshold continuously at a nearly invariant frequency (Benardo and Foster, 1986; Llinás and Yarom, 1986; Bleasel and Pettigrew, 1992) and synchronously when neighboring neurons were recorded (Benardo and Foster, 1986; Llinás and Yarom, 1986). It had been reasonably assumed that electrical synapses between IO neurons were responsible for the synchrony of subthreshold oscillations (STOs). This was confirmed indirectly when it was found that the pharmacological block of T-type calcium channels in the entire IO blocked spontaneous STOs, whereas STOs in an individual IO neuron were unaffected by somatic current injection that depolarized the neuron out of the activation range for T-type calcium channels (Llinás and Yarom, 1986). The finding indicated that STOs could be sustained by currents that were conducted through gap junctions from the surrounding ensemble of oscillating neurons.

Subsequent theoretical and experimental work suggested that electrical synapses had a role that was more fundamental than ensuring STO synchrony—it was proposed that electrical synapses were required for STOs to exist within the IO (Lampl and Yarom, 1997; Manor et al., 1997, 2000). In this view, the transjunctional current mediated by electrical synapses allowed for resonant interactions and equalized the differences in pacemaking currents between coupled neurons, to amplify and distribute the capacity for STOs to all IO neurons throughout the network. The hypothesis predicted that STOs could not exist in the absence of electrical coupling, even though all of the individual neurons possessed the relevant pacemaking conductances. Strong support for the hypothesis had come from a developmental study (Bleasel and Pettigrew, 1992) in which it was demonstrated that STOs in the IO first appeared when electrical synapses were formed (see Fig. 12.5A, B), at approximately P10-P15 for rats (Bourrat and Sotelo, 1983).

Thus, it was surprising when the first electrophysiology of connexin36 knockout mice showed robust STOs in electrically-uncoupled IO neurons (Long et al., 2002). Although STOs of uncoupled IO neurons were neither synchronized nor locked in phase, they were normal in amplitude, frequency, and prevalence, which appeared to disprove the hypothesis that STOs were an emergent property of the electrically coupled network. The conclusion that IO neurons were single-cell oscillators was tempered by a follow-up study of connexin36 knockout mice demonstrating a 34% increase in the T-type calcium current in uncoupled IO neurons, a developmental compensation that potentially explained why STOs occurred in the absence of electrical coupling (De Zeeuw et al., 2003). Dynamic clamp demonstrated that inclusion of the additional T-type current into wild-type IO neurons supported low-threshold spikes at holding potentials more negative than rest, consistent with the possibility of an enhanced contribution of low-threshold currents to STOs of connexin36 knockout mice (De Zeeuw et al., 2003). However, the dynamic clamp experiments could not prove that T-current upregulation was sufficient for IO neurons to show STOs while uncoupled because the tested IO neurons from wild-type mice were integrated within a network of coupled oscillators.

Aware of the potential confounding effects of developmental compensation following embryonic gene deletion (Knobelman et al., 2001), Placantonakis et al. (2006) further explored this issue using viral knock down of Cx36 in IO neurons of young adult rats. The local genetic block of connexin36-mediated coupling between IO neurons occurred without the changes in the electrical properties of IO neurons that followed embryonic gene deletion in mice. The most significant finding was that IO neurons uncoupled from their surrounding ensemble did not show the large-amplitude, continuous STOs observed in normally coupled IO neurons, but instead showed stable resting membrane potential or, at best, very low amplitude ($\sim$1 mV), erratically organized, and discontinuous STOs (Fig. 12.5C, D). The result demonstrated that the large-amplitude and continuous STOs seen in the normal IO were, in fact, an emergent property of the ensemble enabled by electrical coupling. Genetic block of connexin36-mediated coupling within the adult IO recapitulated an early developmental state in which the expression of pacemaking conductances in the absence of coupling was insufficient to generate STOs (Bleasel and Pettigrew, 1992).

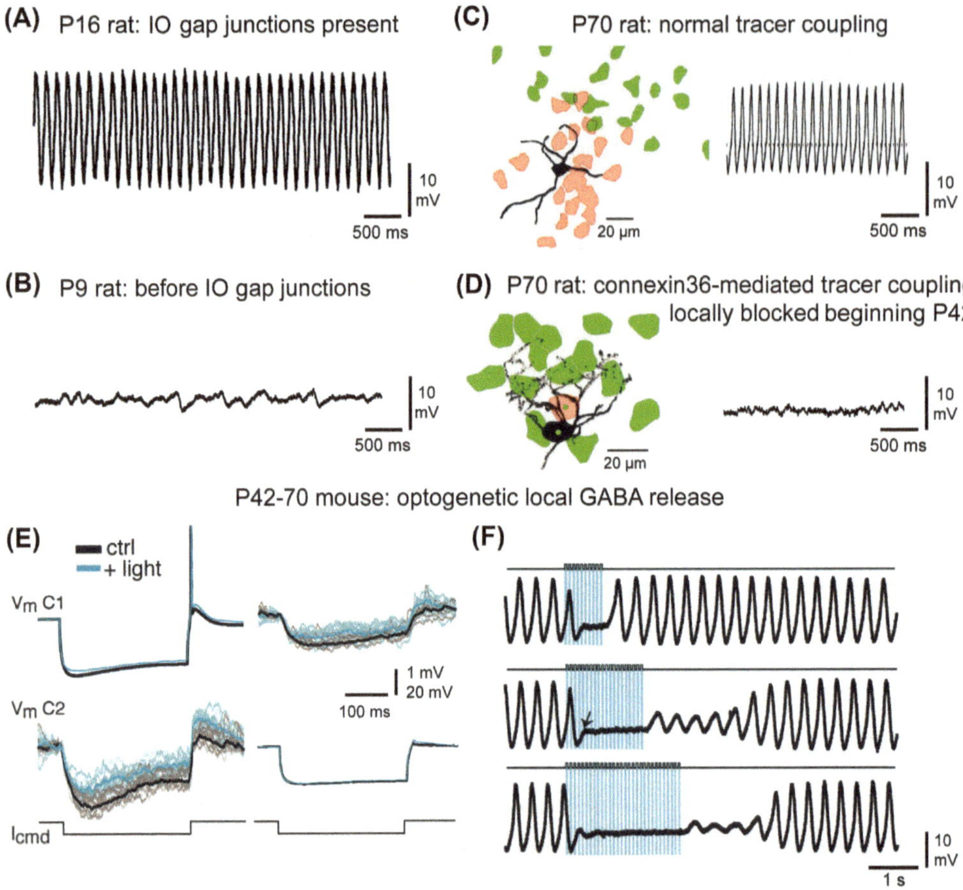

Figure 12.5 Necessity of electrical coupling for subthreshold oscillations (STOs) among inferior olive (IO) neurons. (A, B) Membrane potential of rat IO neurons at P16 when gap junctions are present in glomeruli of IO dendrites (A) and at P9 when gap junctions have not yet developed (B). (C, D) Membrane potential of IO neurons of young adult rats (~P70) in which coupling is normal (C) and nearly completely blocked by local expression of an engineered form of connexin36 that blocks coupling (D). Drawings on the left show the intracellularly recorded neuron (*black*), tracer coupled neurons (*red*), and IO neuron somata that express a mutant connexin36 with dominant-negative effect on wild-type connexin36 plus green fluorescent protein (*green*). (E, F) Light-induced release of GABA reduces electrical coupling of two IO neurons without affecting resting membrane potential (E) and abolishes STOs (F) in young adult mice. *(Modified from (A, B) Bleasel, A.F., Pettigrew, A.G., 1992. Development and properties of spontaneous oscillations of the membrane potential in inferior olivary neurons in the rat. Brain Res. Dev. Brain Res. 65, 43–50, (C, D) Placantonakis, D.G., Bukovsky, A.A., Aicher, S.A., Kiem, H.P., Welsh, J.P., 2006. Continuous electrical oscillations emerge from a coupled network: a study of the inferior olive using lentiviral knockdown of connexin36. J. Neurosci. 26, 5008–5016, and (E, F) Lefler, Y., Yarom, Y., Uusisaari, M.Y., 2014. Cerebellar inhibitory input to the inferior olive decreases electrical coupling and blocks subthreshold oscillations. Neuron 81, 1389–1400.)*

A recent finding provided additional support for the importance of electrical coupling for the manifestation of STOs in IO neurons. Using tissue-specific expression of channelrhodopsin in GABAergic terminals in the IO, Lefler et al. (2014) light evoked GABA release onto IO neurons, which was the mechanism proposed to uncouple IO neurons by current shunting (Llinás, 1974; Lang et al., 1996). Under optimum parameters, light-activated release of GABA reduced the strength of electrical coupling by 26–43% without producing measureable somatic hyperpolarization (Fig. 12.5E). In such preparations, GABA release also blocked STOs for the duration of the light stimulus (Fig. 12.5F). The finding was consistent with an important role of electrical coupling among IO neurons for the expression of STOs and provided the additional insight that the expression of STOs is sensitive to subtotal reductions in the strength of electrical coupling. In sum, multiple lines of evidence now demonstrate that electrical coupling is necessary for STOs to be manifested in the IO network and further demonstrate that electrical synapses are a membrane site at which neuromodulators can control STOs.

3.2 Direct (SLDs) and Indirect (STOs) Effects of Electrical Synapses on Synchronous Spiking Within the IO

Electrical coupling among IO neurons provides two mechanisms that work together to produce synchronous spiking. The first mechanism allows for an SLD in an IO neuron when a coupled neuron fires a spike (Fig. 12.2). It directly contributes to spike synchrony because the SLD can be a spike-triggering event. Yet the ability of an SLD to trigger a spike in a coupled neuron is conditional and requires an additional depolarizing factor. The second mechanism enabled by electrical synapses helps to fulfill that condition by allowing for the emergence of STOs (Fig. 12.5) and is indirect because it provides a facilitating precondition for spike synchrony. Like SLDs, STOs also are not usually sufficiently depolarizing to trigger spikes, but differ from SLDs in providing a continuous backdrop of recurring, gradual approaches to spike threshold upon which a spike can be triggered by an SLD or an excitatory postsynaptic potential (EPSP). Although IO neurons' spikes are not fully constrained in time by STOs, the increased probability of triggering spikes near the crest of the STO underlies spike synchrony only because the STOs themselves are synchronized. The stochastic nature of the interaction between synaptic potentials (chemical or electrical) and STO phase combined with the low coupling strength of IO neurons (Devor and Yarom, 2002; Hoge et al., 2011; Turecek et al., 2014, 2016) explains why the probability of synchronous spiking in vivo under baseline conditions is low, typically less than 10% (Fig. 12.3). The same logic applies for low-threshold spikes generated below mean resting membrane potential and during a phase of the STO that straddles the deinactivation threshold of T-type currents.

The discovery that electrical synapses are required for STOs (Bleasel and Pettigrew, 1992; Placantonakis et al., 2006; Lefler et al., 2014) is fundamentally important for understanding the function of the IO and cerebellum, and suggests that any upward or downward modulation in the strength of electrical coupling can have a profound effect on spike synchrony by producing an analog modulation of STO amplitude. The analog modulation is expressed as a reduction in STO amplitude, or their abolition, when electrical coupling is weakened (Fig. 12.5) or as an enhancement of STO amplitude when electrical coupling is strengthened, via a plasticity process that will be discussed next. Analog control of STO amplitude by modulating electrical coupling has the functional consequence of augmenting or reducing spike synchrony in a graded manner by increasing or reducing the percentage of the STO cycle time at which SLDs or EPSPs can trigger a spike. As we will see next, the study of *N*-methyl-D-aspartate (NMDA) receptor involvement in IO rhythmogenesis contributed to the discovery of strengthening plasticity of electrical synaptic transmission between IO neurons and analog enhancement of STO amplitude.

4. ENHANCING STOs BY UPREGULATING ELECTRICAL COUPLING BY NMDA RECEPTOR ACTIVATION

4.1 NMDA Receptor Contributions to Electrical Rhythmogenesis in the IO

The line of research that led to the finding of strengthening plasticity of electrical synapses in the IO began with the publication of Du and Harvey (1997) which demonstrated that systemic administration of MK-801, an uncompetitive antagonist of the NMDA receptor, blocked harmaline tremor in rabbits. Subsequent in vitro experiments demonstrated that harmaline displaced MK-801 binding in membrane fractions obtained from IO neurons (Du et al., 1997). The finding indicated that harmaline interacted with the phencyclidine binding sites within the NMDA receptor channel where MK-801 was bound. The ability of harmaline to displace MK-801 in vitro and to produce tremor by increasing rhythmic activity in the IO in vivo suggested that harmaline acted as an inverse agonist at the MK-801 binding site to increase the constitutive activity of NMDA receptor-mediated currents (Du and Harvey, 1997; Du et al., 1997). The pharmacological experiments replicated the effect of ibogaine, another tremorogenic beta-carboline, to displace MK-801 binding in caudate and cerebellum (Popik et al., 1994; Mash et al., 1995).

The findings were surprising because they were at odds with the physiology demonstrating that harmaline produced tremor by hyperpolarizing IO neurons and shifting the deinactivation threshold of the T-type calcium channel to increase the amplitude of IO neurons' STOs (Llinás and Yarom, 1986). Larger STOs explained the cellular mechanism of how harmaline induced tremor, because the enhancement of STOs provided a more robust carrier rhythm for generating sustained 10-Hz trains of action potentials in IO neurons and repetitive trains of complex spikes in Purkinje cells (Biscoe et al., 1973; De Montigny and Lamarre, 1975; Headley et al., 1976; Bernard et al., 1984; Sinton et al., 1989). The conflict was that harmaline exerted its tremorogenic action by promoting activity at hyperpolarized membrane potentials (Llinás and Yarom, 1986), whereas the NMDA receptor exerted its influence at depolarized potentials due to the voltage dependence of the magnesium block. Thus it was clear that NMDA receptor involvement in IO neuron rhythmogenesis was not well understood and needed to be examined.

Using intracellular recording of IO neurons in acutely-prepared brainstem slices (Placantonakis and Welsh, 2001), it was found that MK-801 was almost 100% effective in inhibiting STOs in IO neurons, confirming a role for NMDA receptors in IO rhythmogenesis and explaining the effect on tremor (Fig. 12.6A). Because neither blocking the glutamate binding site on the NMDA receptor nor antagonizing α-amino-3-hydroxy-5-methyl-4-isoxazolepropionic acid (AMPA) and kainate glutamate receptors inhibited STOs, the data suggested that constitutive activity of the NMDA receptor was the important contributor to electrical rhythmogenesis in IO neurons. A direct involvement of the NMDA receptor was demonstrated when NMDA receptor activation induced STOs in IO neurons with stable membrane potential, whereas neither AMPA nor kainate receptor activation induced STOs (Placantonakis and Welsh, 2001). STOs occurring during NMDA receptor activation were slower and larger than spontaneous STOs and occurred at depolarized membrane potentials outside of the activation voltage range of T-type calcium channels. When an IO neuron showed STOs spontaneously, NMDA receptor activation depolarized the neuron, increased STO amplitude, and slowed STO frequency by about half (Fig. 12.6B).

As the initial examination of NMDA receptor involvement in STOs was ongoing, Genevieve Yuen in our group examined the localization of NR1 subunits of the NMDA receptor in IO glomeruli. Her work, later published in Turecek et al. (2014), demonstrated the presence of NMDA receptors at postsynaptic densities within IO glomeruli (Fig. 12.6C) and also the presence of extrasynaptic NMDA receptors on spines 10−100 nm away from gap junctions (Fig. 12.6C, D). High-resolution light microscopy demonstrated the colocalization of connexin36-NR1 immunopuncta pairs within the cytosol of IO dendrites and tandems of connexin36 immunopuncta at sites of dendritic adherence, each flanked by an NR1 subunit (Fig. 12.6E, F). The localization of the extrasynaptic NMDA receptors near dendritic gap junctions indicated the potential for a bidirectional interaction between electrical synapses and NMDA receptors to promote rhythmogenesis. In one direction, an inward current through the gap junction sourced by the STO of a neighbor might enhance the receptor's constitutive activity by helping to remove the magnesium block within the NMDA channel via local membrane depolarization. In the other direction, an inward Ca^{2+} current conducted through the NMDA receptor channel might modulate the junctional conductance via a local increase in Ca^{2+} (Peracchia, 1978; Baux et al., 1978) or have an effect on the gap junction via phosphorylation. A third possibility was that the opening of the NMDA receptor channel might reduce the conductance through the gap junction by current shunting (Bennett, 1966).

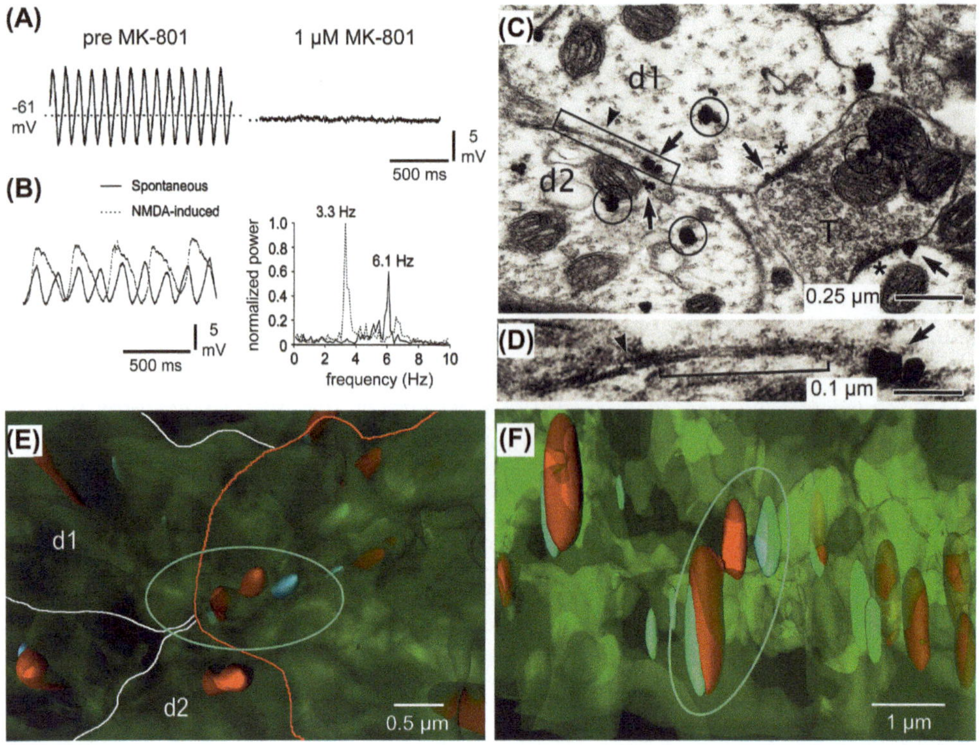

Figure 12.6 The NMDA receptor in inferior olive (IO) function. (A) Intracellular recordings of an IO neuron showing abolition of spontaneous subthreshold oscillations (STOs) by 1 μM MK-801. (B) NMDA receptor activation (100 μM NMDA) increases STO amplitude and reduces STO frequency. (C) Locations of NR1 subunit of the NMDA receptor in an IO glomerulus at synaptic locations on the margin of postsynaptic densities (*arrows adjacent to asterisks*) and extrasynaptically ~50 nm from a gap junction coupling two dendritic spines (*arrow pointing at box*). The arrowhead shows site of a dendrodendritic gap junction. (D) Higher magnification of the boxed region in C showing the NR1 subunit adjacent to the gap junction (*bracket*). The arrowhead shows the margin of the gap junction. NR1 immuonlabeling is indicated by silver-intensified antibody-conjugated microbeads and arrows. (E) Confocal micrograph rendering showing connexin36 (*red*) and NR1 (*blue*) immunopuncta at a site viewed from above where two green fluorescent protein-filled dendrites (d1, d2) merge (*blue oval*). White lines indicate the edges of the two dendrites and the red line indicates their adherence. (F) The same structure viewed from the side showing connexin36-NR1 pairings both at the adherence (*blue oval*) and within the cytosol of the dendrites. *(Modified from (A, B) Placantonakis, D., Welsh, J.P., 2001. Two distinct oscillatory states determined by the NMDA receptor in rat inferior olive. J. Physiol. 534, 123–140 and (C–F)Turecek, J., Yuen, G.S., Han, V.Z., Zeng, X.H., Bayer, K.U., Welsh, J.P., 2014. NMDA receptor activation strengthens weak electrical coupling in mammalian brain. Neuron 81, 1375–1388.)*

There was no evidence to support phosphorylation of neuronal gap junctions in 1997 when we first observed the structural relationship, because connexin36 had not been identified as the primary neuronal gap junction protein. In the meantime, important functional evidence came from work on goldfish.

4.2 NMDA Receptor Activation Strengthens Neuronal Electrical Synapses

Strong indication of a functional interaction between NMDA receptors and electrical synapses came from work on the goldfish eighth-nerve Mauthner cell mixed synapse. The work demonstrated that tetanic stimulation of the goldfish eighth nerve strengthened both the electrical and chemical synaptic components of the mixed synapse, which was found to be due to NMDA receptor activation and an increase in intracellular calcium (Yang et al., 1990). The strengthening of electrical coupling was specifically due to an increase in calcium in the Mauthner cell, implicating a calcium-dependent second messenger molecule that modulated the postsynaptic side of the gap junction (Pereda and Faber, 1996). The molecule was identified as calcium-calmodulin dependent kinase II (CaMKII; Pereda et al., 1998). Across studies, the mean increase in the electrical coupling potential triggered by tetanic stimulation was approximately 35% and could last an hour or more. Freeze-fracture immunolabeling and electron microscopy of the mixed synapse demonstrated that the NMDA receptors in postsynaptic densities of the Mauthner cell were located within 100 nm of connexin35-mediated gap junctions (Pereda et al., 2003), providing a structural correlate to the functional interaction similar to the NMDA receptor localization within IO glomeruli that we observed (Fig. 12.6C, D). A subsequent freeze-fracture double immunolabeling experiment confirmed the close proximity of NR1 subunits and connexin36 within the IO glomerulus (Rash et al., 2004).

After it was discovered that connexin36 was the gap junction protein for neuronal electrical synapses in the mammalian brain (Condorelli et al., 1998), the subsequent sequencing of the connexin36 gene revealed two consensus sites for phosphorylation, one in the cytoplasmic loop and the other in the carboxyl terminal (Alev et al., 2008). Immunohistochemistry detected robust colocalization of CaMKII and connexin36 in IO neurons (Alev et al., 2008), suggesting the interaction that was found in the goldfish for connexin35 (Pereda et al., 1998). The hypothesis of CaMKII modulation of connexin36 was supported when pharmacological blockade of CaMKII and deleting both phosphorylation sites prevented the strengthening of connexin36-mediated conductance in a neuroblastoma expression system (Del Corsso et al., 2012). Subsequent studies of rabbit retina demonstrated colocalized NR1 subunits and connexin36 on AII amacrine cells and demonstrated that an increase in tracer coupling among AII amacrine cells induced by light stimulation required NMDA receptors and the phosphorylation of connexin36 via CaMKII (Kothman et al., 2012; see Ribelayga and O'Brien, 2017 (Chapter 10)).

4.3 NMDA Receptor-Mediated Strengthening of Electrical Synapses in the IO Enhances Network Rhythmogenesis

Because the data from our laboratory obtained by using gene transfer, electrophysiology, and subcellular anatomical imaging demonstrated the necessity of neuronal gap junctions for STOs (Fig. 12.5B), a role for the NMDA receptor in electrical rhythmogenesis of IO neurons (Fig. 12.6A), and a structural correlate of a potential interaction between the two (Fig. 12.6C–F), we examined the effect of NMDA receptor activation on electrical coupling between IO neurons in acutely prepared brainstem slices (Turecek et al., 2014). This research was motivated also by the physiological experiments performed on the goldfish that demonstrated a role for activity-dependent phosphorylation of a neuronal electrical synapse by CaMKII to enhance electrical coupling (Pereda and Faber, 1996; Pereda et al., 1998). Prior to our work, whether the findings from a fish cranial nerve synapse would translate to electrical coupling between central neurons in mammalian brain remained the important outstanding question.

Close examination of the coupling strength of IO neurons of postweaned rats demonstrated a mean coupling coefficient of approximately 1% and that the vast majority of neuron pairs up to 100 μm separation exhibited a junctional conductance between 50 and 200 pS (Turecek et al., 2014). The mean values of junctional conductance that we measured were slightly less than, but showed a similar range to, the values reported in a study of IO neuron pairs separated by up to 50 μm obtained from younger, preweaned rats (Hoge et al., 2011). We found the magnitude of the junctional conductance among IO neurons to be much lower than the means reported for other well-studied systems, such as retinal AII amacrine cells (400 pS, Veruki et al., 2010), the developing reticular thalamic nucleus (up to 1 nS, Parker et al., 2009), retinal ganglion cells (1.35 nS, Hidaka et al., 2004), mesencephalic trigeminal neurons (4.8 nS, Curti et al., 2012), and crayfish septate axons (10 mS, Campos de Carvalho et al., 1984). The weak electrical coupling between IO neurons, largely attributable to the weak junctional conductance, helped explain the small SLDs measured both in vivo and in vitro (Fig. 12.2) and suggested that the high prevalence of electrical coupling distributed over the IO network (Devor and Yarom, 2002) is much more important than the strength of the individual pairwise couplings.

Activation of NMDA receptors by bath application of NMDA and electrical stimulation of glutamatergic axonal terminals strengthened electrical coupling between IO neurons, as measured by an increase in the coupling coefficient, and strengthened electrical synapses, as confirmed by an increase in the junctional conductance. Among weakly and moderately coupled IO neurons, the increase in the junctional conductance produced by NMDA receptor activation scaled positively with an increase in the input conductance of the neurons, indicating a role for the triggered calcium current to strengthen electrical synapses to an extent that countered the shunting inhibition resulting from the increased input conductance. Over equivalent magnitudes of increase of input conductance produced by AMPA and kainate receptor activation, there was a systematic reduction in the junctional conductance consistent with shunting inhibition. Those data provided the first demonstration of strengthening plasticity of electrical synapses in IO neurons in the mammalian brain (Turecek et al., 2014).

An important experimental outcome was that the magnitude of strengthening plasticity of electrical coupling was not uniform in all pairs but varied as a function of the baseline coupling coefficient and the distance between neuronal somata. This observation indicated that a functional consequence of strengthening plasticity of electrical synapses was to expand the distance over which IO neurons were maximally coupled, thereby decreasing the electrotonic distance between neurons and increasing the probability of their influencing one another.

A fundamental property of NMDA-receptor mediated strengthening plasticity was that it did not strengthen electrical coupling beyond the strength of directly adjacent neurons under baseline conditions. The fundamental finding is shown in a representative example (Fig. 12.7A). Here, a pair of IO neurons was bidirectionally coupled with a weak baseline coupling coefficient of 0.5% in both directions. The coupling coefficients were increased to 1.5% and 1.8% by NMDA receptor activation, measured as the elicited STOs were blocked pharmacologically. Fig. 12.7B shows the effect of NMDA receptor activation on coupling strength as a function of intersomatic distance. On average, IO neuron somata less than 25 μm apart had the highest baseline coupling strength and their coupling was not significantly strengthened by NMDA receptor activation. In contrast, IO neurons with somata separated by 25–75 μm had a lower baseline coupling coefficient that was nearly doubled by NMDA receptor activation. Finally, somata separated by 75–100 μm, allowing for only minimal dendritic overlap, showed the lowest baseline coupling coefficients that were only weakly increased by NMDA receptor activation. Expressed as percentage change, the increase in the strength of electrical coupling induced by NMDA receptor activation was 11% at 10 μm, 108% at 35 μm, 122% at 60 μm, and 34% at 88 μm intersomatic distances (Fig. 12.7C). Thus the effect of NMDA receptor activation to strengthen electrical coupling operated primarily on neurons that were both weakly coupled and had nonadjacent somata but were still close enough to allow substantial dendritic overlap and opportunity to form dendrodendritic gap junctions.

The functional consequence of strengthening electrical coupling was to synchronize STOs, increase their amplitude, and reduce their frequency (Fig. 12.7D). Three changes in STO behaviors were found: (1) neighboring IO neurons that did not show STOs before NMDA receptor activation showed synchronized STOs upon NMDA receptor activation (Fig. 12.7Di); (2) neighboring IO neurons that showed asynchronous STOs before NMDA receptor activation were synchronized by NMDA receptor activation (Fig. 12.7Dii); and (3) neighboring IO neurons that showed synchronous STOs before NMDA receptor activation showed larger and slower synchronized STOs during NMDA receptor activation (Fig. 12.6Diii). All of the effects on STOs occurred at depolarized resting membrane potentials at which STOs do not involve T-type calcium currents (Placantonakis and Welsh, 2001). Consistent with the findings in the goldfish, the strengthening plasticity of electrical synapses between IO neurons depended upon

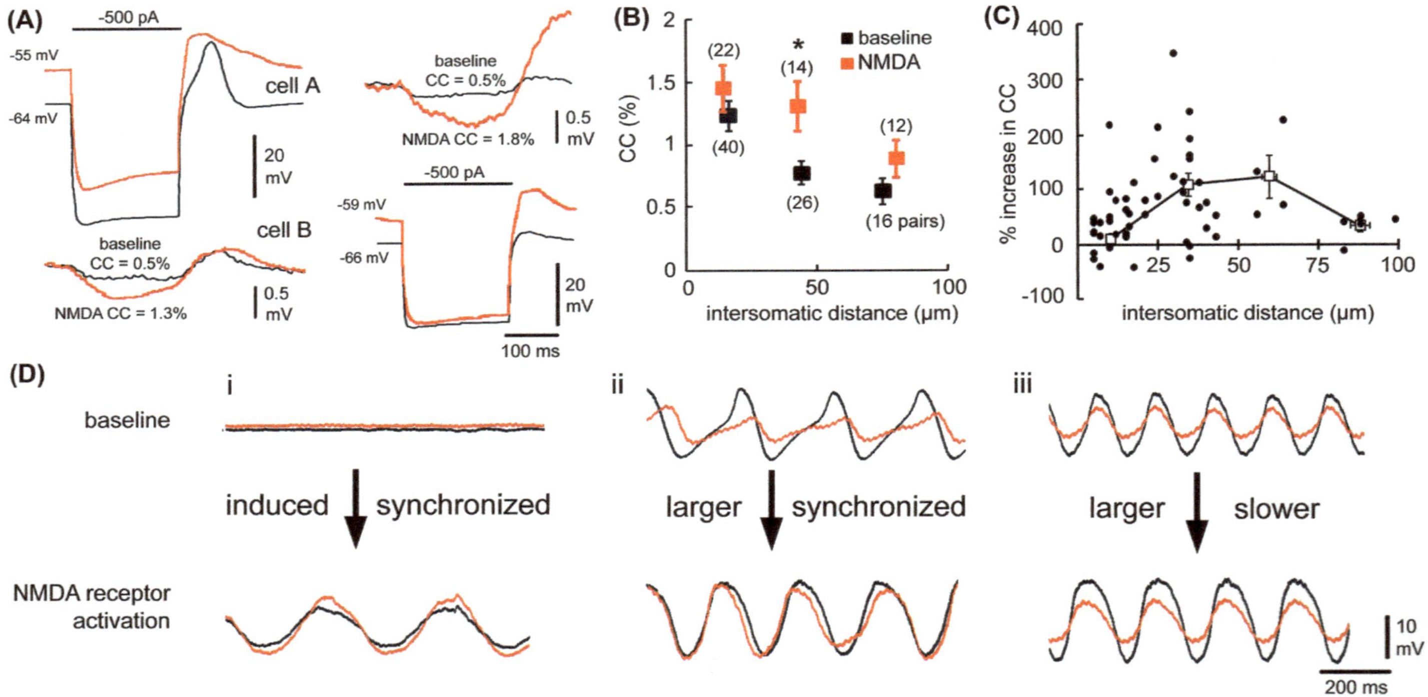

Figure 12.7 NMDA receptor-mediated strengthening plasticity of electrical coupling in the inferior olive (IO). (A) Dual whole-cell patch clamp recordings from two IO neurons show that current injection into cell A or B produces coupled potentials in the other neuron before NMDA (*black*) and larger coupled potentials during NMDA (*red*). (B) Coupling coefficients as a function of intersomatic distance before and during NMDA. *Asterisk* indicates $p < .05$ compared with baseline. (C) Percentage change in coupling coefficient as a function of intersomatic distance produced by NMDA receptor activation. Each *dot* indicates a coupling coefficient measurement ($n = 31$ pairs). The curve shows mean $\pm$ SEM. (D) NMDA receptor activation that strengthens electrical coupling induced synchronized subthreshold oscillations (STOs) (i), synchronized STOs that were previously asynchronous (ii), and enhanced STO amplitude and slowed STO frequency while maintaining synchrony (iii). *(Modified from Turecek, J., Yuen, G.S., Han, V.Z., Zeng, X.H., Bayer, K.U., Welsh, J.P., 2014. NMDA receptor activation strengthens weak electrical coupling in mammalian brain. Neuron 81, 1375–1388.)*

CaMKII, as determined by blocking its activation by Ca^{2+}/calmodulin and by blocking substrate access to the enzyme using intracellular delivery of a short peptide inhibitor (Vest et al., 2007). We hypothesize that the robust STOs in the IO induced by NMDA receptor activation result from the mutual reinforcement of strengthened electrical coupling and slow-frequency oscillations driven by voltage-dependent modulations in NMDA receptor-mediated currents that have been amply documented in other systems (Grillner and Wallen, 1985; Kim and Chandler, 1995; MacLean et al., 1997). In agreement, stable and persistent knock down of the NR1 subunit of the NMDA receptor by short-interfering RNAs blocked the ability of NMDA receptor activation to induce STOs in IO neurons (Zhu et al., 2015).

Fig. 12.8 illustrates the dynamics of the emergent network activity mediated by strengthening electrical coupling to produce synchronized spiking within the IO. Here, two IO neurons with somata adjacent to one another (Fig. 12.8A) showed stable membrane potential and moderate and strong electrical coupling (coupling coefficients of 0.9% and 2.7%, Fig. 12.8B) before NMDA was bath-applied to the IO. The global activation of NMDA receptors in this manner would have been expected to strengthen weak electrical coupling throughout the IO but only moderately in this particular pair due to their adjacency (eg, Fig. 12.7B, C). The electrical coupling strength was not assessed during NMDA receptor activation because both neurons exhibited NMDA-elicited STOs that were not blocked pharmacologically so we could examine the progression of activity below and above the spike threshold. The pair exhibited the STO behavior like the example shown in Fig. 12.7Di.

NMDA elicited STOs at different latencies in the two neurons, with cell A showing STOs prior to cell B. As the STOs of cell A became larger and the membrane depolarized, cell A began to fire spikes. The first five spikes drove SLDs in cell B (down arrows Fig. 12.8C) due to the strong electrical coupling in the direction of cell A to B. Although the SLDs occurred as cell B was developing STOs, their combination was not sufficient to trigger spikes. Only when the STOs of cell B were sufficiently amplified and the membrane sufficiently depolarized, after the sixth SLD, were spikes between the two cells synchronized (Fig. 12.8C, hash marks). The calcium imaging experiments demonstrated increases in fluorescence in varicosities within the dendritic arbor that preceded the onset of synchronous spiking (Fig. 12.8D, AV1, BV1), upon which transient responses relating to the spikes were later superimposed. That finding was important from a functional perspective because dendritic varicosities are the locations from which spines emerge to become invested in glomeruli (De Zeeuw et al., 1998). The calcium increases were specific to the dendritic varicosities, as dendritic shafts were nonresponsive (Fig. 12.8D, AS, BS), thereby implicating a plasticity process whose functional consequence is on dendrodendritic gap junctions but which also involves the varicosity-spine assembly as a potential source of intradendritic calcium to strengthen electrical synapses (Turecek et al., 2014).

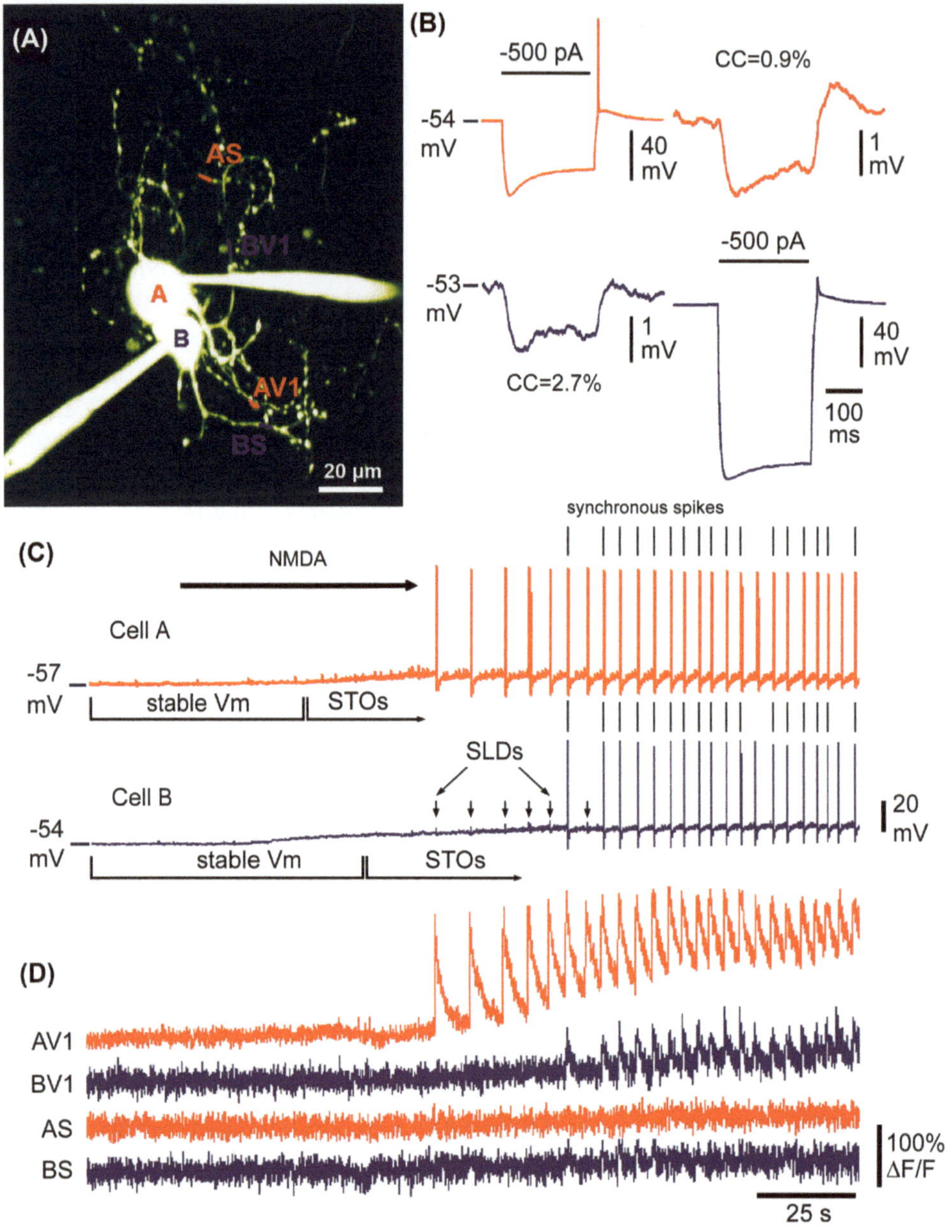

Figure 12.8 Dynamics of subthreshold oscillation (STO) generation and spike synchrony during strengthening electrical synapse plasticity by NMDA receptor activation. (A) A fluorescence image of dual whole-cell patch clamp recordings from two inferior olive (IO) neurons filled with Fluo-4 showing overlapping dendritic arbors and indicating dendritic varicosities (AV1, BV1) and dendritic shafts (AS, BS) whose calcium fluorescence was analyzed during NMDA receptor activation using fast two-photon microscopy. Left and right fluorescence lines are dye-filled patch electrodes. (B) Moderate and strong electrical coupling between cells A (*red*) and B (*blue*) prior to NMDA receptor activation. (C) Effect of bath-applied NMDA (50 μM) on the membrane potential of the two neurons. Hash marks indicate synchronous spikes. (D) Simultaneous two-photon imaging of calcium responses of two dendritic varicosities (AV1, BV1) and dendritic shafts (AS, BS). See text for discussion and Turecek et al. (2014, Supplemental Information) for description of the fast two-photon imaging method.

The example reinforces the following points: (1) electrical coupling among adjacent IO neurons is not sufficient to produce STOs; (2) enhancing weak electrical coupling throughout the IO recruits nonoscillating neurons into the network of oscillators; (3) SLDs mediated by strong electrical coupling are not always sufficient to generate synchronous spiking; and (4) synchronous spiking is due to the emergent STO behavior of the network, which is promoted by strengthening electrical coupling.

5. A HYPOTHESIS OF STRENGTHENING PLASTICITY OF ELECTRICAL SYNAPSES DURING THE LEARNING OF MOTOR SYNERGIES

The olivocerebellar system is required to perform motor skills and may be involved in their learning (Murphy and O'Leary, 1971; Rondi-Reig et al., 1997; Welsh, 2002). Motor skill learning requires assembling the optimum muscle synergies and coordinating their execution in time. This is as true for behaviors as simple as a conditioned eye blink as for complex behaviors that require a wider array of muscles such as a basketball layup, a golf swing, and playing the bass pedal organ. The foregoing evidence suggests that electrical synapses within the IO are important for the execution of motor skills by allowing for the synchronous spiking within groups of IO neurons and Purkinje cells that represent components of muscle synergies. A learning process that facilitates the ability to reconfigure the IO network into different combinations of functional groupings and enhances the precision of their read-out over time may importantly contribute to the acquisition of motor skills.

NMDA receptor strengthening of electrical coupling is a plasticity mechanism that may play an important role in motor skill learning by facilitating the assembly of cell groupings representing components of muscle synergies and enhancing their expression over time via a look-ahead process represented in synchronized STOs. Under baseline conditions, spike synchrony in the IO is strikingly low due to the weak junctional conductance that does not reliably support SLDs of sufficient amplitude to trigger spikes. Yet, the high prevalence of weak electrical coupling within the IO combined with the expression of pacemaking conductances in its neurons allows for the emergence of STOs. The low coupling strength of IO neurons provides a low baseline and makes available a wide dynamic range upon which a strengthening mechanism can have strong functional impact. Until recently, emphasis has been placed on the uncoupling mechanism mediated by GABAergic synapses from the DCN that desynchronizes IO neurons and inhibits STOs. As a complementary mechanism, strengthening plasticity of electrical synapses operating on a low coupling baseline can significantly change the network structure of synchronization in the IO by enhancing STOs.

Strengthening plasticity of electrical synapses in the IO may provide control over STO frequency as well as amplitude. A consequence of strengthening plasticity of electrical synapses in the IO is the slowing of STOs from 6 to 10 Hz into the 2–3 Hz range, which occurs as STOs are synchronized and amplified. This may be at least partially explained by the different kinetics of high-threshold membrane potential oscillations controlled by the NMDA receptor (Grillner and Wallen, 1985; Kim and Chandler, 1995; MacLean et al., 1997), in addition to an increase in membrane capacitance that would be expected to occur with stronger electrical coupling (Turecek et al., 2016). Spatially targeted recruitment of IO glomeruli by glutamatergic inputs originating in higher motor centers may provide the assembly mechanism to establish synchronous groupings in the IO.

A common strategy for learning new motor skills is to practice them slowly to obtain proficiency with the optimum muscle synergies and thereafter increase the speed of execution gradually with repetition and feedback until a desired performance is reached. A second strategy is to break the motor skill into subsequences, or "chunks," and learn them independently at slow speed before concatenating them into a desired motor sequence. The trade-off between movement speed and accuracy is well documented and is greatest during the early stages of motor skill learning (Shmuelof et al., 2012). A hypothesis for future investigation is that early motor skill learning at slow speed involves a transition from low-amplitude, fast STOs within the IO during the naïve state (Fig. 12.9A) to large-amplitude, slow STOs that are a consequence of strengthening plasticity of electrical synapses as it assembles functional cell groupings (Fig. 12.9B). The hypothesis predicts that once such functional cell groupings are established via electrical synapse strengthening as motor skill learning occurs at slow speed, the IO network returns to a state in which STOs are generated at a higher frequency by low-threshold mechanisms, providing a time signal for faster read-out of synchronized activity (Fig. 12.9C). The polysynaptic influence of activity originating in motor cortex during motor learning within IO glomeruli, relayed through nuclei of the mesodiencephalic junction (Fig. 12.1B), is an obvious candidate that may guide strengthening plasticity of electrical synapses as motor skills are being practiced. Inhibition within IO glomeruli must continue to have an important role to allow the coupled network to retain configurational flexibility when motor context changes via the fast and transient uncoupling mechanism of current shunting at the site of electrical synapses by GABAergic synaptic input (Fig. 12.9D). The hypothesis does not predict that strengthening plasticity of electrical synapses in the IO is required for motor skill learning, per se, but rather that it underlies the ability to acquire, through learning, the highest degree of temporal precision in coordinated muscle activity during skilled movement.

OUTSTANDING QUESTIONS

- How general is NMDA receptor-mediated strengthening of neuronal electrical synapses in the mammalian brain? The evidence in teleost and mammalian nervous systems suggests a conserved function over phylogeny. For instance, does NMDA

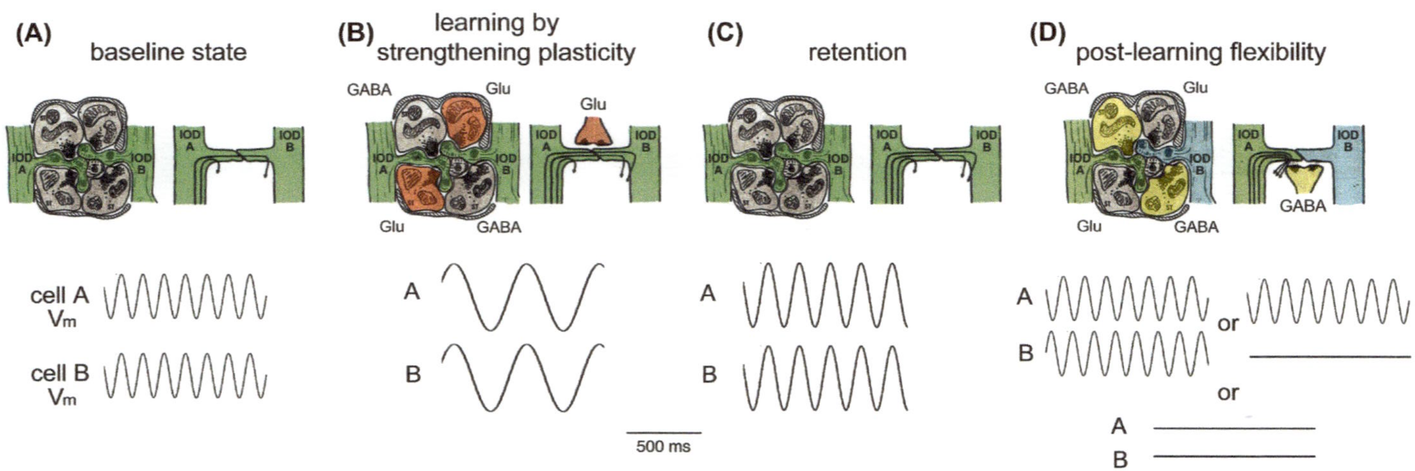

Figure 12.9 Hypothesis of strengthening electrical synapse plasticity in the inferior olive (IO) that contributes to the learned precision of motor skills. (A) During a naïve state prior to skill learning, electrical coupling strength between IO neurons representing the motor groups to be learned to be expressed as a synergy is low but is sufficient to permit synchronized subthreshold oscillations (STOs) driven by low-threshold currents that are reinforced through the coupled network. (B) During motor skill learning, NMDA receptor activation by glutamatergic terminals (*red*) within IO glomeruli strengthens electrical coupling and assembles cell groupings related to coactivated muscles. The STOs of the assembled cell groupings are larger and slower than in the baseline state due to the kinetics of high-threshold currents and increased membrane capacitance. (C) After skill learning, performance is maintained due to retention of the assembled groupings and STOs are driven by low-threshold currents but are larger than in the naïve state due to enhanced electrical coupling. (D) Network flexibility is retained after learning as motor contexts change by the transient uncoupling mechanism provided by GABAergic terminals within the IO glomerulus. The left drawing of each panel shows a simplified glomerulus having two dendritic spines from two IO neurons at its core (cells A and B) surrounded by four synaptic terminals. The right of each panel shows a schematic of the current flow from cell A to B at four different states of modulation by chemical synaptic transmission within the glomerulus. *IOD*, inferior olive dendrite; *red*, glutamatergic terminals activating NMDA receptors; *yellow*, GABAergic terminals activating GABAa receptors. *(The glomerulus drawing is modified from Llinás, R., Baker, R., Sotelo, C., 1974. Electrotonic coupling between neurons in cat inferior olive. J. Neurophysiol. 37, 560–571.)*

receptor-mediated strengthening of electrical synapses contribute the coherence of brain rhythms in the cerebrum?

- Do alterations in NMDA receptor-mediated electrical synapse plasticity contribute to disease states associated with changes in excitatory/inhibitory balance (EIB)? Significant data indicate alterations in NMDA receptor-mediated excitatory transmission in syndromes associated with changes in EIB that are also associated with increases in the coherence of brain activity, such as hypoxic-ischemic encephalopathy, epilepsy, and some forms of intellectual disability.
- Are there changes in the oscillation frequency in the IO during the learning of motor skills related to NMDA receptor activation, as hypothesized? Will characterizing the emergent synchronous activity in the IO supported by electrical coupling during paradigms of motor adaptation that do not explicitly require a change in motor performance, such as eyeblink conditioning and vestibuloocular gain adaptation, help us understand real-world motor skill learning?
- Electrical synapses and synchronized STOs are prevalent in the primate IO (Turecek et al., 2016). Does the strengthening of electrical synapses and enhanced STO coherence by NMDA receptor activation in the primate IO contribute to nonmotor functions, particularly with regard to the hemispheral cerebellum and cognition in primates?

ACKNOWLEDGMENT

This work is supported by United States Public Health Service Grant R01 NS31224 from the National Institute of Neurological Disorders and Stroke.

REFERENCES

Alev, C., Urschel, S., Sonntag, S., Zoidl, G., Fort, A.G., Höher, T., Matsubara, M., Willecke, K., Spray, D.C., Dermietzel, R., 2008. The neuronal connexin36 interacts with and is phosphorylated by CaMKII in a way similar to CaMKII interaction with glutamate receptors. Proc. Natl. Acad. Sci. USA 105, 20964–20969.

Asada, Y., Bennett, M.V., 1971. Experimental alteration of coupling resistance at an electrotonic synapse. J. Cell Biol. 49, 159–172.

Auerbach, A.A., Bennett, M.V., 1969. A rectifying electrotonic synapse in the central nervous system of a vertebrate. J. Gen. Physiol. 53, 211–237.

Baker, R., Llinás, R., 1971. Electrotonic coupling between neurones in the rat mesencephalic nucleus. J. Physiol. 212, 45–83.

Baux, G., Simonneau, M., Tauc, L., Segundo, J.P., 1978. Uncoupling of electrotonic synapses by calcium. Proc. Natl. Acad. Sci. USA 75, 4577–4581.

Belluardo, N., Mudò, G., Trovato-Salinaro, A., Le Gurun, S., Charollais, A., Serre-Beinier, V., Amato, G., Haefliger, J.A., Meda, P., Condorelli, D.F., 2000. Expression of connexin36 in the adult and developing rat brain. Brain Res. 865, 121–138.

Belluardo, N., Trovato-Salinaro, A., Mudò, G., Hurd, Y.L., Condorelli, D.F., 1999. Structure, chromosomal localization, and brain expression of human Cx36 gene. J. Neurosci. Res. 57, 740–752.

Benardo, L.S., Foster, R.E., 1986. Oscillatory behavior in inferior olive neurons: mechanism, modulation, cell aggregates. Brain Res. Bull. 17, 773–784.

Bennett, M.V., 1966. Physiology of electrotonic junctions. Ann. NY Acad. Sci. 137, 509–539.

Bennett, M.V.L., Aljure, E., Nakajima, Y., Pappas, G.D., 1963. Electronic junctions between teleost spinal neurons: electrophysiology and ultrastructure. Science 141, 262–264.
Bennett, M.V., Nakajima, Y., Pappas, G.D., 1967a. Physiology and ultrastructure of electrotonic junctions. I. Supramedullary neurons. J. Neurophysiol. 30, 161–179.
Bennett, M.V., Pappas, G.D., Aljure, E., Nakajima, Y., 1967b. Physiology and ultrastructure of electrotonic junctions. II. Spinal and medullary electromotor nuclei in mormyrid fish. J. Neurophysiol. 30, 180–208.
Bernard, J.F., Buisseret-Delmas, C., Laplante, S., 1984. Inferior olivary neurons: 3-acetylpyridine effects on glucose consumption, axonal transport, electrical activity and harmaline-induced tremor. Brain Res. 322, 382–387.
Bernstein, N.A., 1967. The Coordination and Regulation of Movements. Pergamon, London.
Biscoe, T.J., Duggan, A.W., Headley, P.M., Lodge, D., 1973. Proceedings: rhythmical field potentials induced in the inferior olive complex by iontophoretically applied harmaline and other unrelated alkaloids. Br. J. Pharmacol. 49, 174P–175P.
Bleasel, A.F., Pettigrew, A.G., 1992. Development and properties of spontaneous oscillations of the membrane potential in inferior olivary neurons in the rat. Brain Res. Dev. Brain Res. 65, 43–50.
Blenkinsop, T.A., Lang, E.J., 2006. Block of inferior olive gap junctional coupling decreases Purkinje cell complex spike synchrony and rhythmicity. J. Neurosci. 26, 1739–1748.
Bourrat, F., Sotelo, C., 1983. Postnatal development of the inferior olivary complex in the rat. I. An electron microscopic study of the medial accessory olive. Brain Res. 284, 291–310.
Bruinvels, J., 1969. Comparison of intracisternally and intraperitoneally injected harmaline on body temperature and tremor in the rat. J. Pharm. Pharmacol. 21, 506–508.
Cachope, R., Pereda, A.E., 2015. Opioids potentiate electrical transmission at mixed synapses on the Mauthner cell. J. Neurophysiol. 114, 689–697.
Campos de Carvalho, A., Spray, D.C., Bennett, M.V., 1984. pH dependence of transmission at electrotonic synapses of the crayfish septate axon. Brain Res. 321, 279–286.
Condorelli, D.F., Belluardo, N., Trovato-Salinaro, A., Mudò, G., 2000. Expression of Cx36 in mammalian neurons. Brain Res. Brain Res. Rev. 32, 72–85.
Curti, S., Hoge, G., Nagy, J.I., Pereda, A.E., 2012. Electrical transmission between mammalian neurons is supported by a small fraction of gap junction channels. J. Membr. Biol. 245, 283–290.
Condorelli, D.F., Parenti, R., Spinella, F., Trovato Salinaro, A., Belluardo, N., Cardile, V., Cicirata, F., 1998. Cloning of a new gap junction gene (Cx36) highly expressed in mammalian brain neurons. Eur. J. Neurosci. 10, 1202–1208.
De Gruijl, J.R., Hoogland, T.M., De Zeeuw, C.I., 2014. Behavioral correlates of complex spike synchrony in cerebellar microzones. J. Neurosci. 34, 8937–8947.
De Montigny, C., Lamarre, Y., 1973. Rhythmic activity induced by harmaline in the olivo-cerebello-bulbar system of the cat. Brain Res. 53, 81–95.
De Montigny, C., Lamarre, Y., 1975. Effects produced by local applications of harmaline in the inferior olive. Can. J. Physiol. Pharmacol. 53, 845–849.
De Zeeuw, C.I., Chorev, E., Devor, A., Manor, Y., Van Der Giessen, R.S., De Jeu, M.T., Hoogenraad, C.C., Bijman, J., Ruigrok, T.J., French, P., Jaarsma, D., Kistler, W.M., Meier, C., Petrasch-Parwez, E., Dermietzel, R., Sohl, G., Gueldenagel, M., Willecke, K., Yarom, Y., 2003. Deformation of network connectivity in the inferior olive of connexin 36-deficient mice is compensated by morphological and electrophysiological changes at the single neuron level. J. Neurosci. 23, 4700–4711.
De Zeeuw, C.I., Ruigrok, T.J., Holstege, J.C., Jansen, H.G., Voogd, J., 1990. Intracellular labeling of neurons in the medial accessory olive of the cat: II. Ultrastructure of dendritic spines and their GABAergic innervation. J. Comp. Neurol. 300, 478–494.
De Zeeuw, C.I., Simpson, J.I., Hoogenraad, C.C., Galjart, N., Koekkoek, S.K., Ruigrok, T.J., 1998. Microcircuitry and function of the inferior olive. Trends Neurosci. 21, 391–400.
Del Corsso, C., Iglesias, R., Zoidl, G., Dermietzelm, R., Spray, D.C., 2012. Calmodulin dependent protein kinase increases conductance at gap junctions formed by the neuronal gap junction protein connexin36. Brain Res. 1487, 69–77.
Devor, A., Yarom, Y., 2002. Electrotonic coupling in the inferior olivary nucleus revealed by simultaneous double patch recordings. J. Neurophysiol. 87, 3048–3058.
Dowling, J.E., Boycott, B.B., 1966. Organization of the primate retina: electron microscopy. Proc. R. Soc. Lond B Biol. Sci. 166, 80–111.

Du, W., Aloyo, V.J., Harvey, J.A., 1997. Harmaline competitively inhibits [3H]MK-801 binding to the NMDA receptor in rabbit brain. Brain Res. 770, 26–29.
Du, W., Harvey, J.A., 1997. Harmaline-induced tremor and impairment of learning are both blocked by dizocilpine in the rabbit. Brain Res. 745, 183–188.
Van Der Giessen, R.S., Koekkoek, S.K., van Dorp, S., De Gruijl, J.R., Cupido, A., Khosrovani, S., Dortland, B., Wellershaus, K., Degen, J., Deuchars, J., Fuchs, E.C., Monyer, H., Willecke, K., De Jeu, M.T., De Zeeuw, C.I., 2008. Role of olivary electrical coupling in cerebellar motor learning. Neuron 58, 599–612.
Eccles, J., Llinás, R., Sasaki, K., 1964. Excitation of cerebellar Purkinje cells by the climbing fibers. Nature 203, 245–246.
Eccles, J.C., Llinás, R., Sasaki, K., 1966a. The excitatory synaptic action of climbing fibres on the Purkinje cells of the cerebellum. J. Physiol. 182, 268–296.
Eccles, J.C., Llinás, R., Sasaki, K., Voorhoeve, P.E., 1966b. Interaction experiments on the responses evoked in Purkinje cells by climbing fibres. J. Physiol. 182, 297–315.
Eccles, J.C., Llinás, R., Sasaki, K., 1966c. The action of antidromic impulses on the cerebellar Purkinje cells. J. Physiol. 182, 316–345.
Eccles, J.C., Llinás, R., Sasaki, K., 1966d. Parallel fibre stimulation and the responses induced thereby in the Purkinje cells of the cerebellum. Exp. Brain Res. 1, 17–39.
Fredette, B.J., Mugnaini, E., 1991. The GABAergic cerebello-olivary projection in the rat. Anat. Embryol. Berl. 184, 225–243.
Fuentealba, P., Steriade, M., 2005. The reticular nucleus revisited: intrinsic and network properties of a thalamic pacemaker. Prog. Neurobiol. 75, 125–141.
Furshpan, E.J., 1964. "Electrical transmission" at an excitatory synapse in a vertebrate brain. Science 144, 878–880.
Furshpan, E.J., Potter, D.D., 1959. Transmission at the giant motor synapses of the crayfish. J. Physiol. 145, 289–325.
Grillner, S., Wallén, P., 1985. The ionic mechanisms underlying N-methyl-D-aspartate receptor-induced, tetrodotoxin-resistant membrane potential oscillations in lamprey neurons active during locomotion. Neurosci. Lett. 60, 289–294.
Gwyn, D.G., Nicholson, G.P., Flumerfelt, B.A., 1977. The inferior olivary nucleus of the rat: a light and electron microscopic study. J. Comp. Neurol. 174, 489–520.
Hámori, J., Szentágothai, J., 1966. Identification under the electron microscope of climbing fibers and their synaptic contacts. Exp. Brain Res. 1, 65–81.
Headley, P.M., Lodge, D., Duggan, A.W., 1976. Drug-induced rhythmical activity in the inferior olivary complex of the rat. Brain Res. 101, 461–478.
Hidaka, S., Akahori, Y., Kurosawa, Y., 2004. Dendrodendritic electrical synapses between mammalian retinal ganglion cells. J. Neurosci. 24, 10553–10567.
Hoge, G.J., Davidson, K.G., Yasumura, T., Castillo, P.E., Rash, J.E., Pereda, A.E., 2011. The extent and strength of electrical coupling between inferior olivary neurons is heterogeneous. J. Neurophysiol. 105, 1089–1101.
Ito, M., Yamaguchi, K., Nagao, S., Yamazaki, T., 2014. Long-term depression as a model of cerebellar plasticity. Prog. Brain Res. 210, 1–30.
Ito, M., Yoshida, M., 1966. The origin of cerebral-induced inhibition of Deiters neurones. I. Monosynaptic initiation of the inhibitorypostsynaptic potentials. Exp. Brain Res. 2, 330–349.
Kim, Y.I., Chandler, S.H., 1995. NMDA-induced burst discharge in Guinea pig trigeminal motoneurons in vitro. J. Neurophysiol. 74, 334–346.
King, J.S., 1976. The synaptic cluster (glomerulus) in the inferior olivary nucleus. J. Comp. Neurol. 165, 387–400.
Knobelman, D.A., Hen, R., Blendy, J.A., Lucki, I., 2001. Regional patterns of compensation following genetic deletion of either 5-hydroxytryptamine(1A) or 5-hydroxytryptamine(1B) receptor in the mouse. J. Pharmacol. Exp. Ther. 298, 1092–1100.
Korn, H., Sotelo, C., Crepel, F., 1973. Electronic coupling between neurons in the rat lateral vestibular nucleus. Exp. Brain Res. 16, 255–275.

Kothmann, W.W., Trexler, E.B., Whitaker, C.M., Li, W., Massey, S.C., O'Brien, J., 2012. Nonsynaptic NMDA receptors mediate activity-dependent plasticity of gap junctional coupling in the AII amacrine cell network. J. Neurosci. 32, 6747–6759.

Kriebel, M.E., Bennett, M.V., Waxman, S.G., Pappas, G.D., 1969. Oculomotor neurons in fish: electrotonic coupling and multiple sites of impulse initiation. Science 166, 520–524.

Lamarre, Y., de Montigny, C., Dumont, M., Weiss, M., 1971. Harmaline-induced rhythmic activity of cerebellar and lower brain stem neurons. Brain Res. 32, 246–250.

Lampl, I., Yarom, Y., 1997. Subthreshold oscillations and resonant behavior: two manifestations of the same mechanism. Neuroscience 78, 325–341.

Lang, E.J., Sugihara, I., Llinás, R., 1996. GABAergic modulation of complex spike activity by the cerebellar nucleoolivary pathway in rat. J. Neurophysiol. 76, 255–275.

Lefler, Y., Yarom, Y., Uusisaari, M.Y., 2014. Cerebellar inhibitory input to the inferior olive decreases electrical coupling and blocks subthreshold oscillations. Neuron 81, 1389–1400.

Li, W.-C., Rekling, J.C., 2017. Electrical coupling in the generation of vertebrate motor rhythms. In: Jing, J. (Ed.), Network Functions and Plasticity. Elsevier (Chapter 11).

Llinás, R., 1974. Eighteenth Bowditch lecture. Motor aspects of cerebellar control. Physiologist 17, 19–46.

Llinás, R., Baker, R., Sotelo, C., 1974. Electrotonic coupling between neurons in cat inferior olive. J. Neurophysiol. 37, 560–571.

Llinás, R., Sasaki, K., 1989. The functional organization of the olivo-cerebellar system as examined by multiple Purkinje cell recordings. Eur. J. Neurosci. 1, 587–602.

Llinás, R., Sugimori, M., 1980. Electrophysiological properties of in vitro Purkinje cell dendrites in mammalian cerebellar slices. J. Physiol. 305, 197–213.

Llinás, R., Volkind, R.A., 1973. The olivo-cerebellar system: functional properties as revealed by harmaline-induced tremor. Exp. Brain Res. 18, 69–87.

Llinás, R., Welsh, J.P., 1993. On the cerebellum and motor learning. Curr. Opin. Neurobiol. 3, 958–965.

Llinás, R., Yarom, Y., 1981. Electrophysiology of mammalian inferior olivary neurones in vitro. Different types of voltage-dependent ionic conductances. J. Physiol. 215, 549–567.

Llinás, R., Yarom, Y., 1986. Oscillatory properties of Guinea-pig inferior olivary neurones and their pharmacological modulation: an in vitro study. J. Physiol. 376, 163–182.

Long, M.A., Deans, M.R., Paul, D.L., Connors, B.W., 2002. Rhythmicity without synchrony in the electrically uncoupled inferior olive. J. Neurosci. 22, 10898–10905.

MacLean, J.N., Schmidt, B.J., Hochman, S., 1997. NMDA receptor activation triggers voltage oscillations, plateau potentials and bursting in neonatal rat lumbar motoneurons in vitro. Eur. J. Neurosci. 9, 2702–2711.

Manning, K.A., Evinger, C., 1986. Different forms of blinks and their two-stage control. Exp. Brain Res. 64, 579–588.

Manor, Y., Rinzel, J., Segev, I., Yarom, Y., 1997. Low-amplitude oscillations in the inferior olive: a model based on electrical coupling of neurons with heterogeneous channel densities. J. Neurophysiol. 77, 2736–2752.

Manor, Y., Yarom, Y., Chorev, E., Devor, A., 2000. To beat or not to beat: a decision taken at the network level. J. Physiol. Paris 94, 375–390.

Marshall, S.P., van der Giessen, R.S., de Zeeuw, C.I., Lang, E.J., 2007. Altered olivocerebellar activity patterns in the connexin36 knockout mouse. Cerebellum 6, 287–299.

Mash, D.C., Staley, J.K., Pablo, J.P., Holohean, A.M., Hackman, J.C., Davidoff, R.A., 1995. Properties of ibogaine and its principal metabolite (12-hydroxyibogamine) at the MK-801 binding site of the NMDA receptor complex. Neurosci. Lett. 192, 53–56.

McCormick, D.A., 1995. Motor control. The cerebellar symphony. Nature 374, 412–413.

Middleton, F.A., Strick, P.L., 1997. Cerebellar output channels. Int. Rev. Neurobiol. 41, 61–82.

Murphy, M.G., O'Leary, J.L., 1971. Neurological deficit in cats with lesions of the olivocebellar system. Arch. Neurol. 24, 145–157.

Onodera, S., 1984. Olivary projections from the mesodiencephalic structures in the cat studied by means of axonal transport of horseradish peroxidase and tritiated amino acids. J. Comp. Neurol. 227, 37–49.

Parenti, R., Gulisano, M., Zappala, A., Cicirata, F., 2000. Expression of connexin36 mRNA in adult rodent brain. Neuroreport 11, 1497–1502.
Park, Y.–G., Park, H.–Y., Lee, C.J., Choi, S., Jo, S., Choi, H., Kim, Y.–H., Shin, H.–S., Llinás, R.R., Kim, D., 2010. Cav3.1 is a tremor rhythm pacemaker in the inferior olive. Proc. Natl. Acad. Sci. USA 107, 10731–10736.
Parker, P.R., Cruikshank, S.J., Connors, B.W., 2009. Stability of electrical coupling despite massive developmental changes of intrinsic neuronal physiology. J. Neurosci. 29, 9761–9770.
Peracchia, C., 1978. Calcium effects on gap junction structure and cell coupling. Nature 271, 669–671.
Pereda, A.E., Bennett, M.V.L., 2017. Electrical synapses in fishes: their relevance to synaptic transmission. In: Jing, J. (Ed.), Network Functions and Plasticity. Elsevier (Chapter 8).
Pereda, A.E., Bell, T.D., Chang, B.H., Czernik, A.J., Nairn, A.C., Soderling, T.R., Faber, D.S., 1998. Ca^{2+}/calmodulin-dependent kinase II mediates simultaneous enhancement of gap-junctional conductance and glutamatergic transmission. Proc. Natl. Acad. Sci. USA 95, 13272–13277.
Pereda, A.E., Faber, D.S., 1996. Activity-dependent short-term enhancement of intercellular coupling. J. Neurosci. 16, 983–992.
Pereda, A.E., Nairn, A.C., Wolszon, L.R., Faber, D.S., 1994. Postsynaptic modulation of synaptic efficacy at mixed synapses on the Mauthner cell. J. Neurosci. 14, 3704–3712.
Pereda, A., O'Brien, J., Nagy, J.I., Bukauskas, F., Davidson, K.G., Kamasawa, N., Yasumura, T., Rash, J.E., 2003. Connexin35 mediates electrical transmission at mixed synapses on Mauthner cells. J. Neurosci. 23, 7489–7503.
Placantonakis, D.G., Bukovsky, A.A., Aicher, S.A., Kiem, H.P., Welsh, J.P., 2006. Continuous electrical oscillations emerge from a coupled network: a study of the inferior olive using lentiviral knockdown of connexin36. J. Neurosci. 26, 5008–5016.
Placantonakis, D.G., Bukovsky, A.A., Zeng, X.H., Kiem, H.P., Welsh, J.P., 2004. Fundamental role of inferior olive connexin 36 in muscle coherence during tremor. Proc. Natl. Acad. Sci. USA 101, 7164–7169.
Placantonakis, D., Cicirata, F., Welsh, J.P., 2002. A dominant negative mutation of neuronal connexin 36 that blocks intercellular permeability. Brain Res. Mol. Brain Res. 98, 15–28.
Placantonakis, D., Welsh, J.P., 2001. Two distinct oscillatory states determined by the NMDA receptor in rat inferior olive. J. Physiol. 534, 123–140.
Popik, P., Layer, R.T., Skolnick, P., 1994. The putative anti-addictive drug ibogaine is a competitive inhibitor of [3H]MK-801 binding to the NMDA receptor complex. Psychopharmacol. Berl. 114, 672–674.
Rash, J.E., Pereda, A., Kamasawa, N., Furman, C.S., Yasumura, T., Davidson, K.G., Dudek, F.E., Olson, C., Li, X., Nagy, J.I., 2004. High-resolution proteomic mapping in the vertebrate central nervous system: close proximity of connexin35 to NMDA glutamate receptor clusters and co-localization of connexin36 with immunoreactivity for zonula occludens protein-1 (ZO-1). J. Neurocytol. 33, 131–151.
Rash, J.E., Staines, W.A., Yasumura, T., Patel, D., Furman, C.S., Stelmack, G.L., Nagy, J.I., 2000. Immunogold evidence that neuronal gap junctions in adult rat brain and spinal cord contain connexin-36 but not connexin-32 or connexin-43. Proc. Natl. Acad. Sci. USA 97, 7573–7578.
Ribelayga, C.P., O'Brien, J., 2017. Circadian and light-adaptive control of electrical synaptic plasticity in the vertebrate retina. In: Jing, J. (Ed.), Network Functions and Plasticity. Elsevier (Chapter 10).
Rondi-Reig, L., Delhaye-Bouchaud, N., Mariani, J., Caston, J., 1997. Role of the inferior olivary complex in motor skills and motor learning in the adult rat. Neuroscience 77, 955–963.
Ruigrok, T.J., Voogd, J., 1995. Cerebellar influence on olivary excitability in the cat. Eur. J. Neurosci. 7, 679–693.
Rutherford, J.G., Gwyn, D.G., 1980. A light and electron microscopic study of the inferior olivary nucleus of the squirrel monkey, *Saimiri sciureus*. J. Comp. Neurol. 189, 127–155.
Saint-Cyr, J.A., Courville, J., 1981. Sources of descending afferents to the inferior olive from the upper brain stem in the cat as revealed by the retrograde transport of horseradish peroxidase. J. Comp. Neurol. 198, 567–581.
Sasaki, K., Bower, J.M., Llinás, R., 1989. Multiple Purkinje cell recording in rodent cerebellar cortex. Eur. J. Neurosci. 1, 572–586.
Schild, R.F., 1970. On the inferior olive of the albino rat. J. Comp. Neurol. 140, 255–259.
Shmuelof, L., Krakauer, J.W., Mazzoni, P., 2012. How is a motor skill learned? Change and invariance at the levels of task success and trajectory control. J. Neurophysiol. 108, 578–594.

Sinton, C.M., Krosser, B.I., Walton, K.D., Llinás, R.R., 1989. The effectiveness of different isomers of octanol as blockers of harmaline-induced tremor. Pflugers Arch. 414, 31–36.
Söhl, G., Degen, J., Teubner, B., Willecke, K., 1998. The murine gap junction gene connexin36 is highly expressed in mouse retina and regulated during brain development. FEBS Lett. 428, 27–31.
Sotelo, C., Llinás, R., Baker, R., 1974. Structural study of inferior olivary nucleus of the cat: morphological correlates of electrotonic coupling. J. Neurophysiol. 37, 541–559.
Sotelo, C., Palay, S.L., 1970. The fine structure of the later vestibular nucleus in the rat. II. Synaptic organization. Brain Res. 18, 93–115.
Soto-Treviño, C., Rabbah, P., Marder, E., Nadim, F., 2005. Computational model of electrically coupled, intrinsically distinct pacemaker neurons. J. Neurophysiol. 94, 590–604.
Spira, M.E., Bennett, M.V., 1972. Synaptic control of electrotonic coupling between neurons. Brain Res. 37, 294–300.
Sugihara, I., Lang, E.J., Llinás, R., 1993. Uniform olivocerebellar conduction time underlies Purkinje cell complex spike synchronicity in the rat cerebellum. J. Physiol. 470, 243–271.
Szentágothai, J., Rajkovits, K., 1959. Über den Ursprung der Kletterfasern des Kleinhirns. Anat. Embryol. Berl. 121, 130–141.
Tchumatchenko, T., Clopath, C., 2014. Oscillations emerging from noise-driven steady state in networks with electrical synapses and subthreshold resonance. Nat. Commun. 5, 5512.
Thach Jr., W.T., 1967. Somatosensory receptive fields of single units in cat cerebellar cortex. J. Neurophysiol. 30, 675–696.
Traub, R.D., Pais, I., Bibbig, A., LeBeau, F.E., Buhl, E.H., Hormuzdi, S.G., Monyer, H., Whittington, M.A., 2003. Contrasting roles of axonal (pyramidal cell) and dendritic (interneuron) electrical coupling in the generation of neuronal network oscillations. Proc. Natl. Acad. Sci. USA 100, 1370–1374.
Turecek, J., Han, V.Z., Cuzon Carlson, V.C., Grant, K.A., Welsh, J.P., 2016. Electrical coupling and synchronized subthreshold oscillations in the inferior olive of the rhesus macaque. J. Neurosci. 36, 6497–6502.
Turecek, J., Yuen, G.S., Han, V.Z., Zeng, X.H., Bayer, K.U., Welsh, J.P., 2014. NMDA receptor activation strengthens weak electrical coupling in mammalian brain. Neuron 81, 1375–1388.
Veruki, M.L., Oltedal, L., Hartveit, E., 2010. Electrical coupling and passive membrane properties of AII amacrine cells. J. Neurophysiol. 103, 1456–1466.
Vest, R.S., Davies, K.D., O'Leary, H., Port, J.D., Bayer, K.U., 2007. Dual mechanism of a natural CaMKII inhibitor. Mol. Biol. Cell 18, 5024–5033.
Watanabe, A., 1958. The interaction of electrical activity among neurons of lobster cardiac ganglion. Jpn. J. Physiol. 8, 305–318.
Welsh, J.P., 2002. Functional significance of climbing-fiber synchrony: a population coding and behavioral analysis. Ann. NY Acad. Sci. 978, 188–204.
Welsh, J.P., Harvey, J.A., 1998. Acute inactivation of the inferior olive blocks associative learning. Eur. J. Neurosci. 10, 3321–3332.
Welsh, J.P., Lang, E.J., Sugihara, I., Llinás, R., 1995. Dynamic organization of motor control within the olivocerebellar system. Nature 374, 453–457.
Welsh, J.P., Llinás, R., 1997. Some organizing principles for the control of movement based on olivocerebellar physiology. Prog. Brain Res. 114, 449–461.
Witter, L., Rudolph, S., Pressler, R.T., Lahlaf, S.I., Regehr, W.G., 2016. Purkinje cell collaterals enable output signals from the cerebellar cortex to feed back to Purkinje cells and interneurons. Neuron 91, 312–319.
Yamamoto, T., Fukuda, M., Llinás, R., 2001. Bilaterally synchronous complex spike Purkinje cell activity in the mammalian cerebellum. Eur. J. Neurosci. 13, 327–339.
Yang, X.D., Korn, H., Faber, D.S., 1990. Long-term potentiation of electrotonic coupling at mixed synapses. Nature 348, 542–545.
Zhu, Z., Zeng, X.H., Turecek, J., Han, V.Z., Welsh, J.P., 2015. RNA interference of GluN1 inhibits neuronal rhythmogenesis in the adult inferior olive. J. Mol. Neurosci. 55, 416–429.

CHAPTER 13

Gap Junctions Between Pyramidal Cells Account for a Variety of Very Fast Network Oscillations (>80 Hz) in Cortical Structures

R.D. Traub[1,2], M.A. Whittington[3], A. Draguhn[4]
[1]IBM T.J. Watson Research Center, Yorktown Heights, NY, United States; [2]Columbia University, New York, NY, United States; [3]University of York, York, United Kingdom; [4]Heidelberg University, Heidelberg, Germany

Very fast oscillations ("VFOs," > ~80 Hz), in populations of cortical neurons, occur in a variety of situations; for example, during normal, physiological sharp-wave ripples (SPW-R; Buzsáki et al., 1992; Ylinen et al., 1995); before and during epileptiform events, such as interictal spikes and frank electrographic seizures (Fisher et al., 1992; Traub et al., 2001a,b; Pais et al., 2003); and, on occasion, in isolation in the brains of epileptic patients or epileptic experimental animals, recorded with subdural grids or depth electrodes (Andrade-Valenca et al., 2011; Bragin et al., 1999; Grenier et al., 2003; Jacobs et al., 2008; Schevon et al., 2009). What are the basic cellular mechanisms, and are the mechanisms shared between the different sorts of VFO?

In this chapter, we present the case that populations of pyramidal neurons, electrically coupled through their axons, lie at the "heart" of all of the above-mentioned phenomena; and, by inference, that synaptic currents and fields, recorded during VFO, arise secondarily to events in a plexus of pyramidal cell axons, interconnected through gap junctions that are located on the proximal, unmyelinated, portions of the axons (Schmitz et al., 2001).

To put this chapter in a larger context, two points should be made: first, gap junctions (probably between principal cells) appear to contribute to network oscillations at many different frequencies other than VFO, including alpha (M.A. Whittington et al., unpublished data; Hughes et al., 2011), beta-2 (20–30 Hz; Roopun et al., 2006), delta (Carracedo et al., 2013), gamma (30–70 Hz; Traub et al., 2000; Roopun et al., 2010; Traub and Whittington, 2010), and "fast ripples" (>250 Hz; Engel et al., 2009; Simon et al., 2014). Second, gap junctions exist between other types of neurons besides principal cortical and hippocampal neurons, including but not limited to thalamic nucleus reticularis cells (Lee et al., 2014), inferior olivary neurons (Llinás and Yarom, 1986; Welsh and Turecek, 2017 (Chapter 12)), principal thalamic neurons (Hughes et al., 2011), and cortical interneurons. For the latter, there is an extensive literature concerning both

Network Functions and Plasticity
ISBN 978-0-12-803471-2, http://dx.doi.org/10.1016/B978-0-12-803471-2.00013-8

ultrastructure and electrophysiology (eg, Kosaka, 1983; Amitai et al., 2002), and some analysis of how interneuron gap junctions might influence gamma oscillations (Traub et al., 2001a, 2003b). For reasons of space, and of simplicity, we shall concentrate here on VFOs, which illustrate how oscillations can emerge from electrical coupling alone, without chemical synapses—in other words, in "pure form"; and can emerge by an interesting and novel mechanism, a mechanism apparently different from what is involved in the other examples cited earlier. We shall, however, examine implications of our VFO analysis to selected other sorts of oscillation, in particular to selected oscillations where both gap junctions and chemical synapses cooperate in critical ways.

Another, and even larger, matter of context concerns why neuronal network oscillations are important in the first place, either for physiological or for pathological brain function, besides for such obvious requirements as central pattern generators. This subject is too complex for the present chapter to address properly, and we must refer the reader to a monograph (Traub and Whittington, 2010), and some of its many references, including to the work of Wolf Singer, György Buzsáki, and others.

1. WHERE DID THE IDEA OF AXONAL GAP JUNCTIONS COME FROM?

"Ripples," or VFOs, were described by Buzsáki et al. (1992) as occurring superimposed on physiological sharp waves in the hippocampus, and later in other limbic structures (eg, Chrobak and Buzsáki, 1996). Several findings were noted (see also Ylinen et al., 1995): (1) pyramidal cells were depolarized during the sharp wave, but fired little; (2) fast spiking interneurons fired at much higher rates than pyramidal cells; (3) all firings were phase locked to the rippling field, with pyramidal cells slightly leading interneurons; (4) ripple-frequency synaptic potentials occurred in pyramidal cells, also phase locked to the field, and appeared to largely consist of inhibitory postsynaptic potentials. Subsequently, when an in vitro model of spontaneous SPW-R s was developed (Maier et al., 2003), the in vivo findings were mostly replicated, with a major difference being that pyramidal cells tended to hyperpolarize during the events, rather than depolarize. Remarkably, pyramidal cells might still fire, even during strong hyperpolarizations, with the resulting action potentials being inflected on their rising phase, giving them the appearance of antidromic spikes (see later discussion). In addition, in vitro, it has been shown that synaptic currents in pyramidal cells, during SPW-R, are actually a mixture of excitatory and inhibitory currents (see also later discussion).

In 1998, Draguhn et al. showed that, again, in hippocampal slices, it was possible to observe spontaneous ripples (very-high-frequency field oscillations) without the sharp wave, that is, without the slower, tens of millisecond, field potential (Fig. 13.1). We shall denote such events as "dissociated ripples." (Interestingly, at least to our knowledge, the reverse dissociation, sharp waves without ripples, has never been observed.) Because the "dissociated" ripples were present in slices, it was possible to perform experimental

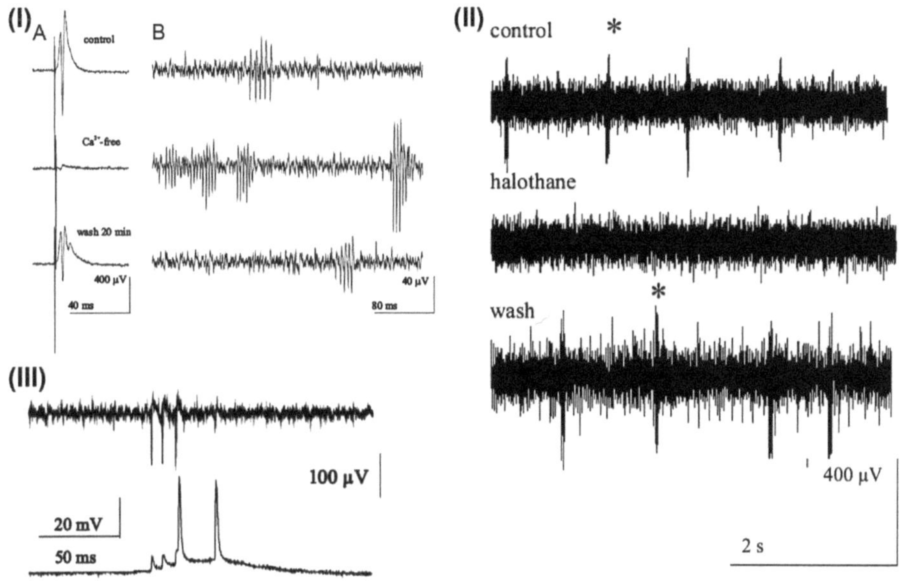

Figure 13.1 ***Ripples without sharp waves in the hippocampal slice, CA1 region.*** (I) field potential ripples occur spontaneously, and are augmented (rather than abolished) in low-calcium media that block synaptic transmission. (II) spontaneous field potential ripples (asterisks) are abolished reversibly by the gap junction blocker halothane. (III) simultaneous recording of local field and patch recording of CA1 pyramidal cell. Each field transient is correlated with an intracellular transient, either a spikelet or a full action potential. The action potentials are inflected and appear antidromic (for comparison, a sharp wave/ripple is shown in Fig. 13.7B). *(From Draguhn, A., Traub, R.D., Schmitz, D., Jefferys, J.G.R., 1998. Electrical coupling underlies high-frequency oscillations in the hippocampus* in vitro. *Nature 394, 189-192., reproduced with permission.)*

manipulations that are difficult, if not impossible, to perform in vivo (Fig. 13.1). Most importantly, the dissociated ripples were actually enhanced by block of chemical synaptic transmission in low-calcium media, rather than suppressed (as one would expect, if the ripple were primarily generated by interneurons and consisted entirely of inhibitory synaptic currents). The dissociated ripples were, however, suppressed by a variety of gap junction blockers (halothane, octanol, carbenoxolone) while being enhanced by alkaline pH, a treatment that increases the conductance of many, but not all, types of gap junction channels (González-Nieto et al., 2008; Palacios-Prado et al., 2009, 2010). Intracellular recordings indicated that the action potentials of an interneuron were not consistently correlated with the field ripple, whereas in a pyramidal cell (Fig. 13.1), each field transient was correlated either with a spikelet or with an action potential.

An initial problem, then, was to formulate a hypothesis that could explain, first, how a population of pyramidal cells, connected by gap junctions, and with chemical synapses blocked, could generate a field ripple—and a field ripple that was associated with spikelets and antidromic spikes in individual pyramidal cells. The model should be able to account for individual ripple waves, as well as the frequency of the ripple.

Our basic hypothesis is that pyramidal cells are electrically coupled between their axons, and that the requisite gap junctions have specific properties:

1. First, each axon should be coupled to more than one other axon, on average. (This is called the "percolation threshold" in the theory of random graphs (Erdös and Rényi, 1960).) Note that the percolation threshold can be reached, in a large population of neurons, at very low (ie, sparse) densities of coupling; for example, in a population of 10,000 cells, if one cell is coupled on average to 0.1% of the others, that would imply each cell couples on average to 10 others, well above the percolation threshold.
2. Gap junctions between axons allow an action potential in one axon to induce an action potential in the other at short latency (submillisecond). (Note that this is a quite different sort of functional gap junctional coupling than the more commonly considered low-pass filtering, but electrical coupling allowing action potentials to induce action potentials, ie, not low-pass filtering, is perfectly consistent with the classic observations of Furshpan and Potter (1959), and with other data shown later.)

It is immediately apparent that in such a case, if spontaneous action potentials occur in the axons at even low rates, then activity will percolate through the plexus (ie, network) of axons, and, by antidromic propagation from the gap junction sites, lead to antidromic spikes and spikelets in somata. It is not at all obvious, however, why a ripple-frequency population oscillation should result. In this context, it is important to note that, so far as is known, pyramidal cell membranes do not possess an intrinsic, subthreshold oscillatory resonance at ripple frequencies. Thus, the population ripple must be emergent in the network and we should not look for analogies in the sinoatrial node of the heart (Vinogradova et al., 2005), or in the inferior olive (Llinás and Yarom, 1986; Devor and Yarom, 2002).

We shall first describe the further network implications of our above-mentioned hypothesis, which was formulated as a network model suitable for computer simulations, and, following that, present data that support the postulates on which the model rests, as well as the predictions that the model makes.

2. IF AXONAL GAP JUNCTIONS EXIST, HOW MIGHT THEY ACCOUNT FOR VFO?

To begin with, let us consider how a single wave of a dissociated in vitro ripple might occur. The idea for how this could happen was suggested by earlier studies of synchronized bursts in the CA3 region of the disinhibited hippocampal slice (Traub and Wong, 1982), where synchronization occurs via excitatory synaptic interactions: it was discovered that a minimal stimulus (eg, a small puff of hypertonic KCl) could elicit a synchronized population burst after a long latent period, many tens of milliseconds or even longer. How to account for this latent period—which will play a role analogous to the oscillation period during a ripple (although the timescales are, of course, quite different)? Intracellular recordings from pyramidal cells showed that most pyramidal cells were silent during the latent interval, only to discharge later with the population, but, on occasion, a cell was found that did burst during the latent period, and that burst again, subsequently, with the population. Such data suggested that there was a prolonged, but still exponential, growth process of activity (ie, bursts of action potentials), which spread from cell to cell. This type of growth of activity is possible, starting with bursts in just a few cells, when each cell synaptically excites more than one other (giving the growth), and when the synaptic connections are powerful enough for a burst to propagate from one cell to another. The duration of the latent period would then depend on two (in principle experimentally determinable) parameters: how long it takes for a burst to propagate from one cell to another, which turned out to be of order 10 ms, and by how many cells any one cell excites monosynaptically (Miles and Wong, 1986, 1987): the larger the synaptic divergence is, the fewer will be the number of steps required to engage the whole population in bursting, hence the shorter will be the latent period.

An analogous model can account for a single wave of a dissociated ripple, in the following way: we suppose that

1. a single wave starts from an ectopic (here, axonal) spontaneous action potential (this corresponds approximately to the exogenous KCl puff used by Traub and Wong (1982));
2. gap junctions between pyramidal cells have the property that a spike in one axon induces, with submillisecond latency, a spike in the coupled cell (analogous to synaptic burst propagation in CA3, but much faster, and operating bidirectionally, rather than the unidirectionality of a chemical synapse);
3. each pyramidal cell is electrically coupled to more than one other cell, on average (again analogous to the proposed, and measured, synaptic interconnectivity of CA3). Just as before, the period of the single wave will be determined by how long it takes for a spike to propagate cell to cell, and by how many "steps" there are until most or all of the cells fire.

Although the aforementioned principles could explain, at least intuitively, how a single ripple wave might occur, with appropriate duration, these principles do not explain (in any obvious way) how the full ripple (or at least dissociated ripple) might occur. To investigate that question, we performed simulations of thousands of multicompartment pyramidal neurons, electrically coupled in axonal compartments, and without chemical synaptic interactions (Traub et al., 1999). Axonal gap junctions were typically located less than about 250 μm from the soma (the experimentally estimated average for CA1 pyramidal neurons is about 80 μm (Schmitz et al., 2001)). Two distinctive global network topologies were used: a binary tree and a random graph (Erdös and Rényi, 1960). Connectivity in the binary tree is of course above the percolation limit, and we used enough connections in the random graphs, so as also to be above the percolation limit. The binary tree would appear to be completely "unnatural," but was used to make a mathematical point: if ripples could emerge on a tree, it must mean that network loops were not involved, or at least not necessary; that is, reentry was not required. The random graph was thought to be a more realistic topology, although, at least in very large networks, a locally random graph might be more appropriate (Lewis and Rinzel, 2000; Traub et al., 2010). Random graphs contain many loops, so additional analysis is required to attempt to separate two possible mechanisms for ripple generation (Lewis and Rinzel, 2000, 2001; Munro and Börgers, 2010; Vladimirov et al., 2012; Simon et al., 2014).

In any case, binary trees of axonally connected neurons (Fig. 13.2), and also random graph topologies, could indeed generate ripple-like oscillations, provided that there was a random background of spontaneous "ectopic" axonal firing. Note the admixture of somatic spikelets and antidromic action potentials in this simulation, as also observed for experimental ripples by Draguhn et al. (1998) and in neocortex by Miles Whittington

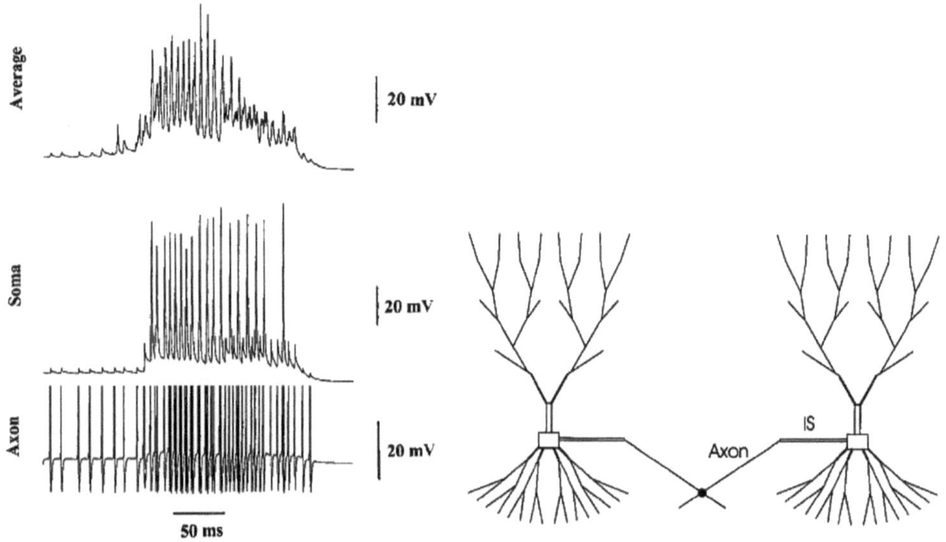

Figure 13.2 ***Demonstration of ripple in a simulated network of 2047 model neurons (hippocampal pyramidal cells), each with soma, axon, and dendrites.*** The only between-cell interactions were through axonal gap junctions (diagrammatically shown for a pair of neurons at right) that could allow spikes to propagate, axon to axon. The cells were arranged into a binary tree, so that there were no network cycles, excluding reentry. Random ectopic (axonal) spikes occurred at mean frequency 1 per 10 s per axon. Activity is organized into network oscillations at ~140–175 Hz. Upper trace is the average somatic potential of 224 cells. Note the high-frequency firing of the axon and the mixture of spikelets and full spikes in the soma. The model cells had intrinsic burst properties, accounting for the slow somatic depolarization. *(From Traub, R.D., Schmitz, D., Jefferys, J.G.R., Draguhn, A., 1999. High-frequency population oscillations are predicted to occur in hippocampal pyramidal neuronal networks interconnected by axoaxonal gap junctions. Neuroscience 92, 407–426, reproduced with permission.)*

(Traub et al., 2010). In the study illustrated in Fig. 13.2, the spontaneous firing in any individual axon could be extremely low, even less than 1 Hz, because a spontaneous spike could, at least sometimes, propagate through most or all of the entire network, an extreme form of amplification. Two features of the network behavior were remarkable: first, that an oscillation occurred at all, over a wide range of ectopic rates—one might imagine that the network activity would simply consist of noise, as indeed it does if ectopic rates are too high. Second, that the population frequency was about right (and not related to axonal refractoriness, although in this model, obviously the population period will not be shorter than axonal refractoriness, and indeed was always much longer than axonal refractory periods). The frequency could be altered by changing network connectivity, with higher connectivity leading to a faster population frequency. Later, it was confirmed, again, as expected, that altering the crossing time (how long it takes a spike to cross from axon to axon) would also control the population period (Cunningham et al., 2012).

Thus, the simulations tell us that axonal coupling, if it really exists and has certain defined properties, could, at least in principle, account for dissociated ripples. We are not aware of alternative explanations, at least plausible ones.

3. PHYSIOLOGICAL EVIDENCE FOR AXONAL GAP JUNCTIONS

Two types of evidence are relevant: first, that whatever type of electrical coupling exists between pyramidal neurons takes place between their respective axons; second, that the coupling is strong enough that an action potential in one cell induces, at short latency, an action potential in the second neuron, something that has been observed in certain invertebrate systems (Furshpan and Potter, 1959).

In the experiment shown in Fig. 13.3 (taken from Schmitz et al. (2001)), presumed axonal stimulation in stratum oriens, in low calcium conditions, led to spikelets in a pyramidal neuron, appearing first in the axon, then in the soma, and finally in the apical dendrite. The stimulus did not appear to be exciting the axon of the recorded neuron directly, as other experiments showed that appearance of the spikelet was suppressed by carbenoxolone (a gap junction blocker), and was also sensitive to intracellular acidification (which closes many, but not all, types of gap junction channel). These data suggest that the occurrence of fast spikelets reflects axonal coupling.

The experiments shown in Figs. 13.4 and 13.5 serve to indicate that coupling between hippocampal pyramidal cells, respectively, prefrontal neocortical layer five pyramidal cells, can be strong enough that a spike in one cell is able to evoke a spike in the second cell, at short (submillisecond) latency. The close apposition of the axons of coupled cells, in Fig. 13.5, is also suggestive that the coupling in question occurs between axons, but this light microscopic image is by no means conclusive.

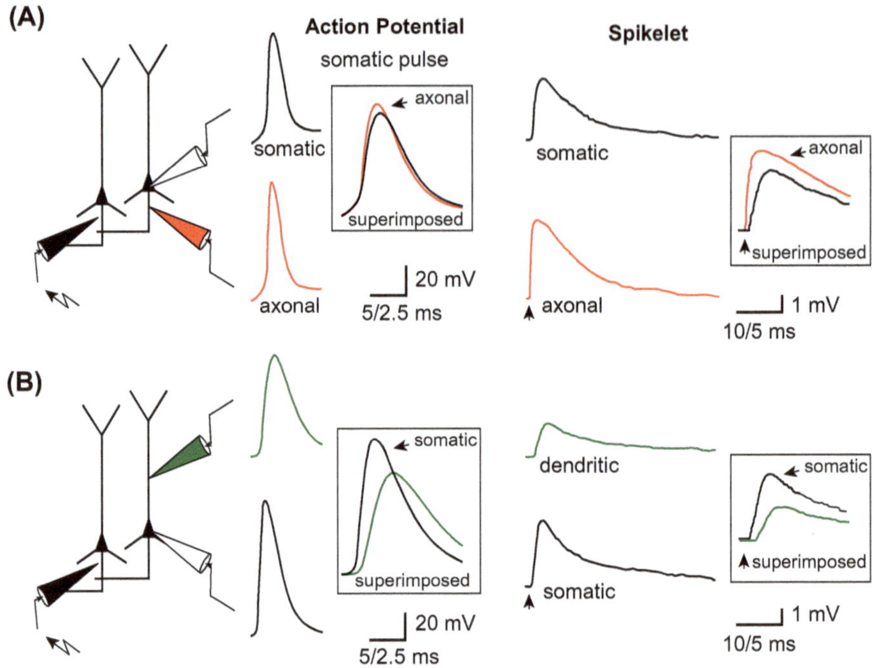

Figure 13.3 ***Somatic spikelets can arise from stimulation of the axon (CA1 pyramidal cell).*** Extracellular stimuli were applied in stratum oriens (left-most electrodes in A and B); simultaneous patch recordings were made on the soma and proximal axon (A); and on the soma and apical dendrite (B). In cases where the stimulus produced a somatic spikelet, the axonal spikelet led the somatic spikelet (A); while the somatic spikelet in turn led the dendritic spikelet (B). *(From Schmitz, D., Schuchmann, S., Fisahn, A., Draguhn, A., Buhl, E.H., Petrasch-Parwez, R.E., Dermietzel, R., Heinemann, U., Traub, R.D., 2001. Axo-axonal coupling: a novel mechanism for ultrafast neuronal communication. Neuron 31, 831–840, reproduced with permission.)*

4. ANATOMICAL EVIDENCE FOR AXONAL GAP JUNCTIONS

The main anatomical issues are these: first, do gap junctions occur on pyramidal cell axons, and, if so, at what locations (ie, axon initial segment, proximal collaterals, nodes of Ranvier, presynaptic terminals)? Second, does each cell couple, on average, to more than one other cell (a critical feature of the model described above)? The anatomical evidence should preferably be ultrastructural in nature; however, to the best of our knowledge, the only telencephalic axonal gap junction demonstrated so far with ultrastructure (specifically, with FRIL, freeze-fracture immunogold labeling) lay on a hippocampal mossy fiber, and contained connexion 36 (Hamzei-Sichani et al., 2007). We shall, therefore, concentrate on dye coupling data. Such data are useful, but cannot eliminate the interposition of a fine intermediate process (such as from an astrocyte), which is below the resolution of light microscopy (Peters et al., 1991).

The data of Church and Baimbridge (1991) indicate that, hippocampal pyramidal neurons do couple to more than one other, on average, provided conditions are sufficiently alkaline; at pH 7.9, one CA1 pyramidal cell coupled to an average of 2.25 others. It should be noted, however, that the experimental use of pH = 7.9 by Church

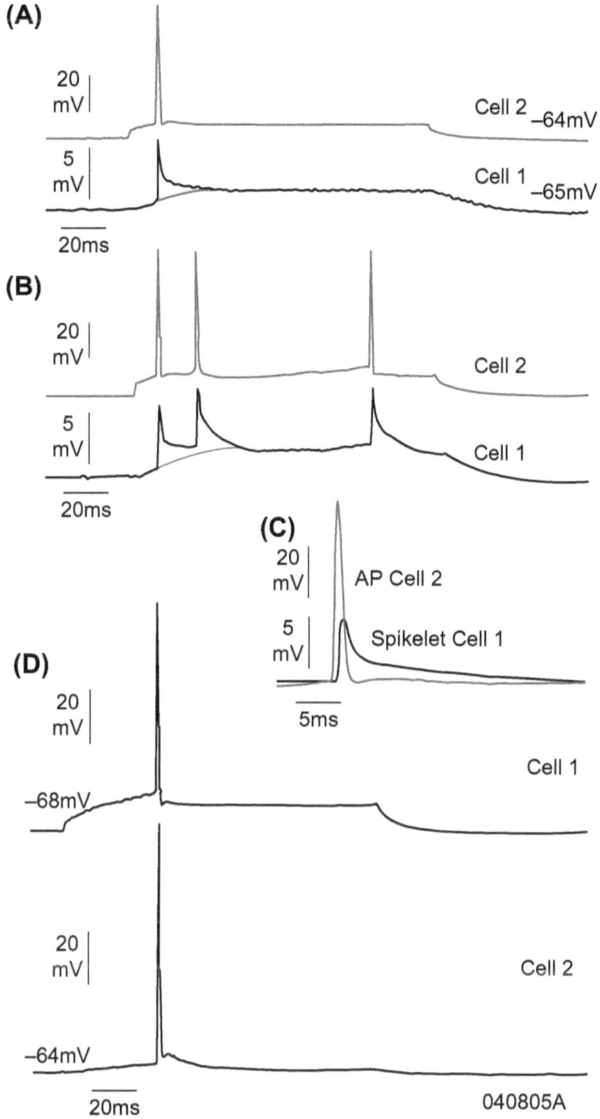

Figure 13.4 ***Strong electrical coupling between pyramidal neurons (CA1 hippocampus).*** In at least some instances, an action potential in one neuron (cell 2 for A–C, cell 1 for D) can elicit either a full action potential (D) or a spikelet (A–C), in a second neuron, at submillisecond latency. *AP*, action potential. *(From Mercer, A., Bannister, A.P., Thomson, A.M., 2006. Electrical coupling between pyramidal cells in adult cortical regions. Brain Cell Biol. 35, 13–27, reproduced with permission.)*

and Baimbridge is probably unphysiological, and whether the necessary degree of coupling occurs at more normal pH levels remains to be demonstrated.

Coupling between hippocampal pyramidal neurons is between axons, at least as determined by dye coupling and light microscopy (confocal imaging) (Fig. 13.6). Such coupling was observed in over 30 instances, in media with either low or normal calcium, and the site of coupling could be identified in four cases, at a mean of about 80 μm from the soma. These anatomical data are thus consistent with the physiological data, and also provide clues as to where immunoflourescent and ultrastructural studies might be directed.

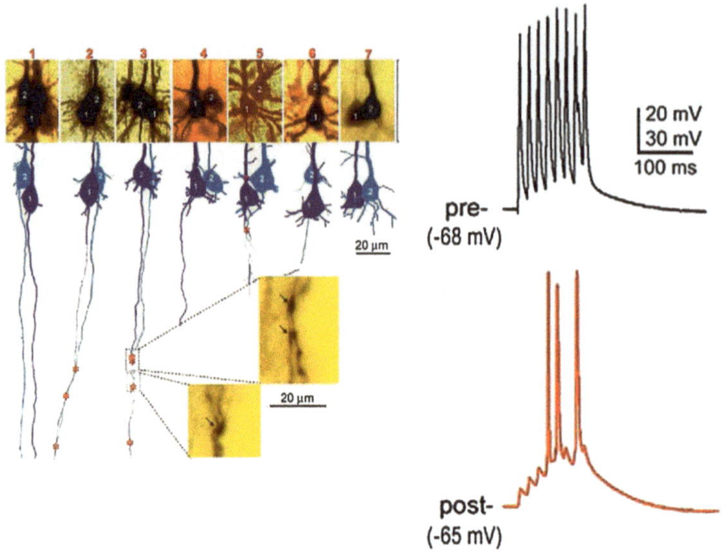

Figure 13.5 ***Strong electrical coupling between large layer five pyramidal neurons in mature rat prefrontal cortex.*** Full action potentials in one cell lead to short-latency action potentials and spikelets in the other cell. In at least some instances, the axons of the two neurons have close appositions (left), whose precise nature is not resolvable by light microscopy. Note that, at least in ferret prefrontal cortex, myelination of layer five pyramidal cell axons can begin hundreds of microns away from the soma (Shu et al., 2007). *(From Wang, Y., Barakat, A., Zhou, H., 2010, Electrotonic coupling between pyramidal neurons in the neocortex. PLoS One 5, e10253, reproduced with permission.)*

5. PREDICTIONS OF THE AXONAL GAP JUNCTION MODEL OF VFO, AND SPECIFICALLY OF RIPPLES

Once a model of dissociated ripples had been developed (Traub et al., 1999), it was natural to extend this model to other sorts of experimentally observed network oscillations, where there were data to suggest that electrical coupling might be important (reviewed in Traub and Whittington, 2010). Examples of such oscillations include "full" sharp wave-ripples occurring either in vivo or in vitro, under conditions where synaptic transmission is intact (Traub and Bibbig, 2000); persistent (ie, drug-induced) gamma oscillations (Traub et al., 2000); and beta-2 (20–30 Hz) oscillations in deep layers of association neocortex (Roopun et al., 2006). Although space does not permit a detailed analysis, we can summarize the situation this way: VFO in the pyramidal cell axonal plexus underlies all three sorts of oscillations, but this VFO interacts with inhibitory processes in somewhat different ways.

Thus, in the case of sharp wave-ripples, the synaptic output of the rapidly oscillating axonal plexus induces high-frequency firing in fast-spiking (and other) interneurons,

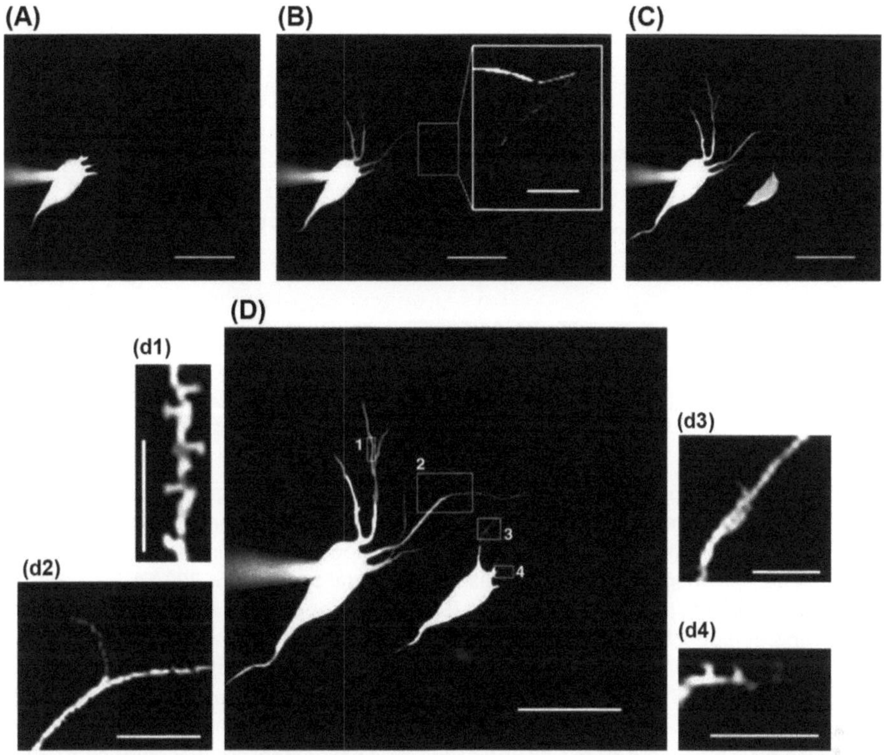

Figure 13.6 ***Dye coupling (rhodamine) between two identified CA1 pyramidal cells (identification by the presence of typical spiny basal dendrites).*** With the use of confocal scanning laser technique, the site of coupling was identified (light microscopically, of course) to be between axons: the coupling neuronal processes lacked spines and had spineless collaterals. The site of coupling, in four well-characterized instances, was 65–117 μm from the soma. Dye coupling was observed in 24 other instances (total of 64 injections) where the site of coupling could not be detected. Artifactual coupling during cell-attached recording was never observed. Dye filling of a pyramidal neuron with rhodamine 123 at increasing time points after establishing whole-cell configuration (dye pipette from left): (A) 30 s (scale bar: 50 μm). (B) 8 min, beginning of axonal and dendritic dye filling. The boxed region and the overframing show a cell process dye filled from the axon (distance between axonal coupling: 91 μm; scale bar: 50 μm; overframing: 10 μm). (C) 11 min, dye-coupling of a second pyramidal neuron. The dye filling of the soma occurs via the cell process, which can now be identified as the axon of the coupled cell (scale bar: 50 μm). (D) 17 min, beginning of dendritic dye filling of the dye-coupled neuron (scale bar: 50 μm). Overframings show details from both neurons: (D1, D4) dendritic segments with spines (scale bar: 5 μm); (D2, D3) axonal segments with branching collaterals (scale bar: 10 μm). *(From Schmitz, D., Schuchmann, S., Fisahn, A., Draguhn, A., Buhl, E.H., Petrasch-Parwez, R.E., Dermietzel, R., Heinemann, U., Traub, R.D., 2001. Axo-axonal coupling: a novel mechanism for ultrafast neuronal communication. Neuron 31, 831–840, reproduced with permission.)*

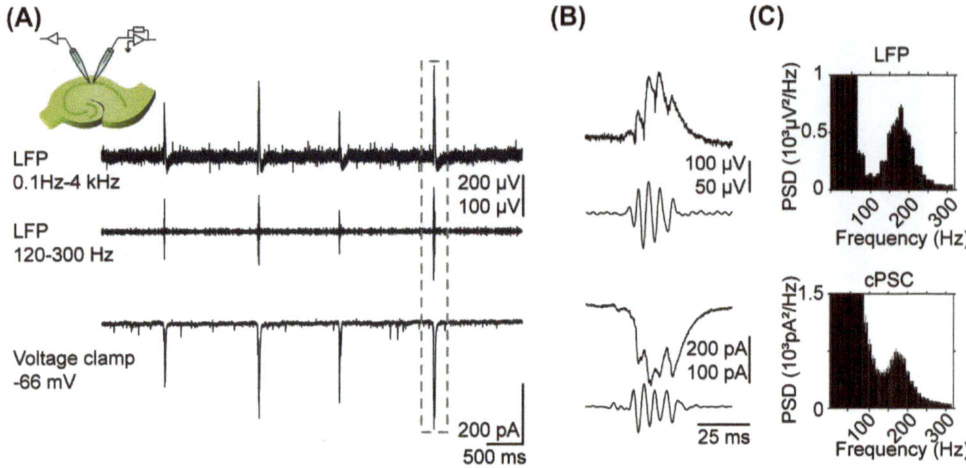

Figure 13.7 ***In vitro sharp wave-ripples are associated with ripple-frequency synaptic currents in pyramidal neurons.*** Here, examples are shown of the field potentials, and of patch recording of a single pyramidal neuron (measuring synaptic currents, below), at slow and faster time-scales (A, B). Power spectra are shown in (C). Note the similarity in spectral peak, about 200 Hz, between the field and the synaptic currents in the single neuron. (Sharp wave-ripples are a normal neuronal network event, occurring in hippocampus and other limbic structures in vivo, associated with awake immobility and slow-wave sleep. They may be important in memory consolidation.) The synaptic currents contain ripple-frequency EPSCs (excitatory postsynaptic currents, shown later in the paper from which this figure was taken). Ripple-frequency EPSCs also occur in CA1 oriens lacunosum/moleculare interneurons, which have little input from CA3, indicating a CA1 axonal source (Pangalos et al., 2013). *LFP*, local field potential; *PSD*, power spectral density. *(From Maier, N., Tejero-Cantero, A., Dorrn, A.L., Winterer, J., Beed, P.S., Morris, G., Kempter, R., Poulet, J.F.A., Leibold, C., Schmitz, D., 2011. Coherent phasic excitation during hippocampal ripples. Neuron 72, 137–152, reproduced with permission.)*

that provides feedback, ripple-frequency inhibitory postsynaptic currents to pyramidal cells, and (at least in vitro) causes deep hyperpolarization of pyramidal neurons (Maier et al., 2003; see also Fig. 13.8, left). In the case of persistent gamma oscillations (Buhl et al., 1998; Fisahn et al., 1998), brief axonal plexus VFO bursts are interrupted by perisomatic inhibition (Traub et al., 2003a). And in the case of beta-2, brief axonal plexus VFO bursts are interrupted by intrinsic K^+ currents, probably generated in the axon itself (Roopun et al., 2006).

The above-mentioned models, regarded as extensions of the axonal coupling model, make a number of specific predictions, some of which we shall illustrate, and for others provide references to the literature:

1. During sharp wave-ripples, ripple-frequency excitatory postsynaptic potentials/excitatory postsynaptic currents should occur in both pyramidal cells and local interneurons, a result of the net orthodromic synaptic output of the rapidly oscillating pyramidal axonal plexus. An example of ripple-frequency synaptic currents in a pyramidal cell, recorded during an in vitro sharp wave-ripple, is shown in Fig. 13.7. The

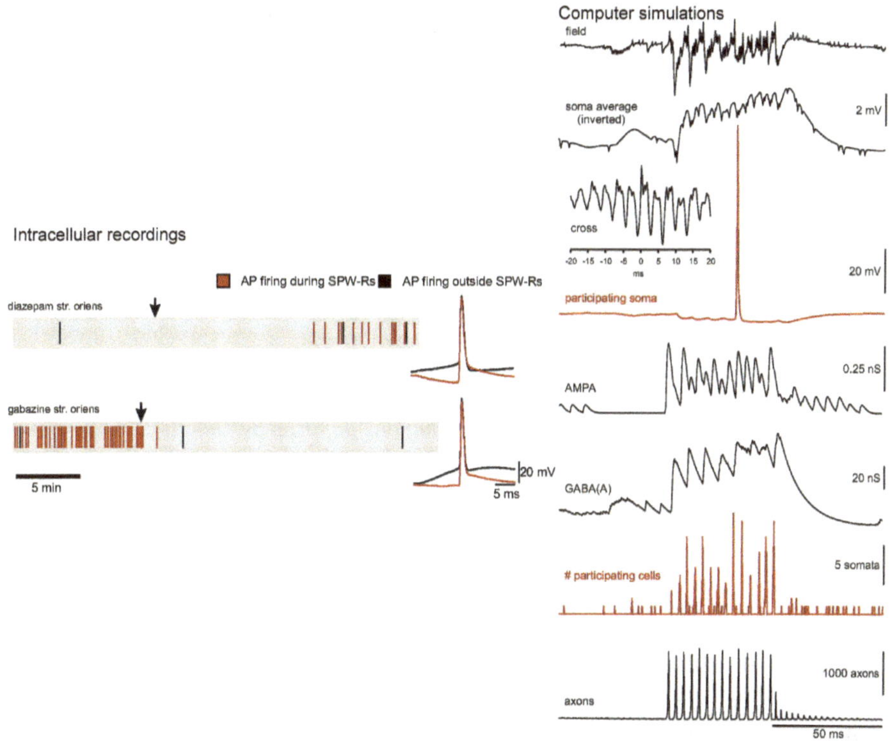

Figure 13.8 ***During in vitro sharp wave-ripples, pyramidal cell action potentials are antidromic, both experimentally (left) and in a network model (right).*** The model contained both pyramidal cells and populations of different interneuron types, with excitatory and inhibitory synaptic actions, along with electrical coupling, included. Electrical coupling was between axons for pyramidal cells, and between dendrites for interneurons. The model network generated ripple-frequency synaptic currents along with antidromic spikes, as rippling was generated by the axonal plexus, with action potentials propagating (line 3 on the right) antidromically to pyramidal cell somata, and (line 4 on the right) orthodromically to synapses on pyramidal cells and interneurons. *AP*, action potential; *AMPA*, α-amino-3-hydroxy-5-methyl-4-isoxazolepropionic acid; *GABA*(A), γ-aminobutyric acid (A); *SPW-R*, sharp-wave/ripple. *(From Bähner, F., Weiss, E.K., Birke, G., Maier, N., Schmitz, D., Rudolph, U., Frotscher, M., Traub, R.D., Both, M., Draguhn, A., 2011. Cellular correlate of assembly formation in oscillating hippocampal networks in vitro. Proc. Natl. Acad. Sci. U.S.A. 108, E607–E616, reproduced with permission.)*

paper from which this figure was reproduced (Maier et al., 2011) demonstrated that the net synaptic currents include EPSCs. Similar ripple-frequency EPSCs have also been recorded in oriens lacunosum/moleculare interneurons during sharp wave-ripples (Pangalos et al., 2013). The ripple-frequency EPSCs can reasonably account for the high-frequency firing of fast spiking interneurons during sharp wave-ripples (Ylinen et al., 1995).

2. During sharp wave-ripples, persistent gamma oscillations, and beta-2 oscillations (in association cortex), pyramidal cell action potentials are, at least in large part,

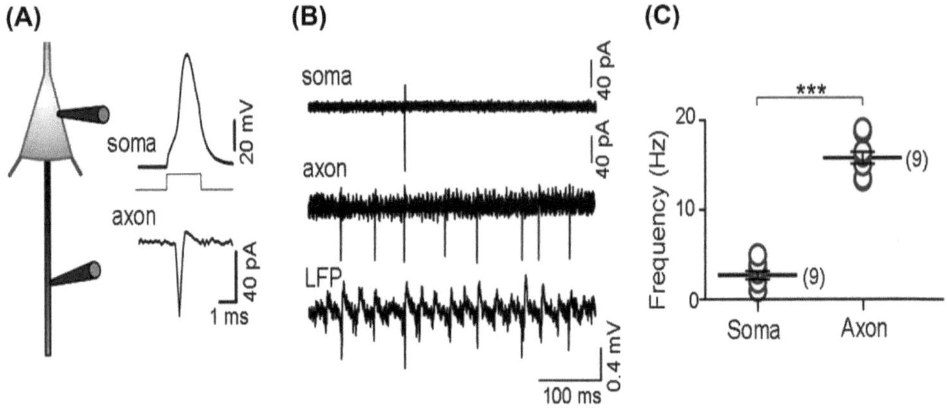

Figure 13.9 ***During in vitro persistent gamma oscillations (induced by kainate), CA3 pyramidal cell axons fire considerably more than do somata, as predicted by the axonal coupling model (Traub et al., 2000).*** (A) Scheme of dual somatic and axonal recording configuration. Action potentials evoked in whole-cell configuration by brief depolarizing current injection into the soma (800 pA) (inset) reliably induced action currents in the axon. (B) Dual somatic and axonal cell-attached recordings directly demonstrate that high-frequency axonal spikes fail to invade the soma during gamma oscillations (LFP). (C) Summary plot shows the highly significant difference in the discharge frequency observed at the soma and proximal axon in dual recordings during network oscillations. *LFP*, local field potential. *(This figure is from Dugladze, T., Schmitz, D., Whittington, M.A., Vida, I., Gloveli, T., 2012. Segregation of axonal and somatic activity during fast network oscillations. Science 336, 1458–1461, reproduced with permission.)*

antidromic, a result of their origin in the pyramidal cell axonal plexus. Fig. 13.8 illustrates action potentials from the real hippocampus in vitro, and from a simulation model of sharp wave-ripples. In the model, all pyramidal cell spikes were antidromic. In the experiments by Bähner et al. (2011), pyramidal cell action potentials that occurred during sharp wave-ripples were all antidromic, whereas action potentials occurring between sharp wave-ripples were not (see also Papatheodoropoulos, 2008). The occurrence of antidromic spikes is an immediate consequence of the axonal plexus model, and not readily explained by models of sharp waves that are generated purely by synaptic interactions. During gamma oscillations in entorhinal cortex and neocortex, spikelets are observed, also indicative of axonal activity (Cunningham et al., 2004a,b).

3. Packets of VFOs should occur at gamma frequency in the local field, during persistent gamma oscillations. This was demonstrated experimentally in hippocampal CA1 by Traub et al. (2003a).
4. During gamma oscillations, pyramidal cell axons should fire at significantly higher rates than somata (Traub et al., 2000). This has been demonstrated by Dugladze et al. (2012) (Fig. 13.9). We predict the same frequency dissociation (axon vs. soma) for sharp wave-ripples and for beta-2 oscillations, but this prediction remains under experimental investigation.

6. WHAT MIGHT THE GAP JUNCTION PROTEIN BE?

Connexins that have been described in neurons and in the retina include connexin36 (Cx36) (Srinivas et al., 1999; Teubner et al., 2000), the most common one in the central nervous system; Cx26 (in developing neocortex (Yu et al., 2012)); Cx45 (Maxeiner et al., 2003); and Cx57 (Palacios-Prado et al., 2009). Pannexin channels also appear to exist (Bruzzone et al., 2003), although the functional significance of pannexin gap junctions between neurons (as opposed to hemichannels) is not clear, to the best of our knowledge (MacVicar and Thomson, 2010).

Whatever the connexin is that presumably mediates axonal coupling, it may not be Cx36, given that alkaline conditions reduce Cx36 channel conductance (Gonzalez-Nieto et al., 2008), whereas axonal coupling and dissociated rippling, as well as dye coupling, are all enhanced by alkaline conditions (Schmitz et al., 2001; Draguhn et al., 1998; Church and Baimbridge, 1991). In addition, knockout of Cx36 does not affect low-calcium ripples in vitro (Hormuzdi et al., 2001), nor SPW-Rs in vivo (Buhl et al., 2003), although there could be compensation in the knockout by one or more other connexins.

Similarly, knockout of Cx45 does not affect in vitro SPW-Rs (Zlomzica et al., 2010), although again there could be compensation. Cx57 participates in axoaxonal coupling in the retina (Janssen-Bienhold et al., 2009), but expression of this protein in the adult brain is, to our knowledge, extremely limited. Immunoflourescent studies of adult brain, searching for Cx26, have also been unrevealing so far (J.I. Nagy, personal communication). Further studies are, obviously, required.

Proper identification of the putative axonal connexin(s) is of utmost importance: (1) it will allow localization of gap junction sites along the axon, including collaterals; (2) it will permit search for axonal coupling in many parts of the nervous system; (3) it may permit development of tools for selective blockade of axonal coupling (as opposed to nonspecific blockade of many sorts of gap junctions, along with unintended side effects); (4) it will greatly facilitate studies of the role of axonal coupling in disease states (such as epilepsy), as well as cognitive disorders. We shall expand on this last point in the next section.

7. CLINICAL IMPLICATIONS

Anatomical or functional abnormalities in neuron (specifically axonal) gap junctions might have two sorts of consequences for human health. We shall discuss these very briefly. As the major connexin mediating axonal coupling has not yet been identified, we shall refer to that connexin as "connexin X."

First, sharp wave-ripples are involved in the processing of experience-induced neuronal firing patterns, and presumably also memory consolidation (Bukalo et al., 2013; Ego-Stengel and Wilson, 2010; Carr et al., 2011). Disruption of gap junction function, certainly its experimental blockade (Maier et al., 2003), will affect sharp wave-ripples, either acutely or chronically. Such disruption might well have cognitive consequences, given the evidence for a connection between sharp wave-ripples and memory consolidation (Eschenko et al., 2008).

Second, there is known to be an intimate association between VFO and the onset of electrographic seizures (Fisher et al., 1992; Traub et al., 2001b). One might expect to find, in inherited epilepsies, molecular alterations or abnormalities (eg, mutations, deletions, nucleotide polymorphisms), either in the gene for connexin X, or for genes encoding for gap junction–associated proteins (eg, zonula occludens-1, AF6, MUPP1, cingulin) (Lynn et al., 2012). Another plausible system that could be involved in inherited epilepsies are genes encoding for receptors and signaling pathways that modulate gap junction conductance (eg, serotonin and dopamine).

Finally, VFO is potentiated by alkaline conditions in experimental protocols (Draguhn et al., 1998; Traub et al., 2001b, 2010). Experimental and clinical febrile seizures appear to be associated with, and probably induced by, respiratory alkalosis (Schuchmann et al., 2006, 2011). It is a classical observation that hyperventilation can induce absence seizures in susceptible individuals, but, in addition, breathing 5% CO_2 diminishes the ability of hyperventilation to induce clinical seizures, and it shortens spike-wave discharges (Yang et al., 2014). Given that, for example, connexins 45 and 57 have channel conductances that are enhanced by alkaline conditions (Palacios-Prado et al., 2009, 2010), it seems likely that gap junction conductance is the causal intermediary between alkaline pH and VFO. It would be most helpful, therefore, to be able to measure pH in the brain itself, simultaneously with electrical potentials, during invasive recordings of the epileptic human brain. A reasonable hypothesis is that astrocytic dysfunction, uncoupling, perhaps, predisposes the epileptic brain to develop alkaline conditions, in turn facilitating VFO and then seizure activity.

To summarize briefly, there is experimental evidence that very fast network oscillations, in cortical structures, can be generated by electrically coupled networks of principal neurons, and the oscillations can even be generated, at least in some circumstances, without the participation of chemical synapses. There is additionally experimental evidence that at least some principal cortical neurons are indeed electrically coupled, as well as being dye coupled. Network simulations have elucidated how the population oscillations can come about, provided that each cell is coupled to more than one other (on average), and provided that the coupling is strong enough for spikes to propagate from one axon to another—conditions that appear to be met experimentally, again under certain circumstances. The most pressing question for the immediate future is to define the requisite connexin(s) involved, as well as the precise anatomical sites of the gap junctions.

REFERENCES

Amitai, Y., Gibson, J.R., Beierlein, M., Patrick, S.L., Ho, A.M., Connors, B.W., Golomb, D., 2002. The spatial dimensions of electrically coupled networks of interneurons in the neocortex. J. Neurosci. 22, 4142–4152.

Andrade-Valenca, L.P., Dubeau, F., Mari, F., Zelmann, R., Gotman, J., 2011. Interictal scalp fast oscillations as a marker of the seizure onset zone. Neurology 77, 524–531.

Bähner, F., Weiss, E.K., Birke, G., Maier, N., Schmitz, D., Rudolph, U., Frotscher, M., Traub, R.D., Both, M., Draguhn, A., 2011. Cellular correlate of assembly formation in oscillating hippocampal networks in vitro. Proc. Natl. Acad. Sci. U.S.A. 108, E607–E616.

Bragin, A., Engel Jr., J., Wilson, C.L., Fried, I., Mathern, G.W., 1999. Hippocampal and entorhinal cortex high-frequency oscillations (100–500 Hz) in human epileptic brain and in kainic acid-treated rats with chronic seizures. Epilepsia 40, 127–137.

Bruzzone, R., Hormuzdi, S.G., Barbe, M.T., Herb, A., Monyer, H., 2003. Pannexins, a family of gap junction proteins expressed in brain. Proc. Natl. Acad. Sci. U.S.A. 100, 13644–13649.

Buhl, E.H., Tamás, G., Fisahn, A., 1998. Cholinergic activation and tonic excitation induce persistent gamma oscillations in mouse somatosensory cortex in vitro. J. Physiol. 513, 117–126.

Buhl, D.L., Harris, K.D., Hormuzdi, S.G., Monyer, H., Buzsáki, G., 2003. Selective impairment of hippocampal gamma oscillations in connexin-36 knock-out mouse in vivo. J. Neurosci. 23, 1013–1018.

Bukalo, O., Campanac, E., Hoffman, D.A., Fields, R.D., 2013. Synaptic plasticity by antidromic firing during hippocampal network oscillations. Proc. Natl. Acad. Sci. U.S.A. 110, 5175–5180.

Buzsáki, G., Horváth, Z., Urioste, R., Hetke, J., Wise, K., 1992. High-frequency network oscillation in the hippocampus. Science 256, 1025–1027.

Carr, M.F., Jadhav, S.P., Frank, L.M., 2011. Hippocampal replay in the awake state: a potential substrate for memory consolidation and retrieval. Nat. Neurosci. 14, 147–153.

Carracedo, L.M., Kjeldsen, H., Cunnington, L., Jenkins, A., Schofield, I., Cunningham, M.O., Traub, R.D., Whittington, M.A., 2013. A neocortical delta rhythm facilitates reciprocal interlaminar interactions via nested theta rhythms. J. Neurosci. 33, 10750–10761.

Chrobak, J.J., Buzsáki, G., 1996. High-frequency oscillations in the output networks of the hippocampal-entorhinal axis of the freely behaving rat. J. Neurosci. 16, 3056–3066.

Church, J., Baimbridge, K.G., 1991. Exposure to high-pH medium increases the incidence and extent of dye coupling between rat hippocampal CA1 pyramidal neurons in vitro. J. Neurosci. 11, 3289–3295.

Cunningham, M.O., Whittington, M.A., Bibbig, A., Roopun, A., LeBeau, F.E.N., Vogt, A., Monyer, H., Buhl, E.H., Traub, R.D., 2004a. A role for fast rhythmic bursting neurons in cortical gamma oscillations in vitro. Proc. Natl. Acad. Sci. U.S.A. 101, 7152–7157.

Cunningham, M.O., Halliday, D.M., Davies, C.H., Traub, R.D., Buhl, E.H., Whittington, M.A., 2004b. Coexistence of gamma and high-frequency oscillations in the medial entorhinal cortex in vitro. J. Physiol. 559, 347–353.

Cunningham, M.O., Roopun, A.K., Schofield, I.S., Whittaker, R.G., Duncan, R., Russell, A., Jenkins, A., Nicholson, C., Whittington, M.A., Traub, R.D., 2012. Glissandi: transient fast electrocorticographic oscillation of steadily increasing frequency, explained by temporally increasing gap junction conductance. Epilepsia 53, 1205–1214.

Devor, A., Yarom, Y., 2002. Generation and propagation of subthreshold waves in a network of inferior olivary neurons. J. Neurophysiol. 87, 3059–3069.

Draguhn, A., Traub, R.D., Schmitz, D., Jefferys, J.G.R., 1998. Electrical coupling underlies high-frequency oscillations in the hippocampus in vitro. Nature 394, 189–192.

Dugladze, T., Schmitz, D., Whittington, M.A., Vida, I., Gloveli, T., 2012. Segregation of axonal and somatic activity during fast network oscillations. Science 336, 1458–1461.

Ego-Stengel, V., Wilson, M.A., 2010. Disruption of ripple-associated hippocampal activity during rest impairs spatial learning in the rat. Hippocampus 20, 1–10.

Engel Jr., J., Bragin, A., Staba, R., Mody, I., 2009. High-frequency oscillations: what is normal and what is not? Epilepsia 50, 598–604.

Erdös, P., Rényi, A., 1960. On the evolution of random graphs. Publ. Math. Instit. Hung. Acad. Sci. 5, 17–61.
Eschenko, O., Ramadan, W., Mölle, M., Born, J., Sara, S.J., 2008. Sustained increase in hippocampal sharp-wave ripple activity during slow-wave sleep after learning. Learn. Mem. 15, 222–228.
Fisahn, A., Pike, F.G., Buhl, E.H., Paulsen, O., 1998. Cholinergic induction of network oscillations at 40 Hz in the hippocampus in vitro. Nature 394, 186–189.
Fisher, R.S., Webber, W.R.S., Lesser, R.P., Arroyo, S., Uematsu, S., 1992. High-frequency EEG activity at the start of seizures. J. Clin. Neurophysiol. 9, 441–448.
Furshpan, E.J., Potter, D.D., 1959. Transmission at the giant motor synapses of the crayfish. J. Physiol. 145, 289–325.
González-Nieto, D., Gómez-Hernández, J.M., Larrosa, B., Gutiérrez, C., Muñoz, M.D., Fasciani, I., O'Brien, J.O., Zappalà, A., Cicirata, F., Barrio, L.C., 2008. Regulation of neuronal connexin-36 channels by pH. Proc. Natl. Acad. Sci. U.S.A. 105, 17169–17174.
Grenier, F., Timofeev, I., Steriade, M., 2003. Neocortical very fast oscillations (ripples, 80–200 Hz) during seizures: intracellular correlates. J. Neurophysiol. 89, 841–852.
Hamzei-Sichani, F., Kamasawa, N., Janssen, W.G.M., Yasamura, T., Davidson, K.G.V., Hof, P.R., Wearne, S.L., Stewart, M.G., Young, S.R., Whittington, M.A., Rash, J.E., Traub, R.D., 2007. Gap junctions on hippocampal mossy fiber axons demonstrated by thin-section electron microscopy and freeze-fracture replica immunogold labeling. Proc. Natl. Acad. Sci. U.S.A. 104, 12548–12553.
Hormuzdi, S.G., Pais, I., LeBeau, F.E.N., Towers, S.K., Rozov, A., Buhl, E.H., Whittington, M.A., Monyer, H., 2001. Impaired electrical signaling disrupts gamma frequency oscillations in connexin 36-deficient mice. Neuron 31, 487–495.
Hughes, S.W., Lörincz, M.L., Blethyn, K., Kékesi, K.A., Juhász, G., Turmaine, M., Parnavelas, J.G., Crunelli, V., 2011. Thalamic gap junctions control local neuronal synchrony and influence macroscopic oscillation amplitude during EEG alpha rhythms. Front. Psychol. 2, 193.
Jacobs, J., LeVan, P., Chander, R., Hall, J., Dubeau, F., Gotman, J., 2008. Interictal high-frequency oscillations (80–500 Hz) are an indicator of seizure onset areas independent of spikes in the human epileptic brain. Epilepsia 49, 1893–1907.
Janssen-Bienhold, U., Trümpler, J., Hilgen, G., Schultz, K., Müller, L.P., Sonntag, S., Dedek, K., Kirks, P., Willecke, K., Weiler, R., 2009. Connexin57 is expressed in dendro-dendritic and axo-axonal gap junctions of mouse horizontal cells and its distribution is modulated by light. J. Comp. Neurol. 513, 363–374.
Kosaka, T., 1983. Gap junctions between non-pyramidal cell dendrites in the rat hippocampus (CA1 and CA3 regions). Brain Res. 271, 157–161.
Lee, S.C., Patrick, S.L., Richardson, K.A., Connors, B.W., 2014. Two functionally distinct networks of gap junction-coupled inhibitory neurons in the thalamic reticular nucleus. J. Neurosci. 34, 13170–13182.
Lewis, T.J., Rinzel, J., 2000. Self-organized synchronous oscillations in a network of excitable cells coupled by gap junctions. Netw. Comput. Neural Syst. 11, 299–320.
Lewis, T.J., Rinzel, J., 2001. Topological target patterns and population oscillations in a network with random gap junctional coupling. Neurocomputing 38–40, 763–768.
Llinás, R., Yarom, Y., 1986. Oscillatory properties of guinea-pig inferior olivary neurones and their pharmacological modulation: an in vitro study. J. Physiol. 376, 163–182.
Lynn, B.D., Li, X., Nagy, J.I., 2012. Under construction: building the macromolecular superstructure and signalling components of an electrical synapse. J. Membr. Biol. 245, 303–317.
MacVicar, B.A., Thomson, R.J., 2010. Non-junction functions of pannexin-1 channels. Trends Neurosci. 33, 93–102.
Maier, N., Nimmrich, V., Draguhn, A., 2003. Cellular and network mechanisms underlying spontaneous sharp wave-ripple complexes in mouse hippocampal slices. J. Physiol. 550, 873–887.
Maier, N., Tejero-Cantero, A., Dorrn, A.L., Winterer, J., Beed, P.S., Morris, G., Kempter, R., Poulet, J.F.A., Leibold, C., Schmitz, D., 2011. Coherent phasic excitation during hippocampal ripples. Neuron 72, 137–152.
Maxeiner, S., Krüger, O., Schilling, K., Traub, O., Urschel, S., Willecke, K., 2003. Spatiotemporal transcription of connexion45 during brain development results in neuronal expression in adult mice. Neuroscience 119, 689–700.

Mercer, A., Bannister, A.P., Thomson, A.M., 2006. Electrical coupling between pyramidal cells in adult cortical regions. Brain Cell Biol. 35, 13–27.

Miles, R., Wong, R.K.S., 1986. Excitatory synaptic interactions between CA3 neurones in the guinea-pig hippocampus. J. Physiol. 373, 397–418.

Miles, R., Wong, R.K.S., 1987. Inhibitory control of local excitatory circuits in the guinea-pig hippocampus. J. Physiol. 388, 611–629.

Munro, E., Börgers, C., 2010. Mechanisms of very fast oscillations in networks of axons coupled by gap junctions. J. Comput. Neurosci. 28, 539–555.

Pais, I., Hormuzdi, S.G., Monyer, H., Traub, R.D., Wood, I.C., Buhl, E.H., Whittington, M.A., LeBeau, F.E.N., 2003. Sharp wave-like activity in the hippocampus in vitro in mice lacking the gap junction protein connexin 36. J. Neurophysiol. 89, 2046–2054.

Palacios-Prado, N., Sonntag, S., Skeberdis, V.A., Bukauskas, F.F., 2009. Gating, permselectivity and pH-dependent modulation of channels formed by connexin57, a major connexin of horizontal cells in the mouse retina. J. Physiol. 587, 3251–3269.

Palacios-Prado, N., Briggs, S.W., Skeberdis, V.A., Pranevicius, M., Bennett, M.V., Bukauskas, F.F., 2010. pH-dependent modulation of voltage gating in connexin45 homotypic and connexin45/connexin43 heterotypic gap junctions. Proc. Natl. Acad. Sci. U.S.A. 107, 9897–9902.

Pangalos, M., Donoso, J.R., Winterer, J., Zivkovic, A.R., Kempter, R., Maier, N., Schmitz, D., 2013. Recruitment of oriens-lacunosum-moleculare interneurons during hippocampal ripples. Proc. Natl. Acad. Sci. U.S.A. 110, 4398–4403.

Papatheodoropoulos, C., 2008. A possible role of ectopic action potentials in the in vitro hippocampal sharp wave-ripple complexes. Neuroscience 157, 495–501.

Peters, A., Palay, S.L., Webster, H.D., 1991. The fine structure of the nervous system: neurons their supporting cells. vol. 3. Oxford University Press, New York, pp. 1–494.

Roopun, A., Middleton, S.J., Cunningham, M.O., LeBeau, F.E.N., Bibbig, A., Whittington, M.A., Traub, R.D., 2006. A beta2-frequency (20–30 Hz) oscillation in non-synaptic networks of somatosensory cortex. Proc. Natl. Acad. Sci. U.S.A. 103, 15646–15650.

Roopun, A.K., Simonotto, J.D., Pierce, M.L., Jenkins, A., Schofield, I., Kaiser, M., Whittington, M.A., Traub, R.D., Cunningham, M.O., 2010. A non-synaptic mechanism underlying interictal discharges in human epileptic neocortex. Proc. Natl. Acad. Sci. U.S.A. 107, 338–343.

Schevon, C.A., Trevelyan, A.J., Schroeder, C.E., Goodman, R.R., McKhann Jr., G., Emerson, R.G., 2009. Spatial characterization of interictal high frequency oscillations in epileptic neocortex. Brain 132, 3047–3059.

Schmitz, D., Schuchmann, S., Fisahn, A., Draguhn, A., Buhl, E.H., Petrasch-Parwez, R.E., Dermietzel, R., Heinemann, U., Traub, R.D., 2001. Axo-axonal coupling: a novel mechanism for ultrafast neuronal communication. Neuron 31, 831–840.

Schuchmann, S., Schmitz, D., Rivera, C., Vanhatalo, S., Salmen, B., Mackie, K., Sipilä, S.T., Voipio, J., Kaila, K., 2006. Experimental febrile seizures are precipitated by a hyperthermia-induced respiratory alkalosis. Nat. Med. 12, 817–823.

Schuchmann, S., Hauck, S., Henning, S., Grüters-Kieslich, A., Vanhatalo, S., Schmitz, D., Kaila, K., 2011. Respiratory alkalosis in children with febrile seizures. Epilepsia 52, 1949–1955.

Shu, Y., Duque, A., Yu, Y., Haider, B., McCormick, D.A., 2007. Properties of action-potential initiation in neocortical pyramidal cells: evidence from whole cell axon recordings. J. Neurophysiol. 97, 746–760.

Simon, A., Traub, R.D., Vladimirov, N., Jenkins, A., Nicholson, C., Whittaker, R., Schofield, I., Clowry, G.J., Cunningham, M.O., Whittington, M.A., 2014. Gap junction networks can generate both ripple-like and fast-ripple-like oscillations. Eur. J. Neurosci. 39, 46–60.

Srinivas, M., Rozental, R., Kojima, T., Dermietzel, R., Mehler, M., Condorelli, D.F., Kessler, J.A., Spray, D.C., 1999. Functional properties of channels formed by the neuronal gap junction protein connexin36. J. Neurosci. 19, 9848–9855.

Teubner, B., Degen, J., Söhl, G., Güldenagel, M., Bukauskas, F.F., Trexler, E.B., Verselis, V.K., De Zeeuw, C.I., Lee, C.G., Kozak, C.A., Petrasch-Parwez, E., Dermietzel, R., Willecke, K., 2000.

Functional expression of the murine connexion 36 gene coding for a neuron-specific gap junctional protein. J. Membr. Biol. 176, 249–262.
Traub, R.D., Bibbig, A., 2000. A model of high-frequency ripples in the hippocampus, based on synaptic coupling plus axon-axon gap junctions between pyramidal neurons. J. Neurosci. 20, 2086–2093.
Traub, R.D., Whittington, M.A., 2010. Cortical Oscillations in Health and Disease. Oxford University Press, New York.
Traub, R.D., Wong, R.K.S., 1982. Cellular mechanism of neuronal synchronization in epilepsy. Science 216, 745–747.
Traub, R.D., Schmitz, D., Jefferys, J.G.R., Draguhn, A., 1999. High-frequency population oscillations are predicted to occur in hippocampal pyramidal neuronal networks interconnected by axoaxonal gap junctions. Neuroscience 92, 407–426.
Traub, R.D., Bibbig, A., Fisahn, A., LeBeau, F.E.N., Whittington, M.A., Buhl, E.H., 2000. A model of gamma-frequency network oscillations induced in the rat CA3 region by carbachol in vitro. Eur. J. Neurosci. 12, 4093–4106.
Traub, R.D., Kopell, N., Bibbig, A., Buhl, E.H., LeBeau, F.E.N., Whittington, M.A., 2001a. Gap junctions between interneuron dendrites can enhance long-range synchrony of gamma oscillations. J. Neurosci. 21, 9478–9486.
Traub, R.D., Whittington, M.A., Buhl, E.H., LeBeau, F.E.N., Bibbig, A., Boyd, S., Cross, H., Baldeweg, T., 2001b. A possible role for gap junctions in generation of very fast EEG oscillations preceding the onset of, and perhaps initiating, seizures. Epilepsia 42, 153–170.
Traub, R.D., Cunningham, M.O., Gloveli, T., LeBeau, F.E.N., Bibbig, A., Buhl, E.H., Whittington, M.A., 2003a. GABA-enhanced collective behavior in neuronal axons underlies persistent gamma-frequency oscillations. Proc. Natl. Acad. Sci. U.S.A. 100, 11047–11052.
Traub, R.D., Pais, I., Bibbig, A., LeBeau, F.E.N., Buhl, E.H., Hormuzdi, S.G., Monyer, H., Whittington, M.A., 2003b. Contrasting roles of axonal (pyramidal cell) and dendritic (interneuron) electrical coupling in the generation of gamma oscillations in the hippocampus in vitro. Proc. Natl. Acad. Sci. U.S.A. 100, 1370–1374.
Traub, R.D., Duncan, R., Russell, A.J.C., Baldeweg, T., Tu, Y., Cunningham, M.O., Whittington, M.A., 2010. Spatiotemporal patterns of electrocorticographic very fast oscillations (>80 Hz) consistent with a network model based on electrical coupling between principal neurons. Epilepsia 51, 1587–1597.
Vinogradova, T.M., Maltsev, V.A., Bogdanov, K.Y., Lyashkov, A.E., Lakatta, E.G., 2005. Rhythmic Ca^{2+} oscillations drive sinoatrial nodal cell pacemaker function to make the heart tick. Ann. N.Y. Acad. Sci. 1047, 138–156.
Vladimirov, N., Tu, Y., Traub, R.D., 2012. Shortest loops are pacemakers in random networks of electrically coupled axons. Front. Comput. Neurosci. http://dx.doi.org/10.3389/fn com.2012.00017.
Wang, Y., Barakat, A., Zhou, H., 2010. Electrotonic coupling between pyramidal neurons in the neocortex. PLoS One 5, e10253.
Welsh, J.P., Turecek, J., 2017. Implications of electrical synapse plasticity in the inferior olive. In: Jing, J. (Ed.), Network Functions and Plasticity. Elsevier (Chapter 12).
Yang, X.F., Shi, X.Y., Ju, J., Zhang, W.N., Liu, Y.J., Li, X.Y., Zou, L.P., 2014. 5% CO_2 inhalation suppresses hyperventilation-induced absence seizures in children. Epilepsy Res. 108, 345–348.
Ylinen, A., Bragin, A., Nádasdy, Z., Jandó, G., Szabó, I., Sik, A., Buzsáki, G., 1995. Sharp wave-associated high frequency oscillation (200 Hz) in the intact hippocampus: network and intracellular mechanisms. J. Neurosci. 15, 30–46.
Yu, Y.-C., He, S., Chen, S., Fu, Y., Brown, K.N., Yao, X.-H., Ma, J., Gao, K.P., Sosinsky, G.E., Huang, K., Shi, S.-H., 2012. Preferential electrical coupling regulates neocortical lineage-dependent microcircuit assembly. Nature 486, 113–117.
Zlomuzica, A., Reichinnek, S., Maxeiner, S., Both, M., May, E., Wörsdörfer, P., Draguhn, A., Willecke, K., Dere, E., 2010. Deletion of connexin45 in mouse neurons disrupts one-trial object recognition and alters kainate-induced gamma-oscillations in the hippocampus. Physiol. Behav. 101, 245–253.

CHAPTER 14

Lineage-Dependent Electrical Synapse Formation in the Mammalian Neocortex

S. He[1], S.-H. Shi[2]
[1]ShanghaiTech University, Shanghai, China; [2]Memorial Sloan Kettering Cancer Center, New York, NY, United States

1. INTRODUCTION

Gap junctions are specialized cell—cell contacts that enable direct intercellular communication. In general, they allow small molecules less than 1000 Da to pass through, which include small ions, such as Na^+, K^+, Ca^{2+}, and Cl^-; metabolites; nutrients; second messengers, such as inositol-1,4,5-trisphosphate and cyclic AMP (cAMP); and small dyes, such as Lucifer yellow, biocytin, and neurobiotin (Evans and Martin, 2002). Owing to the passive diffusion of ions across them, gap junctions are termed "electrical synapses" (Bennett, 1997, 2000) and can be detected by simultaneous recording of two neurons using paired whole-cell electrophysiological recordings. Gap junctions that have no channel activity but maintain cell—cell adhesion are usually classified as adhesive gap junctions (Elias and Kriegstein, 2008; Elias et al., 2007; Lin et al., 2002).

Electrical synapses are widely present in tissues including the cardiovascular system (blood vessels and heart), the digestive system (the gastrointestinal tract, salivary glands, pancreas, and liver), the reproductive system, and the nervous system (Friend and Gilula, 1972; Nielsen et al., 2012). Electrical synapses were first identified in the 1950s between the giant motor neurons of the crayfish using electrophysiological recordings, electron microscopy, and freeze fracture (Robertson, 1953; Sohl et al., 2005). Since then, electrical synapses have been found in the central nervous system of the adult mammalian brain, especially in GABAergic interneurons and glia (Montoro and Yuste, 2004; Wang et al., 2010). Owing to the advances in research techniques including electrophysiology, genetic approaches, and high-resolution cellular imaging, electrical synapses between interneurons have been extensively studied in many regions of the brain leading to significant contributions toward understanding neuronal synchronization (Bennett and Zukin, 2004; Hestrin and Galarreta, 2005; Hormuzdi et al., 2004). Evidently, electrical synapses can function alone or orchestrate with chemical synapses to promote synchronized neuronal activity, thereby promoting direct passage of electrotonic potentials and sharpening neuronal firing. Because of their unique electrophysiological properties, electrical synapses serve as a low-pass filter that allows low-frequency firing to pass

Network Functions and Plasticity
ISBN 978-0-12-803471-2, http://dx.doi.org/10.1016/B978-0-12-803471-2.00014-X

through neurons more easily than high-frequency firing, although this role remains controversial between nonneuronal cells. For example, ultrafast neuronal firing (>150 Hz) is almost completely filtered between neurons coupled by electrical synapses, resulting in their restriction to the starting neurons. Therefore certain frequency ranges of neuronal firing tend to be synchronized into oscillatory neuronal activity within an ensemble of coupled neurons.

Contrary to their prevalence in interneurons, electrical synapses are virtually absent among mature pyramidal cells in the mammalian neocortex (Galarreta and Hestrin, 1999; Gibson et al., 1999) (see Traub et al., 2017 (Chapter 13)). Using multiple-electrode whole-cell electrophysiological recordings, one study showed that only 0.4% of neighboring pyramidal cells (8 of 2000 recorded pairs) were coupled by electrical synapses in the neocortex of ferrets and rats (Wang et al., 2010), and their physiological roles remain elusive. But this is not the case for immature pyramidal cells. Electrical synapses are transiently but abundantly expressed among immature neocortical pyramidal cells (Nadarajah et al., 1997; Peinado et al., 1993; Yuste et al., 1992), and can form extensive "neuronal domains" in the developing mammalian neocortex (Kandler and Katz, 1995, 1998a; Yuste et al., 1995, 1992). The neuronal domains were first characterized in the neonatal neocortex of rats using optical calcium imaging with calcium indicator Fura-2. A neuronal domain consists of a variable number of neurons (from 5 to 50 cells) that display a simultaneous increase in intracellular free calcium concentration ($[Ca^{2+}]_i$). In line with this, single neuronal injection of Neurobiotin at the similar postnatal age revealed similar neuronal columnar domains, in which neurons were radially dispersed from layer II to layer VI. The neuronal domains were not affected by perfusion of a voltage-dependent sodium channel blocker (tetrodotoxin) that specifically blocks chemical, but not electrical, synaptic transmission via preventing action potentials, but were abolished by gap junction blockers halothane and octanol. These data clearly suggest that neuronal domains in the neonatal neocortex consist of an ensemble of neurons interconnected by specific electrical synapses instead of chemical synapses. This further supports the notion that electrical synapses are robustly present between immature pyramidal cells in the developing mammalian neocortex.

Furthermore, electrical synapses were found to be extremely prominent between neuroblasts in the embryonic neocortex. Injection of Lucifer yellow dye into one neuroblast through a recording pipette resulted in a group of cells oriented radially along the ventricle also being labeled (Gutnick et al., 1981; Lo Turco and Kriegstein, 1991), thereby revealing a columnar domain. The dye-labeled cells were neural progenitor cells negative for the neuron-specific class III beta-tubulin (TUJ1), a postmitotic neuronal marker, but positive for a specific radial glial progenitor (RGP) marker (Bittman et al., 1997). The columnar domains can contain up to 90 neuroblasts in the ventricular zone (VZ), a neurogenic region in which RGPs give rise to postmitotic neurons and intermediate progenitor cells, at the early stage of neocortical development.

As development proceeds, the number of labeled cells progressively decreases, despite an increase in the number of migrating postmitotic neurons, suggesting no dye coupling between RGPs and migrating neurons. More than a decade later, adhesive gap junctions that have no channel activity were suggested to be essential for neuronal migration along the radial glial fiber of RGPs toward their final destination in the cortical plate (CP), which forms the future neocortex (Elias et al., 2007).

Besides being abundant in neurons, electrical synapses are also widely found between glia. It has been amply demonstrated that electrical synapses form between astrocytes, and between astrocytes and oligodendrocytes, but not between oligodendrocytes nor between glia and neurons (Massa and Mugnaini, 1982; Nagy et al., 2001). The precise functional roles of electrical synapses between glia remain elusive, although they are generally thought to act as a buffer to prevent the accumulation of intracellular potassium ions in one glial cell after depolarization (Connors et al., 1982; Orkand et al., 1966; Orthmann-Murphy et al., 2008; Ransom et al., 1985). Glial electrical synapses can also play a role in the propagation of calcium waves (Cornell-Bell and Finkbeiner, 1991; Cornell-Bell et al., 1990). Furthermore, gap junction proteins (Connexin 43, Cx43) have been observed to interact with signaling proteins, such as protein kinases and phosphatases, and structural proteins, notably zona occluden-1 and microtubules (Giepmans, 2004), thereby implicating electrical synapses in cellular signaling pathways.

In general, it was widely accepted that electrical synapses tend to form between cells of the same subtype in the central nervous system, but it was unclear whether coupled neurons were likely to be lineage related. A series of studies showed that electrical synapses preferentially form between clonally related cells in the developing mouse neocortex (He et al., 2015; Yu et al., 2012). This chapter focuses on formation of the lineage-specific electrical synapses and their role in the development of the functional neocortex.

2. COMPOSITION OF ELECTRICAL SYNAPSES IN THE MAMMALIAN NEOCORTEX

Electrical synapses are assembled by two hemichannels situated in apposed cell membranes. A hemichannel, also called a connexon, consists of six connexins. At least 20 different connexin genes have been identified in mice and 21 genes in humans (Cina et al., 2007; Nadarajah et al., 1997; Willecke et al., 2002), of which about eight have been characterized in the mammalian neocortex. They include Cx26, Cx29, Cx30, Cx32, Cx36, Cx43, Cx45, and Cx47 (Table 14.1), which are named after their approximate molecular weights (eg, Cx26 has a molecular weight of 26 Da) (Connors and Long, 2004; Nagy and Rash, 2000; Nielsen et al., 2012; Rash et al., 2001).

Cx26, Cx43, and Cx45 are the most abundantly expressed connexins in the developing neocortex at the embryonic and neonatal stages. Cx43 is strongly expressed in

Table 14.1 Expression of Connexins in the Mammalian Neocortex

	Location and Cell Type	Developmental Stage	Functional Compatibility With Other Cxs
Cx26	Immunochemical staining shows broad presence in RGPs and immature excitatory neurons (Nadarajah et al., 1997; Bittman and LoTurco, 1999; Dermietzel et al., 1989); but LacZ reporter line shows Cx26 restricted to meningeal cells (Filippov et al., 2003)	Embryonic and neonatal	Able to form functional heterotypic channels with Cx30, but not with Cx43 (Barrio et al., 1991; White et al., 1995)
Cx29	Predominantly expressed in oligodendrocytes (Montoro and Yuste, 2004; Dermietzel et al., 1997; Nagy et al., 2003a,b)	Adult	Unlikely to form heterotypic channels with other Cxs
Cx30	Neocortical astrocytes (Nagy and Rash, 2000; Nagy et al., 1999)	Adult	Able to form functional heterotypic channels with Cx26, Cx32, and Cx47, but not with Cx43 (Barrio et al., 1991; White et al., 1995) see review by Abrams and Rash, 2009)
Cx32	Neocortical astrocytes and oligodendrocytes (Montoro and Yuste, 2004; Nagy and Rash, 2000; Dermietzel et al., 1989; Li et al., 1997; Dermietzel et al., 1997; Micevych and Abelson, 1991; Scherer et al., 1995; Nagy et al., 2003a,b)	Adult	Able to form heterotypic channels with Cx26, Cx30, and Cx47 (see review Abrams and Rash, 2009)
Cx36	Predominant Cx present in GABAergic interneurons	Adult	Unable to form heterotypic channels with other Cxs
Cx43	Rich in RGPs and newborn neurons	Embryonic and neonatal	Able to form heterotypic channels with Cx47 (White et al., 1995), but not with Cx26 and Cx30 (Barrio et al., 1991; White et al., 1995). Cx30-Cx32 and Cx43-Cx47 are primary components of A-O electrical synapses
Cx45	Colocalization with NeuN (Kruger et al., 2000), most likely in immature excitatory neurons	Embryonic and neonatal	Unable to form heterotypic channels with other Cxs
Cx47	Predominantly expressed in oligodendrocytes (Montoro and Yuste, 2004; Dermietzel et al., 1997; Odermatt et al., 2003; Nagy et al., 2003a,b)	Adult	Able to form heterotypic channels with Cx30 and Cx43 (White et al., 1995)

A-O, astrocyte-oligodendrocyte; *Cx*, connexin; *RGP*, radial glial progenitor.

RGPs and newborn neurons in the embryonic neocortex, and progressively increases in the rodent neocortex during the first postnatal month (Bennett, 1997; Bittman and LoTurco, 1999; Elias et al., 2007; Nadarajah et al., 1997). The expression of Cx45 in the mouse neocortex reaches its peak around birth, after which it progressively declines with neocortical development (Kruger et al., 2000). Cx45 expression is extensively colocalized with NeuN, a marker of neurons, suggesting neuronal expression of Cx45 in the neocortex. Since electrical synapses emerge between interneurons in the second postnatal week, Cx45 is likely expressed in excitatory neurons of the neonatal neocortex. Expression of Cx26 in the neocortex remains controversial. Although immunocytochemical studies showed abundant staining of Cx26 in progenitor cells and neurons of the developing neocortex (Bittman and LoTurco, 1999; Dermietzel et al., 1989; Nadarajah et al., 1997), a study using a LacZ reporter under control of the endogenous Cx26 promoter revealed that expression of Cx26 is restricted to the meningeal cells and absent in neurons and glial cells (Filippov et al., 2003). This inconsistency might be due to the nonspecificity of the antibody or hybridization probes that generated some cross-reactivity. Another possibility could be the relatively low level of Cx26 expression in the heterozygous animals. Future studies are needed to resolve this discrepancy.

Unlike Cxs26, 43, and 45, Cx36 is predominantly expressed in interneurons of the mammalian neocortex. In agreement with this, LacZ gene expression driven by the Cx36 promoter is widely scattered in the adult neocortex (Degen et al., 2004), similar to interneuron distribution. Cx36 expression in the neocortex gradually increases in the first 2 weeks, and thereafter reaches a plateau in the adult brain, in accordance with the timeline for formation of electrical synapses between GABAergic interneurons.

Cx29, Cx32, and Cx47 are mainly found in oligodendrocytes (Dermietzel et al., 1989, 1997; Li et al., 1997; Micevych and Abelson, 1991; Montoro and Yuste, 2004; Nagy et al., 2003a,b; Odermatt et al., 2003; Scherer et al., 1995), whereas Cx30 and Cx43 are expressed in astrocytes in the neocortex (Dermietzel et al., 1989; Micevych and Abelson, 1991; Nagy et al., 1999, 1992; Nagy and Rash, 2000). As astrocytes and oligodendrocytes are primarily produced after birth, these Cxs begin to be expressed at a similar time and progressively increase during postnatal development of the neocortex.

Although numerous subtypes of Cxs have been characterized in the neocortex, electrical synapses are not formed by a random assembly of Cxs. A large body of literature shows that electrical synapses can be homotypic or heterotypic, possibly depending on Cx subtypes (Wong et al., 2008) (Table 14.1). For example, Cx26 and Cx30 can form functional heterotypic channels, but neither of them is able to form functional heterotypic channels with Cx43 (Barrio et al., 1991; White et al., 1995). Cx47 tends to form heterotypic channels with Cx30 or Cx43 (White et al., 1995). These heterotypic channels are often formed between astrocytes and oligodendrocytes owing to expression of different Cx subtypes in these two subtypes of glial cells. Cx36, however, forms homotypic channels between interneurons (Abrams and Rash, 2009; Sohl et al., 2005) (also see the review by

Abrams and Rash, 2009). An electrical synapse contains an aqueous pore, the size of which depends on the Cx subtype. Several studies using fluorescence tracers and uncharged molecules suggest that the permeability of electrical synapses decreases according to the following order of constituent connexins: Cx43 > Cx32 > Cx26 > Cx37 > Cx46 and Cx43 > Cx40 (Harris, 2007).

3. LINEAGE-DEPENDENT SPECIFICITY OF ELECTRICAL SYNAPSES IN THE MOUSE NEOCORTEX

The adult mammalian neocortex is thought to be composed of iteratively arrayed functional columns, in which an ensemble of radially arrayed neurons share/perform similar physiological functions (Hubel and Wiesel, 1959, 1963, 1968; Mountcastle, 1997; Mountcastle et al., 1957). This functional similarity results from neurons within a column being highly interconnected by specific chemical synapses. Based on their anatomical similarity (ie, the vertical arrangement), it has been postulated that the functional column in the adult neocortex originates from the ontogenetic column established during neocortical development. By using retroviruses expressing a fluorescent protein to trace cell lineage, the fate of the progeny of an RGP cell that undergoes a number of rounds of asymmetric divisions can be directly observed. A single RGP can produce a number of excitatory neurons that form a "radial unit," also called an ontogenetic column, in which migrating neurons tightly surround the radial glial fiber of their mother RGP in the embryonic neocortex (Hevner, 2006; Hippenmeyer et al., 2010; Kornack and Rakic, 1995; Luskin et al., 1988; Noctor et al., 2001, 2004; Price and Thurlow, 1988; Rakic, 1995; Ware et al., 1999). In support of the developmental origin of the functional column, recent studies showed that sister excitatory neurons in an ontogenetic clone derived from the same RGP were preferentially connected by chemical synapses in the neocortex (Yu et al., 2009) and shared similar orientation tuning in the mouse visual cortex (Li et al., 2012). Because chemical synapses emerge only in the second postnatal week, how are sister excitatory neurons connected in the first postnatal week? The most plausible explanation is by electrical synapses because previous experiments using dye injection and calcium imaging revealed that electrical synapses are prevalently present between excitatory neurons in the developing neocortex. Electrical synapse coupled neurons appear to form a "neuronal columnar domain," but the precise nature of the relationship between these neurons is unclear. Moreover, before the emergence of chemical synapses, electrical synapses play an essential role in the occurrence of correlated neural activities in the neonatal neocortex during the first postnatal week (Allene and Cossart, 2010). This further implies that electrical synapses may connect sister excitatory neurons in the neonatal neocortex. If so, this raises the questions of when lineage-specific electrical synapses emerge during neocortical development and whether the lineage-specific electrical synapses form between progenitor and progeny.

3.1 Lineage-Dependent Electrical Synapses Between Progenitor and Progeny

Low-titer retrovirus expressing enhanced green fluorescence protein (EGFP) was in utero injected into the lateral ventricle of mouse embryos at embryonic day 12–13 (E12–13) to label isolated individual ontogenetic clones in the neocortex. The infection of individual RGPs results in the labeling of all progeny derived from the EGFP-expressing RGP. During embryonic development, EGFP-expressing radial units that contain an RGP and several daughter cells migrating radially along the mother RGP fiber were frequently observed. Previous dye injection studies revealed extensive gap junction coupling between adjacent progenitor cells, but no dye coupling between progenitor cell and postmitotic neurons (Bittman et al., 1997; Lo Turco and Kriegstein, 1991). Although these dye filling experiments furnished important insights on the development and physiological functions of electrical synapses, their accuracy in revealing gap junction coupling is debated. For instance, leaking dyes from the recording pipette can label unhealthy and dying cells, which confounds interpretation of results. Furthermore, dye coupling may be correlated with the strength of gap junction coupling, with dye coupling observed in strong but not in weak electrical synapses that were recorded by electrophysiological recordings. This brings about a concern that weak electrical synapses with low conductance are present between migrating neurons and the mother glial fiber, but undetected by dye labeling. To circumvent these issues and take advantage of the high sensitivity of electrophysiological recordings, multiple-electrode whole cell recordings were performed to directly probe electrical synapses between the mother radial glial fiber and migrating daughter cells. Two to four cells were simultaneously recorded in the neonatal neocortex by injection of a depolarized or hyperpolarized current into one of them. Electrical coupling is assessed by a relatively smaller depolarization or hyperpolarization in the nonstimulated cells responding to those of the stimulated neuron.

As expected, RGPs formed strong electrical synapses with adjacent RGPs (100% coupled; 44 of 44 neighboring RGP pairs tested), consistent with the previous dye experiments showing robust dye coupling between nearby RGPs (Fig. 14.1). Furthermore, RGPs formed reliable electrical synapses with their daughter cells in the subventricular zone (SVZ) (100% coupled; 43 of 43 progenitor–progeny pairs tested). These SVZ cells expressed voltage-gated sodium channels, which imply a neuronal identity. In contrast, only 16.1% of radially situated RGPs and their nondaughter cell pairs were coupled. Meanwhile, the coupling conductance between an RGP and its daughter cells was substantially larger than that between the RGP and the radially located nondaughter cells, but smaller than that between adjacent RGPs (RGP–RGP, 392 ± 84 pS, range 15–2747 pS, $n = 44$; RGP–daughter cell, 85 ± 12 pS, range 8–78 pS, $n = 43$; RGP–nondaughter cell, 23 ± 3 pS, range 60–2160 pS, $n = 30$). These results suggest that the SVZ daughter cells are preferentially and strongly coupled to their mother RGPs in the VZ/SVZ.

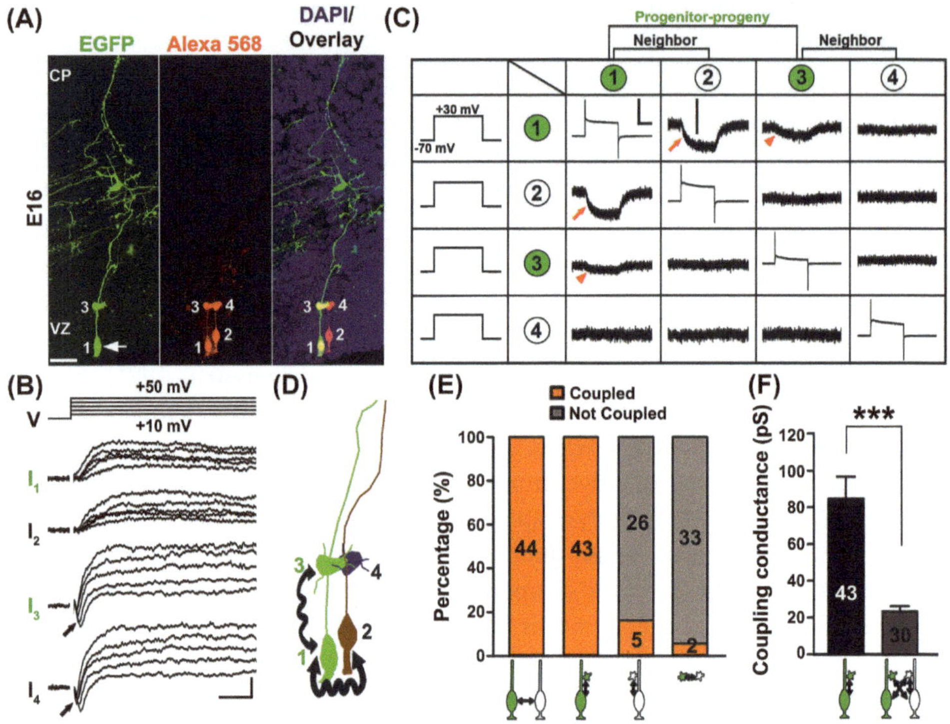

Figure 14.1 ***Preferential formation of strong electrical coupling between radial glial progenitor (RGP) and its progeny.*** (A) Representative images of an EGFP-expressing RGP in the ventricular zone (VZ) (cell 1, arrow) and its progeny in the subventricular zone (SVZ) (cell 3) (*green*), and a nearby non-EGFP-expressing RGP in the VZ (cell 2) and a nearby non-EGFP-expressing SVZ cell (cell 4) on the same side in the E16 neocortex. Cells were subjected to quadruple whole-cell recording, filled with Alexa Fluor 568 hydrazide (*red*) and stained with 4,6-diamidino-2-phenylindole (DAPI, *blue*). Scale bar: 30 μm. (B) Sample current traces of the four cells evoked by 10-mV step changes in membrane potential from +10 to +50 mV. Note that inward voltage-dependent sodium channel–mediated currents are elicited in cells 3 and 4 (*arrows*), but not in cells 1 and 2. Scale bars: 50 pA, 5 ms. (C) Sample traces of currents elicited by sequential depolarization of one of the four cells from −70 to +30 mV. Green circles indicate an EGFP-expressing RGP and its progeny in the SVZ, whereas white circles indicate a nearby non-EGFP-expressing RGP and a nearby non-EGFP-expressing SVZ cell. The average traces are shown in corresponding rectangles with the driver cell at the main diagonal. *Red arrows* and *arrowheads* indicate detectable junctional currents between RGPs and between RGP and SVZ progeny, respectively. Similar symbols and panel display are used in subsequent figures. Scale bars: 500 pA, 100 ms, 5 pA. (D) Morphological reconstruction of the four cells. The wavy lines with arrows indicate electrical coupling and the thickness of the line reflects the coupling strength. (E) Summary of the rate of electrical coupling observed between nearby RGPs, between EGFP-expressing RGPs and their progeny, between non-EGFP-expressing RGPs and SVZ cells adjacent to EGFP-expressing cells on the same side, and between nearby nonsister SVZ cells. (F) Quantification of the coupling conductance observed between EGFP-expressing RGP and progeny pairs and between RGP and nonprogeny SVZ cell pairs at E16-18. Data are presented as mean ± s.e.m; ***, $p < .001$. *(Adapted from He, S., Li, Z., Ge, S., Yu, Y.C., Shi, S.H., 2015. Inside-out radial migration facilitates lineage-dependent neocortical microcircuit assembly. Neuron 86 (5) 1159–1166. doi:10.1016/j.neuron.2015.05.002 with permission.)*

3.2 Lineage-Dependent Specificity of Electrical Synapses Between Excitatory Neurons in the Mouse Neocortex

As development proceeds, neurons progressively migrate to their final destination to constitute the neocortex. The majority of clonally related excitatory neurons are radially situated in the neocortex after they have completed migration. Interestingly, about 38.9% (14 of 36 pairs tested) and 35.3% of radially situated sister pairs (12 of 34 pairs tested) were coupled by electrical synapses at postnatal day 1 (P1) and P2, respectively. The coupling rate was reduced to approximately 25% at P3-4 (9 of 39 pairs tested at P3, and 9 of 35 pairs tested at P4) and 10% at P5-6 (5 of 48 pairs tested). After P6, electrical synapses were rare with only 3 of 193 pairs coupled, similar to that between more mature neocortical excitatory neurons (Galarreta and Hestrin, 1999; Gibson et al., 1999; Wang et al., 2010) (Fig. 14.2). Unlike the neocortex, sister excitatory neurons in the hippocampus were negligibly coupled throughout brain development (Xu et al., 2014), suggesting that the robust lineage-dependent coupling occurs specifically in the neocortex. In addition to a progressive decrease in the coupling rate, the coupling strength of electrical synapses between sister excitatory neurons in the neocortex was also progressively reduced as development proceeded. These findings suggest transient electrical synapses between sister excitatory neurons during cortical development, consistent with the

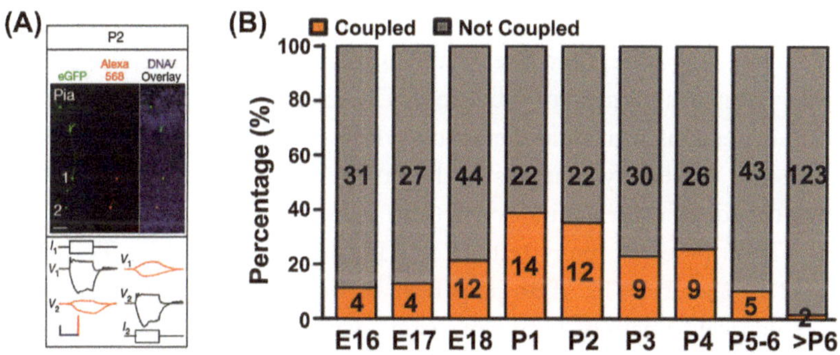

Figure 14.2 ***Progressive development of electrical synapses between sister excitatory neurons in the neocortex.*** (A) Dual whole-cell recordings of sister excitatory neurons in ontogenetic columns at neonatal stage P2. Top, images of sister excitatory neurons expressing EGFP (*green*) in an ontogenetic column. Scale bar, 50 mm. Bottom, simultaneous depolarization or hyperpolarization of two sister excitatory neurons (V1 and V2) when a positive or negative current (1100 or 2100 pA) was injected into only one of the neurons (I1 or I2) under current-clamp conditions, indicating electrical coupling. Subsequent figures use a similar panel layout. Scale bars, 50 mV (*black*), 10 mV (*red*), and 200 ms (*blue*). Pia, pial surface. (B) Summary of the frequency of electrical coupling between sister excitatory neurons in the ontogenetic columns at different developmental stages. *(Adapted from Yu, Y.-C., He, S., Chen, S., Fu, Y., Brown, K.N., Yao, X.-H., Shi, S.-H, 2012. Preferential electrical coupling regulates neocortical lineage-dependent microcircuit assembly. Nature 486 (7401) U113–U139 and He, S., Li, Z., Ge, S., Yu, Y.C., Shi, S.H., 2015. Inside-out radial migration facilitates lineage-dependent neocortical microcircuit assembly. Neuron 86 (5) 1159–1166. doi:10.1016/j.neuron.2015.05.002 with permission.)*

development of electrical coupling between random neurons in the neonatal neocortex (Kandler and Katz, 1998b; Montoro and Yuste, 2004). By performing quadruple whole-cell recordings of two sister neurons in an individual ontogenetic clone and two nearby nonsister neurons on the same side, we found that electrical synapses preferentially formed between sister excitatory neurons in the developing neocortex. About 28.2% (49 of 174 pairs tested) of sister pairs were coupled, whereas only 2.6% (8 of 303 pairs tested) of nonsister pairs were coupled at P1-P6 (Fig. 14.3). In addition, the coupling between sister excitatory neurons was substantially stronger than that between nonsisters (sisters, 430 ± 50 pS, range 60−2160 pS, $n = 85$; nonsisters, 130 ± 30 pS, range 60−810 pS, $n = 24$). These findings suggest lineage-dependent specificity of electrical synapses between excitatory neurons in the developing neocortex.

4. PROGRESSIVE DEVELOPMENT OF ELECTRICAL SYNAPSES BETWEEN SISTER EXCITATORY NEURONS

Electrical synapses are transiently expressed between sister excitatory neurons in the developing neocortex, starting at the late embryonic stage and reaching a maximal level at P1. Normally, the neocortex is organized into six layers (layer I−VI) in a neuronal birth date−dependent "inside-out" manner. That is, the early-born neurons occupy layers V−VI, whereas the late-born neurons constitute layers II−IV. An RGP typically gives rise to about eight to nine neurons that normally expand from layer II to layer VI (Gao et al., 2014), suggesting that sister neurons in an ontogenetic clone are produced at different time points via asymmetric divisions. Consistent with the orderly production of clonally related neurons, multiple electrode whole-cell electrophysiological recordings on newborn sister excitatory neurons in the embryonic neocortex revealed that 11.4% (4 of 35 pairs tested) of sister excitatory neuron pairs were coupled by electrical synapses at E16 and 12.9% (4 of 31 pairs tested) at E17 (He et al., 2015). The coupling rate increased to 21.4% at E18 (12 of 56 pairs tested) (Fig. 14.2), suggesting progressive formation of electrical synapses between sister excitatory neurons during embryonic neocortical development. More importantly, compared with the nearby nonsister neurons that were virtually not coupled by electrical synapses, sister excitatory neurons were preferentially coupled by electrical synapses in the embryonic neocortex, suggesting that lineage-specific electrical synapses emerge at the beginning of neocortical development.

5. MECHANISMS UNDERLYING LINEAGE-DEPENDENT ELECTRICAL SYNAPSE FORMATION

Despite the discovery of electrical synapses in crayfish over 50 years ago, the mechanisms of electrical synapse formation remain obscure. It has been shown that the interplay of

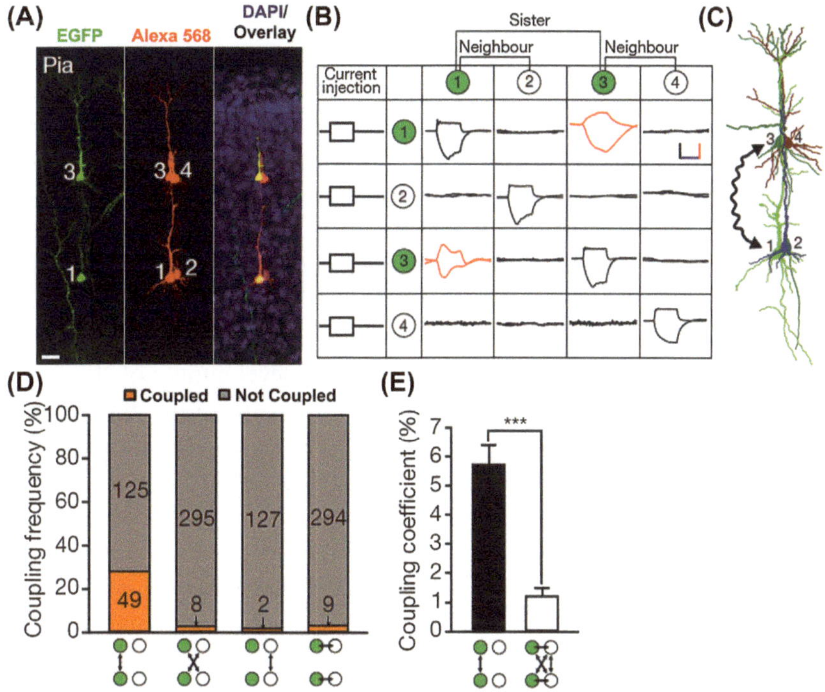

Figure 14.3 ***Preferential formation of strong electrical coupling between sister excitatory neurons in neonatal neocortical ontogenetic columns.*** (A) Image of a quadruple whole-cell recording of two EGFP-expressing sister excitatory neurons (neurons 1 and 3, *green*) in an ontogenetic column and two non-EGFP-expressing excitatory neurons (neurons 2 and 4) adjacent to the sisters on the same side. Scale bar, 20 μm. (B) Sample traces of voltage changes in the four neurons in response to sequential current injection into one of the four neurons. *Green circles* indicate EGFP-expressing sister excitatory neurons, and *white circles* indicate non-EGFP expressing neighboring excitatory neurons. The average traces are shown in each table cell. Scale bars, 50 mV (*black*), 5 mV (*red*), and 200 ms (*blue*). (C) Morphological reconstruction of the four neurons in the quadruple recording. The *wavy arrow* indicates reciprocal electrical coupling between sister excitatory neurons 1 and 3. (D) Summary of the frequency of electrical coupling observed between sister excitatory neurons in the ontogenetic columns and their adjacent nonsister excitatory neurons at P1-P6. (E) Summary of the coupling coefficient (sisters, $n = 74$; nonsisters, $n = 24$; $***<1 \times 10^{-8}$) at P1-P6. *DAPI*, 4,6-diamidino-2-phenylindole. *(Adapted from Yu, Y.-C., He, S., Chen, S., Fu, Y., Brown, K.N., Yao, X.-H., Shi, S.-H., 2012. Preferential electrical coupling regulates neocortical lineage-dependent microcircuit assembly. Nature 486 (7401) U113–U139 with permission.)*

metabotropic glutamate receptors (mGluRs) and gamma-aminobutyric acid A receptor ($GABA_AR$) is involved in the development of electrical synapses (Park et al., 2011). Activation of group II mGluR augmented a progressive increase in the expression of Cx36 protein and coupling strength of Cx36-containing electrical synapses in the rat hypothalamus during brain development, whereas inactivation prevented this increase. In contrast, activation of $GABA_AR$ prevented, but inactivation augmented, the developmental increase of Cx36 protein. A developmental change in Cx36 expression was regulated by group II mGluRs and $GABA_ARs$ via different downstream pathways. The regulation of Cx36 protein by group II mGluRs was via cAMP/protein kinase A -dependent signaling, whereas the regulation by $GABA_ARs$ was by way of Ca^{2+}/L-type voltage-gated calcium channel/protein kinase C-dependent signaling. Unlike with Cx36, alterations in these signalings had no effect on expression of Cx43 in the developing brain, which is thought to be predominantly present in glial cells and immature excitatory neurons, thereby illustrating the specificity of the different pathways involved in developmental regulation of Cx36 expression. Because Cx43 and Cx26 are mainly expressed in immature excitatory neurons of the developing neocortex, these channel activity-mediated pathways may not be involved in progressive development of electrical synapses between sister excitatory neurons in the embryonic neocortex. Our study might provide crucial insights into understanding how preferential electrical synapses progressively form between sister excitatory neurons in the developing neocortex (He et al., 2015).

It is well known that nearly all neocortical excitatory neurons are derived from RGPs in the VZ at the embryonic stage. An RGP is characterized by a bipolar morphology with a short apical end foot attached to the ventricular basement membrane and a long basal glial fiber that radially extends to the pial surface. At the early embryonic stage, RGPs are subject to symmetric divisions that expand the progenitor pool. As development proceeds, RGPs switch to asymmetric divisions to directly produce neurons and intermediate progenitor cells, which then give rise to neurons through symmetric divisions in the SVZ (Haubensak et al., 2004; Hevner, 2006; Miyata et al., 2004; Noctor et al., 2004). In addition to intermediate progenitor cells, RGPs can also indirectly produce cortical excitatory neurons through other transient amplifying progenitor cells including outer SVZ RGPs (oRGPs) (Fietz et al., 2010; Hansen et al., 2010; Martinez-Cerdeno et al., 2012; Reillo et al., 2011; Shitamukai et al., 2011; Wang et al., 2011) and short fiber neural precursors (Gal et al., 2006). Newborn excitatory neurons then migrate radially along the radial glial fibers of their mother RGPs to reach their destination, the CP. About 50 years ago, researchers found that the late-born neurons have to migrate past the early-born neurons to constitute the superficial layers in the mammalian neocortex. This was termed a birth date–dependent "inside-out" radial migration. Although the early-born neurons occupy neocortical layers V–VI, the late-born neurons constitute layers II–IV. Although the classic "inside-out" migration was established 50 years ago, little is known about the significance of this orderly radial migration for functional development of the mammalian neocortex.

As discussed earlier, lineage-dependent electrical synapses emerge between progenitor and progeny as newborn neurons are produced. This coupling is reliable (coupling rate, 100%) when newborn neurons are located in the VZ/SVZ. As newborn neurons migrate along the radial glial fibers of their mother RGPs toward the CP, electrical synapses between progenitor and progeny become progressively less prominent. The coupling rate was reduced to ~35% (10 of 28 tested) when migrating neurons were situated at the intermediate zone (IZ). No functional coupling was observed between progenitor and progeny once newborn neurons migrated to the CP. Because neuronal coupling occurs shortly after migration, the events necessary for formation of electrical synapses between sister excitatory neurons might have taken place during neuronal migration. Sister neurons are derived from the same RGPs and migrate along a similar path of the radial glial fibers of their mother RGPs in a birth date—dependent "inside-out" manner. That is to say that cell—cell encounter likely occurs between sister excitatory neurons at the time the late-born neurons migrate past the early-born neurons. To test this, we took advantage of the *Reeler* mouse that lacks REELIN, a large secreted protein that is involved in neuronal migration and positioning in the developing brain (D'Arcangelo et al., 1995; Hirotsune et al., 1995; Ogawa et al., 1995). Importantly, the late-born neurons are unable to migrate past the early-born neurons in the *Reeler* neocortex, although the radial migration along the glial fibers of their mother RGPs is intact. Contrary to the inside-out organization in the wild-type mouse neocortex, the early-born neurons occupy the superficial layers (layers II—IV), whereas the late-born neurons constitute the deep layers (layers V—VI) in the *Reeler* neocortex. The "outside-in" pattern with inverted cortical layers was identified by specific markers for neocortical layers. Cut-like homeobox 1 (CUX1) and chicken ovalbumin upstream promoter transcription factor-interacting protein 2 (CTIP2) are markers for neurons in layers II—IV and layers V—VI in the wild-type neocortex, respectively. However, in the *Reeler* neocortex CUX1-positive neurons were mainly found in the bottom layers V—VI, whereas CTIP2 was expressed in neurons located in the superficial layers II—IV, implying inverted cortical layers. Furthermore, using 5-Bromo-2′-deoxyuridine (BrdU) and 5-Ethynyl-2′-deoxyuridine (EdU) to trace the birth dates of neurons in an ontogenetic clone, we found that late-born neurons were located below the early-born neurons in the *Reeler* neocortex, contrary to that in the wild-type mouse neocortex (Fig. 14.4).

Electrical synapses between sister excitatory neurons were examined in the *Reeler* neocortex using quadruple whole-cell recordings. It is well documented that a fair amount of neurons display defective morphology with disoriented processes in the *Reeler* neocortex (Goffinet, 1979; Pinto et al., 1979). Therefore sister excitatory neurons with pia-directed principle dendrites were chosen for recordings to rule out the effect of abnormal morphology on the formation of electrical synapses. Intriguingly, preferential electrical synapses between sister excitatory neurons were abolished in the *Reeler* neocortex in spite of a large overlap of their neuronal morphology. The rate of electrical coupling between sister excitatory neurons was similar to that of nonsisters in the *Reeler* neocortex at both embryonic (E17-18: sisters, 3.7% coupled, 2 of 54 pairs; nonsisters, 1.6% coupled, 3 of 184 pairs) and neonatal stages (P1-4: sisters, 3.7% coupled, 4 of 107 pairs; nonsisters, 5.3% coupled, 14 of 265 pairs) (He et al., 2015) (Fig. 14.4). Compared with that in the wild-type neocortex, the coupling rate was dramatically reduced for sister excitatory neurons, but not affected for nonsisters in the *Reeler*

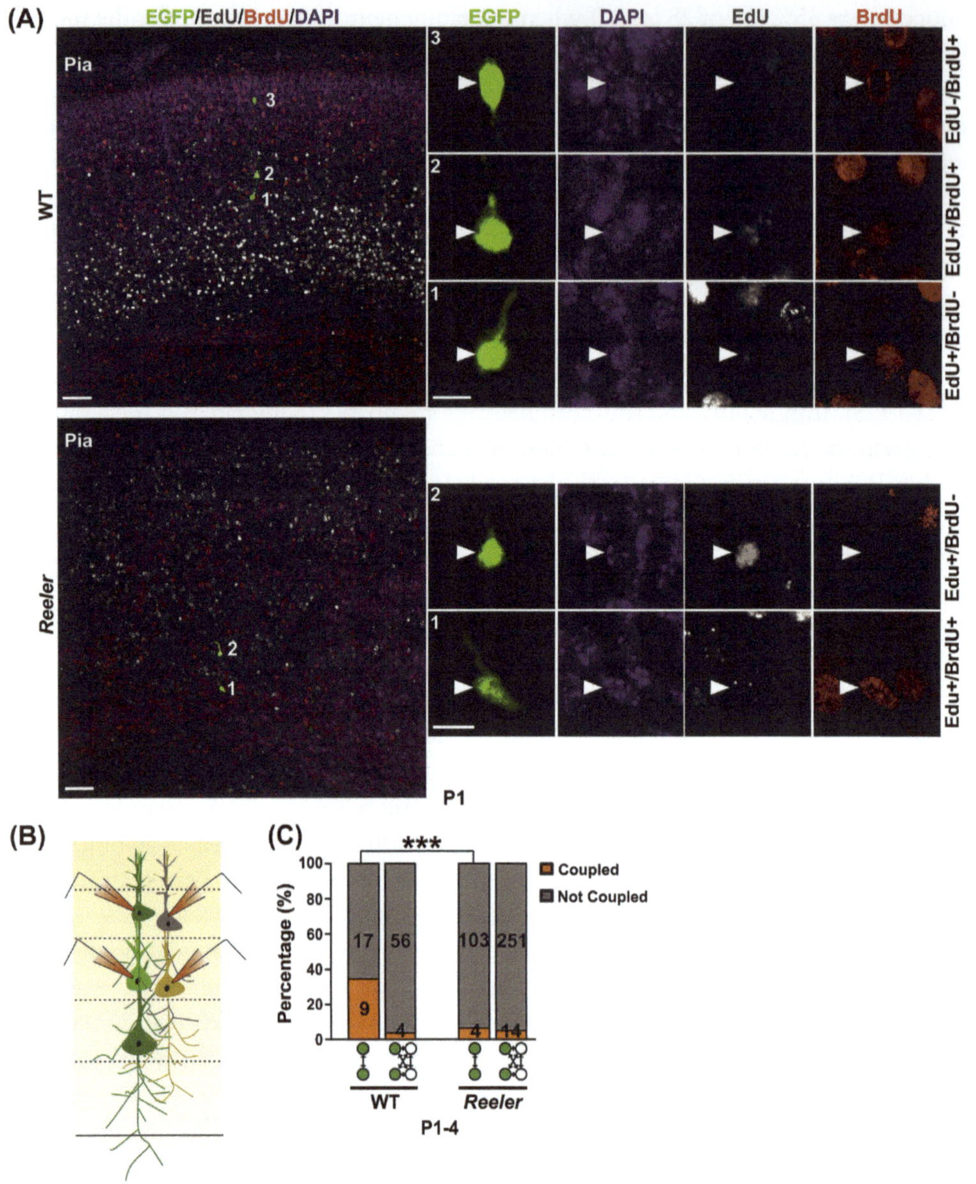

Figure 14.4 ***Birth date–dependent lamination is largely inverted at the clonal level in the* Reeler *neocortex.*** (A) Representative images of the P1 wild type (WT) (*top*) and *Reeler* (*bottom*) neocortices that received intraventricular injection of EGFP-expressing retroviruses (*green*) at E12, EdU (*white*) and BrdU (*red*) administration at E12 and E14, and stained for 4,6-diamidino-2-phenylindole (DAPI) (*blue*). High-magnification images of isolated EGFP-expressing sister excitatory neurons (*arrowheads*) in a clone are shown to the right. Note that in the WT neocortex, an EdU+ early-born neuron (1) is located deeper than two BrdU+ late-born neurons (2 and 3). In contrast, in the *Reeler* neocortex, the BrdU+ late-born neuron (1) is located deeper than the EdU+ early-born neuron (2). In addition,

neocortex. The preferential coupling between progeny and progenitor, however, remained unchanged in the *Reeler* neocortex. This abolishment of the lineage-specific coupling was not due to alterations in neuronal migration or production because neither the neuronal migration guided by radial glial fibers nor the number of neurons in an ontogenetic clone were changed in the *Reeler* neocortex. The lateral distances between progeny and the glial fibers of their mother RGP and the dispersion width of an ontogenetic clone in the *Reeler* neocortex were comparable with those in the wild-type neocortex at the embryonic stage, consistent with previous studies showing that the REELIN pathway is not required for glia-guided neuronal migration (Britto et al., 2011; Franco et al., 2011; Jossin and Cooper, 2011; Sanada et al., 2004). In addition, lineage progression and neuronal differentiation are not changed because the resting membrane potentials of RGPs and excitatory neurons throughout development were not altered in the *Reeler* neocortex. Taken together, these results suggest that disruption of the radial "inside-out" neuronal migration impairs the preferential electrical coupling between radially aligned sister excitatory neurons in the *Reeler* neocortex.

As neuronal morphology and lamination are globally affected in the *Reeler* neocortex, it raises the question whether disruption of the preferential electrical coupling between sister excitatory neurons is indeed due to impaired "inside-out" radial migration at the clonal level. Retroviruses expressing EGFP together with short hairpin RNAs against *Dabled-1* (*Dab 1*) were used to specifically knockdown the DAB1 protein, which is an essential downstream effector of the REELIN pathway (Howell et al., 1997; Rice et al., 1998). As expected, knockdown of DAB1 expression specifically at the clonal level merely abolished the preferential electrical coupling between sister excitatory neurons, but had no effect on the rate of electrical coupling between nonsisters in the developing neocortex (He et al., 2015). This effect is not due to an alteration in neuronal maturation because RMPs in neocortical neurons were not changed following knockdown of DAB1. Several lines of evidence show that removal of DAB1 does not affect the radial glia-guided neuronal migration, but results in the inverted neocortical layers similar to that of the *Reeler* neocortex (Franco et al., 2011; Sanada et al., 2004). The evidence further suggests that the "inside-out" radial migration along the glial fibers of RGPs plays

BrdU+ late-born neurons are in general superficial to EdU+ early-born neurons in the WT neocortex, consistent with an inside-out radial migration and lamination pattern. In contrast, the overall birth date–dependent lamination is disrupted in the *Reeler* neocortex. Scale bars: 50 and 10 μm. (B) Schematic drawings of a quadruple whole-cell recording of two sister excitatory neurons (*green*) in an ontogenetic column and two non-EGFP-expressing excitatory neurons (*nongreen*) adjacent to the sisters on the same side. (C) Summary of the rate of electrical coupling observed between sister excitatory neurons and their adjacent nonsister excitatory neurons in the WT and *Reeler* neocortices at P1-4. ***, $p < .001$. *(Adapted from He, S., Li, Z., Ge, S., Yu, Y.C., Shi, S.H., 2015. Inside-out radial migration facilitates lineage-dependent neocortical microcircuit assembly. Neuron 86 (5) 1159–1166. doi: 10.1016/j.neuron.2015.05.002 with permission.)*

an essential role in the formation of electrical coupling between sister excitatory neurons in the developing neocortex.

To disrupt the cell—cell encounters, retroviruses were used to overexpress Ephrin-A 5 (EFNA5) or Ephrin-A receptor (EPHA7) specifically in sister excitatory neurons within an ontogenetic column (He et al., 2015). In line with a previous study (Torii et al., 2009), overexpression of EFNA5 or EPHA7 led to lateral/tangential dispersion. The lateral width of an ontogenetic clone expressing EGFP together with EFNA5 or EPHA7 was significantly larger than that expressing EGFP alone. Furthermore, the rate of electrical coupling between radially situated sister excitatory neurons expressing EGFP-EFNA5/EPHA7 was substantially reduced in the neonatal neocortex (P1-4) when compared with that between the control sister neuron pairs expressing EGFP alone (He et al., 2015). The rate of electrical coupling between EGFP-EFNA5/EPH7—expressing neurons and their nearby non-EGFP-expressing neurons was similar to that between EGFP-expressing neurons and their nearby non-EGFP-expressing neurons. These data suggest that the preferential coupling between sister excitatory neurons is abolished following overexpression of EFNA5 or EPHA7. This impairment is not due to altered neuronal maturation because RMPs were not changed in neurons expressing EFNA5 or EPHA7. Taken together, these results suggest that the lateral dispersion impairs formation of the preferential electrical synapses between radially situated sister neurons. This further supports the notion that the "inside-out" radial migration facilitates formation of electrical synapses between sister excitatory neurons in the developing neocortex.

A neuron may also encounter numerous nonclonally related neurons during its radial migration, raising the question of how a neuron recognizes its sister neurons to form electrical synapses with them. The chemoaffinity hypothesis to explain the specificity of synapse formation proposed by John Langley in 1895 (Langley, 1895) and later formulated by Roger Sperry (Sperry, 1943, 1944, 1963) might shed some light on the formation of lineage-specific electrical synapses. Neuronal surface transmembrane proteins encoded by a complex genomic locus serve as chemoaffinity molecules to determine synaptic specificity via homophilic binding, in which two matching surface molecules are expressed in axons and their targets (postsynaptic sites, usually dendritic spines) and bind to each other with high affinity (Zipursky and Sanes, 2010). It is possible, therefore, that sister excitatory neurons are more likely to express two identical chemoaffinity molecules. The lineage-specific surface molecules may promote distinguishment of sister neurons from nonsisters encountered during migration.

6. SIGNIFICANCE OF LINEAGE-DEPENDENT ELECTRICAL SYNAPSES IN NEOCORTICAL MICROCIRCUIT ASSEMBLY

The concept of functional columns, which are the basic units of the neocortex, was proposed by Mountcastle in 1957, and later substantiated by Hubel and Wiesel in the

1960s. A functional column is a relatively closed microcircuit in which an ensemble of radially situated neurons is highly connected by specific chemical synapses. Despite the discovery of functional columns that are essential for information processing in the neocortex over 50 years ago, precisely how the microcircuit develops is an outstanding question in the neuroscience field. About 40 years ago, Hubel and Wiesel (1974) postulated, "It may be that there is a great developmental advantage in designing such machinery once only, and repeating it over and over monotonously, like a crystal." This indicates that the developmental origin might matter for the formation of functional columns.

Owing to the discovery of fluorescent proteins and the development of molecular genetics, researchers have found that an ontogenetic column derived from a single progenitor cell forms a radially situated column, resembling a functional column in the anatomical distribution. It has, therefore, been speculated that these two types of columns are developmentally related. Indeed, compared with nearby nonsister neurons, sister excitatory neurons preferentially form chemical synapses with each other in the mouse neocortex (Yu et al., 2009). Moreover, using in vivo two-photon calcium imaging to measure neuronal activities of sister excitatory neurons responding to different orientation bars in layers II/III of the mouse visual cortex, sister excitatory neurons are found to share similar orientation tuning property (Li et al., 2012) (Fig. 14.6). Another study reported a similar observation that "clonally related neurons" in a very large clone display similar orientation tuning property in the mouse visual cortex (Ohtsuki et al., 2012). These two studies suggest that the preferential chemical connectivity might contribute to functional similarity for clonally related excitatory neurons in the neocortex. Taken together, these results further support the concept that clonally related specific microcircuits are the fundamental basis of functional columns in the adult neocortex.

Actually, before the emergence of chemical synapses, preferential electrical synapses transiently develop between sister excitatory neurons in the neonatal neocortex. To examine whether preferential electrical and chemical synapses are related to each other during cortical development, retroviruses expressing EGFP together with dominant-negative Cx26 subunits containing a mutant amino acid (Thr135Ala), which cause electrical synapses containing mutant Cx26 to close, were delivered into the lateral ventricle by in utero injection to specifically disrupt electrical synapses between sister excitatory neurons. Notably, this point mutation has been reported not to affect the gap junction–mediated adhesion force that is required for neuronal migration along the radial glial fibers (Elias et al., 2007). As expected, although neuronal position and laminar distribution remained intact in the neocortex, the rate of electrical coupling between sister neurons expressing mutant Cx26 was significantly reduced compared with that between sister pairs expressing EGFP alone or wild-type Cx26 (Fig. 14.5). More importantly, disruption of preferential electrical synapses prevented the development of preferential chemical synapses and functional similarity between sister excitatory

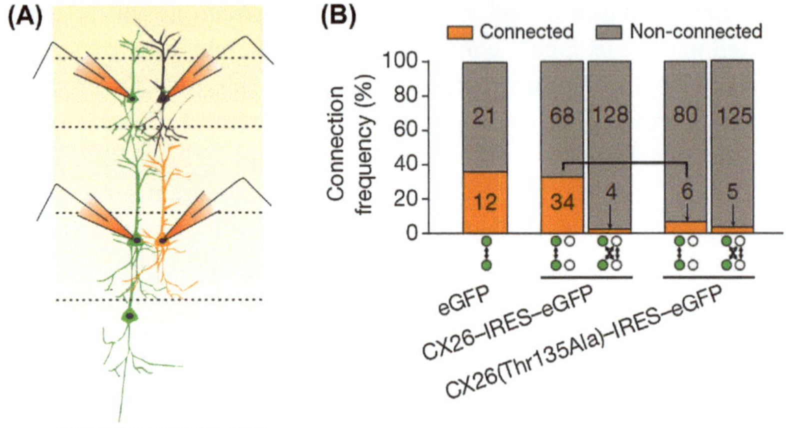

Figure 14.5 ***Preferential electrical coupling is required for chemical synapse formation between sister excitatory neurons in neocortical ontogenetic columns.*** (A) Schematic drawings of a quadruple whole-cell recording of two sister excitatory neurons (*green*) and two adjacent non–EGFP-expressing excitatory neurons (*nongreen*). (B) Summary of the frequency of chemical synapse formation between sister excitatory neurons in ontogenetic columns expressing EGFP, Cx26–IRES–EGFP, or Cx26(Thr135Ala)–IRES–EGFP and adjacent non-EGFP-expressing excitatory neurons. The bracket highlights the difference in the connectivity of Cx26–IRES–EGFP-expressing sister neuron pairs and Cx26(Thr135Ala)–IRES–EGFP-expressing sister neuron pairs. *(Adapted from Yu, Y.-C., He, S., Chen, S., Fu, Y., Brown, K.N., Yao, X.-H., Shi, S.-H, 2012. Preferential electrical coupling regulates neocortical lineage-dependent microcircuit assembly. Nature 486 (7401) U113–U139 with permission.)*

neurons (Li et al., 2012; Yu et al., 2012). Only 7.0% (6 of 86 pairs tested) of sister pairs expressing mutant Cx26 were connected by chemical synapses, which was significantly lower than the 33–36% of sister neuron pairs expressing EGFP alone (12 of 33 pairs tested) or wild-type Cx26 (34 of 102 pairs tested). Moreover, the orientation tuning similarity between sister excitatory neurons was also disrupted by the closed channel mutant of Cx26 (Fig. 14.6). This effect of dominant-negative Cx26 expression on the similarity of visual selectivity could be mimicked by daily administration of a specific blocker of electrical synapses during the first postnatal week. These results together demonstrated that lineage-dependent electrical synapses play a crucial role in the subsequent formation of chemical synapses between sister excitatory neurons as well as the assembly of precise microcircuits involving excitatory neurons in the neocortex.

As electrical and chemical synapses are present between sister excitatory neurons at different developmental stages, how electrical synapses at the early stage regulate subsequent formation of chemical synapses remains an open question. In 1949, Donald Hebb postulated a theory regarding cell assembly, "A cell assembly is a group of neurons clustered together functionally because of a past history of being stimulated together. Their main characteristic is that they are capable of acting together for a time as a closed

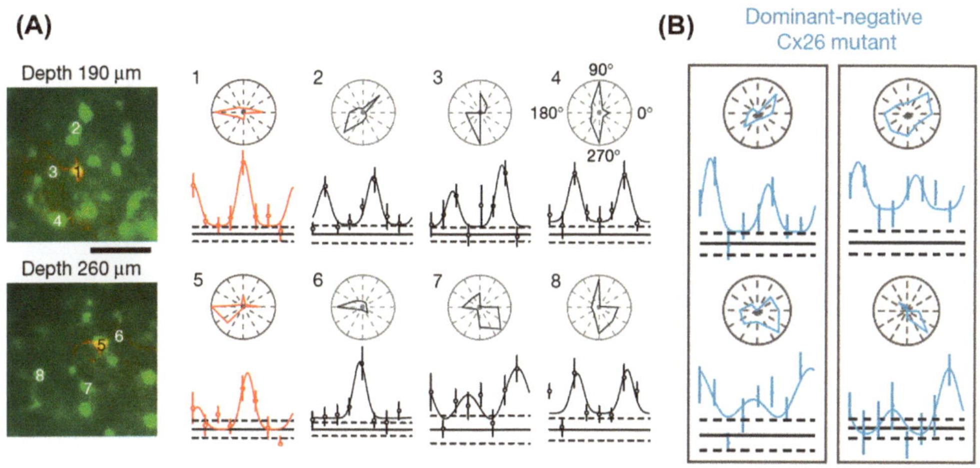

Figure 14.6 ***Expression of dominant-negative Cx26 mutant disrupted orientation and direction preference of sister excitatory neurons in the visual cortex.*** (A) Orientation tuning of the (green fluorescence protein) GFP pair (*red*) and several nearby cells (indicated by numbers) in the experiment shown in Fig. 14.1A, C, and D. *Circle*, mean; *vertical line*, ± s.e.m; *curve*, fitted double Gaussian function. *Horizontal lines*, baseline (*solid*) ± s.e.m (*dashed*). Inset above, polar plot of orientation tuning. Scale bar, 50 μm. (B) Two sample experiments showing tuning curves of sister pairs (each pair in a box) in mice injected with retrovirus containing Cx26(T135A)-T2A-EGFP. *(Adapted from Li, Y., Lu, H., Cheng, P.L., Ge, S., Xu, H., Shi, S.H., Dan, Y., 2012. Clonally related visual cortical neurons show similar stimulus feature selectivity. Nature 486 (7401) 118–121 with permission.)*

system." Subsequently, researchers have found that correlated neuronal activity promotes the formation of chemical synapses and the assembly of neuronal microcircuits (Hebb, 1949; Katz and Shatz, 1996). Corroborating the role of electrical synapses in the synchronized neuronal firing of interneurons (Galarreta and Hestrin, 1999; Gibson et al., 1999; Gibson et al., 2005), strong electrical communication between sister excitatory neurons does indeed promote the generation of action potentials and synchronous neuronal firing (Fig. 14.7), leading to neuronal microcircuit assembly in the developing neocortex (Yu et al., 2012). Because lineage-specific coupling is short lived in the mouse neocortex, it raises the question whether early coupling between sister excitatory neurons uncouples early as well. If so, the rate of coupling between sister excitatory neurons is underestimated in the developing mouse neocortex and might be substantially higher than the rate of chemical connectivity between them in the adult neocortex (~30%). This implies that sister excitatory neurons coupled by strong electrical synapses tend to develop chemical synapses between them by way of synchronous neuronal firing.

Lineage-dependent electrical synapse formation among excitatory neurons may serve as one of the first steps in the development of functional columns in the neocortex. Our studies suggested that preferential electrical synapses begin to form between sister excitatory neurons at the early stage of neocortical development just as they emerge in the

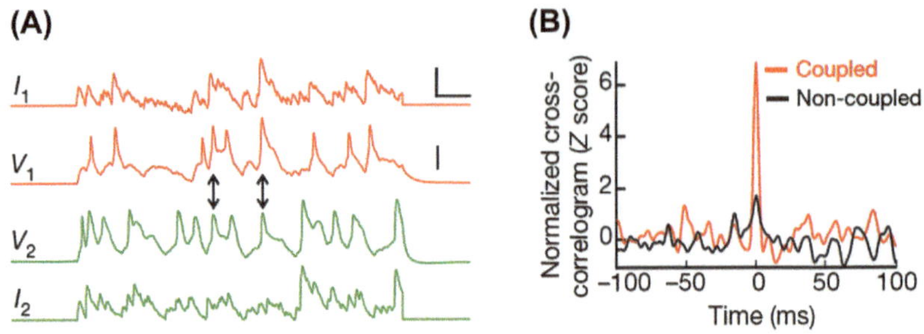

Figure 14.7 ***Electrical coupling promoting synchronous firing of sister excitatory neurons in ontogenetic columns in response to uncorrelated native neuronal activity.*** (A) Sample traces of voltage changes in electrically coupled sister excitatory neurons. *Arrows* indicate the spikes that occur in both neurons within a 1-ms window. Scale bars, 100 pA, 25 mV, and 100 ms. (B) Normalized cross-correlogram analysis. The bin size is 1 ms. Note that the firing frequency is significantly increased near 0 ms for coupled sister excitatory neuron pairs (*red*) but not for noncoupled sister excitatory neuron pairs (*black*), indicating synchronous firing. *(Adapted from Yu, Y.-C., He, S., Chen, S., Fu, Y., Brown, K.N., Yao, X.-H., Shi, S.-H., 2012. Preferential electrical coupling regulates neocortical lineage-dependent microcircuit assembly. Nature 486 (7401) U113—U139 with permission.)*

neocortex. Interestingly, lineage-dependent electrical synapses are first observed between progenitor and progeny in the VZ, where the progeny are produced. Although gap junction—mediated adhesive force rather than channel activity is essential for radial migration of postmitotic neurons along the glial fibers of their mother RGPs, it is electrical synapses (channel activity) that preferentially connect progeny and their mother RGPs at the beginning of neuronal radial migration at the VZ/SVZ. Moreover, the classic "inside-out" radial neuronal migration guides the formation of preferential electrical synapses between sister excitatory neurons, which plays a fundamental role in precise microcircuit assembly in the neocortex. These findings bridge the early neuronal developmental processes (neurogenesis and neuronal migration) and precise microcircuit assembly, and provide important insights into the construction of iteratively arrayed functional columns in the adult mammalian neocortex during brain development.

7. CONCLUSIONS

This chapter summarizes the developmental and functional implications of lineage-dependent electrical synapses among excitatory neurons in the mouse neocortex as well as the importance of early developmental processes, such as neurogenesis and neuronal migration, for precise microcircuit assembly. An emerging theme for the development of a functional neocortex in the mouse is related to cell lineage (Gao et al., 2013). Lineage-specific electrical synapses emerge between progenitor and progeny immediately after neuronal birth, but their functional roles remain elusive. Subsequently, sister excitatory neurons begin to form preferential electrical synapses in the mouse neocortex, and thereby promote the formation of preferential chemical synapses between them at the later stage of cortical development. For excitatory neurons, lineage appears to become a pivotal factor for instructing neuronal migration, spatial organization and development of highly specific connections between them, which together contribute to precise microcircuit assembly for information processing in the rodent neocortex (Fig. 14.8).

Studies have shown that neocortical expansion in higher mammals resulted from enlarged germinal zones containing a repertoire of diverse progenitor cells and rearranged

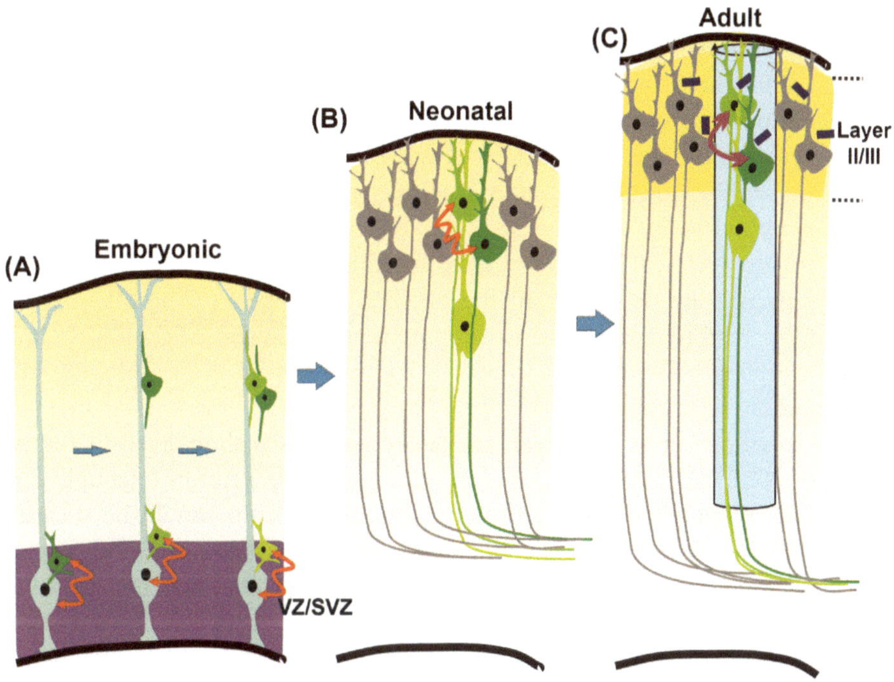

Figure 14.8 ***Schematic summary of early developmental processes including neurogenesis and neuronal migration facilitating precise lineage-dependent microcircuit assembly in the neocortex.*** (A) A progenitor cell sequentially gives rise to three sister excitatory neurons at the embryonic stage. The late-born neuron migrates past the early-born sister neurons along a similar path of the radial glial fiber of their mother radial glial progenitor. The *red wavy arrows* indicate electrical synapses between progenitor and progeny in the embryonic neocortex. (B) The inside-out radial migration facilitates formation of electrical synapses between sister excitatory neurons in the neocortex. The *red wavy arrow* indicates an electrical synapse between sister excitatory neurons at the neonatal stage. (C) Electrical synapses that occur at the neonatal stage promote formation of preferential chemical synapses and functional similarity of orientation tuning response between sister excitatory neurons at the adult stage. The *purple arrows* indicates a chemical synapse between sister excitatory neurons. The *blue bars* indicate orientation responses for neurons in layers II/III of the visual neocortex. *VZ/SVZ*, ventricular zone/subventricular zone.

scaffolds for supporting neuronal radial migration (Florio and Huttner, 2014; Lui et al., 2011). It will, therefore, be of tremendous interest to understand the relationship between early developmental processes, such as neurogenesis and neuronal migration, and lineage-dependent microcircuit assembly in higher mammals including humans under normal and disease conditions. This might provide the basis for the development of clinical therapeutic treatments for some neurological diseases.

OUTSTANDING QUESTIONS

- Lineage-dependent electrical synapses are present between progenitor and progeny in the SVZ. Before newborn neurons reach their final destinations, they have to migrate past three cortical regions including the SVZ, IZ, and CP. Migrating neurons undergo differential morphological changes in these regions. This brings about more questions unanswered. For example, do lineage-specific electrical synapses exist between progenitor and progeny as well when migrating neurons are in the IZ and CP? What are the functional roles of lineage-specific coupling between progenitor and progeny?
- In addition to ventricular RGPs, a large number of cortical excitatory neurons in higher mammals are derived from the oRGPs. The fibers of the oRGPs serve as the scaffolds that are rearranged to support neuronal radial migration. Is lineage-specific coupling present between oRGP and progeny in the developing primate neocortex?
- Lineage-specific coupling between excitatory neurons is short-lived in the mouse neocortex. Can early coupling between sister excitatory neurons uncouple early? If so, the rate of coupling between sister excitatory neurons should be substantially higher than the rate of chemical connectivity between them in the adult neocortex. This raises a new question of whether sister excitatory neurons coupled by strong electrical synapses tend to develop chemical synapses between them by way of synchronous neuronal firing?
- It is known that excitatory neurons in different regions including the neocortex, hippocampus, and thalamus are derived from different progenitor cell pools. This raises some new questions. For example, is lineage a common factor for functional organization of excitatory neurons located throughout the brain or specifically in the neocortex? Can clonally related clusters of excitatory neurons in different regions be preferentially connected, such as the long-range projections that include thalamocortical and corticothalamic connections?
- Unlike the mouse, the primate neocortex has gyrus. Can lineage-dependent microcircuit assembly apply to the primate neocortex that consists of numerous gyri? Furthermore, does lineage-dependent microcircuit assembly contribute to the formation of functional columns in the primate neocortex?

ACKNOWLEDGMENTS

We thank Dr. Yvette Chin and Dr. Guisheng Zhong for critical reading of this manuscript. Our research was supported by the grants from the Simons Foundation (to S.-H.S.), the National Institutes of Health (R01DA024681 and R01MH101382 to S.-H.S.), R21MH083624 (S.-H.S. and Kun Huang) and R01GM065947 (Gina E. Sosinsky), the McKnight Foundation and the March of Dimes Foundation (S.-H.S.), the Tri-Institutional Stem Cell Initiative STARR Stem Cell Fellowship (to S.H.), and by the ShanghaiTech University start-up funds (to S.H.).

REFERENCES

Abrams, C.K., Rash, J.E., 2009. Connexins in the nervous system. In: Harris, A., Locke, D. (Eds.), Connexins: A Guide. Humana Press, a Part of Springer Science, pp. 323–357.

Allene, C., Cossart, R., 2010. Early NMDA receptor-driven waves of activity in the developing neocortex: physiological or pathological network oscillations? J. Physiol. 588 (Pt 1), 83–91.

Barrio, L.C., Suchyna, T., Bargiello, T., Xu, L.X., Roginski, R.S., Bennett, M.V., Nicholson, B.J., 1991. Gap junctions formed by connexins 26 and 32 alone and in combination are differently affected by applied voltage. Proc. Natl. Acad. Sci. U.S.A. 88 (19), 8410–8414.

Bennett, M.V., Zukin, R.S., 2004. Electrical coupling and neuronal synchronization in the mammalian brain. Neuron 41 (4), 495–511.

Bennett, M.V., 1997. Gap junctions as electrical synapses. J. Neurocytol. 26 (6), 349–366.

Bennett, M.V., 2000. Electrical synapses, a personal perspective (or history). Brain Res. Brain Res. Rev. 32 (1), 16–28.

Bittman, K.S., LoTurco, J.J., 1999. Differential regulation of connexin 26 and 43 in murine neocortical precursors. Cereb. Cortex 9 (2), 188–195.

Bittman, K., Owens, D.F., Kriegstein, A.R., LoTurco, J.J., 1997. Cell coupling and uncoupling in the ventricular zone of developing neocortex. J. Neurosci. 17 (18), 7037–7044.

Britto, J.M., Tait, K.J., Johnston, L.A., Hammond, V.E., Kalloniatis, M., Tan, S.S., 2011. Altered speeds and trajectories of neurons migrating in the ventricular and subventricular zones of the reeler neocortex. Cereb. Cortex 21 (5), 1018–1027.

Cina, C., Bechberger, J.F., Ozog, M.A., Naus, C.C., 2007. Expression of connexins in embryonic mouse neocortical development. J. Comp. Neurol. 504 (3), 298–313.

Connors, B.W., Long, M.A., 2004. Electrical synapses in the mammalian brain. Annu. Rev. Neurosci. 27, 393–418.

Connors, B.W., Ransom, B.R., Kunis, D.M., Gutnick, M.J., 1982. Activity-dependent K^+ accumulation in the developing rat optic nerve. Science 216 (4552), 1341–1343.

Cornell-Bell, A.H., Finkbeiner, S.M., 1991. Ca^{2+} waves in astrocytes. Cell Calcium 12 (2–3), 185–204.

Cornell-Bell, A.H., Finkbeiner, S.M., Cooper, M.S., Smith, S.J., 1990. Glutamate induces calcium waves in cultured astrocytes: long-range glial signaling. Science 247 (4941), 470–473.

D'Arcangelo, G., Miao, G.G., Chen, S.C., Soares, H.D., Morgan, J.I., Curran, T., 1995. A protein related to extracellular matrix proteins deleted in the mouse mutant reeler. Nature 374 (6524), 719–723.

Degen, J., Meier, C., Van Der Giessen, R.S., Sohl, G., Petrasch-Parwez, E., Urschel, S., Willecke, K., 2004. Expression pattern of lacZ reporter gene representing connexin36 in transgenic mice. J. Comp. Neurol. 473 (4), 511–525.

Dermietzel, R., Traub, O., Hwang, T.K., Beyer, E., Bennett, M.V., Spray, D.C., Willecke, K., 1989. Differential expression of three gap junction proteins in developing and mature brain tissues. Proc. Natl. Acad. Sci. U.S.A. 86 (24), 10148–10152.

Dermietzel, R., Farooq, M., Kessler, J.A., Althaus, H., Hertzberg, E.L., Spray, D.C., 1997. Oligodendrocytes express gap junction proteins connexin32 and connexin45. Glia 20 (2), 101–114.

Elias, L.A., Kriegstein, A.R., 2008. Gap junctions: multifaceted regulators of embryonic cortical development. Trends Neurosci. 31 (5), 243–250.

Elias, L.A., Wang, D.D., Kriegstein, A.R., 2007. Gap junction adhesion is necessary for radial migration in the neocortex. Nature 448 (7156), 901–907.

Evans, W.H., Martin, P.E., 2002. Gap junctions: structure and function (review). Mol. Membr. Biol. 19 (2), 121–136.

Fietz, S.A., Kelava, I., Vogt, J., Wilsch-Brauninger, M., Stenzel, D., Fish, J.L., Huttner, W.B., 2010. OSVZ progenitors of human and ferret neocortex are epithelial-like and expand by integrin signaling. Nat. Neurosci. 13 (6), 690–699.

Filippov, M.A., Hormuzdi, S.G., Fuchs, E.C., Monyer, H., 2003. A reporter allele for investigating connexin 26 gene expression in the mouse brain. Eur. J. Neurosci. 18 (12), 3183–3192.

Florio, M., Huttner, W.B., 2014. Neural progenitors, neurogenesis and the evolution of the neocortex. Development 141 (11), 2182–2194.

Franco, S.J., Martinez-Garay, I., Gil-Sanz, C., Harkins-Perry, S.R., Muller, U., 2011. Reelin regulates cadherin function via Dab1/Rap1 to control neuronal migration and lamination in the neocortex. Neuron 69 (3), 482–497.

Friend, D.S., Gilula, N.B., 1972. Variations in tight and gap junctions in mammalian tissues. J. Cell Biol. 53 (3), 758–776.

Gal, J.S., Morozov, Y.M., Ayoub, A.E., Chatterjee, M., Rakic, P., Haydar, T.F., 2006. Molecular and morphological heterogeneity of neural precursors in the mouse neocortical proliferative zones. J. Neurosci. 26 (3), 1045–1056.

Galarreta, M., Hestrin, S., 1999. A network of fast-spiking cells in the neocortex connected by electrical synapses. Nature 402 (6757), 72–75.

Gao, P., Sultan, K.T., Zhang, X.J., Shi, S.H., 2013. Lineage-dependent circuit assembly in the neocortex. Development 140 (13), 2645–2655.

Gao, P., Postiglione, M.P., Krieger, T.G., Hernandez, L., Wang, C., Han, Z., Shi, S.H., 2014. Deterministic progenitor behavior and unitary production of neurons in the neocortex. Cell 159 (4), 775–788.

Gibson, J.R., Beierlein, M., Connors, B.W., 1999. Two networks of electrically coupled inhibitory neurons in neocortex. Nature 402 (6757), 75–79.

Gibson, J.R., Beierlein, M., Connors, B.W., 2005. Functional properties of electrical synapses between inhibitory interneurons of neocortical layer 4. J. Neurophysiol. 93 (1), 467–480.

Giepmans, B.N., 2004. Gap junctions and connexin-interacting proteins. Cardiovasc. Res. 62 (2), 233–245.

Goffinet, A.M., 1979. An early development defect in the cerebral cortex of the reeler mouse. A morphological study leading to a hypothesis concerning the action of the mutant gene. Anat. Embryol. (Berl.) 157 (2), 205–216.

Gutnick, M.J., Connors, B.W., Ransom, B.R., 1981. Dye-coupling between glial cells in the guinea pig neocortical slice. Brain Res. 213 (2), 486–492.

Hansen, D.V., Lui, J.H., Parker, P.R., Kriegstein, A.R., 2010. Neurogenic radial glia in the outer subventricular zone of human neocortex. Nature 464 (7288), 554–561.

Harris, A.L., 2007. Connexin channel permeability to cytoplasmic molecules. Prog. Biophys. Mol. Biol. 94 (1–2), 120–143.

Haubensak, W., Attardo, A., Denk, W., Huttner, W.B., 2004. Neurons arise in the basal neuroepithelium of the early mammalian telencephalon: a major site of neurogenesis. Proc. Natl. Acad. Sci. U.S.A. 101 (9), 3196–3201.

He, S., Li, Z., Ge, S., Yu, Y.C., Shi, S.H., 2015. Inside-out radial migration facilitates lineage-dependent neocortical microcircuit assembly. Neuron 86 (5), 1159–1166.

Hebb, D.O., 1949. The Organization of Behavior. Wiley.

Hestrin, S., Galarreta, M., 2005. Electrical synapses define networks of neocortical GABAergic neurons. Trends Neurosci. 28 (6), 304–309.

Hevner, R.F., 2006. From radial glia to pyramidal-projection neuron: transcription factor cascades in cerebral cortex development. Mol. Neurobiol. 33 (1), 33–50.

Hippenmeyer, S., Youn, Y.H., Moon, H.M., Miyamichi, K., Zong, H., Wynshaw-Boris, A., Luo, L., 2010. Genetic mosaic dissection of Lis1 and Ndel1 in neuronal migration. Neuron 68 (4), 695–709.

Hirotsune, S., Takahara, T., Sasaki, N., Hirose, K., Yoshiki, A., Ohashi, T., et al., 1995. The reeler gene encodes a protein with an EGF-like motif expressed by pioneer neurons. Nat. Genet. 10 (1), 77–83.

Hormuzdi, S.G., Filippov, M.A., Mitropoulou, G., Monyer, H., Bruzzone, R., 2004. Electrical synapses: a dynamic signaling system that shapes the activity of neuronal networks. Biochim. Biophys. Acta 1662 (1–2), 113–137.

Howell, B.W., Hawkes, R., Soriano, P., Cooper, J.A., 1997. Neuronal position in the developing brain is regulated by mouse disabled-1. Nature 389 (6652), 733–737.

Hubel, D.H., Wiesel, T.N., 1959. Receptive fields of single neurones in the cat's striate cortex. J. Physiol. 148, 574–591.

Hubel, D.H., Wiesel, T.N., 1963. Shape and arrangement of columns in cat's striate cortex. J. Physiol. 165, 559–568.

Hubel, D.H., Wiesel, T.N., 1968. Receptive fields and functional architecture of monkey striate cortex. J. Physiol. 195 (1), 215–243.

Hubel, D.H., Wiesel, T.N., 1974. Uniformity of monkey striate cortex: a parallel relationship between field size, scatter, and magnification factor. J. Comp. Neurol. 158 (3), 295–305.

Jossin, Y., Cooper, J.A., 2011. Reelin, Rap1 and N-cadherin orient the migration of multipolar neurons in the developing neocortex. Nat. Neurosci. 14 (6), 697–703.

Kandler, K., Katz, L.C., 1995. Neuronal coupling and uncoupling in the developing nervous system. Curr. Opin. Neurobiol. 5 (1), 98–105.

Kandler, K., Katz, L.C., 1998a. Coordination of neuronal activity in developing visual cortex by gap junction-mediated biochemical communication. J. Neurosci. 18 (4), 1419–1427.

Kandler, K., Katz, L.C., 1998b. Relationship between dye coupling and spontaneous activity in developing ferret visual cortex. Dev. Neurosci. 20 (1), 59–64.

Katz, L.C., Shatz, C.J., 1996. Synaptic activity and the construction of cortical circuits. Science 274 (5290), 1133–1138.

Kornack, D.R., Rakic, P., 1995. Radial and horizontal deployment of clonally related cells in the primate neocortex: relationship to distinct mitotic lineages. Neuron 15 (2), 311–321.

Kruger, O., Plum, A., Kim, J.S., Winterhager, E., Maxeiner, S., Hallas, G., Willecke, K., 2000. Defective vascular development in connexin 45-deficient mice. Development 127 (19), 4179–4193.

Langley, J.N., 1895. Note on regeneration of prae-ganglionic fibres of the sympathetic. J. Physiol. 18 (3), 280–284.

Li, J., Hertzberg, E.L., Nagy, J.I., 1997. Connexin32 in oligodendrocytes and association with myelinated fibers in mouse and rat brain. J. Comp. Neurol. 379 (4), 571–591.

Li, Y., Lu, H., Cheng, P.L., Ge, S., Xu, H., Shi, S.H., Dan, Y., 2012. Clonally related visual cortical neurons show similar stimulus feature selectivity. Nature 486 (7401), 118–121.

Lin, J.H., Takano, T., Cotrina, M.L., Arcuino, G., Kang, J., Liu, S., Nedergaard, M., 2002. Connexin 43 enhances the adhesivity and mediates the invasion of malignant glioma cells. J. Neurosci. 22 (11), 4302–4311 doi:20026450.

Lo Turco, J.J., Kriegstein, A.R., 1991. Clusters of coupled neuroblasts in embryonic neocortex. Science 252 (5005), 563–566.

Lui, J.H., Hansen, D.V., Kriegstein, A.R., 2011. Development and evolution of the human neocortex. Cell 146 (1), 18–36.

Luskin, M.B., Pearlman, A.L., Sanes, J.R., 1988. Cell lineage in the cerebral cortex of the mouse studied in vivo and in vitro with a recombinant retrovirus. Neuron 1 (8), 635–647.

Martinez-Cerdeno, V., Cunningham, C.L., Camacho, J., Antczak, J.L., Prakash, A.N., Cziep, M.E., Noctor, S.C., 2012. Comparative analysis of the subventricular zone in rat, ferret and macaque: evidence for an outer subventricular zone in rodents. PLoS One 7 (1), e30178.

Massa, P.T., Mugnaini, E., 1982. Cell junctions and intramembrane particles of astrocytes and oligodendrocytes: a freeze-fracture study. Neuroscience 7 (2), 523–538.

Micevych, P.E., Abelson, L., 1991. Distribution of mRNAs coding for liver and heart gap junction proteins in the rat central nervous system. J. Comp. Neurol. 305 (1), 96–118.

Miyata, T., Kawaguchi, A., Saito, K., Kawano, M., Muto, T., Ogawa, M., 2004. Asymmetric production of surface-dividing and non-surface-dividing cortical progenitor cells. Development 131 (13), 3133–3145.

Montoro, R.J., Yuste, R., 2004. Gap junctions in developing neocortex: a review. Brain Res. Brain Res. Rev. 47 (1–3), 216–226.

Mountcastle, V.B., Davies, P.W., Berman, A.L., 1957. Response properties of neurons of cat's somatic sensory cortex to peripheral stimuli. J. Neurophysiol. 20 (4), 374–407.

Mountcastle, V.B., 1997. The columnar organization of the neocortex. Brain 120 (Pt 4), 701–722.

Nadarajah, B., Jones, A.M., Evans, W.H., Parnavelas, J.G., 1997. Differential expression of connexins during neocortical development and neuronal circuit formation. J. Neurosci. 17 (9), 3096–3111.

Nagy, J.I., Rash, J.E., 2000. Connexins and gap junctions of astrocytes and oligodendrocytes in the CNS. Brain Res. Brain Res. Rev. 32 (1), 29–44.

Nagy, J.I., Yamamoto, T., Sawchuk, M.A., Nance, D.M., Hertzberg, E.L., 1992. Quantitative immunohistochemical and biochemical correlates of connexin43 localization in rat brain. Glia 5 (1), 1–9.

Nagy, J.I., Patel, D., Ochalski, P.A., Stelmack, G.L., 1999. Connexin30 in rodent, cat and human brain: selective expression in gray matter astrocytes, co-localization with connexin43 at gap junctions and late developmental appearance. Neuroscience 88 (2), 447–468.

Nagy, J.I., Li, X., Rempel, J., Stelmack, G., Patel, D., Staines, W.A., Rash, J.E., 2001. Connexin26 in adult rodent central nervous system: demonstration at astrocytic gap junctions and colocalization with connexin30 and connexin43. J. Comp. Neurol. 441 (4), 302–323.

Nagy, J.I., Ionescu, A.V., Lynn, B.D., Rash, J.E., 2003a. Connexin29 and connexin32 at oligodendrocyte and astrocyte gap junctions and in myelin of the mouse central nervous system. J. Comp. Neurol. 464 (3), 356–370.

Nagy, J.I., Ionescu, A.V., Lynn, B.D., Rash, J.E., 2003b. Coupling of astrocyte connexins Cx26, Cx30, Cx43 to oligodendrocyte Cx29, Cx32, Cx47: implications from normal and connexin32 knockout mice. Glia 44 (3), 205–218.

Nielsen, M.S., Axelsen, L.N., Sorgen, P.L., Verma, V., Delmar, M., Holstein-Rathlou, N.H., 2012. Gap junctions. Compr. Physiol. 2 (3), 1981–2035.

Noctor, S.C., Flint, A.C., Weissman, T.A., Dammerman, R.S., Kriegstein, A.R., 2001. Neurons derived from radial glial cells establish radial units in neocortex. Nature 409 (6821), 714–720.

Noctor, S.C., Martinez-Cerdeno, V., Ivic, L., Kriegstein, A.R., 2004. Cortical neurons arise in symmetric and asymmetric division zones and migrate through specific phases. Nat. Neurosci. 7 (2), 136–144.

Odermatt, B., Wellershaus, K., Wallraff, A., Seifert, G., Degen, J., Euwens, C., Willecke, K., 2003. Connexin 47 (Cx47)-deficient mice with enhanced green fluorescent protein reporter gene reveal predominant oligodendrocytic expression of Cx47 and display vacuolized myelin in the CNS. J. Neurosci. 23 (11), 4549–4559.

Ogawa, M., Miyata, T., Nakajima, K., Yagyu, K., Seike, M., Ikenaka, K., Mikoshiba, K., 1995. The reeler gene-associated antigen on Cajal-Retzius neurons is a crucial molecule for laminar organization of cortical neurons. Neuron 14 (5), 899–912.

Ohtsuki, G., Nishiyama, M., Yoshida, T., Murakami, T., Histed, M., Lois, C., Ohki, K., 2012. Similarity of visual selectivity among clonally related neurons in visual cortex. Neuron 75 (1), 65–72.

Orkand, R.K., Nicholls, J.G., Kuffler, S.W., 1966. Effect of nerve impulses on the membrane potential of glial cells in the central nervous system of amphibia. J. Neurophysiol. 29 (4), 788–806.

Orthmann-Murphy, J.L., Abrams, C.K., Scherer, S.S., 2008. Gap junctions couple astrocytes and oligodendrocytes. J. Mol. Neurosci. 35 (1), 101–116.

Park, W.M., Wang, Y., Park, S., Denisova, J.V., Fontes, J.D., Belousov, A.B., 2011. Interplay of chemical neurotransmitters regulates developmental increase in electrical synapses. J. Neurosci. 31 (16), 5909–5920.

Peinado, A., Yuste, R., Katz, L.C., 1993. Extensive dye coupling between rat neocortical neurons during the period of circuit formation. Neuron 10 (1), 103–114.

Pinto Lord, M.C., Caviness Jr., V.S., 1979. Determinants of cell shape and orientation: a comparative Golgi analysis of cell-axon interrelationships in the developing neocortex of normal and reeler mice. J. Comp. Neurol. 187 (1), 49–69.

Price, J., Thurlow, L., 1988. Cell lineage in the rat cerebral cortex: a study using retroviral-mediated gene transfer. Development 104 (3), 473–482.
Rakic, P., 1995. Radial versus tangential migration of neuronal clones in the developing cerebral cortex. Proc. Natl. Acad. Sci. U.S.A. 92 (25), 11323–11327.
Ransom, B.R., Yamate, C.L., Connors, B.W., 1985. Activity-dependent shrinkage of extracellular space in rat optic nerve: a developmental study. J. Neurosci. 5 (2), 532–535.
Rash, J.E., Yasumura, T., Davidson, K.G., Furman, C.S., Dudek, F.E., Nagy, J.I., 2001. Identification of cells expressing Cx43, Cx30, Cx26, Cx32 and Cx36 in gap junctions of rat brain and spinal cord. Cell Commun. Adhes. 8 (4–6), 315–320.
Reillo, I., de Juan Romero, C., Garcia-Cabezas, M.A., Borrell, V., 2011. A role for intermediate radial glia in the tangential expansion of the mammalian cerebral cortex. Cereb. Cortex 21 (7), 1674–1694.
Rice, D.S., Sheldon, M., D'Arcangelo, G., Nakajima, K., Goldowitz, D., Curran, T., 1998. Disabled-1 acts downstream of Reelin in a signaling pathway that controls laminar organization in the mammalian brain. Development 125 (18), 3719–3729.
Robertson, J.D., 1953. Ultrastructure of two invertebrate synapses. Proc. Soc. Exp. Biol. Med. 82 (2), 219–223.
Sanada, K., Gupta, A., Tsai, L.H., 2004. Disabled-1-regulated adhesion of migrating neurons to radial glial fiber contributes to neuronal positioning during early corticogenesis. Neuron 42 (2), 197–211.
Scherer, S.S., Deschenes, S.M., Xu, Y.T., Grinspan, J.B., Fischbeck, K.H., Paul, D.L., 1995. Connexin32 is a myelin-related protein in the PNS and CNS. J. Neurosci. 15 (12), 8281–8294.
Shitamukai, A., Konno, D., Matsuzaki, F., 2011. Oblique radial glial divisions in the developing mouse neocortex induce self-renewing progenitors outside the germinal zone that resemble primate outer subventricular zone progenitors. J. Neurosci. 31 (10), 3683–3695.
Sohl, G., Maxeiner, S., Willecke, K., 2005. Expression and functions of neuronal gap junctions. Nat. Rev. Neurosci. 6 (3), 191–200.
Sperry, R.W., 1943. Visuomotor coordination in the newt (*Triturus uiridescens*) after re-generation of the optic nerve. J. Comp. Neurol. 79, 33–55.
Sperry, R.W., 1944. Optic nerve regeneration with return of vision in anurans. J. Neurophysiol. 7, 57–69.
Sperry, R.W., 1963. Chemoaffinity in the orderly growth of nerve fiber patterns and connections. Proc. Natl. Acad. Sci. U.S.A. 50, 703–710.
Traub, R.D., Whittington, M.A., Draguhn, A., 2017. Gap junctions between pyramidal cells account for a variety of very fast network oscillations (>80 Hz) in cortical structures. In: Jing, J. (Ed.), Network Functions and Plasticity. Elsevier (Chapter 13).
Torii, M., Hashimoto-Torii, K., Levitt, P., Rakic, P., 2009. Integration of neuronal clones in the radial cortical columns by EphA and ephrin-A signalling. Nature 461 (7263), 524–528.
Wang, Y., Barakat, A., Zhou, H., 2010. Electrotonic coupling between pyramidal neurons in the neocortex. PLoS One 5 (4), e10253.
Wang, X., Tsai, J.W., LaMonica, B., Kriegstein, A.R., 2011. A new subtype of progenitor cell in the mouse embryonic neocortex. Nat. Neurosci. 14 (5), 555–561.
Ware, M.L., Tavazoie, S.F., Reid, C.B., Walsh, C.A., 1999. Coexistence of widespread clones and large radial clones in early embryonic ferret cortex. Cereb. Cortex 9 (6), 636–645.
White, T.W., Paul, D.L., Goodenough, D.A., Bruzzone, R., 1995. Functional analysis of selective interactions among rodent connexins. Mol. Biol. Cell 6 (4), 459–470.
Willecke, K., Eiberger, J., Degen, J., Eckardt, D., Romualdi, A., Guldenagel, M., Sohl, G., 2002. Structural and functional diversity of connexin genes in the mouse and human genome. Biol. Chem. 383 (5), 725–737.
Wong, R.C., Pera, M.F., Pebay, A., 2008. Role of gap junctions in embryonic and somatic stem cells. Stem Cell Rev. 4 (4), 283–292.
Xu, H.-T., Han, Z., Gao, P., He, S., Li, Z., Shi, W., Shi, S.-H., 2014. Distinct lineage-dependent structural and functional organization of the hippocampus. Cell 157 (7), 1552–1564.
Yu, Y.C., Bultje, R.S., Wang, X., Shi, S.H., 2009. Specific synapses develop preferentially among sister excitatory neurons in the neocortex. Nature 458 (7237), 501–504.

Yu, Y.-C., He, S., Chen, S., Fu, Y., Brown, K.N., Yao, X.-H., Shi, S.-H., 2012. Preferential electrical coupling regulates neocortical lineage-dependent microcircuit assembly. Nature 486 (7401), U113–U139.

Yuste, R., Peinado, A., Katz, L.C., 1992. Neuronal domains in developing neocortex. Science 257 (5070), 665–669.

Yuste, R., Nelson, D.A., Rubin, W.W., Katz, L.C., 1995. Neuronal domains in developing neocortex: mechanisms of coactivation. Neuron 14 (1), 7–17.

Zipursky, S.L., Sanes, J.R., 2010. Chemoaffinity revisited: dscams, protocadherins, and neural circuit assembly. Cell 143 (3), 343–353.

GLOSSARY

To facilitate reading of the book, we generated a Glossary. In particular, we include those terms that may be unfamiliar or confusing to some readers. Italicized words/phrases are terms in this Glossary. Most of the terms appear in the Index where you may find the locations of in-depth discussions in the main text.

Asymmetric coupling When *coupling ratios* are measured between neuron A and neuron B, if the coupling ratio from neuron A to neuron B is not the same as that from neuron B to neuron A, then the *electrical coupling* is considered to be asymmetric. Asymmetric coupling could result from differences between the *intrinsic properties* of neuron A and neuron B and/or a *rectifying electrical synapse* between the two neurons [see Chapter 5, p. 89].

Central pattern generator A network of neurons in the nervous system with defined chemical and electrical connections that produce a specific activity pattern between circuit elements over time, typically in a *rhythmic* fashion (see *Pattern generation*.) A central pattern generator (CPG) is a type of a *neural circuit/ microcircuit* often present in, but not limited to, a motor circuit. Circuit elements in a CPG are called pattern-generating neurons. In addition, some circuit elements either within or outside of a CPG can be classified as *command neurons* or *modulatory neurons*. Notably, one of the most fundamental discoveries in contemporary circuit research is that most, if not all, CPGs are **multifunctional**, ie, they produce more than one type/version of activity pattern. Most commonly, multifunctionality of a CPG results from actions of *neuromodulators*. It is likely that, in addition to CPGs, multifunctionality also applies to other types of neural circuits.

Chemical connection, Chemical synapse, Chemical transmission A mode of synaptic transmission between two neurons whereby an electrical signal is transmitted indirectly (compare *Electrical synapse*) through release of a chemical messenger/transmitter from a presynaptic neuron that binds to a postsynaptic receptor and ultimately induces a voltage/conductance change in the postsynaptic neuron. A hallmark of chemical transmission is that transmitter release requires Ca^{2+} influx in the presynaptic neuron. In addition, the chemical signal is transmitted unidirectionally, ie, from the presynaptic to the postsynaptic neuron. Typically, an action potential in the presynaptic neuron is also required for chemical transmission to occur (but see *Graded transmission*).

Classical conditioning, Pavlovian conditioning An associative learning procedure whereby an animal learns a new association between two stimuli. After a conditioned stimulus (CS) is repeatedly paired with an unconditioned stimulus (US), the animal's response to the CS will gradually change. For example, if sound (CS) is paired with food (US), an animal can learn to salivate when sound is applied on its own.

Command interneuron, Command neuron, Command-like neuron A neuron or a group of neurons in a motor circuit that receive appropriate sensory inputs, and when active, can drive a specific motor pattern underlying behavior.

Connectome A comprehensive map of neural connections (including *chemical* and *electrical connections*) in the brain, which may be thought of as its "wiring diagram." More broadly, a connectome would include the mapping of all neural connections within an organism's nervous system (adapted from Wikipedia, https://en.wikipedia.org/wiki/Connectome).

Connexin *Gap junction* proteins in chordates.

Coupling coefficient, Coupling rate, Coupling ratio For two neurons connected with an *electrical synapse*, a ratio of voltage change in the postsynaptic neuron over voltage change in the presynaptic neuron when the presynaptic neuron is injected with either a positive or a negative current. Note that the amplitude of the coupling ratio can be affected not only by *electrical coupling* (see *Junctional conductance*), but also by the location of the *electrical synapse* and/or *intrinsic properties* of the two coupled neurons.

Dye coupling When a live neuron is injected with a fluorescent dye, such as Lucifer yellow, other neurons not injected with the dye also exhibit fluorescence after some amount of time (typically hours or longer). This phenomenon is called dye coupling. Neurons with *electrical coupling/gap junctions* usually show dye coupling; thus, dye coupling may be used as morphological evidence for the presence of electrical coupling.

Ectopic electrical synapse Genetic insertion of vertebrate *connexins* in invertebrates, and invertebrate *innexins* in vertebrates. These techniques are used in experiments that determine the role of *electrical coupling* in a *neural circuit/microcircuit* (see Chapter 1).

Electrical connection, Electrical coupling, Electrical synapses, Electrical transmission A mode of synaptic transmission between two neurons where an electrical signal is transmitted directly, without involvement of a chemical messenger/transmitter (see *Chemical synapse*). A hallmark of electrical coupling is that the synaptic transmission does not require Ca^{2+} influx in the presynaptic neuron (compare *Chemical synapse*). In addition, electrical coupling typically allows bidirectional transmission of an electrical signal (but see *Rectifying electrical coupling*). In *neural circuits* without *graded transmission*, electrical coupling can usually be identified using electrophysiological recordings by injecting negative or positive current into a presynaptic neuron, and observing smaller voltage changes with the similar polarity in the postsynaptic neuron. However, when the electrical synapse is located far away from the recording site, eg, the soma of a presynaptic neuron with a long axon or axons (especially a *projection neuron*), the absence of voltage change in the postsynaptic neuron cannot be taken as evidence against electrical coupling (see Section 2.2.3 of Chapter 5).

Gap junction Synaptic structure that connects two neurons to allow direct current flow between them. It is the morphological basis of *electrical coupling*. Each gap junction is composed of two hemichannels, which consist of homo- or heterohexameric arrays of gap junction proteins, and the hemichannel in one plasma membrane docks end to end with a hemichannel in the membrane of a closely opposed cell. The hemichannel is made of six subunits. Gap junction proteins are termed *innexins* in prechordates, and *connexins* in chordates. Innexins and connexins are not homologous. In chordates, proteins that are homologous to innexins are present and termed *pannexins*, but typically do not mediate electrical coupling.

Graded postsynaptic potential, Graded potentials Postsynaptic potentials that are induced between neurons with *graded transmission*.

Graded synapse, Graded transmission A form of *chemical transmission* in some neurons, typically *non-spiking neurons*, of some animals (eg, invertebrates: nematodes, crustaceans, insects; vertebrates: retina, cortex, cerebellum), whereby presynaptic spikes are not required to induce transmitter release. Rather, a transmitter is released in a graded fashion as the presynaptic membrane is depolarized.

Heterogeneous elements Neuronal circuit elements with different synaptic connectivity, different morphology, and/or different *intrinsic properties*. Thus, heterogeneous elements in a *neural circuit/microcircuit* can be considered to be neurons of different types. See *Homogenous elements*.

Heterologous coupling *Electrical coupling* between neurons of different types.

Heterologous gap junctions Similar to *heterotypic gap junctions*.

Heterotypic channel, Heterotypic electrical synapse, Heterotypic gap junction *Gap junctions* in which gap junction proteins in the presynaptic neuron differ from gap junction proteins in the postsynaptic neuron (see Section 3.3 of Chapter 2).

Homogenous (Homogeneous) elements, Homogenous neurons Neuronal circuit elements with identical synaptic connectivity, morphology, and *intrinsic properties* in an animal. Thus, homogeneous elements in a *neural circuit/microcircuit* can be considered to be neurons of the same type. Note that homogenous neurons are different from *homologous neurons*. See *Heterogeneous elements*.

Homologous coupling *Electrical coupling* between neurons of the same type.

Homologous gap junctions Similar to *homotypic gap junctions*.

Homologous neurons Between different species within a class, such as mollusks or insects, individual neurons that show similar locations, morphology, transmitter phenotypes, synaptic connectivity, and functions within a specific *neural circuit* can be considered homologous. Note that homologous neurons are different from *homogenous neurons.*

Homotypic channel, Homotypic electrical synapse, Homotypic gap junction *Gap junctions* in which gap junction proteins in the presynaptic neuron are the same as gap junction proteins in the postsynaptic neuron.

Innexin *Gap junction* proteins in prechordates.

Intrinsic properties of a neuron Membrane properties of a neuron that include membrane capacitance and membrane conductance. The latter is due to the presence of a unique set of ionic channels in a particular neuron. In addition, for a spiking neuron, the threshold to spiking, ie, the minimum current required to induce a spike, is also considered an intrinsic property. Other intrinsic properties of a neuron include, but are not limited to, the spontaneous firing frequency; afterhyperpolarization (amplitude and delay), rheobase, and sag.

Junctional conductance Synaptic conductance between two electrically coupled neurons that can be measured under voltage clamp. Note that the junctional conductance is not affected by the *intrinsic properties* of the pre- and postsynaptic neurons, whereas the *coupling ratio* is.

Local circuit neurons Neuronal circuit elements with somata and axons within one brain region or a ganglion (see *Projection neurons*).

Microcircuit Similar to *neural circuit*, but generally refers to a subset of neurons within a neural circuit with a more defined function(s).

Modulation Same as *neuromodulation.*

Modulators Same as *neuromodulators.*

Modulatory neurons A neuron or a group of neurons within a circuit that contain and release a *neuromodulator* or neuromodulators. Neuromodulators function to change circuit properties (ie, *intrinsic properties* of circuit elements and the pattern of *chemical* and *electrical synaptic connections* between them). Modulatory neurons can be present in any *neural circuit*, including a motor circuit. In a motor circuit, these neurons contrast with *command neurons* in that they do not drive a specific motor pattern when active.

Modules (building blocks) A subset of neurons, often interneurons, with a specific function or a set of functions within a *neural circuit/microcircuit.*

Motif see *Network motif.*

Motivated behaviors Behaviors whose expression is influenced by internal factors, ie, *motivation.* These include, but are not limited to, feeding, drinking, locomotion, avoidance, and reproductive behavior.

Motivation A term used to refer to internal factors that may account for variations in behavior. These internal factors include, but are not limited to, hunger/satiation, thirst, and arousal. Most commonly, changes in internal state may be mediated, at least in part, by the circuit actions of *modulatory neurons*, which release a *neuromodulator* or neuromodulators. Furthermore, changes in the internal state may be mediated by actions of hormones. Other factors that may influence behavior are sensation and learning.

Neural circuit, Neural network A general term for any group of neurons within the brain (or the nervous system) that interact with each other with a specific pattern of *chemical* and *electrical synaptic* connections and perform one or more functions (eg, sensory, motor and/or higher-order functions) (see *Microcircuit*). Notably, the synaptic connection between any two neurons in a circuit can be chemical, electrical, or both. The activity of a neural circuit is determined by the *intrinsic properties* of circuit elements and the pattern of *chemical* and *electrical synaptic connections* between them. A subset of circuit elements may form a *module* or *network motif.* One prototypical neural circuit is a *central pattern generator.*

Neuromodulation Alterations in the network/circuit properties (including the *chemical* and *electrical synaptic* connections between circuit elements and *intrinsic properties* of specific circuit elements) resulting from the release of *neuromodulator*(s) or *modulator*(s). Most commonly, neuromodulators (such as biogenic amines and neuropeptides) activate G-protein-coupled receptors (or more generally, metabotropic receptors) to induce slower and often persistent changes in postsynaptic neurons.

Neuromodulators Neurotransmitters that act on postsynaptic neurons typically through activation of G-protein coupled receptors (or more generally, metabotropic receptors) to induce slower and often persistent changes in postsynaptic neurons. Examples include biogenic amines (eg, dopamine, noradrenaline (norepinephrine), serotonin, and histamine) and peptides (or neuropeptides). Neuropeptides are relatively large transmitter molecules composed of 3 or more (typically less than 40) amino acids, and contrast with small molecule neurotransmitter. Examples of the latter include glutamate, acetylcholine, GABA, and glycine, as well as biogenic amines. With the exception of biogenic amines, small molecule neurotransmitters often act on ionotropic receptors to induce fast changes in postsynaptic neurons. This is not always the case, however. When small molecule neurotransmitters act on metabotropic receptors, they are classified as neuromodulators.

Network motif A network of three or more neurons with a specific connectivity pattern within a *neural circuit/microcircuit* (see Chapter 1).

Nonspiking neurons Neurons that do not generate action potentials or spikes. *Chemical transmission* for nonspiking neurons is *graded*.

Operant conditioning An associative learning procedure in which the execution of a behavior (so-called operant; eg, biting) is modified by presentation of a reinforcing stimulus (a reward or punishment; eg, food or electric shock). The probability or intensity of the conditioned behavior changes as a result of the operant conditioning. Operant conditioning is also called reinforcement learning or instrumental conditioning.

Oscillation This can refer to the membrane potential or the spiking activity of a single neuron, or it can describe network activity. In general, it refers to repetitive variation, typically in time. Note that if a network of neurons is involved, the oscillation between the neurons may or may not be in *synchrony*.

Pannexin Chordate proteins that are homologous to innexins, but typically do not mediate electrical coupling.

Pattern generation A network of neurons that produce a specific activity pattern between circuit elements over time. When the pattern is repeated, it becomes a *rhythmic* pattern (see *Central pattern generator*).

Plasticity, Neural plasticity Network/circuit properties, including the *chemical* and *electrical synaptic* connections between circuit elements and *intrinsic properties* of specific circuit elements, that change over time. Changes can result from short- or long-term activity within the network, the release of *neuromodulator(s)*, and/or learning and memory training procedures (see *Classical conditioning* and *Operant conditioning*). Persistent plasticity in a *neural circuit* is the neural basis for learning and memory. In the context of the goals in this book, it should be emphasized that learning and memory may involve not only alterations in chemical synaptic connections between circuit elements (as is commonly described in the literature), but also alterations in electrical synaptic connections between them (see Chapters 3, 6, 10, and 12). In addition, they may involve alterations in the intrinsic properties of specific circuit elements.

Projection neurons Neuronal circuit elements with somata in one brain region or a ganglion, and an axon or axons projecting to a different brain region or a ganglion (see *Local circuit neurons*).

Rectification of an electrical synapse The property of a *rectifying electrical synapse*.

Rectifying electrical synapse, Rectifying gap junction, Rectifying junction, Rectifying synapse An *electrical synapse* between neuron A and B is considered to be rectifying if current passes better in one direction than the other. One of its hallmarks is that for the two coupled neurons, depolarizing current injected into A will produce larger responses in B than hyperpolarizing current injected into A. Moreover, hyperpolarizing current injected into B will produce larger responses in A than depolarizing current injected in B (see Section 2.1.3 of Chapter 5). Rectifying electrical coupling likely results from *heterotypic gap junctions* between the two coupled neurons, where the two hemichannnels in presynaptic and postsynaptic neurons have different conductances (see Section 3.3 of Chapter 2). Importantly, rectifying electrical coupling enables effective unidirectional signal transmission, which is also a characteristic of chemical transmission.

Rhythm Similar to *oscillation*, but this term typically refers to the spiking activity of neurons.

Symmetric coupling When *coupling ratios* are measured between two neurons (eg, neuron A and neuron B), if the coupling ratio from neuron A to neuron B is the same as that from neuron B to neuron A, then the electrical coupling is considered to be symmetric. Symmetric coupling can result from a *nonrectifying or homotypic electrical synapse* between two neurons with similar *intrinsic properties*.

Synchrony When two or more neurons generate action potentials at about the same time, their activity is synchronous. Note that neurons that fire in synchrony may or may not fire in *oscillation* or in a *rhythmic* fashion.

INDEX

'*Note*: Page numbers followed by "f" indicate figures, "t" indicate tables.'

D

E

H

N

S

T

U

V

FSC
www.fsc.org
MIX
Papier | Fördert
gute Waldnutzung
FSC® C083411

Kolibri 360 GmbH
Ferdinand-Jühlke-Straße 7
99095 Erfurt, Deutschland
produktsicherheit@kolibri360.de